Inverse Methods in Electromagnetic Imaging

Part 1

This Advanced Research Workshop was co-sponsored by:
NATO, Scientific Affairs Division
DFVLR, Institute for Radio Frequency Technology
U.S.-ARO, European Branch Office
SIEMENS A.G., Medical Division
UIC/EECS, Communications Laboratory

NATO ASI Series

Advanced Science Institutes Series

A series presenting the results of activities sponsored by the NATO Science Committee, which aims at the dissemination of advanced scientific and technological knowledge, with a view to strengthening links between scientific communities.

The series is published by an international board of publishers in conjunction with the NATO Scientific Affairs Division

A	Life Sciences	Plenum Publishing Corporation
B	Physics	London and New York
C	Mathematical and Physical Sciences	D. Reidel Publishing Company Dordrecht, Boston and Lancaster
D	Behavioural and Social Sciences	Martinus Nijhoff Publishers
E	Engineering and Materials Sciences	The Hague, Boston and Lancaster
F	Computer and Systems Sciences	Springer-Verlag
G	Ecological Sciences	Berlin, Heidelberg, New York and Tokyo

Series C: Mathematical and Physical Sciences Vol. 143 - Part 1

Inverse Methods in Electromagnetic Imaging

Part 1

edited by

Wolfgang-M. Boerner

Communications Laboratory, EECS Department,
University of Illinois at Chicago, U.S.A.

Hans Brand

High Frequency Engineering Laboratories,
Universität Erlangen-Nürnberg, F.R.G.

Leonard A. Cram

Thorn EMI, Electronics Ltd.,
U.K.

Dag T. Gjessing

ESTP, Royal Norwegian Council for Industrial
and Scientific Research, Kjeller, Norway

Arthur K. Jordan

Space Sciences Division,
Naval Research Laboratory, U.S.A.

Wolfgang Keydel

Institüt für HF-Technik,
DFVLR, F.R.G.

Günther Schwierz

Siemens Medical Division,
Siemens AG, Erlangen, F.R.G.

Martin Vogel

Institüt für HF-Technik,
DFVLR, F.R.G.

Springer-Science+Business Media, B.V.

Proceedings of the NATO Advanced Research Workshop on
Inverse Methods in Electromagnetic Imaging
Bad Windsheim, Franconia, F.R.G.
18-24 September 1983

Library of Congress Cataloging in Publication Data

NATO Advanced Research Workshop on Inverse Methods in Electromagnetic Imaging
(1983: Bad Windsheim, Germany)
Inverse methods in electromagnetic imaging.

(NATO ASI series. Series C, Mathematical and physical sciences; vol. 143)
"Proceedings of the NATO Advanced Research Workshop on Inverse Methods in
Electromagnetic Imaging, Bad Windsheim, Franconia, F.R.G., September 18–24, 1983"–
T.p. verso.
"Published in cooperation with NATO Scientific Affairs Division."
Bibliography: p.
Includes index.
1. Imaging systems–Congresses. 2. Electromagnetic waves–Congresses.
I. Boerner, Wolfgang M. II. North Atlantic Treaty Organization, Scientific Affairs
Divison. III. Title. IV. Series: NATO ASI series. Series C, Mathematical and physical
sciences; no. 143.
TK8315.N36   1983   621.36'7   84–22843

ISBN 978-94-010-9446-7   ISBN 978-94-010-9444-3 (eBook)
DOI 10.1007/978-94-010-9444-3

TABLE OF CONTENTS (Part 1)   v

TOPIC 0 - MEMORIAL PAPER AND OVERVIEW
(Papers are cross-referenced with the **Final Technical Program Out-line.** Given are session and sequence of presentation, i.e., OS.3)

TOPIC I - MATHEMATICAL INVERSE METHODS AND TRANSIENT TECHNIQUES
(Papers are cross-referenced with the **Final Technical Program Out-line.** Given are session and sequence of presentation, i.e., OS.3)

TOPIC II – NUMERICAL INVERSION METHODS
(Papers are cross-referenced with the **Final Technical Program Outline**. Given are session and sequence of presentation, i.e., OS.3)

**TOPIC III   –   POLARIZATION UTILIZATION IN THE ELECTROMAGNETIC
VECTOR INVERSE PROBLEM**
(Papers are cross-referenced with the **Final Technical Program Out-
line.** Given are session and sequence of presentation, i.e., OS.3)

CONTENTS OF PART 2

TOPIC III (continued)
(Papers are cross-referenced with the **Final Technical Program Outline.** Given are session and sequence of presentation, i.e., OS.3)

TOPIC IV - **IMAGE QUALITY AND IMAGE RESOLUTION IN REMOTE SENSING
             AND SURVEILLANCE**
(Papers are cross-referenced with the **Final Technical Program Outline.** Given are session and sequence of presentation, i.e., OS.3)

TOPIC VI – WORKING DISCUSSION GROUP REPORTS

ADDRESS OF THE HOST NATION

It is an honour and pleasure for me to welcome the experts of inverse methods in electromagnetic imaging in Bad Windsheim, which is located in Franconia, the historical heart of Germany. If more than 100 experts from 11 countries representing the key persons on this important and fundamental science area meet for a workshop, then the scientific community will expect relevant results. I hope this little town, with its quiet and easy-going atmosphere, will provide the environment for fruitful discussions and for stimulating scientific interaction. The inverse methods in electromagnetic imaging are fundamental tools for remote sensing and they should find application in several military and civilian areas. However, they are in wide parts not yet developed completely. A lot of scientific and technical problems are still to be solved and I want to express my hope that the results of this workshop will contribute to the further advancement of the state of the art in this scientific field. Therefore, I request that we teach, learn and work together in order to bring the different inverse methods in electromagnetic imaging a step further into the direction of operational applicability.

A more general text and source book on this subject matter does not exist. It seems an impossible task for one man to write a suitable book because of the great variety of subjects and the wide range of applications; even if one or a group of experts could have produced a book, it would certainly have removed them from active research for several years. Therefore, in preparing the Proceedings of this Workshop, several review papers have been included which together with more specific papers should provide a good overview on the subject matter. Thus, may the resulting Proceedings serve as a source and reference book where none existed before.

Dr. Wolfgang Keydel, Director
Institute for Radio Frequency Technology
German Aerospace Research Establishment
DFVLR-Oberpfaffenhofen, FR Germany

IN MEMORIAM
EDWARD M. KENNAUGH
PROFESSOR EMERITUS
1922 - 1983

The Electrical Engineering profession and the electromagnetics discipline lost a great luminary and teacher with the death of Professor Edward M. Kennaugh at age 60 on March 11, 1983. Beyond the borders of the campus of his alma mater and that of the U.S.A he became known and highly admired, specifically through his contributions to NATO Advanced Study Institutes and AGARD Workshops.

Coming to The Ohio State University in 1946 he stayed as a graduate student under Professor Victor Rumsey, and labored for 35 yrs. on electromagnetic scattering theory, bringing international fame to himself and honor to the ElectroScience Laboratory in which he worked. His fundamental theory of the polarizing properties of radar scatterers, accomplished in the early 1950's while he was still a student, remains today the definitive work on this subject. His introduction in 1958 of the "Impulse Response" concept for three dimensional scattering obstacles brought new importance to the time-domain viewpoint in scattering theory and opened avenues of research which continue to branch even today. The large body of literature which has been built upon this one concept is overwhelming witness to the depth of Professor Kennaugh's insight into the scattering process. For twenty years, he and his students labored to refine the theory and contribute to this literature while proceeding to develop basic ideas in other directions. In the early 1960's he laid the foundations of antenna scattering theory and called attention to the importance of antennas in the modification and control of radar scattering. This work was in advance of its time and came to be widely appreciated only with the advent of low cross section aircraft. In the middle 1960's he inspired and fostered the development of modal descriptions of scattering by obstacles of arbitrary shape. In this fundamental work he displayed his genius for seeing common things in an uncommon way by bringing together the familiar ideas of power factor and eigenvectors, a union which continues to bear fruit even today. In the early 1970's, Professor Kennaugh took on the directorship of the ElectroScience Laboratory and remained at that post for three years while continuing his researches in scattering theory. His "retirement" in 1977 was one in name and not in fact, for he proceeded in the six years given to him since then to develop the so-called "K-pulse" concept which he formally introduced in 1981 and which he was carrying forward at the time of his death. This concept crowns Professor Kennaugh's years of research in time-domain methods and, with its singular originator now gone, its development remains a formidable challenge to present and future investigators.

PREFACE

This NATO Advanced Research Workshop on "Inverse Methods in Electromagnetic Imaging" has come at a time of greatly increased interest
in using electromagnetic imaging methods as tools for remote sensing, material testing and medical diagnosis. Spurred by a combination of recent advances in sensor and device technology spanning the
entire m-to-sub-mm wavelengths, the infra-red, optical, ultraviolet
and x-ray regions of the electromagnetic spectrum, the rapid development of radar polarimetry including the deployment of active/pas-
sive sensors on spacecraft platforms, it has become evident that
these advanced imaging techniques can only then become useful when
supported by well-developed inverse scattering theories for proper
consistent and efficient data interpretation. However, these data
interpretative techniques, which are urgently needed in several military and civilian areas, are in many aspects not yet completely developed. Thus, the main purpose of this NATO-ARW-IMEI-1983 was to
bring together internationally renowned key experts of this new
scientific discipline of Inverse Methods in Electromagnetic Imaging
and have them prepare well-written overviews, treatises on new
areas, papers on special topics; have them deliberate on still unre-
sovled problems in working discussion groups, and then report back
on their important findings to the entire NATO Scientific Community
in dealing with these exciting current developments.

The scientific and technical aspects of the organization, the struc-
turing and grouping of seventy-five papers into the seven topics of
the Proceedings are presented in the lead article on "Inverse Methods in Electromagnetic Imaging - An Overview of and Introduction to
the Workshop Papers with a State-of-the-Art-Review", prepared in
collaboration with Dr. Arthur K. Jordan.

All participants of this NATO-ARW-IMEI-1983 were actively and con-
structively engaged in the Working Discussion Group Program. All
invitees were provided with clear and rather detailed instructions,
forwarded with the pre-workshop information package, explaining why
these working discussion group activities are most essential to the
success of any NATO-ASI and/or ARW. Originally, we had planned five
separate groups, which increased by two, and which met separately
each day for about two-to-three hours. The well-written informative
and stimulating reports, published in Part 2, Topic VI, of the Proceedings, give witness to the fact that although we were pressed for
time, the adopted plan worked and was very successful in spite of
these unavoidable time limitations and the relatively extensive
working discussion group material that had to be absorbed, filtered,
and put to use in formulating recommendations for the NATO Scientific Affairs Division, ASI/ARW Selection Panel Committees. We thank
all discussion group participants, and particularly, commend the
hard work done by the coordinators, moderators, advisors and especially the reporters for their tireless work.

After inspection of the headings and the memorial address you will observe that our workshop has been dedicated to the late Professor Edward Morton Kennaugh, who departed from us on March 11, 1983, at the age of 60. This dedication was made because of his extensive and numerous contributions to the field of inverse methods in electromagnetic imaging for which reason he had originally been chosen as one of the main speakers and as moderator and advisor for the working discussion group programs. On this occasion we are happy to report that Ed's dear wife, Mrs. Mary Kennaugh, was present during the dedication address at the Sunday evening, September 18, 1983 Opening Session and that she was able to join us for the duration of the workshop; and so we know that Professor Kennaugh's invigorating spirit and humor will be still alive and carried among us and with us into the future.

Finally, it is instructive to know that NATO Advanced Research Workshops are primarily sponsored with contributions from the United Kingdom. In retrospect, we would like to honor these financial commitments of its Ministry of Defense to the programmes of the NATO Scientific Affairs Division. In this NATO-ARW-IMEI-1983 the British contributions were doubled by the active participation of some of its finest engineering scientists as was so strongly reflected in the working discussion group activities. Our British colleagues, especially Mr. Len Cram, Prof. Alan P. Anderson, Dr. Roderick Logan, Prof. John O. Thomas, Dr. Ed Pike, together with Dr. Dag T. Gjessing, Dr. Tilo Kester, Dr. Wolfgang Keydel and Dr. Martin Vogel, punctually at 7:00 every morning sat down with us for breakfast to preview the day's scheduled events. Recalling the active participation of the latter German colleagues, it is also gratefully recorded here that they made possible the badly needed financial "fill-in" support from their Institute budget within DFVLR, Oberpfaffenhofen, to keep the workshop budget afloat. Here we also wish to acknowledge the financial contributions made by the US Army Research Office, European Branch, Edison House, London, for their assistance, and for having Dr. Karl Steinbach of MERADCOM, Ft. Belvoir, present it. Many of us were totally or partly supported by our home base institutes and we thank the Department Heads, Deans, Division Directors, and the Graduate Research Colleges for their willingness to support us. We thank them all for their contributions and the NATO Scientific Affairs Division for awarding this prestigious ARW-Award No. SA.5.2.04(0009/82)191. And last but not least, the financial sacrifices of every single participant are acknowledged, since without such individual cost sharing, only a trickle of all the key experts and their accompanients would have entered the appended list of participants, and without their presence, this NATO-ARW-IMEI-1983 just would not have taken place.

W-M. Boerner
Workshop Director

NATO-ARW-IMEI-1983
Group of Participants
present during break of
FINAL WORKING GROUP SESSION
KuK Hotel Residenz, Bad Windsheim
Friday, Sept. 23, 1983, 15:30
- Exit to the Kurpark -
"Auf der Brücke"

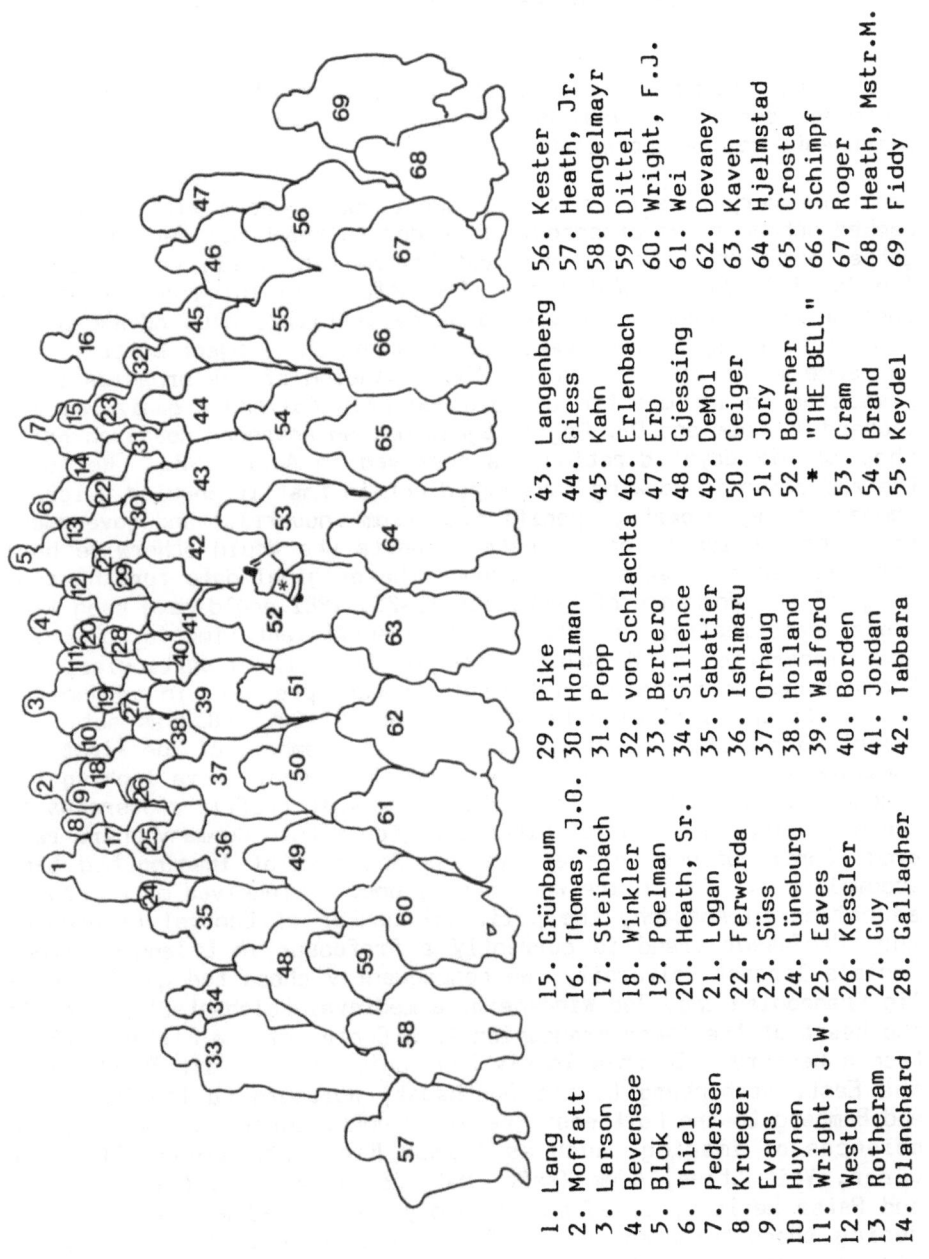

1. Lang
2. Moffatt
3. Larson
4. Bevensee
5. Blok
6. Thiel
7. Pedersen
8. Krueger
9. Evans
10. Huynen
11. Wright, J.W.
12. Weston
13. Rotheram
14. Blanchard
15. Grünbaum
16. Thomas, J.O.
17. Steinbach
18. Winkler
19. Poelman
20. Heath, Sr.
21. Logan
22. Ferwerda
23. Süss
24. Lüneburg
25. Eaves
26. Kessler
27. Guy
28. Gallagher
29. Pike
30. Hollman
31. Popp
32. von Schlachta
33. Bertero
34. Sillence
35. Sabatier
36. Ishimaru
37. Orhaug
38. Holland
39. Walford
40. Borden
41. Jordan
42. Tabbara
43. Langenberg
44. Giess
45. Kahn
46. Erlenbach
47. Erb
48. Gjessing
49. DeMol
50. Geiger
51. Jory
52. Boerner
  * "THE BELL"
53. Cram
54. Brand
55. Keydel
56. Kester
57. Heath, Jr.
58. Dangelmayr
59. Dittel
60. Wright, F.J.
61. Wei
62. Devaney
63. Kaveh
64. Hjelmstad
65. Crosta
66. Schimpf
67. Roger
68. Heath, Mstr.M.
69. Fiddy

DIRECTOR'S FOREWORD

This foreword deals exclusively with the planning, organization and execution of the Workshop's scientific, as well as cultural programs. An informal discussion about some of the organizational details, may recapture the spirit of this NATO Advanced Research Workshop and will reveal how successful it was and how its accomplishments may be appreciated.

From the very outset, a rather ambitious program with a densely packed schedule was prepared. In order to reach our set goals, together with Prof. Hans Brand and Dr. Martin Vogel, we approached (in April 1981) the NATO Scientific Affairs Division and requested funding for a combined Advanced Study Institute plus Advanced Research Workshop of two weeks' duration on the subject matter. After some minor alterations and the expansion of the organizational committee to include two experienced NATO ASI/ARW organizers, Mr. Leonard A. Cram and Dr. Dag T. Gjessing, an Advanced Research Workshop of six days' duration was approved in April 1982. We found the reduction in working days expedient in that it enabled a larger number of key experts, specifically from industrial and governmental laboratories to participate; experts who would otherwise have not been able to attend. Although the original date for the staging of the Workshop of September 12-25, 1982 could have been feasible logistically, in order to allow sufficient time for planning and preparation of Workshop arrangements and, in particular, for the authors to prepare print-ready manuscripts well in advance of the Workshop, a final date of September 18-24, 1983 was set. In selecting a suitable Workshop site we, the members of the first Organizing Committee — originating from FR Germany, were looking for a modern Congress Center located off the main traffic crossroads in the historical heart of a cultural region within Germany where relevant contributions to the Arts and Sciences of Imaging had been accomplished. The choice was an easy one. I received my secondary education in Ansbach, the political center of Central Franconia, and, Professor Brand is currently a Professor in Erlangen, Franconia's largest University, we consequently chose the quiet, idyllic Franconian spa, Bad Windsheim, a medieval Reichsstadt placed in the heart of the famed Franconian Arts Center of the Fifteenth/Sixteenth Century. In this lovely Franconian region with Nürnberg to the East, Rothenburg to the Southwest, Würzburg to the Northwest and Bamberg to the Northeast, renowned contributors to the arts and sciences of imaging, such as Tilman Riemenschneider, Albrecht Dürer, Veit Stoss, Adam Kraft, Matthias Grünewald, Martin Behaim and Peter Henlein, lived and strongly interacted with one another during those turbulent years of the reformation. It was this region of Germany, which at that time, was the cultural heart of the Roman Empire of Germanic Nations North of the Alps; where the cultural and industrial crossroads of Europe, "South ↔ North", "Southwest ↔ Northeast", "Northwest ↔ Southeast", met, bringing great wealth, industrial and scholarly craftsmanship to this part of Franconia. This tradition of Franconia being a cultural and scientific center of the arts and sciences in imaging has regained new impetus and value today with the famed Siemens Zentral Laboratorium and Siemens Medical Division having their research centers in Erlangen, where exciting R&D&T advances in zeugmatography, nuclear magnetic resonance imaging are accomplished. Therefore, it was not a mere coincidence that the NATO-ARW-IMEI-1983 should take place in this historical heart of Germany. In Bad Windsheim, about forty km straight west from Erlangen, we found the ideal workshop center in the Kur und Kongress Hotel Residenz, and all those who actively participated in that Workshop will agree with us that it was a choice location. The center is embedded in the lovely Kurpark and is close to that dreamy, restful, pretty township, remote from modern life's hustle and well removed from rustling tourist traps.

It was the intent of the Workshop Organizing Committee to provide a forum for internationally renowned key experts to expand the interaction and to enlarge the scope of their activities in pursuit of promoting this rapidly growing new engineering sciences discipline of "Inverse Methods in Electromagnetic Imaging", which covers mathematical, metrological, numerical, signal processing, statistical, and also technical apparatus design topics in the engineering, physical and aeronomic sciences. As a consequence, the list of interested key experts became excessively large and, because the NATO-Scientific Affairs Division requested the total number of active participants not to exceed sixty, by too many, a careful choice in the selection of specific topic areas and of invitees had to be made so that a meaningful program and harmonious execution of the adopted approach could be guaranteed. We note here that by early September 1983, the number of key experts requesting late acceptance for participation rose steadily and reached twice as many as we were able to accommodate with the room capacity provided by the KuK Hotel Residenz. We regret that so many other valuable contributions to this, otherwise, well functioning and highly productive workshop did not materialize. The final total of accepted invited scientific participants was one hundred twenty-five, and that of accompanients was twenty-four, with approximately twenty no-shows due to illness or other sudden priority commitments. It was the determined position of the Organizing Committee to limit the total number of participants to that not exceeding the maximum available hotel capacity, and not to allow any participants to board in neighboring hotels. Thus, we achieved the assembly of all participants under one roof so as to allow optimum direct interaction and the ultimate use of a day's working hours (from 7:00 to 22:00 hours). The Workshop Director requested early (6:00) daily wake-up calls to be executed by a hotel bellboy for all participants, which indeed caused some rumbling complaints during the first two mornings of the NATO-ARW-IMEI-1983 conference, without which, however, we would otherwise not have successfully completed a long week of hard work and inspiring interaction.

The Workshop Organizing Committee had, at one time, considered the possibility of reducing the number of papers selected for oral presentation to approximately twenty and to collect the remaining papers in a Poster Session. The Workshop Director dropped this plan and every invited paper was presented and is included in the Workshop Proceedings. We note that invited papers which could not be presented due to illness or other priority commitments are also formally included in the Workshop Proceedings, except for those not cleared for publication by their administrations. In total some seventy-five papers were presented and, as a consequence, the duration for oral paper presentation and subsequent open discussions was regretfully limited. Yet, it was a pleasure to witness that the speakers, with the exception of some inexperienced junior members, obeyed the time limitations strictly. We also wish to mention that there were no simultaneous lectures scheduled nor presented; and that the language of paper presentations was exclusively English. Because all papers were of direct input to the working discussion group activities, the main purpose of oral paper presentation was to declare it an open forum for identifying unresolved questions which resulted from the research to be discussed in more detail in the Proceedings papers. Therefore, speakers were strictly advised to make optimum use of the alloted time, i.e. for a normal paper with twenty minutes time alloted for presentation, the first three minutes were to identify the specific new research under consideration, to summarize succinctly the research results of their Proceedings' paper during the next ten minutes; to use the next three to four minutes for clearly amplifying and stating the particular research problems that are unresolved; and the remainder was intended for questions from the audience. This procedure certainly required a great deal of concentration on the part of all participants, as well as strict adherence to the set session schedules which was admirably executed by all session chairmen. The Workshop Director permitted himself the liberty of using the KuK-Congress-Center-bell, whenever "Call to Order" was required, to get us back to the pre-set time schedule. In retrospect, everyone will agree that this approach not only worked, but, as so many stated in their enthusiastic, positive thank you notes, it also contributed to the optimum information exchange, stimulation of new ideas, and the creation of new interactions among participants who only knew of each other by names on publications, and will now desire to meet again in the near future. In concluding, the Organizing Committee wishes to thank all speakers and session chairmen for the demonstrated harmonious collaboration, for the excellence of paper content, as well as paper presentation and open-minded question handling.

In enriching the overall program, cultural events were scheduled for the evenings on a daily basis in addition to the two main cultural scientific events of Wednesday, September 21 and Saturday,

September 24, 1983. Furthermore, deliberately on late call, a rather functional ladies program was intelligently improvised with the dear assistance of Ms. Ursula Allmendinger, Verkehrsamt, Bad Windsheim, by Mrs. Barbara Kester, Mrs. Mary Jordan and Mrs. Eileen Boerner, whose inspiration made the cultural ladies tours, sightseeing, walking and biking tours a memorable event for the accompanients and also some scientists. The two cultural/scientific tour events were mainly planned, organized and executed by Prof. Dr.-Ing. Hans Brand and his pleasant institute staff members, Mrs. Margarete Geiger, Dipl-Ing. Siegfried Osterrieder and Dipl-Ing. Roman Glöckler who deserve our highest praise and admiration. The cultural evening programs were arranged with the assistance of Mr. Rolph K. Erlenbach, the Hotel Manager, Mr. Hans-Dieter Erb, the Congress Center Manager, and Ms. Ursula Allmendinger who went out of their way to help, advise and assist us in any way they could. Therefore, we wish to use this ocassion to express, on behalf of all participants, our admiration and gratitude for the thorough and sincere assistance we received during our stay in Bad Windsheim, not only from its friendly citizens during our walks and shopping sprees through their inviting township but, in particular, to the staff and management of the KuK Hotel Residenz, specifically, the head receptionist, Mrs. Rossmann and her friendly assistants, the most impressive, capable master of the Restaurant "die Brücke", Herr Dick, with his well selected team of friendly, attentive waiters, and not to forget our friend "Bruno". It was a pleasure to be their guests, because the excellent service we received and the peaceful atmosphere of the restaurant enabled us to work hard, yet at the same time, eat well and relax. Then, after "DAY'S END", some of us disappeared downstairs into the hotel lounge to enjoy the relaxed atmosphere, created by lively Fräulein Kleine. The KuK Hotel Residenz with its recreational facilities and the adjacent Kurpark certainly provided an ideal stimulating, yet pleasant environment for such a work-loaded Advanced Research Workshop, lying as it does in the historical heart of the modern German culture, Franconia.

In the following, some of the important highlights of our scientific/cultural events will be summarized with the intention of concluding the description of our NATO-ARW-IMEI-1983 which was a successful, well-rounded scientific as well as cultural experience.

Monday Evening Event (September 19, 1983)
We recall with pleasure the leisurely walk from the KuK Hotel Center through the Kurpark and downtown Bad Windsheim to the City Hall, where we were given a reception planned by the First Mayor Mr. Otmar Schaller, with the Second Mayor, Mr. Josef Heinrich present to receive us in the Historische Rathaussaal. For those U.S. citizens of German descent it was an historical event in that Francis Daniel Pastorius (1651-1719), the son of a former Bad Windsheim Mayor Pastorius (1675-1691), was raised and received his outstanding early education in this city, passed his inaugral disputa-

tion (doctorate of law) at the age of 25 in Altdorf bei Nürnberg, the famed medieval Franconian University which in the 18th century was moved to Erlangen, preparing him to become the co-founder of the first official German Settlement in the United States of America on October 6, 1683 in Germantown, PA. It is noteworthy to mention that it was Francis Daniel Pastorius the well travelled, highly respected lawyer and scholar, who formulated the first decree in Northern America banning slavery and racial segregation, in 1697, which later on became one of Abraham Lincoln's most admired documents on the subject matter. It was a pleasant evening and made everyone relax after having survived the first day of our "Arbeitszusammentreffen" which, indeed, was tough for many who arrived the day before from North America.

Tuesday Evening Event (September 20, 1983)

A total surprise was our pleasant reception in the Franconian Open Air Museum, where on a perfect "Indian-Summer" evening, we viewed original small villages and farms from the Fifteenth to Eighteenth Century in Franconia which were brought from the surrounding countryside, salvaged from complete destruction, and reassembled in this lovely location. The delicious carp dinner (the traditional Franconian fall festivity) was followed with the superb presentation by the "Bad Windsheimer Sänger", so masterfully conducted by Professor Horst Steinmetz, who, with his capable artists, introduced us to Franconian folk and art songs of the Fifteenth/Sixteenth Century including many enjoyable cultural originals created within the vicinity of Bad Windsheim. This lively event provided us with the pleasure to get to know many new and dear friends, to appreciate the deep humorous and relaxed character of the Franconians, and set the foundation for a harmonious working environment for the remainder of the Workshop.

Wednesday, Scientific/Cultural Tour (September 21, 1983)

By now, all of the participants of this NATO-ARW-IMEI-1983 were able to appreciate that not only are the scientific Workshop programs on schedule, but so are the cultural ones, and this according to a motto: "hard work deserves hard play"! This Wednesday Scientific/Cultural Tour in itself must have made it worth coming to Franconia, and our praise, admiration and thanks go equally to the Siemens Medical Division, Siemens AG., and the High Frequency Engineering Laboratories, UEN, both in the famed Hugenottenstadt, Erlangen. While Dr. Manfred Pfeiler, Dr. Werner J. Haas. and Dr. Günter Schwierz provided the financial support for the elegant Workshop portfolio, the bus tour, the scientific visits and most of the contributions towards the overhead of other cultural events; Prof. Dr.-Ing. Hans Brand and his institute staff planned these tour programs. Dipl-Ing. Roman Glöckler, who was responsible for organizing the tours, was unable to attend due to illness and other priority commitments. We all commend the pleasant, dear, intelligent and polite manner in which Dipl-Ing. Siegfried Osterrieder substituted and executed the tour programs so efficiently.

After the late morning lectures on Tomographic Imaging and the excellent overview of Dr. Günter Schwierz on the subject matter, introducing us to one of the most exciting fields of "Inverse Methods in Electromagnetic Imaging", namely Zeugmotography or Nuclear Magnetic Resonance Imaging, we were introduced to the development of this exciting new medical imaging technology which sooner or later may replace hazardous, detrimental x-ray tomographic methods. However, the ultimate deployment of these not only very demanding advanced imaging techniques, side-by-side, will also require a very expansive high technology base plus expertise in "Inverse Methods in Electromagnetic Imaging", at a cost level of about three to five million dollars per final imaging unit which includes very large, super-cooled magnets (5m x 3m x 3m), very sensitive RF receivers, fifth generation parallel processing computers equal to or probably larger than the CRAY-type, and many other extensive software supporting programs. To the participants of this workshop it was very interesting to be informed that as large and extensive a global enterprise as the Siemens, AG. is, the small and disadvantaged business component suppliers are now from almost all NATO-member countries, due to their highly selective technological expertise, providing another reason for our close future collaboration and the further expansion of the ASI/ARW programmes of the NATO-Scientific Affairs Division. It was also interesting to note that a great demand on improved "Inverse Methods in Electromagnetic Imaging" is still required to make Magnetic Resonance Imaging become of real practical use in everyday medical diagnostic imaging. We express our fond gratitude to all of the scientists and engineers who guided us and replied so knowledgeably and enthusiastically to our questions all pertaining to and demonstrating the rapid advancement of the new engineering sciences discipline of "Inverse Methods in Electromagnetic Imaging" of which Erlangen now has become one of the most important R&D&T centers, certainly another valid reason to have our NATO-ARW-IMEI-1983 take place adjacent to its centers of research activities in this new exciting engineering sciences discipline.

The subsequent visit of the "Germanische National Museum", where concurrently, the "500 Anniversary Exhibition Martin Luther (1483-1546)" and the "450 Anniversary Exhibition Veit Stoss (1440-1533)" were taking place, and the late evening "Franconian Smorgasbord at the Heiliggeist Spital" were well organized and culturally inspiring events. At the museum we split up into several groups, all visiting the Veit Stoss Exhibition, and unfortunately, missing the Martin Luther Anniversary, which took place in that Section of the Museum where Dürer's, Behaim's, Kepler's, Henlein's and all the other great artist's and scientists' of the Fifteenth/Sixteenth Century contributions to Imaging of Nature, the Human Mind, and the Extraterrestrial Bodies are collected. Yet, there is so much to be comprehended on the subject matter; and, our guided tours through

the Veit Stoss Exhibition certainly were gratifying in that the interaction of all of the contemporaries of this culmination point of
medieval Trans-European culture was so vividly demonstrated to us.
So, it is interesting to note, that Albrecht Dürer, the painter;
Veit Stoss, the painter/sculptor; and Tilman Riemenschneider, the
sculptor/artist/statesman had produced combined works of art in
Imaging, not only of the lively body, but also of the human mind,
certainly the most daring and demanding subfield of expertise in
the general sciences of inverse methods. In concluding this exciting visit to the Germanische National Museum in Nürnberg, it is
worthwhile mentioning that Albrecht Dürer set some formal foundations of tomographic imaging in his treatise on stereographic projections. Thus, all those participants and others who are going to
revisit Nürnberg, don't miss the Germanische National Museum and
the Dürer Haus.

This exciting day was concluded with an elegant, neatly prepared,
Franconian Smorgasbord Buffet at the Heiliggeist-Spital, located on
a bridge across the river Pegnitz, which provided another event of
socializing and of finding new friends and scientific interaction.
By suprise, the traditional "Altstadt Fest: Die Pegnitz brennt!"
took place with fireworks right below the restaurant on the river,
an historic event which dates back to the middle ages. By this
time, many of those who wanted to depart and disappear in tourist
traps, hung on to us, with the understanding and conviction that
cultural/scientific offerings of a NATO-ARW, such as ours, could in
no way be surpassed or substituted by individually arranged side-
stepping tours, i.e. we were successful in aiding a flock of approximately one-hundred-and-thirty individualistically minded pro-
fessors and engineering scientists to actively live together and
interact with one another until the very end of the Workshop.

Friday Evening Workshop Dinner (September 23, 1983)
After the closing down of the Scientific Paper Presentations and
completing the Working Discussion Group Reports, it was a delight
to sit down, relax and get prepared for the final event, the introduction to the Life of Tilman Riemenschneider, Artist/Scholar/-
Statesman. Before our invited dinner speaker, Mr. William McKee
Wisehart, was given access to the podium, Dr. Dag T. Gjessing, Dr.
Tilo Kester and Mr. Leonard A. Cram discharged the Workshop Director from his duty and made him replace the KuK Congress Center
(school) bell by a tiny mountain bell (with Saxionian pitch: very
high in tone, clear and not too loud). With pleasure he accepted
this relief and handed over the loud instrument which was used for
"Call to Order", to the next NATO-ARW Director, Dr. Robert C.
Worrest, of the Oregon State University in Corvallis, who was thus
given the maddening instrument to punctually commence with his
workshop on Monday morning, September 26, 1983 (who knows whether
he needed the bell for some thirty or less participants). Here, I
also wish to thank my friend, Prof. Walter K. Kahn, for participat-

ing in the Workshop, and for making himself feel at home again in the beloved homeland of his honorable ancestors who had contributed so much to its wealth of knowledge, scholarship, and industry. Also, to say thank you to all the participants for some lively days of hard scientific work and friendly, accomodating and intelligent interaction.

The post-dinner speech of Bill McKee Wisehart was by no means a standard one, and with intelligent, sensitive elaborations, he introduced us, most of whom were not capable of speaking the German Language, nor familiar with its way of life, into the Franconian sphere of life at the turn of the Fifteenth/Sixteenth Century and then, step-by-step, made us appreciate the artistic and scholarly achievements of one of mankind's greatest craftsman, Tilman Riemenschneider. Bill had done a marvelous job and no Professional Scholar of the Historical Fine Arts could easily have replaced him. It is important to note that with his fine selection of slides he was able to trace the connections and interactions of Tilman Riemenschneider with Veit Stoss and Albrecht Dürer and to explain why Tilman's contribution became almost forgotten. In the early 1500's Bad Windsheim also sponsored and displayed one of Riemenschneider's Works of Art, which in reproduction has been recently reinstalled in its original shrine in the Stadtkirche.

In the final presentation of the evening, Prof. Brand and Dipl-Ing. Osterrieder were kind enough to provide us with a preview of the Workshop Concluding Scientific/Cultural Tour planned for Saturday which was centered on the artistic productions of the sculptor, Tilman Riemenschneider. Every intermediate stop and the highlights of that program were clearly identified. Both the presentations of Bill McKee Wisehart and of Prof. Brand, with Dipl-Ing. Osterrieder complimented one another very well, and got us all excited about participating in the Final Workshop Close-Down Tour Activities. Equal thanks are extended to all three.

Saturday Cultural/Scientific Close-Down Tour (September 24, 1983)
As requested by the KuK Hotel Residenz management, by 9:00 AM all participants of this NATO-ARW-IMEI-1983 had vacated the hotel. Approximately sixty participants were transported by bus, others followed in their rental/own cars and departed Bad Windsheim, guided across the Frankenhöhe to Rothenburg by Bill, who, himself a U.S. citizen and fine scholar, carefully explained all the specific differences in village structure, farming, forresting, etc. that could be observed during our tour, which was highly appreciated by those not acquainted with Franconia and/or Central Europe.

In Rothenburg, a totally rebuilt city, everyone was on his own to view this "idol" of a medieval township: city walls, narrow "Gassen", medieval wood-framed housefronts, old cathedrals and also a Judaic cemetery rebuilt next to the house from which the famed scholar of the Talmud, and temporary citizen of Rothenburg, Meier Ben Baruch (1215-1293), fought against the unsuccessful attempt by

Emperor Rudolf to have the Jews pay taxes, resulting in his impri-
sonment (1286) and death in the Emperial prison at Worms. The main
event though, was to view one of Tilman Riemenschneider's master-
pieces, the Heiligblutaltar in the St. Jakob's Kirehe, Rothenburg
ob der Tauber. After viewing the many memorabilia of the middle
ages, most of us must have been grateful that the Workshop took
place in Bad Windsheim and not in Rothenburg; otherwise, many of us
would have simply gone astray in this quaint and inviting tourist
trap which has so many exciting views to offer.

Bill guided us during the next part of the bus tour to Creglingen,
where we admired the Marienaltar, which by many occidental and ori-
ental fine arts historians is considered to be the culmination and
absolute perfection in wood carving, demonstrating the grandeur of
that Franconian phase of high cultural, but also high technological, activities (e.g. the colour woodprinting process invented by
Albrecht Dürer, the first pocket-watch instrumented by Peter
Henlein, the first globe created by Martin Behaim, etc.)

In leaving Creglingen, we soon departed from the deep-rutted Tauber
valley and ended up near the Käpelle above Würzburg for lunch under
open blue skies to enjoy the view of the River Main, and the famed
baroque spires of Würzburg, being overpowered by the mighty fort-
ress, Marienburg, encroaching the river and the city on the south
side. We mention here that the 450 Anniversary Exhibition of
Tilman Riemenschneider (1460-1531) took place in the Mainfränkische
Museum in the Marienburg in, 1981, setting the prerogatives of the
Workshop Director in selecting this cultural/scientific program.
After viewing the baroque Käpelle or the Roccocco Residenz, with
the famed wall paintings and frescoes by the Venetian Master,
Giovanni Batista Tiepolo (1696-1770), and its famous fountain, "Der
Künstler Brunnen", depicting the contributions of three of Main-
franken's world-famed artists: Walther von der Vogelweide (1170-
1230), Matthias Grünewald (1465-1535), and Tilman Riemenschneider,
the tour bus returned along the main valley road and then via the
Autobahn to the Nürnberg-Airport and to Nürnberg Hauptbahnhof,
where the Workshop came to a fulfilled end. The remaining participants dispersed for a relaxing weekend either in Nürnberg, in nearby München and/or the Alps and we hope that they all returned home
safely with the understanding that there is a good reason why we
need to have a North Atlantic Treaty Organization and that its Scientific Affairs Division serves the noble purpose of reminding us
that there is a lot to be protected in any one of our many NATO
member countries, rich in history, talent, and free societies.

In retrospect, it had been the intention, that next to accomplish-
ing outstanding technical and scientific workshop results, to introduce the participants of this NATO-ARW-IMEI-1983 to the rich
cultural and scientific life of one of the most productive periods
of German and European history. Although born in Austral-Asia, of
Saxonian descent, the Workshop Director made Franconia his

"Wahlheimat" (motherland of choice) and prefers to speak its dialect rather than use the official instrument of German linguistic interaction "Hochdeutsch". The reason for this laudable action is his early realization that this part of Germany had discovered a lot of basic truths, produced honest statesmenship and generated great artistic values which still are not fully comprehended, but had so much to offer to make real day-to-day life more livable and easier to appreciate for all of mankind. Hence, the Workshop Director personally hopes that all participants acquired deep respect for this dreamy, well-tempered stronghold of German medieval culture which had so much to offer toward obtaining a more realistic view, down-to-earth pure image of our terrestrial and celestial environment.

In preparing such an extensive NATO-ARW, enormous secretarial assistance, organizational ability and know-how was required. The prime source of such contributions came from within the Communications Laboratory, Department of Electrical Engineering & Computer Science, Univeristy of Illinois at Chicago, and all our thanks go directly to Richard and Deborah Foster, without whose talents and dedication the pre-workshop and post-workshop processing of all relating material, particularly producing the Workshop Proceedings could not have been accomplished. Here, we would like to thank Drs. Sujeet K. Chaudhuri, Koroda Umashankar and Anthony J. Devaney for their assistance in proof reading the manuscripts.

On this occasion, I would like to express my very special, warm and hearty thanks to Dr. Mario Di Lullo and Dr. Craig Sinclair, the Directors of the NATO-Scientific Affairs Division, ASI/ARW Programmes, within the NATO Headquarters in Brussels; together with Dr. Tilo Kester and his capable, dear wife, Barbara Kester, of the NATO Publications Coordination Office, for their generous advice and many discussion hours so freely offered to me during the pre-planning and organization phase of this NATO-ARW-IMEI-1983.

Last, but not least, my understanding wife, Eileen Annette, deserves my deep appreciation for her patient tolerance of the many long nights and lost weekends of extra work during the entire pre to post workshop engagement.

Wolfgang-M. Boerner
The Workshop Director

NATO-ARM-IMEI-1983
Friday, Sept. 23, 1983, 18:30
KuK Hotel Residenz, Bad Windsheim, FR Germany
The Ladies Group

1. Barbara Kester
2. Joan Rotheram
3. Vila Eaves
4. Virginia Jory
5. Jackie Weston
6. Mary Jordan
7. Susan Moffatt

8. Mae Bevensee
9. Mary Kennaugh
10. Aartji Block
11. Dianne Logan
12. Cindy Manson
13. Phylene Raven
14. Carli Poelman

15. Dixie Carpenter
16. Rie Fewerda
17. Ruth Wright
18. Etty Huynen
19. Christine de Mol
20. Avril Guy
21. Maureen Anderson
22. Eileen Boerner

Photographed by: Bruce Z. Hollmann
Portrayed by   : Richard W. Foster
Reproduced by : John Darby
Available from : Communications Lab/EECS Dept., M/C 154
                 University of Illinois at Chicago
                 P.O. Box 4348, Chicago, IL 60680 USA

ORGANIZATIONAL COMMITTEE

PLANNING COMMITTEE

DIRECTOR: Professor Wolfgang-M. Boerner
               CL-EECS, University of Illinois at Chicago
               Chicago, IL 60680 USA

CO-DIRECTORS:

Professor Hans Brand
Universität Erlangen-Nürnberg
Erlangen, FR GERMANY

Mr. Leonard A. Cram
THORN EMI, Radar Electronics
Wells/Somerset, UK

Professor Dag T. Gjessing
ESTP, Royal Norwegian Council
for Industrial and Scientific
Research, Kjeller, NORWAY

Dr. Wolfgang Keydel
DFVLR, Institute for Radio
Frequency Technology
Oberpfaffenhofen, FR GERMANY

Dr. Günter Schwierz
SIEMENS A.G., Medical Division
Erlangen, FR GERMANY

Dr. Martin Vogel
DFVLR, Institute for Radio
Frequency Technology
Oberpfaffenhofen, FR GERMANY

NATO-ADVISORS

Dr. Mario Di Lullo
ARW/ASI Programmes, Scientific
Affairs Division
NATO HQ, Brussels BELGIUM

Dr. Tilo Kester
NATO-Publications
Coordination Office
Overyse, BELGIUM

Professor Anton de Hoop
Delft University of Technology
Delft, THE NETHERLANDS

Dr. Craig Sinclair
ARW Programme, Scientific
Affairs Division
NATO HQ, Brussels, BELGIUM

TECHNICAL COMMITTEE AND EDITORIAL BOARD

Professor Wolfgang-M. Boerner
Professor Hans Brand
Dr. Günter Schwierz
Dr. Arthur K. Jordan
SSD, Naval Res. Lab., USA

Mr. Leonard A. Cram
Dr. Wolfgang Keydel
Professor Dag T. Gjessing
Dr. Martin Vogel

COORDINATORS OF SCIENTIFIC/CULTURAL TOURS

Dipl.-Ing. Roman Glöckler
Dipl.-Ing. Siegfried Osterrieder
UEN, Erlangen, FR GERMANY

Professor Hans Brand
Dr. Günter Schwierz
Professor Wolfgang-M. Boerner

HOSTESSES OF THE LADIES/CULTURAL PROGRAMMES

Frau Margarete Geiger
UEN, Erlangen, FR GERMANY
Mrs. Eileen A. Boerner
Northbrook, IL, USA

Frau Ursula Allmendinger
Verkehrsamt, Bad Windsheim, FRG
Frau Barbara Kester
Overyse, BELGIUM

WORKSHOP PROGRAM AND PROCEEDINGS DEVELOPMENT AND PREPARATION

Mr. Richard W. Foster   Mrs. Deborah A. Foster
CL-EECS, University of Illinois at Chicago, Chicago IL, USA

0.1 (OS.3)

TRANSIENT CURRENT DENSITY WAVEFORMS ON A PERFECTLY
CONDUCTING SPHERE

E.M. Kennaugh
D.L. Moffatt

The Ohio State University
Department of Electrical Engineering
2015 Neil Avenue
Columbus, Ohio 43212

FOREWORD

The transient current densities induced on a conducting
spherical scatterer by an incident plane wave with impulsive
time dependence, while in existence for some time at the
ElectroScience Laboratory, have never been formally published.
The time-dependent current density waveforms are an elegant
example of how the impulse response concept can illuminate and
simplify relatively complex relationships. Much of the time
domain research by Prof. E.M. Kennaugh, together with his
students and colleagues, has appeared in the literature. This
paper, while it is not his final work, serves to further
complete a report of his unique contributions. I am indebted
to Prof. W-M Boerner for the opportunity to include this paper
in the Workshop Proceedings.

ABSTRACT

The time-dependent surface current density waveforms at
various points on the surface of a perfectly conducting sphere
illuminated by a plane electromagnetic wave are presented. The
incident plane wave has an impulsive or shock-type time
dependence. Current density waveforms in the E-plane (strong
creeping wave) and H-plane on both the "lit" and shadowed sides
of the sphere are shown. On the illuminated side of the sphere,
removal of the Kirchhoff or physical optics approximation for
the surface current density permits a detailed examination

1

W.-M. Boerner et al. (eds.), Inverse Methods in Electromagnetic Imaging - Part 1, 1–31.
© *1985 by D. Reidel Publishing Company.*

of the deficiencies of this approximation for short times (high frequencies). From these results the form of a first order correction to the Kirchhoff current is suggested. On the shadowed side of the sphere, the changing character of the current waveform is noted.

INTRODUCTION

Development of the impulse response concept in 3-dimensional electromagnetic scattering problems[1,2] was materially aided by a Fourier synthesis procedure whereby harmonically related complex scattering data were used to synthesize approximate far-zone canonical response waveforms[2,3,4,5]. The same procedure permits one to generate the time-dependent surface current density waveforms on a scatterer or radiator. A comparison of Fourier synthesis and space-time integral equation calculations has been made[6]. Both procedures have an inherent resolution limitation dictated either by a finite summation (Fourier synthesis) or a Gaussian pulse excitation (space-time integral equation), but it is not correct to suggest that an infinite summation[7] or equivalently an impulsive excitation must be used. However, some care must be exercised in interpreting singularities and/or jump discontinuities particularly if they occur simultaneously.

Impetus for this paper stems from a depolarization correction to physical optics derived by Bennett[8] and utilized by Boerner[9]. We show that even for an object where no depolarization occurs (the conducting sphere), the form of a basic correction to the Kirchhoff current can be deduced. It is suggested that similar first order corrections will improve the Kirchhoff approximation for a number of different object geometries.

Surface Current Density Waveforms

Mie series calculations of the tangential magnetic field and radial electric field at the surface of a conducting sphere of radius a immersed in an incident plane wave have been made for sphere circumferences in wavelengths of 0.2 (0.2)20.0. Locations on the sphere surface in both the E-plane and H-plane for theta angles of 0 (15) 180 degrees were used with 0 degrees corresponding to the specular point. The real, time-dependent surface current density waveforms synthesized from these data are shown in Figures 1, 2 and 3. Figure 1 combines the E-plane (positive) and H-plane (negative) waveforms on the illuminated side of the sphere. The abscissa scale is in units of transit time for the sphere diameter. On the illuminated side of the sphere, the weighted quasi-impulses at the origin of each waveform illustrate the resolution obtained (approximately o.75a). On the illuminated side of the sphere, the H-plane waveforms have a sign reversal. The abscissa scales are in units of transit time for the sphere diameter. On the illuminated side of the sphere, the current density is primarily impulsive; details for short times are best seen after the Kirchhoff approximation for the current is removed. There is clear evidence of the creeping wave contributions. The surface current density waveforms on the shadowed side of the sphere are shown in Figure 2 (E-plane) and Figure 3 (H-plane). Note that the ordinate scales for H-plane are not constant. On the shadowed side of the sphere note that the character of the current density at the onset of the waveform is still apparently impulsive near the shadow boundary (theta less than 120 degrees) in the H-plane but not in the E-plane. There is a strong creeping wave contribution in the E-plane and some evidence of a very weak creeping wave contribution in the H-plane.

In Figure 4 the surface current density waveforms on the illuminated side of the sphere are shown after the Kirchhoff approximation to the current density has been removed. The H-plane waveforms are dashed. It is very evident that as the observation point progresses from the specualr point to the shadow boundary, the Kirchhoff approximation becomes progressively too large in the E-plane and too small in the H-plane. At a given angle, the magnitude of the errors in the E-plane and H-plane are approximately equal. This would seem to imply that on a cut at 45 degrees from the E-plane and H-plane the Kirchhoff approximation to the current may be quite good even near the shadow boundary. In the E-plane at theta equal 60 degrees and 75 degrees one can also clearly discern two distinct creeping wave contributions, both launched from the shadow boundary and delayed from the waveform onset by different travel times on the sphere surface.

Correction

Based on the results shown for the conducting sphere, the following formulas for correcting the physical optics estimates for the induced currents are suggested for smooth (no edge) scatterers. In the E-plane modify the physical optics currents by the multiplicative factor

$$[1 - K \sin\theta] \quad . \tag{1}$$

In the H-plane modify the physical optics currents by the multi-plicative factor

$$[1 + K \frac{\sin\theta}{\cos\theta}] \quad , \tag{2}$$

where θ is the angle between the outward normal to the surface of scatterer and the backscatter direction. The H-plane correction becomes infinite at the shadow boundary ($\theta=90°$) but when multiplied by the physical optics current estimate ($\cos\theta$ factor) yields a finite current estimate. The specular error (exact minus physical optics) before correction is shown in Figure 5. Figures 6 and 7 show the effect of the correction term for values of the constant K of 0.2 and 0.1, respectively. For the simple form of the correction suggested, a value K near 0.1 appears optimum. Other forms for the correction, possibly exponential, may suggest themselves to the reader. Our purpose here, however, was to show that simple corrections are possible.

Conclusions

It has been demonstrated that relatively simple formulas can be developed for correcting the physical optics estimates of the current densities induced on a conducting spherical scatterer. It has also been suggested that similar simple corrections could be made for other "smooth" scatterers. It is noted that the corrections indicated here are not related to depolarization properties of the scatterer.

The surface current density waveforms on the shadowed side of the sphere have been included here for the possible benefit of other researchers. The physical optics estimate has long been a favorite direct scattering solution for inverse scattering because the estimate can be generally related to a target cross sectional area function on the illuminated side of the scatterer. A similar relationship (general) for the shadowed side of the scatterer would greatly improve imaging techniques based on the physical optics estimate.

REFERENCES

[1]   E.M. Kennaugh and R.L. Cosgriff, "The Use of Impulse Responses in Electromagnetic Scattering Problems," 1958 IRE Nat'l Conv. Rec., Pt. 1, pp. 72-77.

[2]   E.M. Kennaugh and D.L. Moffatt, "Transient and Impulse Response Approximations," Proc. IEEE, Vol. 53, No. 8, pp. 893-901, August, 1965.

[3]   E.M. Kennaugh, "The Scattering of Short Electromagnetic Pulses by a Conducting Sphere," Proc. IRE (Correspondence) Vol. 49, p. 380, January, 1961.

[4]   D.L. Moffatt and E.M. Kennaugh, "The Axial Echo Area of a Perfectly Conducting Prolate Spheroid," IEEE Trans. on Antennas & Propagation, Vol. AP-13, pp. 401-409, May, 1965.

[5]   E.M. Kennaugh and D.L. Moffatt, "On the Axial Echo Area of the Cone Sphere Shape," Proc. IRE (Correspondence) Vol. 50, p. 199, February, 1962; see also K.M. Siegel, et. al., "Radar Cross Section of a Cone Sphere," Proc. IRE (Correspondence) Vol. 51, pp. 231-232, January, 1963.

[6]   C.L. Bennett, "The Numerical Solution of Transient Electromagnetic Scattering Problems," SCRC-RP-77-55, Sperry Research Center, August, 1977.

[7]   E.M. Kennaugh and D.L. Moffatt, Comments on "Impulse Response of a Conducting Sphere Based on Singularity Expansion Method," Proc. IEEE, Vol. 70, No. 3, pp. 294-295, March, 1982.

[8]   C.L. Bennett, A.M. Auckanthaler, R.S. Smith and J.D. DeLorenzo, "Space Time Integral Equation Approach to the Large Body Scattering Problem," Sperry Rand Research Center, Sudbury, Mass., May 1973. (AD 763794) (RADC-CR-73-70)

[9]   S.K. Chaudhuri and W-M. Boerner, "A Monostatic Inverse Scattering Model Based on Polarization Utilization," Applied Physics, Vol. 11, pp. 337-350, 1976.

ADDITIONAL REFERENCES

A relatively complete list of publications and reports of The Ohio State University's ElectroScience Laboratory on the application of time domain concepts to electromagnetic scattering and radiation problems is given here.

Publications

[1]   E.M. Kennaugh and R.L. Cosgriff, "The Use of Impulse Responses in Electromagnetic Scattering Problems," 1958 National Convention IRE Record, Pt. 1, pp. 72-77.

[2]   E.M. Kennaugh, "The Scattering of Short Electromagnetic Pulses by a Conducting Sphere," Proc. IRE (Correspondence), Vol. 49, 1961, p. 380.

[3]   E.M. Kennaugh and D.L. Moffatt, "On the Axial Echo Area of the Cone Sphere Shape," Proc. IRE (Correspondence) Vol. 50, 1962, p. 199. Also see Proc. IRE (Correspondence) Vol. 51, 1963, p. 232.

[4] D.L. Moffatt and E.M. Kennaugh, "The Axial Echo Area of a Perfectly Conducting Prolate Spheroid," Proc. IEEE (Correspondence), Vol. 52, 1964, pp. 1252-1253.

[5] D.L. Moffatt and E.M. Kennaugh, "The Axial Echo Area of a Perfectly Conducting Prolate Spheroid," IEEE Trans. on Antennas and Propagation, Vol. AP-13, 1965, pp. 401-409.

[6] E.M. Kennaugh and D.L. Moffatt, "Transient and Impulse Response Approximations," Proc. IEEE, Vol. 53, 1965, pp. 893-901.

[7] D.L. Moffatt, "The Echo Area of a Perfectly Conducting Prolate Spheroid," IEEE Trans. on Antennas and Propagation, Vol. AP-17, No. 3, 1969, pp. 299-307.

[8] D.L. Moffatt, "Impulse Response Waveforms of a Perfectly Conducting Right Circular Cylinder," Proc. of IEEE, Vol. 57, No. 5, 1969, pp. 816-817.

[9] D.L. Moffatt, R.H. Paul and R.A. Voss, "The Echo Area of a Perfectly Conducting Prolate Spheroid," IEEE Trans. on Antennas and Propagation, AP-21, No. 2, p. 231, March 1973.

[10] J.D. Young, D.E. Svoboda and W.D. Burnside, "A Comparison of Time- and Frequency-Domain Measurement Techniques in Antenna Theory," IEEE Trans., Vol. AP-21, No. 4, 1973, pp. 581-583.

[11] D.L. Moffatt and R.K. Mains, "Detection and Discrimination of Radar Targets," IEEE Trans. on Antennas and Propagation, Vol. AP-23, No. 3, 1975, pp. 78-83.

[12] J.D. Young, "Radar Imaging from Ramp Response Signatures," IEEE Trans. on Antennas and Propagation, Vol. AP-24, No. 3, May 1976.

[13] D.L. Moffatt and R.J. Puskar, "A Subsurface Electromagnetic Pulse Radar," Geophysics, Vol. 41, No. 3, 1976, pp. 506-518.

[14] C.W. Chuang and D.L. Moffatt, "Natural Resonances Via Prony's Method and Target Discrimination," IEEE Trans. Aerospace Electronic Systems, Vol. 12, No. 5, 1976.

[15] K.A. Shubert, J.D. Young and D.L. Moffatt, "Synthetic Radar Imagery," IEEE Transactions on Antennas and Propagation, Vol. AP-25, No. 4, 1977.

[16] D.L. Moffatt and K.A. Shubert, "Natural Resonances Via Rational Approximants," IEEE Transactions on Antennas and Propagation, Vol. AP-25, No. 5, 1977.

[17] L.C. Chan, D.L. Moffatt and L. Peters, Jr., "A Character-
      ization of Subsurface Radar Targets," Proceedings of the
      IEEE, Vol. 67, No. 7, 1979.

[18] L.C. Chan, D.L. Moffatt and L. Peters, Jr., "Estimation
      of the Complex Natural Resonances from a Class of Sub-
      surface Targets," in Acoustic, Electromagnetic and Elastic
      Wave Scattering - Focus on the T-Matrix Approach. Edited
      by V.K. Varadan, Pergamon Press, 1980, pp. 463-482.

[19] J.D. Young, "Approximate Image Reconstruction from Tran-
      sient Signature," Part 9 of Acoustic, Electromagnetic
      and Elastic Scattering - Focus on the T-Matrix Approach,
      (edited by Varadan and Varadan), Pergamon Press, 1980.

[20] E.M. Kennaugh, "Opening Remarks," IEEE Transactions on
      Antennas and Propagation, Vol. AP-29, No. 2, 1981.

[21] D.L. Moffatt, J.D. Young, A.A. Ksienski, H.C. Lin and
      C.M. Rhoads, "Transient Response Characteristics in Iden-
      tification and Imaging," IEEE Trans. on Antennas and
      Propagation, Vol. AP-29, No. 2, 1981. Invited.

[22] E.M. Kennaugh, "The K-Pulse Concept," IEEE Trans. on
      Antennas and Propagation, Vol. AP-29, No. 2, 1981.

[23] D.L. Moffatt, "Ramp Response Radar Imagery Spectral Con-
      tent," IEEE Trans. on Antennas and Propagation, Vol.
      AP-29, No. 2, March, 1981.

[24] L.C. Chan, D.L. Moffatt and L. Peters, Jr., "Improved
      Performance of a Subsurface Radar Target Identification
      System Through Antenna Design," IEEE Trans. on Antennas
      and Propagation, Vol. AP-29, No. 2, 1981. Invited.

[25] E.M. Kennaugh, "Polarization Dependence of RCS - A
      Geometrical Interpretation," IEEE Trans. on Antennas
      and Propagation, Vol. AP-29, No. 2, 1981.

[26] L.C. Chan, D.L. Moffatt and L. Peters, Jr., "Subsurface
      Radar Target Imaging Estimates," IEEE Trans. on Antennas
      and Propagation, Vol. AP-29, No. 2, March, 1981.

[27] J.D. Young and J.L. Volakis, "Phase Linearization of a
      Broad-Band Antenna Response in Time Domain," IEEE Trans.
      Antennas and Propagation, Vol. AP-30, No. 2, March, 1982.

[28] D.L. Moffatt and C.M. Rhoads, "Radar Identification of
      Naval Vessels," IEEE Transactions on Aerospace and
      Electronic Systems, Vol. AES-18, No. 2, March, 1982.

[29] T.W. Johnson and D.L. Moffatt, "Electromagnetic Scattering
      by an Open Circular Waveguide," Radio Science, Vol. 17,
      No. 6, pp. 1547-1536, November-December, 1982.

[30] E.M. Kennaugh and D.L. Moffatt, Comments on "Impulse
      Response of a Conducting Sphere Based on Singularity
      Expansion Method," Proceedings of the IEEE, Vol. 70,
      No. 3, March, 1982.

[31] J.L. Volakis and L. Peters, Jr., "Improved Identification
      of Underground Targets Using Video Pulse Radars by
      Elimination of Undesired Natural Resonances," IEEE Trans.
      Antenna Propagation, Vol. AP-31, No. 2, 1983.

[32] D.L. Moffatt, C-Y. Lai and T.C. Lee, "Time Domain Electro-
      magnetic Scattering by Open Ended Circular Waveguide and
      Related Structures," To be published in WAVE MOTION.

Reports

MONOSTATIC AND BISTATIC MEASUREMENT OF SCATTERING
SHAPE AND SYNTHESIS OF SCATTERING SHAPES
Air Force Cambridge Research Center
Contract No. AF 19(604)-6157
Project No. 1073

J.W. Eberle and R.W. St. Clair, "Echo Area of Combinations of
     Cones, Spheroids, and Hemispheres as a Function of Bi-
     static Angle and Target Aspect," 1073-1, 30 June 1960.
     (AD 242472) (AFCRC-TN-60-795)

R. Tsu, "The Theory and Application of the Scattering Matrix
     for Electromagnetic Waves," 1073-2, 1 August 1960.
     (AD 243689) (AFCRC-TN-60-950)

R. Tsu, "The Evaluation of Incomplete Normalization Integrals
     and Derivatives with Respect to the Order of Associated
     Legendre Polynomials," 1073-3, 1 April 1960. (AD 238953)
     (AFCRC-TN-60-376)

"Problems in Electromagnetic Scattering Analysis," Final Report,
     1073-4, 1 January 1961. (AD 255839) (AFCRL-193)

RESEARCH ON RADAR CAMOUFLAGE AND ANTENNA ECHOS
Aeronautical Systems Division
Contract No. AF 33(616)-8039
Project No. 1223

D.L. Moffatt and S.A. Redick, "Memorandum on a Camouflaged Model
     of the B-47 Aircraft," 1223-1, 1 August 1961. (AD 326301)

D.L. Moffatt, "Selective Camouflage of Spheres and Cylinders," 1223-2, 15 October 1961. (AD 327636)

R.J. Garbacz, "Electromagnetic Scattering by Radially Inhomogeneous Spheres," 1223-3, 9 January 1962. (AD 275370)

D.L. Moffatt, "Low Radar Cross Sections, The Cone-Sphere," 1223-5, 1 June 1962. (AD 283338)

Annual Summary Engineering Report, 1223-6, 1 January 1962. (AD 328318)

R.J. Garbacz, "A Memorandum on a Measurement Technique for Low Cross Sections," 1223-7, 1 May 1962. (AD 276102)

R.J. Garbacz, "The Determination of Antenna Parameters by Scattering Cross-Section Measurements. I. Antenna Impedance," 1223-8, 30 September 1962. (AD 286760)

R.J. Garbacz, "The Determination of Antenna Parameters by Scattering Cross-Section Measurements. II. Antenna Gain," 1223-9, 30 November 1962. (AD 297953)

R.J. Garbacz, "The Determination of Antenna Parameters by Scattering Cross-Section Measurements. III. Antenna Scattering Cross Section," 1223-10, 30 November 1962. (AD 295031)

E.M. Kennaugh and D.L. Moffatt, "A Memorandum of Antenna Camouflage," 1223-11, 31 December 1963. (AD 347113)

D.L. Moffatt, "Determination of Antenna Scattering Properties from Model Measurements," 1223-12, 1 January 1964. (AD 441216)

Annual Summary Engineering Report, 1223-13, 31 December 1962. (AD 335006)

D.L. Moffatt, "Camouflage of the Parabolic Reflector Antenna," 1223-14, 1 April 1963.

D.L. Moffatt and R.H. Ott, "Echo Patterns of Spheres with Radial Stubs," 1223-15, 10 May 1963. (AD 337610)

E.M. Kennaugh and R.H. Ott, "Fields in the Focal Region of a Parabolic Receiving Antenna," 1223-16, 31 August 1963. (AD 420437)

R.B. Green, "The General Theory of Antenna Scattering," 1223-17, 31 August 1963. (AD 429186)

S.J. Skarote, "A Parabolic Reflector Receiving Antenna Utilizing Heterodyne Conversion and Control at the Focus," 1223-18, 1 January 1964. (AD 351271)

Final Engineering Report, 1223-19, 1 January 1964. (AD 348098)

RESEARCH ON RADAR ECHO REDUCTION STUDIES
Air Force Avionics Laboratory
Wright-Patterson Air Force Base, Ohio
Contract No. AF 33(615)-1318
Project No. 1774

E.M. Kennaugh and D.L. Moffatt, "The Axial Echo Area of a Perfectly Conducting Prolate Spheroid," 1774-1, 15 June 1964 (AD 446398)

J.H. Richmond, "Narrow-Band Window Design for Antenna Camouflage," 1774-2, 1 September 1964. (AD 354599)

R.H. Ott, "Camouflage of the Parabolic Reflector Type Antenna: II. Use of Tuned Surfaces," 1774-3, 1 September 1964. (AD 354979)

R.H. Ott, "Design Curves for Series RLC Tuned Surfaces," 1774-4, 22 January 1965. (AD 457198)

R.E. Ryan, Jr., "Camouflage of the Parabolic Reflector Type Antenna: III. The Concave Fresnel Reflector," 1774-5, 26 Janaury 1965. (AD 358349)

R.H. Ott and S.A. Redick, "Camouflage of the Parabolic Reflector Type Antenna: IV. The OSU Tuned Reflector," 1774-6, 5 January 1965. (AD 358348)

D.L. Moffatt, R.H. Ott and C.E. Ryan, Jr., "Internal Memorandum, A Memorandum on Test Results of a Camouflaged Parabolic Reflector Type Antenna," 1774-7, 22 April 1965.

Annual Summary Engineering Report, 1774-8, 31 December 1964 (AD 358054)

J.H. Richmond and R.H. Ott, "A Linear-Equation Solution for Scattering by a Periodic Plane Array on Thin Conducting Plates," 1774-9, 1 May 1965.

R.H. Ott and S.A. Redick, "Internal Memorandum, Camouflage of the Parabolic Reflector Type Antenna: V. Monostatic Echo Area Measurements on the OSU Tuned Reflector," 1774-10, 5 May 1965.

SCATTERING OF ELECTROMAGNETIC ENERGY FROM
HIGHLY CONDUCTING BODIES
Air Force Cambridge Research Laboratories
Office of Aerospace Research
Contract No. F19628-67-C-0239
Project No. 2415

D.L. Moffatt, "Interpretation and Application of Transient and
Impulse Response Approximations in Electromagnetic
Scattering Problems," 2415-1, 27 March 1968. (AD 668124)
(AFCRL-67-0690)

R. Schaffer, "Transient Currents on a Perfectly Conducting
Cylinder Illuminated by Unit-Step and Impulsive Plane
Waves," 2415-2, 3 May 1968. (AD 668536) (AFCRL-67-0691)

C.E. Ryan, Jr., "Diffraction Analysis of Scattering by a Cube
with Application to the Time Response Waveforms," 2415-3,
13 March 1970. (AD 702874) (AFCRL-69-0490)

D.B. Hodge, "The Calculation of the Spheroidal Wave Equation
Eigenvalues and Eigenfunctions," 2415-4, 22 September
1969. (AD 694711) (AFCRL-69-0359)

D.B. Hodge, "Spectral and Transient Response of a Circular Disk
to Plane Electromagnetic Waves," 2415-5, 28 May 1970.
(AD 707493) (AFCRL-70-0237)

D.L. Moffatt, Final Report: "Scattering of Electromagnetic
Energy From Highly Conducting Bodies," 2415-6, 7 September
1970. (AD 875533) (AFCRL-70-0238)

RESEARCH IN THE RADAR REFLECTING PROPERTIES OF DISCRETE
TARGET IN THE PRESENCE OF AN INTERFERING MEDIUM
Air Force Cambridge Research Laboratories
Office of Aerospace Research
United States Air Force
Contract No. F44620-67-C-0095

Li-Jen Du, "Rayleigh Scattering from Leaves," 2467-1, January
1969. (AD 682335) (AFCRL-68-0454)

D.A. Hill and D.L. Moffatt, "The Transient Fields of Dipole
Antennas in the Presence of a Dielectric Half-Space,"
2467-2, 13 March 1970. (AD 707106) (AFCRL-70-0016)

J.D. Young, D.L. Moffatt and E.M. Kennaugh, "Time-Domain Radar
Signature Measurment Techniques," 2467-3, 30 July 1969.
(AFCRL-69-0202)

D.L. Moffatt, "Characteristic Waveforms of Metallic Targets
    in the Presence of Background Signal," 2467-4, 13 March
    1970. (AD 870168) (AFCRL-69-0527)

Final Summary Report, 2467-5, 13 March 1970. (AD 870169)
    (AFCRL-69-0402)

ELECTROMAGNETIC SCATTERING RESEARCH
Air Force Systems Command
Laurence G. Hanscom Field
Contract No. F19628-70-C-0125
Project No. 2971

D.A. Hill, "Electromagnetic Scattering Concepts Applied to the
    Detection of Targets Near the Ground," 2971-1, 4 September
    1970. (AD 875889) (AFCRL-70-0250)

D.L. Moffatt, Final Report: "Time Domain Electromagentic
    Scattering from Highly Conducting Objects," 2971-2,
    May 1971. (AD 885883) (AFCRL-71-0319)

ELECTROMAGNETIC PULSE SOUNDING FOR GEOLOGICAL
SURVEYING WITH APPLICATION IN ROCK MECHANICS
AND RAPID EXCAVATION PROGRAM
Bureau of Mines, Department of Interior (ARPA)
Contract No. H0230009
Project No. 3408-A1

D.L. Moffatt, L. Peters, Jr., and R.J. Puskar, Semiannual
    Report: "Electromagnetic Pulse Sounding for Geological
    Surveying with Application in Rock Mechanics and the
    Rapid Excavation Program," 3408-1, April 1973.

D.L. Moffatt, R.J. Puskar and L. Peters, Jr., Final Report:
    "Electromagnetic Pulse Sounding for Geological Surveying
    with Application in Rock Mechanics and the Rapid
    Excavation Program," 3408-2, September 1973. (AD 772065)

INVESTIGATION OF THE MULTI-FREQUENCY RADAR
REFLECTIVITY OF RADAR TARGET
Dept. of the Air Force, Hanscom AFB, MA
Contract No. F19628-72-C-0203
Project No. 3424

R.K. Mains and D.L. Moffatt, "Complex Natural Resonances of an
    Object in Detection and Discrimination," 3424-1, June
    1974. (AFCRL-TR-74-0282)

S.C. Lee, "Control of Electromagnetic Scattering by Antenna Impedance Loading," 3424-2, July 1974. (AFCRL-TR-74-0426)

C.W. Chuang and D.L. Moffatt, "Complex Natural Resonances of Radar Targets Via Prony's Method," 3424-3, April 1975. (AFCRL-TR-75-0203)

J. Aas, "Control of Electromagnetic Scattering from Wing Profiles by Impedance Loading," 3424-4, August 1973. (AFCRL-TR-75-0463)

D.L. Moffatt, R.C. Rudduck, C.W. Chuang and J.A. Aas, "Continuation of the Investigation of Multi-Frequency Radar Reflectivity and Radar Target Identification," 3424-5, July 1975. (AFCRL-TR-75-0417)

C.W. Chuang, "Radiation of Backscatter by Impedance Loading," 3424-6, October 1976. (RADC-TR-76-376) (AD/A-035 509)

C.W. Chuang, D.L. Moffatt, L. Peters, Jr. and K.A. Shubert, "Continuation of the Investigation of the Multi-Frequency Radar Reflectivity and Radar Target Identification," 3424-7, Final Report, March 1977. (RADC-TR-77-115)

THREE-D IMAGES FOR SPECTRAL SIGNATURE DATA
Rome Air Development Center
Contract No. F30602-74-C-0170

J. D. Young, "Three-D Images for Spectral Signature Data," 3919-1, August 1974.

J.D. Young, Quarterly Report: "Three-D Images for Spectral Signature Data," 3919-2, November 1974.

K.A. Shubert and D.L. Moffatt, "Three-D Images for Spectral Signature Data," 3919-3, March 1975.

K.A. Shubert and D.L. Moffatt, "Three-D Images for Spectral Signature Data," 3919-4, May 1975.

K.A. Shubert and D.L. Moffatt, "Three-D Images for Spectral Signature Data," 3919-5, August 1975.

K.A. Shubert, D.L. Moffatt, J.D. Young and W.E. Cory, "Multiple Frequency Scattering Data," 3919-6, October 1975.

K.A. Shubert and J.D. Young and D.L. Moffatt, "Target Imaging from Multiple Frequency Radar Data," 3919-7, Final Report, November 1975.

S.C. Lee, "Control of Electromagnetic Scattering by Antenna
    Impedance Loading," 3424-2, July 1974. (AFCRL-TR-74-0426)

C.W. Chuang and D.L. Moffatt, "Complex Natural Resonances of
    Radar Targets Via Prony's Method," 3424-3, April 1975.
    (AFCRL-TR-75-0203)

J. Aas, "Control of Electromagnetic Scattering from Wing Profiles
    by Impedance Loading," 3424-4, August 1973. (AFCRL-TR-
    75-0463)

D.L. Moffatt, R.C. Rudduck, C.W. Chuang and J.A. Aas, "Continu-
    ation of the Investigation of Multi-Frequency Radar
    Reflectivity and Radar Target Identification," 3424-5,
    July 1975. (AFCRL-TR-75-0417)

C.W. Chuang, "Radiation of Backscatter by Impedance Loading,"
    3424-6, October 1976. (RADC-TR-76-376) (AD/A-035 509)

C.W. Chuang, D.L. Moffatt, L. Peters, Jr. and K.A. Shubert,
    "Continuation of the Investigation of the Multi-Frequency
    Radar Reflectivity and Radar Target Identification,"
    3424-7, Final Report, March 1977. (RADC-TR-77-115)

THREE-D IMAGES FOR SPECTRAL SIGNATURE DATA
Rome Air Development Center
Contract No. F30602-74-C-0170

J. D. Young, "Three-D Images for Spectral Signature Data,"
    3919-1, August 1974.

J.D. Young, Quarterly Report: "Three-D Images for Spectral
    Signature Data," 3919-2, November 1974.

K.A. Shubert and D.L. Moffatt, "Three-D Images for Spectral
    Signature Data," 3919-3, March 1975.

K.A. Shubert and D.L. Moffatt, "Three-D Images for Spectral
    Signature Data," 3919-4, May 1975.

K.A. Shubert and D.L. Moffatt, "Three-D Images for Spectral
    Signature Data," 3919-5, August 1975.

K.A. Shubert, D.L. Moffatt, J.D. Young and W.E. Cory, "Multiple
    Frequency Scattering Data," 3919-6, October 1975.

K.A. Shubert and J.D. Young and D.L. Moffatt, "Target Imaging
    from Multiple Frequency Radar Data," 3919-7, Final
    Report, November 1975.

DEVELOPMENT OF A SUBTERRANEAN RADAR FOR TUNNEL LOCATION
U.S. Army Mobility Equipment Research
& Development Command
Ft. Belvoir, VA
Contract DAAG53-76-C-0179
Project 4460

G.A. Burrell and B.A. Munk, "The Array Scanning Method and
Applying it to Determine the Impedance of Linear Antennas
in a Lossy Half Space," 4460-1, October 1976.

G.A. Burrrell, L. Peters, Jr. and A.J. Terzuoli, Jr., "The
Propagation of Electromagnetic Video Pulses with Application to Subsurface Radar for Tunnel Application,"
4460-2, December 1976.

L. Peters, Jr., G.A Burrell and H.B. Tran, "A Scattering Model
for Detection of Tunnels Using Video Pulse Systems,"
4460-3, February 1977.

C.A. Tribuzi, "An Antenna for Use in an Underground (HFW) Radar
System," 4460-4, November 1977.

D.O. Stapp, "Method for Gray Scale Mapping of Underground
Obstacles Using Video Pulse Radar Return," 784460-5,
December 1978.

CHARACTERIZATION OF SUBSURFACE ELECTROMAGNETIC SOUNDINGS
National Sciene Foundation
Grant ENG76-04-04344
Project 784490

D.L. Moffatt, "Characterization of Subsurface Electromagnetic
Soundings," 784490-1, May 1977.

D.L. Moffatt and L.C. Chan, "Characterization of Subsurface
Electromagnetic Soundings," Final Report, 784490-2.

RADAR IDENTIFICATION OF NAVAL VESSELS
Office of Naval Research
Arlington, VA
Contract N00014-76-C-1079
Project 784558

D.L. Moffatt and C.M. Rhoads, "Radar Identification of Naval
Vessels," 784558-1, December 1977. (AD/A050 047)
(AD/A071 921)

D.L. Moffatt and C.M. Rhoads, "Radar Identification of Naval
Vessels," Final Report, 784558-2, April 1979.

A STUDY OF RADAR TARGET IDENTIFICATION BY THE
MULTIPLE-FREQUENCY RESONANCE AND TIME DOMAIN POLE METHOD
Deputy for Electronic Technology (RADC/ETER)
Hanscom AFB, MA 01731
Contract F19628-77-C-0125
Project 784677

K.A. Shubert and D.L. Moffatt, "Swept Frequency Scattering
Measurements of Aircraft," 784677-1, January 1979.
(AD/A071 474) (RADC-TR-79-110)

D.L. Moffatt, K.A. Shubert and E.M. Kennaugh, "Radar Target
Identification," Final Report, 784677-2. (AD/A070 793)
(RADC-TR-79-118)

BASIC RESEARCH IN 3-DIMENSIONAL IMAGING
FOR TRANSIENT RADAR SCATTERING SIGNATURES
Ballistic Missile Defense Systems Command
Contract DASG60-77-C-0133
Project 784785

J.D. Young, R.A. Day, F.R. Gross and E.K. Walton, "Basic
Research in the Three-Dimensional Imaging from Transient
Radar Scattering Signatures," Annual Report, 784785-1,
July 1978.

R.A. Day, "Automated Imaging of Cone-Like Targets from Transient
Radar Signature Data," 784785-2, March 1979.

F. Gross, "Application of Physical Optics Inverse Diffraction
to the Identification of Cones from Limited Scattering
Data," 784785-4, June 1979.

J.D. Young, "Basic Research in Three-Dimensional Imaging from
Transient Radar Scattering Signatures," Final Report,
784785-5. (AD/A079 423)

JOINT SERVICES ELECTRONICS PROGRAM
Dept. of the Navy
Office of Naval Research
Arlington, Virginia 22217
Contract N00014-78-C0049
Project 710816

"Joint Services Electronics Program," First Annual Report,
Annual Report, 710816-1, December 1978. (AD/A064 791)

D.B. Hodge, "The Calculation of Far Field Scatter by a Circular Metallic Disk," 710816-2, Feburary 1979. (AD/A068 208)

D.P. Mithouar and D.B. Hodge, "Electromagnetic Scattering by a Metallic Disk," 710816-3, September 1979. (AD/A080 441)

I.L. Volakis, "Improved Identification of Underground Targets Using Video-Pulse Radars by Elimination of Undesired Natural Resonances," 710816-4, October 1979. (AD/A078 328)

"Joint Services Electronics Program," Second Annual Report, 710816-6, December 1979. (AD/A087 061)

T.W. Johnson and D.L. Moffatt, "Electromagnetic Scattering by Open Circular Waveguides," 701816-9, December 1980.

"Joint Services Electronics Program," Third Annual Report, 710816-10, December 1980.

"Joint Services Electronics Program," Fourth Annual Report, 710816-11, December 1981.

"Joint Services Electronics Program," Fifth Annual Report, 710816-12, December 1982.

RESEARCH IN HIGH RESOLUTION RADAR TARGET POLARIMETRY

Polarization Studies for Echo Characteristics
Rome Air Development Center
Contract No. AF 28(099)-90
Project No. 389

389-1 to 389-15, and 398-17 to 389-24
Quarterly Progress & Final Reports, "EFFECT OF TYPE OF POLARIZATION ON ECHO CHARACTERISTICS, Sept., 1949 to Oct., 1954

Landing System Antennas and Echo Enhancement Problems
Rome Air Development Center
Griffiss Air Force Base, New York
Contract No. AF 30(635)-2811
Project No. 612

612-1 to 612-16
Quarterly Progress and Interim & Final Reports, "POLARIZATION DEPENDENCE OF RADAR ECHOES", January, 1955 to January, 1957

List of Reports on Polarimetric Properties of Radar Targets is in "Proceedings of the Second Workshop on Polarimetric Radar Technology", Conducted at the U.S. Army Missile Command, DRSMI-RN, Redstone Arsenal, AL 35898, 3-5 May, 1983, Vol. 1, pp. 178-180. A commemorative collection of unpublished notes and reports in 4 volumes will be made available by the ElectroScience Lab., June 1984.

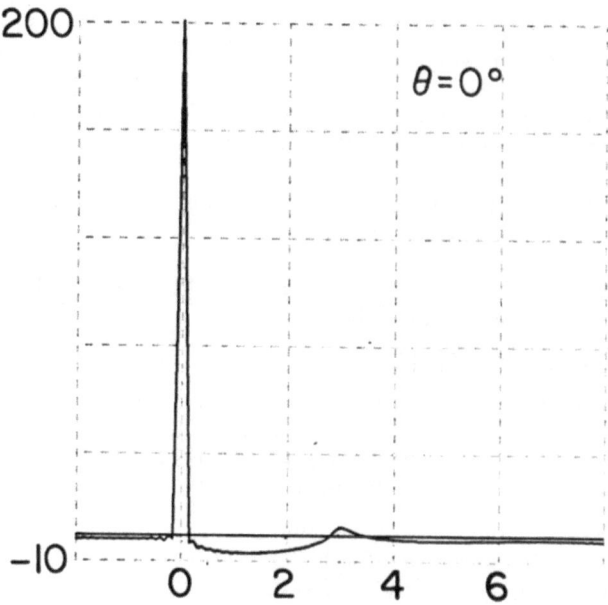

a. θ = 0° (specular)

Figure 1. E-plane (positive) and H-plane (negative) surface
current density waveforms induced by an impulsive
plane wave on a perfectly conducting sphere.
Illuminated side of sphere.

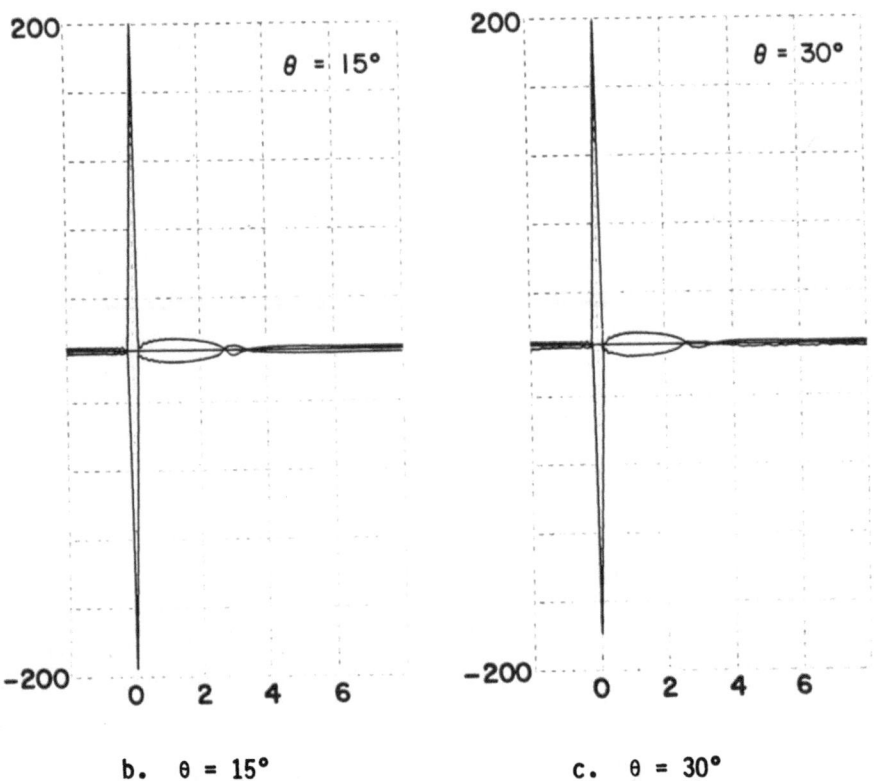

Figure 1. E-plane (positive) and H-plane (negative) surface
current density waveforms induced by an impulsive
plane wave on a perfectly conducting sphere.
Illuminated side of sphere.

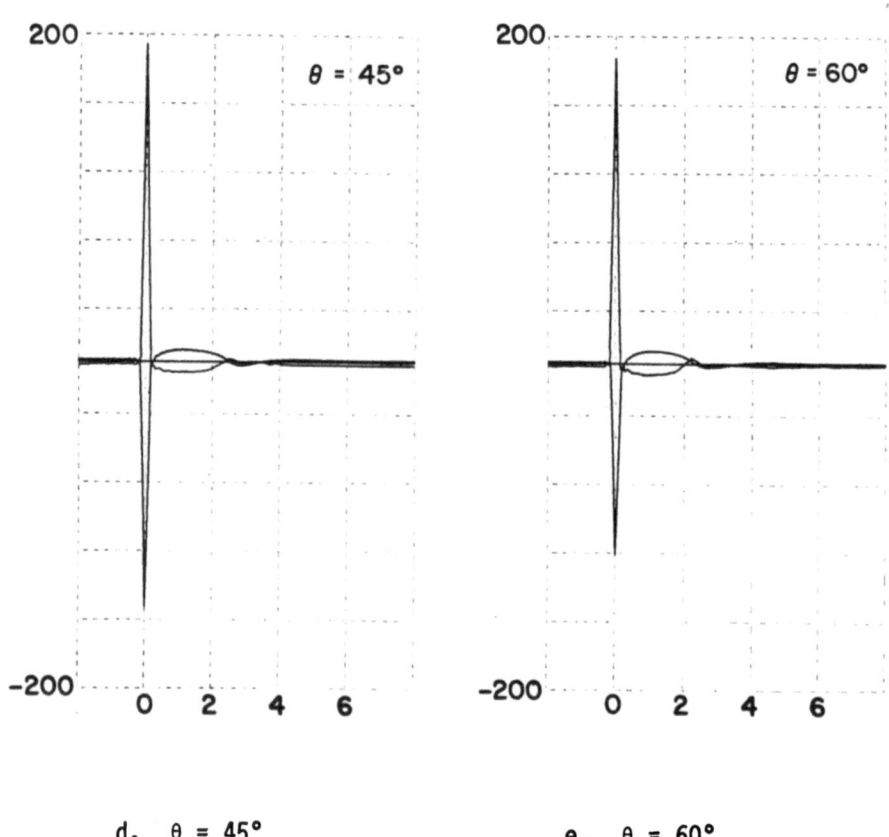

d.   θ = 45°   e.   θ = 60°

Figure 1.   E-plane (positive) and H-plane (negative) surface
current density waveforms induced by an impulsive
plane wave on a perfectly conducting sphere.
Illuminated side of sphere.

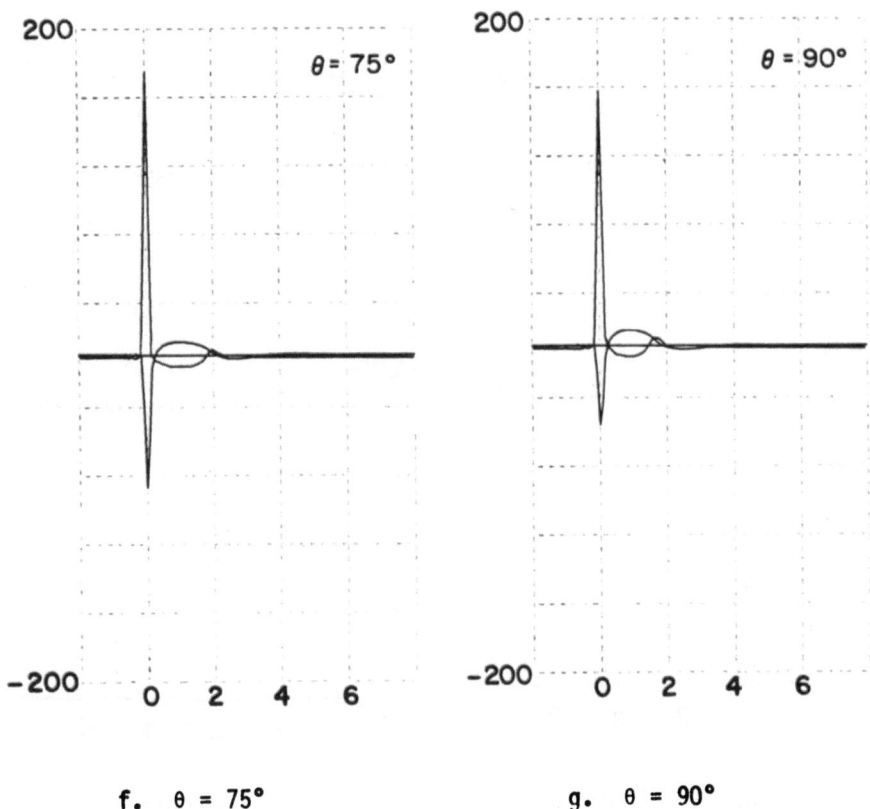

Figure 1. E-plane (positive) and H-plane (negative) surface
current density waveforms induced by an impulsive
plane wave on a perfectly conducting sphere.
Illuminated side of sphere.

a. θ = 105°

b. θ = 120°

c. θ = 135°

d. θ = 150°

Figure 2. E-plane surface current density waveforms induced by
an impulsive plane wave on a perfectly conducting
sphere. Shadowed side of sphere.

e. θ = 165°

f. θ = 180°

Figure 2. E-plane surface current density waveforms induced by an impulsive plane wave on a perfectly conducting sphere. Shadowed side of sphere.

a.   θ = 105°

Figure 3.   H-plane surface current density waveforms induced by an impulsive plane wave on a perfectly conducting sphere. Shadowed side of sphere.

b.   θ = 120°

c.   θ = 135°

d.   θ = 150°

f.   θ = 165°

Figure 3. H-plane surface current density waveforms induced by
           an impulsive plane wave on a perfectly conducting
           sphere. Shadowed side of sphere.

a. θ = 0°

Figure 4. E-plane (solid) and H-plane (dashed) surface current
density waveforms induced by an impulsive plane wave
on a perfectly conducting sphere after the physical
optics approximation for the surface current density
has been removed. Illuminated side of sphere.

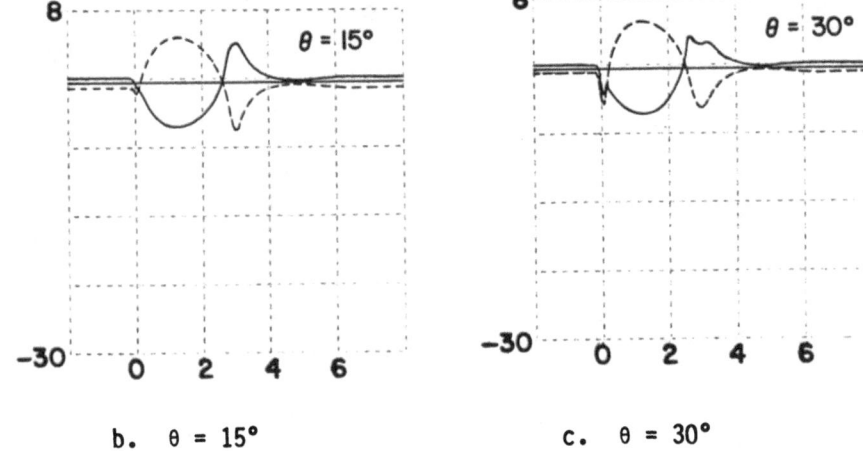

b.   θ = 15°   c.   θ = 30°

d.   θ = 45°

Figure 4.   E-plane (solid) and H-plane (dashed) surface current
            density waveforms induced by an impulsive plane wave
            on a perfectly conducting sphere after the physical
            optics approximation for the surface currrent density
            has been removed. Illuminated side of sphere.

e. θ = 60°   f. θ = 75°

g. θ = 90°

Figure 4. E-plane (solid) and H-plane (dashed) surface current
           density waveforms induced by an impulsive plane wave
           on a perfectly conducting sphere after the physical
           optics approximation for the surface current density
           has been removed. Illuminated side of sphere.

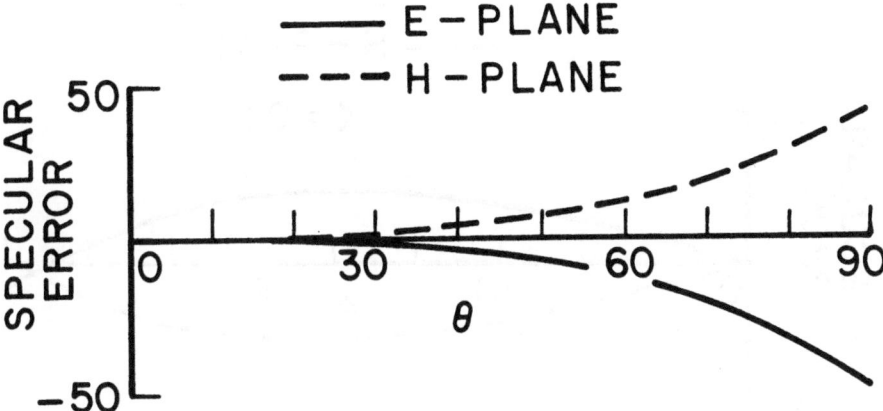

Figure 5. E-plane (solid) and H-plane (dashed) error for
specular term (exact minus physical optics) for
surface current density induced by an impulsive
plane wave on a perfectly conducting sphere.

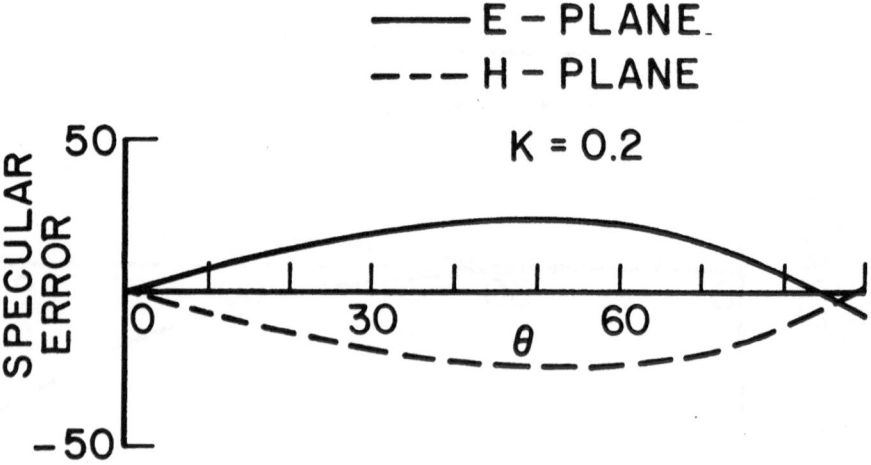

Figure 6. E-plane (solid) and H-plane (dashed) error for
specular term (exact minus corrected physical
optics, value of K constant is 0.2) for surface
current density induced by an impulsive plane
wave on a perfectly conducting sphere.

Figure 7. E-plane (solid) and H-plane (dashed) error for specular term (exact minus corrected physical optics, value of K constant is 0.2) for surface current density induced by an impulsive plane wave on a perfectly conducting sphere.

0.2 (OS.4)

INVERSE METHODS IN ELECTROMAGNETIC IMAGING

Arthur K. Jordan

Space Science Division,
Naval Research Laboratory,
Washington, DC 20375

Wolfgang-M. Boerner

Communications Laboratory
EECS Department
University of IL at Chicago
Chicago, IL 60680

Abstract

Imaging can be described as the generation of the image field I from an object field O by means of appropriate electromagnetic and computational systems L, so that $I = L \cdot O$. In remote sensing, non-destructive evaluation and similar applications, the object field O is not known explicitly and must be estimated by an inverse method $O = L^{-1} \cdot I$ subject to some constraint $\| I - O \| < \varepsilon$, which results in inherent numerical instabilities. These problems in general are "ill-posed" and are treated by a variety of methods: constraints on the solution for O, use of *a priori* knowledge of O, relaxation of the error limit ε on image quality, addition of more data in the image I, etc. Using this rationale, the seventy-five papers comprising these Proceedings are summarized and organized into five Topics: Topic I was concerned with fundamental inversion theories; Topic II with numerical instabilities; Topic III with the utilization of polarization information to improve high resolution imaging; Topic IV with the effects of environmental noise and clutter on image quality; and Topic V with holographic and tomographic imaging and their associated phase problems. This General Review concludes with a summary of recent relevant advances.

1. Introduction

In general, images are representations of objects that are indirectly or remotely sensed. Since indirect sensing is used, inverse methods are necessary to reconstruct images of these objects. This NATO Advanced Research Workshop was concerned with imaging by electromagnetic waves in two or three space dimensions and with the mathematical and experimental implementation of inverse methods for

W.-M. Boerner et al. (eds.), Inverse Methods in Electromagnetic Imaging - Part 1, 33–42.
© *1985 by D. Reidel Publishing Company.*

imaging systems.

The complete proceedings of this NATO Advanced Research Workshop
on Inverse Methods in Electromagnetic Imaging are collected in two
parts. A perusal of the Table of Contents reveals the diversity
of topics presented, ranging from abstract mathematical discussions
to practical designs and experimental results. For such a rapidly
developing research area, it is necessary to classify the progress
reported here approximately according to topics. This arrangement
will enable the reader to become acquainted with the current research trends and to appreciate the fundamental unity of these inverse methods. It has not been possible to include many important
contributions that were not presented at this ARW. However, most
of the papers in these Proceedings contain extensive references
which serve to extend the reader's understanding. This general
review section also summarizes recent related monographs that contain reviews with extensive references; recent relevant NATO Advanced Study Institutes and Workshops Proceedings are also listed.

Since the participants in the ARW presented a wide-ranging variety
of papers, it will be worthwhile to give a general background and
description of imaging and why inverse methods are essential to
any practical imaging system. It is convenient to organize this
overview by first considering a mathematical representation of the
imaging process which serves to highlight the roles of inverse
theory, image quality and image reconstruction techniques. This
problem definition also serves to organize these Proceedings into
six topics with their corresponding Working Groups. The Proceedings are then summarized with one or two key sentences abstracted
from each paper (identified by the authors' names). We conclude
with a brief review of references on imaging and inverse methods,
on progress in related sciences and technologies, and also related
NATO Advanced Study Institutes.

2. Problem Definition

For the purposes of this discussion, imaging can be described as
the generation of the image field, I, from an object field, O, by
means of appropriate electromagnetic and computational systems L:

$$I = L \cdot O$$

The image quality is measured (this subjective process is, of
course, difficult to quantify) by the "closeness" of I to O in
some appropriate mathematical representation, i.e.,

$$\| I - O \| < \varepsilon$$

In remote sensing, nondestructive evaluation, medical imaging, and
similar applications, the object field O is not known explicitly
and must be estimated by indirect or inverse methods, L^{-1}, from

data contained in the image fields I. This procedure is conveniently written as:

$$O = L^{-1} \cdot I \quad ,$$

although we may not be able to define L^{-1} mathematically. Inverse
problems of this type have inherent numerical instabilities, so
that small variations in the image data, I, can cause large errors
in the reconstructed object, O. These problems are called "ill-
posed" and are treated by a variety of methods: constraints on
the solution for O, use of *a priori* knowledge of O, relaxation of
the error limit ε on image quality, addition of more data in the
image I, etc. Using this rationale we have organized (somewhat
arbitrarily) the papers comprising these Proceedings into five
topics, roughly corresponding to the five Working Groups, and
these papers are briefly summarized in the next section.

3. Synopsis of Proceedings of the Workshop

3.0 The memorial paper by Kennaugh & Moffatt also contains extensive lists of references of the unpublished research reports by
the late Prof. Edward M. Kennaugh and his coworkers on transient
electromagnetic problems, those on the polarization properties of
radar targets are now available in four volumes (O).

3.1 The papers of Topic 1 were concerned with the fundamental inversion theories required to obtain the object from the image. In
the first paper Sabatier, provides an analysis of the common
structure of inverse methods and demonstrates this by examples of
three important aspects of inverse methods:

   a. the importance of *a priori* information,

   b. the necessary use of geometrical methods in function
       spaces,

   c. the interest in these methods from the algorithmic points
       of view.

Basic topological considerations are used by Güttinger to classify
singularities in waves and rays in order to study the formation of
patterns. Three-dimensional, time-dependent inverse problems are
considered by Moses,.who obtains solutions for the current distributions that are not unique, but conditions are given that make
them so. Langenberg et al consider two types of transient methods: broadband signals, such as the singularity expension method,
and synthetic aperture methods with band-limited data in K-space.
Roger describes a numerical method (Newton-Kantorovich iterative
algorithm) that demonstrates a characteristic instability of inverse problems. Krueger et al use time-domain methods to reconstruct both the permittivity and conductivity profiles of a one-
dimensional dissipative medium: four numerical examples are given.

Jordan et al extend inverse scattering theory to include profiles that are almost-periodic functions; applications to modulated in-homogeneous and non-deterministic media are indicated. Several time-domain methods are compared by Blok & Tijhuis: direct-ap-proach which involves a step-by-step determination of the unknown profile, iterative approach for successive approximations to the unknown constitutive coefficient, optimization approach by mini-mization of the mean square error. Approaches to electromagnetic inverse problem are given by Eftimiu by using analogies to the quantum mechanical inverse problem. The Abel integral is applied to a variety of remote sensing problems by Süss; a combined analy-tic-numeric inversion algorithm has relatively high stability. The importance of group representations, i.e., symmetry analysis, in scattering problems, direct and inverse, is reviewed by Wacker. Borden has used partial scattering data to obtain the shape of a convex body from a variational solution to an inverse scattering problem. Ferwerda et al presented a method of profile reconstruc-tion based on the Gelfand-Levitan-Marchenko approach; bound states were accounted for.

3.2 The papers of Topic II were concerned with the numerical in-stabilities inherent in inverse problems. One cause of numerical instability is the lack of sufficient data; this was discussed by Weston for large perturbations so a nonlinear iteration approach is used. Grünbaum considers the problem of recovering a real func-tion, e.g. the density, from bandlimited data, the projections; ex-tension to the hyperbolic case enables the use of prolate spheroid-al wave functions. Crosta analyzes the modeling procedures in di-rect and inverse problems. Singular function expansions are used by Bertero et al in two papers to obtain stable solutions and to define resolution limits. Conjugate gradient methods are used by Sarkar to obtain stable results as long as the number of effective bits in the signal is correctly chosen. Jory uses approximation schema for the Moore-Penrose generalized inverse to obtain regular-ization operators. Several maximum entropy methods are reviewed by Bevensee and fundamental limitations on the resolution on various radiating sources are discussed. Complex electromagnetic problems are treated by Taflove & Umashankar using a combined finite-differ-ence time-domain method; accurate results have been obtained for three-dimensional structures as large as 20 wavelengths. Geometri-cal and quasi-geometrical imaging are considered by Danglmayr with special attention to the caustics associated with topological sing-ularities of the distance function. Dittel has investigated three algorithms for the deconvolution of radiometry data using *a priori* information. Time-domain inverse scattering methods have been re-viewed by Tabbara. Complex natural resonances of an object have been extracted by Moffatt & Lee using rational function approxim-ants; these can be used to correct and supplement poles obtained by a K-pulse technique.

3.3   The papers of Topic III were concerned with the appropriate utilization of polarization information to improve high resolution target imaging. The tools used for handling polarization problems are reviewed by Deschamps. Suggestions are made to account for symmetries and their use in inverse problems. Utilization of polarization information by radar systems is reviewed by Poelman. Thiel, briefly, notes the transformation of the Stokes' vectors. After a review of fundamentals, Grüner proposes a correlation measuring method, using polarization switching for use in microwave radiometry. Three-dimensional radar images are constructed by polarization-dependent methods by Heath; efficient Fourier or Radon transforms are used. The phenomenological theory of radar targets is reviewed by Davidovitz & Boerner and the optimum polarization case is extended to asymmetric polarization scattering matrices. The graphical display of polarization parameters was discussed by Raven and polarization nulls of simulated, complex reflectors were displayed. Meteorological information is obtained from polarimetric measurements by Nespor & Boerner. The related problem of polarization glint has been considered from the view point of the measuring system by Stock. The use of polarization signatures for target angle tracking systems has been considered in Carpenter, whereas, it is shown in Manson & Boerner how high resolution scattering matrix pulse compression radar data can be used for reconstructing the three-dimensional target image downrange using Kennaugh's target characteristic operator and Huynen's Mueller matrix decomposition concept. Agreement between experiment and simulation provides support for the single-photon coherence zone concept of Wei et al. The criteria for reducing errors in recovering polarized scattering parameters are provided by Blanchard & Newton. Five conceptual issues relating to polarization processing are discussed by Vannicola & Lis.

3.4   The papers of Topic IV were concerned with the effects of environmental noise and clutter on image quality and methods for improvement of imagery. A fundamental approach to electromagnetic imaging is radiative transfer theory. Ishimaru has reviewed this theory for arbitrary polarized electromagnetic waves; limitations due to multiple scattering effects are also discussed. Image quality also depends upon the perception of the data by the observer; Huynen reviews the methods of data presentation of objects based upon fields and powers. Radar imagery is reviewed by Orhaug from the viewpoint of image quality, image encoding, and computer analysis. Controlled experiments to assess and develop inverse processing algorithms have been described by Cram. Optimum methods of detection and identification of surface chemicals (e.g. paints) and target shapes are discussed by Gjessing et al. Inverse methods are used by Rotheram & Macklin to apply linear demodulation and speckle removal to synthetic aperture radar images. Rough surface properties have been obtained by inversion of scattering data by Fung. SAR imagery was also discussed by Larson et al with respect to fine

resolution imagery of stationary scenes and imaging of a moving
surface.   Radio exploration of glaciers was reviewed by Walford
with respect to the inverse problems for polar and temperate glaciers. Results from fast millimeter wave imaging for traffic studies were reviewed by Brand. Low-frequency imaging algorithms were
presented by Chaudhuri, using ellipsoidal models. Dielectric inhomogeneities have been reconstructed by Chaloupka, using the Born
approximation. Experimental results from a millimeter-wave imaging
system were presented by Detlefsen.

3.5 The papers of Topic V were concerned with holographic and tomographic imaging and the phase problems that are critical for successful imaging. Lohmann reviewed the basic holographic principle,
pointing out that information can be lost to phase-, color-, and
polarization blindness. Medical tomographic imaging was reviewed
by Schwierz from algebraic techniques to iterative and integral
transform methods. Using circuit theory analogies, Kahn considers
inverse methods using resonance conditions, transformation invari-
ants, and physical symmetry. Anderson & Adams have produced satisfactory images using fast convenient FFT-based algorithms. Devaney
has reviewed the theoretical foundations of computed tomography
using diffracting wavefields (diffraction tomography) and contrasted them with the foundations of conventional computed tomography.
Born and Rytov approximations were presented as special cases of a
general view of diffraction tomography by Kaveh & Soumekh. An approach to extend scalar propagation tomography to the electromagnetic vector case is presented in James, Yang & Boerner. Techniques
to increase the number of degrees-of-freedom in image processing
were reviewed by Szu. The iterative method for image restoration
has been reviewed by Gori & Wabnitz and the discrete and continuous
cases contrasted. Cases where mono-pulse techniques can be applied
to synthetic aperture radar imagery were proposed by Gough. Methods for probing dielectric radomes were reviewed by Tricoles.
Methods for constructing and retrieving authenticity features were
reviewed by Baltes & Huiser. Methods for improving target recognition from microwave imagery were presented by Gniss & Magura. The
inverse problem of far-field-to-near-field transformation in spherical coordinates was considered by Hess incorporating many concepts
discussed in previous papers.

3.6 Five Working Discussion Groups evaluated the state-of-the-art
of each of these Topics and made specific recommendations for future work. Their reports, which are collected at the end of Part
2, provide a concise summary of the present status and the important future emphasis of inverse methods applied to electromagnetic
imaging. In addition, Working Group VI considered the methods of
enhancing the interactions between workers and institutions concerned with this multi-disciplinary technology.

4. Reviews of Related Progress

There have been several recent, comprehensive reviews with voluminous references to various aspects of imaging and inverse methods.
Several of these are: Colin (1), which has reviews of the state-
of-the-art of inverse scattering circa 1972; a survey of the literature up to 1978 has been presented by Boerner (2); Boerner,
Jordan & Kay (3), which reviews progress in inverse methods in
electromagnetics up to 1981; Baltes (4,5), which reviews progress
in inverse methods in optics up to 1978 and 1981; Sabatier (6) reviews applied inverse problems and electromagnetic inverse scattering (12).

Several important topics were not discussed or were only briefly
mentioned in this lead article. Zuev & Naats (7) review inverse
problems of lidar (laser radar) sensing; Wilde & Barrett (8) review the mathematical basis for digital image processing and
Devaney (9) reviews related optical inverse methods. Mathematical
methods for inverse problems of remote sensing are treated by
Twomey (10). A comparison of phase retrieval methods, which topic
was considered by Working Group V, has been provided by Fienup (11).

Nonlinear processes and their effects on imaging merit considerable attention. Sabatier (12), Scott et al (13), Dodd et al (14),
among others, discuss these problems and their applications to
solitons. Remote sensing of the environment uses digital imaging
techniques; Gjessing (15), Bernstein (16), Lillesand (17), Ulaby &
Moore (18), and Rosenfeld (19) have made extensive reviews. Laser
beam propagation was only briefly considered in this overview;
Ishimaru (20), Zuev (21) and Hinkley (22) provide further information. Artificial intelligence, robot vision and related topics
inherently depend upon inverse methods for imaging; Winston (23)
has reported on the progress in this rapidly developing field.
Problems of perception and psycho-physics are important for the
subjective interpretation of image quality; Geissler (24) has reported on recent developments. Imaging methods for radiology have
been extensively reviewed by Barrett and Swindell (25).

NATO Advanced Study Institutes and Research Workshops have provided a continuing forum for fundamental results and latest developments. Several recent and relevant conferences are:

Physics and Engineering of Medical Imaging (26)

Industrial Robotic Vision (27)

Vision & Image Understanding (28)

Pictorial Data Analysis (29)

Diagnostic Imaging in Medicine (30)

Atmospheric Effects on Radar Target Identification and
Imaging (31)

Surveillance of Environmental Pollution and Resources by
Electromagnetic Waves (32)

Theoretical Methods for Determining the Interaction of
Electromagnetic Waves with Structures (33)

Pattern Recognition Theory and Applications (34)

Remote Sensing Applications in Marine Science and
Technology (35)

Optical Metrology (36)

The Application of Laser Light Scattering to the Study of
Biological Motion (37)

Nonlinear Phenomena at Phase Transitions and
Instabilities (38)

In conclusion, this NATO Advanced Research Workshop has demonstrated both the basic unity of inverse methods and the diversity of
their applications to imaging. The Proceedings of this Workshop
represent, however, only one sampling of the development of this
active area of science and technology. These Proceedings should
provide workers with sufficient reference materials and suggestions for new research topics to further their continuing development.

5. References

0.   Kennaugh, E.M., Contributions to the Polarization Properties
     of Radar Targets, Commemorative Collection of Unpublished
     Research Reports in Four Volumes, ElectroScience Lab., Dept.
     of Electrical Engineering, Ohio State University, Columbus,
     OH 43212, June 1984.

1.   Colin, L., ed., Mathematics of Profile Inversion,
     NASA TMX-62, 150; Aug., 1972.

2.   Boerner, W.-M., Detailed State of the Art Review, Communica-
     tion Lab., Univ. of Ill. (Chicago), Report 78-3; Oct., 1978.

3.   Boerner, W.-M., Jordan, A. K., & Kay, I., Special Issue on
     Inverse Methods in Electromagnetics, IEEE Trans. Ant. & Prop.
     AP-29; March, 1981.

4.   Baltes, H. P., Inverse Source Problems in Optics, Springer
     Verlag; 1978.

5.   Baltes, H. P., Inverse Scattering Problems in Optics,
     Springer Verlag; 1981.

6.   Sabatier, P. C., Applied Inverse Problems, Springer Verlag;
     1978.

7.   Zuev, V. E., & Naats, I. E., Inverse Problems of Lidar
     Sensing of the Atmosphere, Springer Verlag; 1983.

8.   Wilde, C. O., & Barrett, E., Image Science Mathematics,
     Western Periodicals Co., 1977.

9.   Devaney, A. J., Inverse Optics, SPIE Proceedings, vol. 413;
     1983.

10. Twomey, S., Introduction to the Mathematics of Inversion in
     Remote Sensing and Indirect Measurements, Elsevier Scientific
     Publ. Co.; 1977.

11. Fienup, J. R., "Phase retrieval algorithms: a comparison",
     Appl. Optics 21, 2758; 1982.

12. Sabatier. P. C., "Theoretical considerations for inverse
     scattering", Radio Science 18. 1-18; 1983.

13. Scott, A. C. et al, "The soliton: a new concept in applied
     science", Proc. IEEE 61, 1443-1483; 1973.

14. Dodd, R. K. et al, Solitons and Nonlinear Wave Equations,
     Academic Press; 1982.

15. Gjessing, Dag T., Adaptive Radar in Remote Sensing, Ann Arbor
     Publishers; 1981.

16. Bernstein, R., ed., Digital Image Processing for Remote
     Sensing, J. Wiley Sons; 1978.

17. Lillesand, T. M., Remote Sensing and Image Interpretation,
     J. Wiley Sons; 1979.

18. Ulaby, F. T. & Moore; R. K., Microwave Remote Sensing: Active
     and Passive, Addison-Wesley Publishing Co., 1981.

19. Rosenfeld, A., Digital Picture Processing, Academic Press;
     1982.

20. Ishimaru, A., Wave Propagation and Scattering in Media,
     Vols. 1 & 2, Academic Press; 1978.

21. Zuev, V. E., Laser Beam Propagation in the Atmosphere,
     Consultants Bureau; 1982.

22. Hinkley, E. D., ed., Laser Monitoring of the Atmosphere,
     Springer Verlag; 1976.

23. Winston, P. H., Artificial Intelligence, Addison-Wesley; 1984.

24. Geissler, H.-G., Modern Issues in Perception, North-Holland;
     1983.

25. Barrett, H.H. & Swindell, W., Radiological Imaging, Vols. 1&2,
     Academic Press; 1981.

26. Physics & Engineering of Medical Imaging, NATO Advanced Study
     Institute, 23 Sept.-5 Oct. 1984; Maratea, Italy (R. Guzzardi).

27. Industrial Robotic Vision, NATO Advanced Study Institute,
     6-17 August 1984; Leuven, Belgium (A. Oosterlinck).

28. Vision and Image Understanding, NATO Advanced Study Institute,
     1-12 July, 1984; Erice, Italy (A. Borsellino).

29. Haralick, R.M., ed., Pictorial Data Analysis, Proc. of ASI,
     Springer Verlag; 1983.

30. Reba, R.C., Goodenough, D.J., Diagnostic Imaging in Medicine,
     Proc. of ASI, M. Nijhoff Publishers; 1983.

31. Jeske, H., ed., Atmospheric Effects on Radar Target
     Identification and Imaging, Proc. of ASI, D. Reidel
     Publishing Co.; 1976.

32. Lund, T., ed., Surveillance of Environmental Pollution and
     Resources by Electromagnetic Waves, Proc. of ASI, D. Reidel
     Publishing Co.; 1979.

33. Skwirzynski, J.F., ed., Theoretical Methods for Determining
     the Interaction of Electromagnetic Waves with Structures,
     Proc. of ASI, M. Nijhoff Publishers; 1981.

34. Kittler, J., Fu, K., Pau, L., eds., Pattern Recognition
     Theory and Applications, Proc. of ASI, D. Reidel Publishing
     Co.; 1982.

35. Cracknell, A.P., ed., Remote Sensing Applications in Marine
     Science and Technology, Proc. of ASI, Reidel Publishing Co.;
     1983.

36. Optical Metrology, NATO Advanced Study Institute, 16-27 July,
     1984, Viana do Castelo, Portugal.

37. Earnshaw, J.C., ed., The Application of Laser Light Scatter-
     ing to the Study of Biological Motion, NATO Advanced Study
     Institute, Plenum Publishing Co.; 1983.

38. Riste, T., ed., Nonlinear Phenomena at Phase Transitions and
     Instabilities, NATO Advanced Study Institute, Plenum Press;
     1982.

I.1 (TS.1)

CRITICAL ANALYSIS OF THE MATHEMATICAL METHODS USED IN ELECTRO-
MAGNETIC INVERSE THEORIES: A QUEST FOR NEW ROUTES IN THE SPACE
OF PARAMETERS

Pierre C. SABATIER

Département de physique mathématique
Université des sciences et Techniques du Languedoc
34060   MONTPELLIER   CEDEX - FRANCE

It has been remarked by Lanczos that no mathematical trick can
remedy a lack of information. This point is illustrated in two
kinds of problem belonging to Inverse Scattering Theory. The
first one is the one dimensional inverse potential scattering, as
it is used for instance in electron-density profile reconstructions). The smoothness of the surface is discussed as a possible
source of uncertainties. By using a generalized study of the
Schrödinger equation with rational reflection coefficients, one
produces a simple method to take it into account. In turn, this
method suggests a new way to explore the space of parameters, by
which paths are followed through geometrical transforms. This
way enables one to study possible bifurcations of solutions and
exact ambiguities of the one dimensional impedance inversion, which
may yield several different physical impedances for one set   of
reflection and transmission coefficients. It is proved that the
previous results showing these ambiguities are not isolated in
the parameters space but are only points on continuous paths of
equivalent solutions. This non uniqueness of the one-dimensional
Inverse Scattering Theory for impedances illustrates in an illuminating way the limitations of most mathematical methods currently used in inverse theories : for obvious reasons related with
the existence of algorithms, they are local methods in the space
of solutions, and therefore they cannot deal with this kind of
problems.

W.-M. Boerner et al. (eds.), Inverse Methods in Electromagnetic Imaging - Part 1, 43–64.
© *1985 by D. Reidel Publishing Company.*

INTRODUCTION

The word "critical study" is not used here for giving a
title to a more or less censorious review of other methods, with
some "good" remarks on their relevance and their usefulness. We
do not feel competent or interested in this kind of work,
although it may prove to be very useful. Rather, we would like
to show that the common structure of all current methods, ie the
obvious fact that they are local methods in the spaces of parameters and datas, is neither a necessary or a particularly good
feature of inverse methods to be used in electromagnetic inverse
problems. Indeed, the usual analyses only try to locate inside
the space $\mathcal{C}$ of parameters an element that "<u>reasonably</u>" corresponds
to the given "data" in the space $\mathcal{E}$ of results. To be more specific, let us assume that it has been possible to define distances
in $\mathcal{C}$ and $\mathcal{E}$ and that the physical problem has been translated
into mathematical words by defining a mapping $\mathcal{M}$ of $\mathcal{C}$ into $\mathcal{E}$.
One must keep in mind that $\mathcal{E}$ has to contain both $\mathcal{M}(\mathcal{C})$ and all
data with all possible errors. Now, the word "reasonably" can
imply several meanings. If $\mathcal{M}$ is a bicontinuous bijection between
$\mathcal{C}$ and $\mathcal{E}$ ("well-posed problem"), certainly one can call "reasonable" the true solution of the problem. However, suppose that
$\mathcal{M}^{-1}$ is not Holder-continuous, and suppose that small errors in $\mathcal{E}$
give rise to large shifts inside $\mathcal{C}$, which may get at very un-
likely parameters. Then using the exact solution may be a not
very reasonable way of interpreting datas. In slightly more gene-
ral cases, $\mathcal{M}(\mathcal{C})$ does not cover $\mathcal{E}$, but there is still a continuous mapping of $\mathcal{M}(\mathcal{C})$ back into $\mathcal{C}$. Then it is sound to define
a "quasisolution" corresponding to an arbitrary element e $\in \mathcal{E}$
as an element c of $\mathcal{C}$ such that $\mathcal{M}(c)$ is close to e. This "quasisolution" is "reasonable" if choosing $\mathcal{M}(c)$ can be done by a
continuous mapping of $\mathcal{E}$ into $\mathcal{M}(\mathcal{C})$. e.g. the orthogonal projection if $\mathcal{E}$ is a Hilbert space and $\mathcal{M}(\mathcal{C})$ a convex compact set.
In other cases, the quasisolution may be not reasonable and the
only acceptable analyses define "approximate solutions" for which
a trade-off is achieved between the "quality of the fit" and the
"likelihood of the parameter". No matter how this likelihood is
defined (distance to a reference point in $\mathcal{C}$, best estimator in
given statistics, maximum entropy, singular values analyses, etc),
the structure of this analysis is necessarily "local" in the
metric space $\mathcal{C}$. Furthermore, the "best" algorithms are those
which are confined in the smallest subsets of $\mathcal{C}$.
        As we already said, we are only interested by the general
features of the mathematical methods which introduce local
algorithms. Their most obvious quality is their flexibility.
Most of these methods can be adapted to many kinds of inverse
problems, with almost the same algorithms. However, they have
also two defects. First, if there are several branches of solutions, these methods are unable to discover more than one of them.

Next, they are often sensitive to "a priori" assumptions, which may be clear cut, but which are in most cases hidden in the very structure of the algorithms.

The only way to implement a creative criticism is by giving an example of what one might wish to obtain. Hence, we would like to give here an example of a "global" method for exploring the parameter space, and connecting the paths we follow in $\mathcal{C}$ with the corresponding ones in $\mathcal{E}$. The word "global" is of course understood as an opposite word to "local", but it is also reminiscent of the way we explore our natural world, the surface of the Earth. In the naval History, there has been hardly any approach of a port by an algorithmic like approach, sort of spiralling around and down into the harbour. Captains prefer knowing the harbour coordinates and following a cross-coordinated route. Admittedly, a similar method cannot exist in most inverse problems, and its main defect is a complete lack of flexibility. But we can present it in some of the best known problems, and in particular in the scalar one-dimensional inverse scattering problem for waves governed by the Schrödinger equation or by the telegraph equation. This problem is met in several geophysical situations (ionospheric soundings, electromagnetic soundings of the Earth, elastic sounding of layered media by parallel plane waves, etc.). The paths we follow in $\mathcal{C}$ and $\mathcal{E}$ are geometrical paths along which the parameter undergoes "Darboux transforms" which preserve the wave equation structure, whereas the reflection and transmission coefficients are simply multiplied by phase factors. Following a given path, we always get at what we call an "undressed parameter", which is "trivial" in some way, and from which one can construct the desired parameter by following backwards the same path.

In section 1, we recall the elements of the one-dimensional scattering, and show how the geometrical transforms can span the spaces $\mathcal{C}$ and $\mathcal{E}$. In this section we restrict $\mathcal{C}$ and $\mathcal{E}$ to the sets of parameters and data that satisfy the so-called "Deift-Trubowitz" conditions, for which $\mathcal{M}$ is a bicontinuous bijection. Admittedly, even with this assumption, the net of our paths does not completely cover $\mathcal{C}$ and $\mathcal{E}$, since one only gets at coefficients that are rational fractions of k. But the subspaces which are thus generated are everywhere dense in $\mathcal{C}$ and $\mathcal{E}$, since any reflection or transmission coefficient can be arbitrarily approximated by a rational fraction (e.g. a Padé approximant) and $\mathcal{M}^{-1}$ is continuous. Then we show that this "global" method gives us a simple way to study a practical problem : in plasma soundings, we study profile inversion uncertainties related with uncertainties of the plasma surface position.

In section 2, giving up the Deift-Trubowitz conditions enables us to demonstrate ambiguities in $\mathcal{C}$. An "ambiguity" shows up when two or more parameters exactly correspond to the same data. In the one-dimensional Schrödinger inverse problem some examples of ambiguities were recently given by Abraham, de Facio, Moses[15],

Brownstein[16], Moses[17], but they can be objected on physical
grounds. These objections break down in the one-dimensional inverse problems where the physical parameter is the impedance or
an equivalent parameter. If this parameter is twice differentiable,
the problem is equivalent to the Schrödinger problem. But physics
only guarantees that the impedance is piecewise continuous and
ambiguities are allowed. In this paper, we like better giving
methods for constructing ambiguities rather than showing a few
examples like in previous studies of the subject. It is also
shown for the first time that ambiguities are not isolated but
follow paths in the parameter space, where they "travel" through
Darboux transforms. Two paths can cross each other at a point
where they stop, because the two equivalent parameters become
equal (bifurcation point). A path can also "stop" when a new
transform yields a "singular" potential, that goes out of $\mathscr{C}$.
It is important to notice that deriving these results is done
by means of a new transform, which generalizes the Darboux one,
and fits the requirements of piecewise continuous impedances. This
transform can also be used to undress (of their background) impedances that are made of layered jump discontinuities and of a
background that can be represented by rational coefficients.

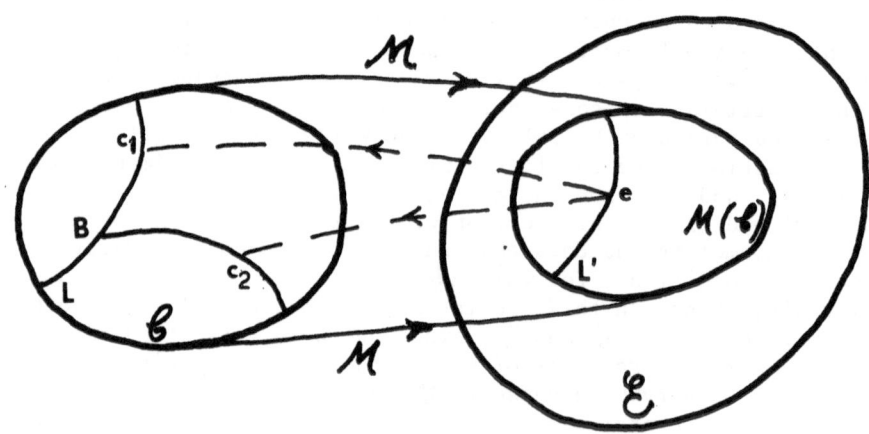

This figure shows the sets $\mathscr{C}$, $\mathscr{E}$, the mapping $\mathscr{M}$, and an
example of ambiguity : 2 parameters c_1 and c_2 correspond to the
same element e of $\mathscr{M}(\mathscr{C})$. When e follows in $\mathscr{M}(\mathscr{C})$ the line L'
whose elements differ from each other by a multiplicative phase
factor only, the corresponding parameters c_1 and c_2 follow in $\mathscr{C}$
the graph L, which shows at B a bifurcation point, and shows
also three end-points at the boundaries of $\mathscr{C}$. Is it necessary

to emphasize the fact that local methods are not able to deal with this kind of problems ?

In a concluding section, we survey some other aspects of a global approach, and some open questions.

SECTION 1. ONE DIMENSIONAL SCATTERING AND RATIONAL COEFFICIENTS

We first recall the mathematical problem, next we briefly survey known results for rational reflection coefficients and their connection with the surface position. Then we present our own methods, which are partly founded on the Darhoux method or on the "Wronskian method" used by Calogero and Degasperis in their analysis of non linear partial differential equation, and we briefly discuss the surface problem in this framework.

1.1. The mathematical problem

In its simplest form, the one-dimensional inverse scattering problem reduces to the Schrödinger equation :

$$[-\frac{d^2}{dx^2} + V(x)]\varphi(k,x) = k^2\varphi(k,x) \qquad (x \in \mathbb{R}) \qquad (1)$$

where we shall definitely assume that V belongs to the class C of functions defined by

$$\int_{-\infty}^{+\infty} dx[1 + |x|] \, | \, V(x) \qquad \infty \qquad (2)$$

For real k, it is possible to define the (unique) solution of (1) asymptotic to $\exp[ikx]$ for $x \to +\infty$, say, $f_+(k,x)$, and that asymptotic to $\exp[-ikx]$ for $x \to -\infty$, say $f_-(k,x)$. They are called the Jost solutions. Their Fourier transforms are traveling waves, which enable us to identify the physical meaning of coefficients in the formula :

$$T(k)f_-(k,x) = f_+(-k,x) + R_+(k)f_+(k,x) \qquad (3)$$

T(k) is the transmission coefficient, $R_+(k)$ is the reflection coefficient from right to left. The direct scattering problem (V → R) and the inverse one (R → V) have been the object of a huge body of studies. With the only assumption (2), Faddeyev [2] obtained a wide set of results. He had used for that only usual techniques of inverse scattering (Chadan-Sabatier [3]). With the help of higher mathematics, Deift and Trubowitz [4] were able to complete this study. However, a pedagogical presentation was needed, at least for the direct part and the reconstruction part of the problem, that are most useful in so many problems. Such a presentation obviously needs minimizing the use of high mathematics

and the length of proofs. The first goal was already attained by
Atkinson[5] and we shall use the general lines of his lecture.
We think that the proofs we have introduced[6] and which are
reproduced here achieve the second goal. Setting
$F_+(k,x) = \exp[-ikx] f_+(k,x)$ (resp. $F_- = \exp[+] f_-$), one readily
derives from (1) the Volterra equations :

$$F_+(k,x) = 1 - \int_x^\infty dy \, L_+(x,y) F_+(k,y) V(y) \tag{4a}$$

$$F_-(k,x) = 1 - \int_{-\infty}^x dy \, L_-(x,y) F_-(k,y) V(y) \tag{4b}$$

where $L_+(x,y) = L_-(y,x) = -k^{-1} \sin[k(y-x) \exp ik(y-x)]$ (5)

In order to continue $F_\pm(k,x)$ into the upper half plane, all
authors have used the following bound, which holds for any $y \geqslant x$,
and where $\sigma(k) = |Imk| - Imk$:

$$\exp[-(y-x)\sigma(k)] |L_+(x,y)| \leqslant 2\inf\{|k|^{-1}, y - x\} \tag{6}$$

This is used in two steps, with first keeping $2|k|^{-1}$ only at the
right hand side of (6) (we call it the first choice in (6)),
then using $(y-x)$ instead. However, this last bound is not convenient, and led to some errors. Our improvement was to replace
$(y-x)$ in the right hand side of (6) by $[1+|x|\theta(-x)][1+|y|\theta(y)]$,
where θ is the Heaviside function (for $L_-(y,x)$, and $x \geqslant y$, it
suffices to permute x and y). This way of proceeding is called
below the second choice. Hence the kernel of (4a) (resp. 4b)
can be absolutely majorized by either of two separable kernels
of the general form $\alpha(x)\beta(y)$. For instance, with the second choice,
and with corresponding labels $\pm$ put at the same level, we have :

$$\alpha_\pm(x) = \exp[\mp x\sigma(k)] \{1+|x|\theta(\mp x)\}$$

$$\beta_\pm(x) = 2\exp[\pm x\sigma(k)] |V(x)| \{1+|x|\theta(\pm x)\} \tag{7}$$

The same formulas hold for the first choice, with k^{-1} instead of
$[1+|x|\theta]$. The interest of such separable kernels is that the
corresponding Volterra equation can be exactly solved and its
solution expansion yields uniform absolute bounds for corresponding terms in the solution expansion to (4a) or (4b). Since each
term obviously is an entire function of k, we readily prove that
$F_\pm(k,x)$ is holomorphic in the half plane $Imk \geqslant 0$ and bounded as
follows :

$$|F_+(k,x)-1| \leqslant \alpha_+(x) \int_x^\infty du \, \beta_+(u) \exp[2\int_x^u dt |V(t)| \gamma(t)]$$

$$|F_-(k,x)-1| \leqslant \alpha_-(x) \int_-^x du \, \beta_-(u) \exp[2\int_u^x dt |V(t)| \gamma(t)] \tag{8}$$

where $\gamma(t)$ is equal to $|k|^{-1}$ for our "first choice", to $(1+|t|)$ for our "second choice". By the way, one sees here the interest of assumption (2). It follows that $|F_+(k,x)-1|$ is $O(k^{-1})$ as $|k| \to \infty$, $\mathrm{Im}\, k \geqslant 0$. Hence we can define, at least in L_2, the following Fourier transforms :

$$K_\pm(x,y) = (2\pi)^{-1}\int_{-\infty}^{+\infty} dk \, \exp[\mp ik(y-x)]\{F_\pm(k,x)-1\} \qquad (9)$$

By iterating (4) once, and inserting the result into (9), one easily sees that the integral in (9) uniformly converges for $y \neq x$. By closing the contour in the upper-half k-plane, we deduce from (9) that $K_+(x,y)$ vanishes for $x > y$ and that so does $K_-(x,y)$ for $x < y$. Hence, the inverse Fourier transform yields, at least in the L_2 sense, the transformation formulas :

$$F_+(k,x)-1 = \int_x^\infty dy \, \exp[ik(y-x)]K_+(x,y) \qquad (10a)$$

$$F_-(k,x)-1 = \int_{-\infty}^x dy \, \exp[-ik(y-x)]K_-(x,y) \qquad (10b)$$

By substituting (4) into (9), then using (10) to eliminate $F_\pm$ in favor of $K_\pm$, and performing the Fourier transforms, we obtain the integral equations :

$$K_+(x,y) = \frac{1}{2}\int_{\frac{1}{2}(x+y)}^\infty ds\{V(s)+2\int_0^{\frac{1}{2}(y-x)} dr V(s-r)K_+(s-r,s+r)\} \qquad (11a)$$

$$K_-(x,y) = \frac{1}{2} \int_{-\infty}^{\frac{1}{2}(x+y)} ds\{V(s)+2\int_{\frac{1}{2}(y-x)}^0 dr \, V(s-r)K_-(s-r,s+r)\} \qquad (11b)$$

It follows from (11) that $K_+(x,x)$ and $K_-(x,x)$ are trivially related to V. In fact, the formulas (10) and (11) show that the transformation kernels $K_\pm$ are the key to the direct problem. They can be constructed from V by integrating (11), and, in turn, they readily yield V and F through (10), ie the parameter and the solutions of (1). We now go to the inverse problem, which can proceed either from $R_+(k)$ or from the reflection coefficient from left to right, $R_-(k)$, here defined by

$$f_-(-k,x) + R_-(k)f_-(k,x) = T(k)f_+(k,x) \qquad (12)$$

Some algebraic derivations of wronskians yield the properties :

$$T(k) = 2ik[\Delta(k)]^{-1} \qquad (13a)$$
$$R_\pm(k) = \pm[\Delta(k)]^{-1}W[f_\pm(-k,\cdot),f_\mp(k,\cdot)] \qquad (13b)$$
$$\text{where} \quad \Delta(k) = W f_-(k,\cdot),f_+,\cdot) \qquad (13c)$$
$$\text{and} \quad R_+(k)T(-k) + \bar R_-(-k)T(k) = 0 \qquad (14a)$$
$$R_+(k)\bar R_+(-k)+T(k)T(-k) = R_-(k)R_-(-k)+T(k)T(-k) = 1 \qquad (14b)$$

Because of the reality of $V(x)$, we have in addition for real k :

$$R_\pm(-k) = [R_\pm(k)] \qquad\qquad T(-k) = [T(k)] \tag{15}$$

The so called "Jost solution" $\Delta(k)$ is holomorphic in $\text{Im}k \geqslant 0$ and therefore so is $T(k)$, except may be at the zeros of $\Delta(k)$. Now it can be proved that these zeros correspond to bound states ("trapped modes") and we can exclude them in a study intended for applications in electromagnetic scattering. Assuming this is done, we can give a method of reconstruction of $V(x)$ from $R(k)$. Indeed, the analytic and asymptotic properties of $F_\pm(k,.)$ in $\text{Im}k \geqslant 0$ enable us to write down the Cauchy formula

$$F_\pm(k,x)-1 = \lim_{\varepsilon \to 0^+} \frac{1}{2\pi i} \int_{-\infty}^{+\infty} dk' \frac{F_\pm(k',x)-1}{k'-k-i\varepsilon} \tag{16}$$

which can be transformed by making $x \to -x$ and by inserting (3) or (12), to yield

$$f_\pm(k,x) \exp[\mp ikx -1 = \lim_{\varepsilon \to 0^+} \{$$
$$\frac{1}{2\pi i} \int_{-\infty}^{+\infty} dk' \frac{1-T(k')f_\mp(k',x)\exp[\pm ik'x]}{k'+k+i\varepsilon}$$
$$+ \frac{1}{2\pi i} \int_{-\infty}^{+\infty} dk' \frac{R(k')f_\pm(k',x)\exp[\pm ik'x]}{k'+k+i\varepsilon} \} \tag{17}$$

Using (13 a,c), the asymptotic behavior of $F_\pm$, and the holomorphy assumption for $T(k)$ in $\text{Im}k \geqslant 0$, we can show that the first term in the right-hand side of (17) vanishes. Hence we obtain an equation that is convenient for the desired reconstruction :

$$F_\pm(k,x)-1 = \frac{1}{2\pi i} \lim_{\varepsilon \to 0^+} \int_{-\infty}^{+\infty} dk' \frac{R_\pm(k')F_\pm(k',x)\exp[\pm 2ik'x]}{k'+k+i\varepsilon} \tag{18}$$

The interest of equations like (18) for reconstruction problems has been shown by Karlsson[7]. In fact, (18) is equivalent to the Marchenko equation, which is readily obtained by Fourier-transforming (18) :

$$K_\pm(x,y)+M_\pm(x+y)+\int_{-\infty}^{+\infty} dz\theta(z\mp x)M(y+z)K_\pm(x,z) \tag{19}$$

where

$$M_\pm(x) = (2\pi)^{-1} \int_{-\infty}^{+\infty} dk\, R_\pm(k)\exp[ikx] \tag{20}$$

The characterization problem in the class defined by (2) is still open. In a slightly narrower class, Deift and Trubowitz[4] gave existence and uniqueness theorems. However, we are interested here in a still narrower class. Thus we shall not recall these complicated results but only refer to the original paper when necessary.

1.2. On "rational" reflection coefficients

Reflection coefficients that are rational fractions of k already appear in the very first important paper on the inverse scattering problem on the line, an unpublished report by Kay[8], who recognized the closed form of potentials corresponding to a coefficient R(k) holomorphic everywhere but a finite number of poles in the lower half-plane. Kay[9,10] gave then a complete study of these "rational" reflection coefficients for potentials vanishing identically after a finite distance on one side of the real axis ("cut off potentials"), whereas applications of reflectionless potentials were sought in electromagnetic propagation[11]. From that time on, and although the method given by Kay could have been used for the general case, almost all authors only considered cut off potentials (say $V(x) = 0$ for $x < 0$). (See for instance Reilly & Jordan[12], Pechenick and Cohen[13], and the references listed therein).Cut potentials are studied because they naturally appear in plasma sounding. Exact calculations with two or three poles in the reflection coefficient were performed, either with Kay's method or with a method beginning by a Laplace transforming the Marchenko equation, or with one using a linear differential operator to cancel R(x) in the Marchenko equation and derive instead a linear differential equation. Concerning full range potentials, let us cite the papers of Moses[14] and of Abraham, de Facio, Moses[15], in which some assumptions of §(1.1) are given up in order to get singular examples. Studies of non linear partial differential equations (NLPDE connected with the inverse scattering problem (ISP) involved the studies of N-soliton solutions, for which R(k) has a finite number of simple poles on the upper half imaginary axis. They gave rise, among others, to a large number of papers on rational solutions of NLPDE (rational functions of x and t). However, our knowledge of reflection coefficients that exhibit poles in the lower half plane, the most interesting in scattering studies, was hardly improved by them. Hence we have undertaken a systematic study of rational reflection coefficients in the ISP. Here we shall only present some results of interest for electromagnetic inverse scattering. To derive them, we introduce two new methods. The first one is a generalization of Kay's method. The second one, founded on the wronskian method, is completely different and yields a sequential algorithm for solving the ISP with rational reflection coefficients. Throughout these methods[6] new distinction between poles has been suggested to us by our current interest in isospectral transformations and has proved to be essential <u>also</u> in the inverse scattering problem.

1.3. The methods

We start with the following assumptions :
(a) T(k) is a rational fraction, with poles located only in the lower half-plane, goes to 1 for $|k| \to \infty$, vanishes on the real axis at k = 0 only, where it has a simple pole, and satisfies (15).
(b) $R_\pm(k)$ are rational fractions, go to zero as $0(|k|^{-2})$ for $|k| \to \infty$, go to -1 for $|k| \to 0$, and satisfy the formulas (14) and (15).
(c) Either $R_+(k)$ or $R_-(k)$ is given, consistently with Assumptions (a) and (b).
Thanks to the assumptions (a) and (b), it is possible to check the Deift and Trubowitz conditions that are sufficient to guarantee the existence and uniqueness of a real potential in the class defined by (2), with reflection coefficients $R_\pm(k)$, transmission coefficient T(k), and for which all the formulas in (1.1) do hold. These formulas involve products like R(k)R(-k), which are invariant under inserting into R(k) a "phase factor" P(k)/P(-k). This remark, and a few algebraic manipulations, show that the only forms consistent with no bound state are

$$T(k) = \frac{\prod\limits_{i=1}^{q} (\kappa_i + k)}{\prod\limits_{j=1}^{q} (\lambda_j - k)} = \frac{N(k)}{D(k)} \tag{21}$$

$$R_-(k) = \frac{P(k)}{\prod\limits_{j=1}^{q} (\lambda_j - k)} \prod\limits_{i=q+1}^{r} \frac{\kappa_i - k}{\kappa_i + k} \prod\limits_{\ell=p+1}^{s} \frac{\lambda_\ell - k}{\lambda_\ell + k} \tag{22}$$

$$R_+(k) = \frac{P(-k)}{\prod\limits_{j=1}^{q} (\lambda_j - k)} \prod\limits_{i=2}^{r} \frac{\kappa_i + k}{\kappa_i - k} \prod\limits_{\ell=p+1}^{s} \frac{\lambda_\ell + k}{\lambda_\ell - k} \tag{23}$$

where the κ_i's and λ_ℓ's are complex numbers such that $\kappa_1 = 0$, Im $\kappa_i > 0$ (i > 1), Im $\lambda_i < 0$, and $p \leqslant q \leqslant r,s$. The polynomial $P(k)$ degree is at most q-1. Some factors may disappear if these forms are algebraically reduced. In any case, the poles of $R_-(k)$ in Imk > 0 belong to two sets, $\mathcal{P}_s$ and $\mathcal{P}_I^-$, with $\mathcal{P}_s^- \subset$ $\subset \{-\lambda_\ell | \ell \in (p+1,q)\}$, $\mathcal{P}_I^- \subsetneq \{-\lambda_\ell | \ell \in (q+1,s)\}$. The poles of $R_+(k)$ in Imk > 0 belong to $\mathcal{P}_I^+$ and $\mathcal{P}_s^+$, with $\mathcal{P}_s^+ \subset \{\kappa_i | i \in (2,q)\}$, $\mathcal{P}_I^+ \subset \{\kappa_i | i \in (q+1,r)\}$. Together with T(k), the sets labelled s correspond to the spectral properties of our problem. Those labelled I do not depend on T and correspond to isospectral transformations of R_+. The whole sets $\mathscr{S}^+ (= \mathscr{S}_I^+ \cup \mathscr{S}_s^-)$ and (similarly) $\mathscr{S}^-$ contribute in the derivation of V from $R^{\pm}$. Equations (18) are very convenient for solving this problem. In the simple case when all poles are simple, with residues R_+^i and R_-^j, we readily derive for $x > 0$

$$F_+(k,x) - 1 = \sum_{\kappa_i \in \mathscr{S}^+} \frac{R_+^i \, F_+(\kappa_i, x) \exp[\, 2i\kappa_i x]}{k + \kappa_i} \qquad (24)$$

and for $x < 0$

$$F_-(k,x) - 1 = \sum_{-\lambda_j \in \mathscr{S}^-} \frac{R_-^i \, F_-(-\lambda_j, x) \exp[\, 2i\lambda_j x]}{k - \lambda_i} \qquad (25)$$

We can identify $K_{\pm}(x,y)$ for instance by comparing (24-25) and (10a-10b) :

$$K_+(x,y) = -i \sum_{\kappa_i \in \mathscr{S}^+} R_+^i F_+(\kappa_i, x) \exp[\, i\kappa_i (x+y)] \qquad (26a)$$

$$K_-(x,y) = -i \sum_{-\lambda_j \in \mathscr{S}^-} R_-^i \, F_-(-\lambda_j, x) \exp[\, i\lambda_j (x+y)] \qquad (26b)$$

Let k run through $\mathscr{S}_+$ in (24), through $\mathscr{S}_-$ in (25). We hereby obtained two linear systems whose solution yields everything in our problem through (26a) and (26b). It is interesting to notice that $R(k)$ being a rational fraction depends on the position of the origin. Changing this position would insert in $R(k)$ a factor $\exp[-2ika]$. Besides, the two linear systems are independent, so that taking care of the phase isospectral factor enables us to modify almost independently the two sides ($x \gtrless 0$) of the potential. Hence, this way of processing is convenient for a study of the smoothing of the "surface" of $V(x)$ at $x = 0$. However, there is a better method, that fully makes use of the isospectral transformations properties. This method(6) could be based either on the Darboux theorem or on the generalized Wronskian formula. Here we use the Darboux theorem. We start with a potential V_1 that already corresponds to rational coefficients. Hence (24) and (25) hold. They mean that $F_{\pm}(k,x)$ and the relations (3) or (12) can be continued everywhere in the k-plane except at isolated poles. Let us now consider a solution u_o of

$$u_o'' + (\lambda_o^2 - V_1)u_o = 0 \qquad (27a)$$

where, say, $\operatorname{Im} \lambda_o > 0$. Let f be a solution of

$$f'' + (k^2 - V_1)f = 0 \qquad (27b)$$

Following Darboux, we easily check that the function ψ defined by

$$\psi = f' - f \, u_o'/u_o \tag{28}$$

is a solution of the equation

$$\psi'' + (k^2 - V_2)\psi = 0 \tag{29}$$

with the modified potential

$$V_2 = V_1 - 2 \frac{d}{dx}(u_o'/u_o) \tag{30}$$

Let us now derive the corresponding reflection and transmission coefficients, setting

$$u_o = \alpha \, f_-^{(1)}(\lambda_o, x) + \beta \, f_-^{(1)}(-\lambda_o, x) \tag{31}$$

where the upper label $^{(1)}$ refers to V_1. A study of (3) and (12) shows the possible cases :

(1) $\beta \neq 0$, any α but $\alpha = \beta \, R_-(\lambda_o)$

In this case, u_o'/u_o is asymptotic to $i\lambda_o$ for $x \to -\infty$ $^{(1)}$, and to $-i\lambda_o$ for $x \to +\infty$. Starting from the solution $f \equiv f_-^{(1)}(k,x)$ of (27b), we can write down the formula analogous to (3) and show by identification that

$$T_2(k) = -\frac{\lambda_o + k}{\lambda_o - k} \, T_1(k) \quad ; \quad R_2^+(k) = \frac{\lambda_o + k}{\lambda_o - k} \, R_1^+(k) \tag{32}$$

(2) $\beta \neq 0$, $\alpha = \beta \, R_-(\lambda_o)$. Then

$$T_2(k) = T_1(k) \quad ; \quad R_2^+(k) = \frac{\lambda_o - k}{\lambda_o + k} \, R_1^+(k) \tag{33}$$

(3) $\beta = 0$, λ_o not a bound state. Then

$$T_2(k) = T_1(k) \quad ; \quad R_2^+(k) = \frac{\lambda_o + k}{\lambda_o - k} \, R_1^+(k) \tag{34}$$

(4) $\beta = 0$, λ_o bound state. Then

$$T_2(k) = -\frac{\lambda_o - k}{\lambda_o + k} \, T_1(k) \quad ; \quad R_2^+(k) = \frac{\lambda_o - k}{\lambda_o + k} \, R_1^+(k) \tag{35}$$

Only the cases (2) and (3) correspond to isospectral transformations. They enable one to insert into $R(k)$ a phase factor with a pole located either in the upper half plane or in the lower half plane. This possibility of sequentially inserting the poles of $R(k)$ can be used to write down a new algorithm. In this algorithm[19], the coefficients $R_\pm(k)$ are written as products of "undressed" coefficients $R_\pm^o(k)$ without poles in the $\pm$ half k-plane respectively, times finitely many phase factors. These "routes" in $\mathcal{C}$ correspond to "routes" in $\mathcal{C}$ starting at "undressed" $v_\pm^{(0)}$

which (see (24),(25)), resp. vanish on $x \gtrless 0$, and one travels along these routes by Darboux transforms.

As an application, we start from a coefficient $R_+(k)$ which corresponds to a potential cut-off at $x = 0$, and we represent uncertainties on a data almost equal to $R_+(k)$ by introducing an isospectral extra factor $(i\mu-k)/(i\mu+k)$, with $\mu \gg 0$, so as to appraise the uncertainties on the surface position due to data errors. For a more specific example, let $R_-(k)$ have only two poles in $\mathrm{Im}\ k < 0$.

$$R_-(k) = (a^2+b^2)[\,(k+ib)^2-a^2]^{-1} \tag{36}$$

where a, b, > 0, $b^2 > a^2$. The corresponding other coefficients are

$$T(k) = k(k+i\kappa)[\,(k+ib)^2-a^2]^{-1} \tag{37}$$

$$R_+(k) = -\frac{k+i\kappa}{k-i\kappa}\ \frac{a^2+b^2}{(k+ib)^2-a^2} \tag{38}$$

where $\kappa = [\,2(b^2-a^2)]^{1/2}$. The corresponding potential is

$$V(x) = \begin{cases} 0 & \text{for}\ \ x < 0 \\[2mm] 8\kappa^2\dfrac{c^2\exp[-2\kappa x]}{\{1-c^2\exp[-2\kappa x]\}^2} & \text{for}\ x > 0 \end{cases} \tag{39}$$

where $c^2 = [\,c\,(\kappa)]^2 = [\,(b+\kappa)^2+a^2]^{-1}(a^2+b^2)$ \hfill (40)

Assume now the uncertainties on $R_-(k)$ are such that the isospectral phase factor $(i\mu-k)/(i\mu+k)$ in front of $R_+(k)$ can represent them, with $\mu \gg \mu_o$. Then the uncertainty on $V(x)$ is seen in the formula

$$V_\mu(x) = \begin{cases} 8\mu^2 d^2\exp[2\mu x]\{1-d^2\exp[2\mu x]\}^{-2} & (x < 0) \\[2mm] -8\kappa^2 c^2\exp[-2\kappa x]\dfrac{\kappa^2-\mu^2}{\{(\kappa+\mu)+(\kappa-\mu)c^2\exp[-2\kappa x]\}^2} & (x > 0) \end{cases} \tag{41}$$

where $d^2 = c^2(\mu)$.

It clearly shows that for $\mu/\kappa \gg 1$, the uncertainty for $x > 0$ is much weaker than the one for $x < 0$. Set for instance $a^2 = 1$, $b^2 = 1,125$, so that $\kappa = 0,5$, $c^2 \sim 0,6$, $\mu = 10$, $d^2 \sim 0,016$. Then $V(x)$ has values between $0,1$ and 1 for x between 0 and 2. The perturbation due to the phase factor lies for $x > 0$ all along $V(x)$ with a relative value not larger than $1/200$, whereas for $x < 0$, it is approximately $13\exp[20x]$, which has values larger than 0.2 for $x > -0.2$. Hence these uncertainties on $R_+(k)$ result into an uncertainty on the position $x = 0$ rather than an uncertainty on the shape of $V(x)$.

SECTION 2.

Let us now deal with exact ambiguities. We know that the inverse problem on the line is well-posed if V is sought in the set L_2' where $\int_{-\infty}^{+\infty} dx(1+|x|^2)|V(x)|$ is finite and if R(k) satisfies the Deift-Trubowitz conditions :

(a) T(k), $R_+(k)$, $R_-(k)$, satisfy on the real axis (14) and (15)

(b) T(k) is holomorphic in Imk > 0, continuous down to the real axis

(c) $T(k)-1 = 0(|k|^{-1})$ (Imk ≥ 0) ; $R_\pm(k) = 0(|k|^{-1})$ (k ∈ ℝ)

(d1) $|T(k)| > 0$ (Imk ≥ 0, k ≠ 0) .

(d2) either $|T(k)| > 0$ (Imk ≥ 0) or there exists $\bar{T}$ and ρ_+ such that $T(k)-k\bar{T} = 0(k)$, (Imk ≥ 0) and $1+R_+(k)-k\rho_+ = 0(k)$, (k ∈ ℝ)

(e) $M_+(x)$ is absolutely continuous, and there exists c(a) finite such that $\int_{-\infty}^{+\infty} dx \, \theta[\pm(x-a)]|M_+'(x)|(1+x^2) \leq c(a)$ for any real a.

In order to show exact ambiguities in potential reconstructions, one must violate at least one of these conditions. This has been done by Abraham, de Facio and Moses[15] whose example involves a δ-function in V, and violates conditions (d) and (e). In particular, T(k) vanishes at k = 0, but $R_+(0)$ is equal to +1 (this feature corresponding to a Jost solution singular at k = 0). Other examples were given by Brownstein[16] and by Moses[17]. None of these authors described how they discovered their examples. It is certainly not a systematic way of exploring the subject. Here we shall try to give one of the possible ways which can be applied in the cases of interest arising in the one-dimensional inverse scattering problem. Thus we study the equation

$$A^{-1} \frac{d}{dx} A \frac{d}{dx} p + k^2 p = 0 \qquad (42)$$

where A may be the admittance, p the potential, k the frequency. This equation arises in various electromagnetic and geophysical problems (see for example references 18 and 19). It must be solved in such a way that p and $A \frac{dp}{dx}$ are continuous. A must be non negative but is not necessarily continuous. We studied elsewhere[19] the case where A shows an arbitrary number of simple discontinuities (jumps). So as to get a chance of discovering a simple way towards ambiguities, we assume here that A is twice differentiable everywhere except at the point x = 0, where A and $\frac{dA}{dx}$ may show simple discontinuities (jumps). Let V be the function defined at each point x ≠ 0 by

$$V(x) = [A(x)]^{-1/2} \frac{d^2}{dx^2} [A(x)]^{1/2} \qquad (43)$$

We assume that V belongs to L_2'. This guarantees the existence of the "Jost solutions" $p_\pm(k,x)$ of (42), that go respectively to $[A(x)]^{-1/2} \exp[\pm ikx]$ as x goes to ±∞. The reflection and transmission coefficients are defined by the relations

$$\hat{T}(k)p_-(k,x) = p_+(-k,x) + \hat{R}_+(k)p_+(k,x) \tag{44}$$

$$p_-(-k,x) + \hat{R}_-(k)p_-(k,x) = \hat{T}(k)p_+(k,x) \tag{45}$$

It is easy to see that $\hat{T}(k)$ and $\hat{R}_+(k)$ satisfy (14) and (15). Now, both for $x > 0$ and for $x < 0$, the equation (42) is equivalent to the Schrödinger equation (1), with $V(x)$ as in (43). The function $p_\pm(k,x)$ are related to the ordinary Jost solutions $f_\pm(k,x)$ of (1) on both sides of $x = 0$. We can write down for instance

$$p_-(k,x) = u_\pm[A(x)]^{-1/2}f_+(-k,x) + v_\pm[A(x)]^{-1/2}f_+(k,x) \tag{46}$$

where the coefficients u_+ and v_+ hold for $x \geq 0$, u_- and v_- for $x < 0$. They are trivially related with $R_+, T, \hat{R}_+, \hat{T}$. Writing down the continuity relations at $x = 0$ we obtain :

$$u_+ = [\hat{T}(k)]^{-1} = [T(k)]^{-1}[\alpha(k) + \beta(k)R_+(k)] \tag{47}$$

$$v_+ = [\hat{T}(k)]^{-1}\hat{R}_+(k) = [T(k)]^{-1}[\beta(-k) + \alpha(-k)R_+(k)] \tag{48}$$

where

$$\alpha(k) = t^{-1}\{1 + s(ik)^{-1}f_+(k)f_+(-k) + r(2ik)^{-1}[f_+'(k)f_+(-k) + f_+'(-k)f_+(k)]\} \tag{49}$$

$$\beta(k) = (ikt)^{-1}\{s[f_+(k)]^2 + r\ f_+(k)f_+'(k)\} \tag{50}$$

$$f_\pm(k,0) \equiv f_\pm(k) \tag{51}$$

$$t^{-1} = \frac{1}{2}\{[A(0^+)/A(0^-)]^{1/2} + [A(0^-)/A(0^+]^{1/2}\} \tag{52}$$

$$r/t = \frac{1}{2}\{[A(0^+)/A(0^-)]^{1/2} - [A(0^-)/A(0^+)]^{1/2}\} \tag{53}$$

$$s/t = -\frac{1}{4}[A(0^+)A(0^-)]^{-1/2}[A'(0^+) - A'(0^-)] \tag{54}$$

We can obtain in the same way the relations

$$T(k)/\hat{T}(k) = \bar{\alpha}(-k) - \bar{\beta}(k)R_-(k) \tag{55}$$

$$\hat{R}_-(k)T(k)/\hat{T}(k) = -\bar{\beta}(-k) + \bar{\alpha}(k)R_-(k) \tag{56}$$

where

$$t\ \bar{\alpha}(k) = 1 - (ik)^{-1}sf_-(k)f_-(-k) - (2ik)^{-1}r[f_-'(k)f_-(-k) + f_-'(-k)f_-(k)] \tag{57}$$

$$-ikt\ \bar{\beta}(k) = s[f_-(k)]^2 + r\ f_-(k)f_-'(k) \tag{58}$$

We can work either on $R_+(k)$ and $T(k)$, using the system (47)-(48), or $R_-(k)$ and $T(k)$, using the system (57)-(58). Once R_+ and T

are fixed, R_- is fixed by (14) (or conversely), so that the two
approaches are equivalent. In the following, we choose working
on $R_+(k)$, and drop everywhere the index +. We look for two diffe-
rent admittances A_1 and A_2 such that $\hat{R}_1(k) = \hat{R}_2(k)$ (with $\hat{R}$ stan-
ding for $\hat{R}_+$). Our working assumption is that the reflection and
transmission coefficients are rational fractions of k. Now it
follows from the algebraic proofs recalled in (1.3) that giving
$\hat{R}(k)$ determines $\hat{T}(k)$ up to the "bound state factor", (not written
in (21)), which we shall ignore in this first study, since
it is related to unphysical zeros of the admittance[20]. Hence
we are led also to impose $\hat{T}_1(k) = \hat{T}_2(k)$. Thus it is natural to
take as an additional working assumption that $T_1(k) = T_2(k), s_1 = s_2$.
With these three equalities, it follows from (47) and (48) that
$R_1(k)$ and $R_2(k)$ satisfy the system

$$\left.\begin{array}{l} \beta_1(k)R_1(k) - \beta_2(k)R_2(k) + \alpha_1(k) - \alpha_2(k) = 0 \\[2mm] \alpha_1(-k)R_1(k) - \alpha_2(-k)R_2(k) + \beta_1(-k) - \beta_2(-k) = 0 \end{array}\right\} \qquad (59)$$

We shall now reduce our study to the case where the impedance is
continuous, ie t = 1, r = 0. Provided $\alpha_1(k) \neq \alpha_2(k)$, and
$\beta_1(k) \neq \beta_2(k)$, the system (59) is equivalent to the system

$$R_1(k) = -\psi(k)/m(k) \qquad (60)$$

$$R_2(k) = -\psi(-k)/m(k) \qquad (61)$$

where

$$\psi(k) = sf_2(k)f_1(-k)k^{-1}[f_2(k)f_1(-k) - f_2(-k)f_1(k)] + i[f_1(k)f_1(-k) -$$
$$f_2(k)f_2(-k)] \qquad (62)$$

and

$$m(k) = sf_1(k)f_2(k)k^{-1}[f_2(k)f_1(-k) - f_2(-k)f_1(k)] + i\{[f_1(k)]^2 - [f_2(k)]^2\} \qquad (63)$$

Hence $R_1(k)/R_2(k)$ is a "phase factor". The first factor in the
right-hand side of (23) is the same one for $R_1(k)$ and $R_2(k)$, a
result which is fortunately consistent with our working assumption $T_1(k) = T_2(k)$. Hence one goes from V_1 to V_2 by means of one
or several Darboux transforms of the form (33) or (34).
Referring to (23), let us define $R_o(k)$ as a common factor of
$R_1(k)$ and $R_2(k)$ which has no pole in Imk > 0 and also (so as to
guarantee its uniquess), no zero in Imk > 0. Setting

$$R_1(k) = R_o(k)\psi_1(k)/\psi_1(-k) \quad R_2(k) = R_o(k)\psi_2(k)/\psi_2(-k) \qquad (64)$$

we see that the Darboux transform which are associated with these
two factors give rise to an ambiguity if two conditions are

fulfilled :
(a) there exists a rational function of k^2, $C(k^2)$, such that

$$\psi_1(k)\psi_2(-k) = C(k^2)\psi(k) \tag{65}$$

(b) $\quad R_o(k) = -\psi_1(-k)\psi_2(-k)[C(k^2)]^{-1}[m(k)]^{-1} \tag{66}$

Indeed, the condition (65) guarantees that $R_1(k)/R_2(k) = \psi(k)/\psi(-k)$, and the condition (66) guarantees that $R_1(k)+R_2(k)$ is consistent with this condition and (60)-(61). The functions $\psi_i(k)$ are polynomials of the form $\prod_{i=1}^{n_i} (\mu_i - k)$, and the (n_1+n_2) parameters μ_i completely describe the problems corresponding to R_1 and R_2. Using the results of §(1.3) enables one to construct $f_1(k)$ and $f_2(k)$ and hence $\psi(k)$ and $m(k)$ in terms of the μ_i's and of the numbers $R_o(\mu_i)$. The parity condition (65) imposes then constraints on the μ_i's and $R_o(\mu_i$'s). $R_o(k)$ can then be constructed from (66) and must satisfy the consistency condition on its zeros and poles location in the lower half plane. We believe that these constraints can in general be fulfilled for whatever choice of the degrees of ψ_1 and ψ_2 (ie n_1 and n_2). Our guess is supported on one hand by a counting of the number of (non linear) equations, constraints, and variables, on the other hand by the existence of a solution in the very simplest and most constrained case where ψ_1 is of degree 0 and ψ_2 of degree 1. Let us now give this example :

Example
Let $\psi_1(k) = 1$, $\psi_2(k) = i\gamma s + k$, with $s > 0$, $0 < \gamma < 2(\sqrt{2}-1)$. The method yields two "equivalent potentials", say $-2s\delta(x)-8\theta(x)\gamma^2 s^2[1+(\gamma-1)e^{-2\gamma sx}]^{-2}(\gamma-1)e^{-2\gamma sx}$ and the symmetrical one, with the same reflection and the same transmission coefficients :

$$\widetilde{T}(k) = k(k+i\gamma s)(k-i\gamma s)^{-1}(k+is)^{-1} \tag{67}$$

$$\widetilde{R}(k) = is(k+i\gamma s)(k-i\gamma s)^{-1}(k+is)^{-1} \tag{68}$$

These potentials have a bound state so that the corresponding impedance has a zero and is not physical. The method is consistent because the "background" potential (ie without the δ function) has no bound state.
In the case of continuous admittances, the formula (43) defines $V(x)$ as a generalized function which contains at $x = 0$ the distribution $-2s\delta(x)$. It is easy to see that the Schrödinger equation (1), with the thus defined function $V(x)$, and the equation (42) are equivalent. They yield the same reflection and transmission coefficients, and the Jost solutions of (42) are those of the Schrödinger equation multiplied by $[A(x)]^{-1/2}$. The Darboux transforms apply without modification to the Schrödinger

equation with V(x) involving δ-functions. Let us call "equivalent" two parameters $\tilde{V}$ which correspond to the same scattering coefficients $\tilde{R}$ and $\tilde{T}$. If a given Darboux transform is applied to two "equivalent" parameters, the corresponding scattering coefficients are multiplied by the same phase factors, and hence the two new parameters V are equivalent. It follows that each pair of equivalent parameters V generates a pair of continuous paths of equivalent parameters in the space of parameters. One can do it as well by starting from the example cited above as by starting from the pairs (or multiplets) introduced by Abraham, de Facio, Moses[14,15] or Brownstein[16]. Some non-physical features are kept in the transformations (e.g. $R_+(0) \neq -1$). Others may be introduced (e.g. zeros of the admittances or double poles of the potential on real x). In fact, there may be several points of special interest on two paths of equivalent parameters. Look for instance on the example given above. We may find it interesting to remove the bound state by using the Darboux transform (35). This is easy but the result is the potential $2s \delta (x)$ for whatever potential V_1 or V_2 we were starting from. Hence the ambiguity disappears beyond this point in our travel along the paths. In other words, this is a multiple point of the transform (see (32)). We can also find it interesting to introduce arbitrary phase factors, for instance factors $(i\beta s-k)^{-1}(i\beta s+k)$, which keep V real. Then we see that for a range of values of β, the Darboux transform introduces singularities (double poles on the real axis coming from the zeros of f_-). In the space of scattering data, the reflection coefficients violate the constraint $|R_+(k)| < 1$ on the real axis. This constraint actually restricts the domain of allowed pole positions in the complex k-plane, as noticed by several authors[13]. Hence the paths of equivalent parameters may also stop at the boundary of the set of allowed parameters.

In the general case (piecewise continuous impedance), the equation (42) is no longer equivalent to the Schrödinger equation with a potential. Nevertheless, since we have still more freedom in the choice of parameters (see (49),(50)), it is clear that ambiguities still exist. The problem is to know whether a generalized transform still enables us (or not) to construct paths in the parameter space that correspond to multiplying by phase factors in the spaces of scattering coefficients. We are able to exhibit such a transform :

Let A be a piecewise continuous function, asymptotically constant at $x = \pm\infty$, bounded away from zero, and which has only a finite set of (jump) discontinuities. A is the "parameter" in the equation (42). Let $p_k \equiv p(k,x)$ be a continuous solution of (42), whose definition is achieved by imposing $Ap'_k \equiv A \frac{d}{dx} p(k,x)$ absolutely continuous, and (arbitrary) convenient asymptotic conditions. Let $p_\mu \equiv p(\mu,x)$ be an absolutely continuous solution of the equation

$$A^{-1} \frac{d}{dx} A \frac{dp_\mu}{dx} + \mu^2 p_\mu = 0 \qquad (69)$$

with Ap'_μ absolutely continuous, and convenient asymptotic conditions. Now, we claim
(1) that the function A^T defined by

$$(A^T)^{1/2} = -A^{1/2} p'_\mu / p_\mu \qquad (70)$$

is piecewise continuous, with a finite set of jump discontinuities, and is asymptotically constant at $\pm\infty$.
(2) that the function

$$p_k^T = p_k - p'_k \, p_\mu / p'_\mu \qquad (71)$$

is a continuous solution of the equation

$$(A^T)^{-1} \frac{d}{dx} A^T \frac{d}{dt} p_k + k^2 p_k = 0 \qquad (72)$$

such that $A^T \frac{d}{dx} p_k$ is absolutely continuous. The proof of these results is easy. We only give to the reader an intermediate result:

$$\frac{d}{dx} p_k^T = Ap_\mu (Ap'_\mu)^{-2} [k^2 p_k Ap'_\mu - \mu^2 \, p_\mu Ap'_k] \qquad (73)$$

Now, choosing the asymptotic conditions like in (31) to (35), enables us to transform parameters in such a way that the corresponding scattering coefficients are multiplied by phase factors like in (32), (33), (34), and (35). In other words, this generalized transform exactly works like the Darboux transform. In particular, if two admittances are equivalent, the generalized transform (70) enables us to construct two paths of equivalent admittances.

Needless to say, all the other applications of these transforms are also generalized. In particular, suppose an admittance is made of jump discontinuities, with homogeneous and constant parts between discontinuities (e.g. Goupillaud or Claerbout models in geophysics). Such a problem is well-known and can be dealt with by matrix methods. Suppose now the parts between discontinuities are no longer homogeneous constant but are such that their effect can be represented[19] by a rational phase factor in the scattering coefficients. Then it is possible to use the generalized transform to undress the problem, reducing it to the simplest one.

Conclusions

So as to conclude our "critical study", it is interesting now to care for the physical "teachings" of a global approach to electromagnetic inverse problems.

(a) the first point which has been made shows the importance of
the (generally concealed) a priori assumptions. We saw in
section 1 that choosing the gap position of a truncated impedance, and allowing or not some diffuseness of this surface may
have large consequences on the whole reconstruction. We saw in
section 2 that allowing surface diffuseness can lead to ambiguities in the reconstruction, may-be (hopefully) discarded by
convenient physical requirements. Other authors have discussed
for example the importance of the no absorption assumption, either
when it is related to the potential [13] or to the spectral
problem itself [20]. But this assumption usually is not forgotten.
(b) the second point is the necessary use of special geometrical
methods for constructing all solutions of the inverse problem.
Of course, geometrical methods in function spaces are well-known:
they enable us to define most quasisolutions or approximate solutions. But here we sought the exact solutions that correspond to
results which are rational fractions of k (and any result can be
arbitrarily approached by a rational fraction of k). And remarkab-
ly enough, the only way to explore completely the function space
containing these solutions is by using Lie-Backlund transforms,
more specially the Darboux ones which are described in § 1. The
result is very elegant.
(c) last but not least, the third point is the particular interest
of these methods from the algorithmic point of view. We have not
given a detailed study of this problem in the present paper, but
we did it elsewhere [19]. Of course the reader can imagine that
using a cross-coordinated route in the parameter space is the
shortest way to go from one point to another one. The value of this
guess   for a large number N of poles has been proved [19] :
inversion algorithms based on Darboux transforms involve $O(N^2)$
operations only, in contrast with $O(N^3)$ for other algorithms.

Connected with these physical points there is still a number of open problems. Let us say honestly that the first one is
how the scattering coefficients, with their zeros and poles, can
be constructed from the experimental data. More theoretical open
questions are connected with the ambiguities : is the number of
their paths finite ?, and, if so, is it possible to construct
all of them - what are the minimum requirements on A such that
there is no ambiguity - and what is the influence of absorption ?

References

(1) Calogero F. and Degasperis A.   Spectral transform and solitons I. North-Holland 1982.

(2) Faddeyev L.D.   Properties of the S-matrix of the one-dimensional Schrödinger equation, Trudy Mat. Inst. Steklov 73, 1964 pp. 314-333 ; AMST 2, 65, 139-166.

(3) Chadan K., and Sabatier P.C.   Inverse Problems in Quantum Scattering Theory, Springer-Verlag, New York 1977.

(4) Deift P., and Trubowitz E.   Inverse Scattering on the Line Comm. Pure Appl. Math. 32, 121-251 (1979).

(5) Atkinson D.   Marchenko in one dimension. Internal Rpt Inst. for Theoretical Physics Groningen, Netherlands (1979).

(6) Sabatier P.C.   Rational reflection coefficients and Inverse Scattering on the line, to be published in Nuovo Cimento.

(7) Karlsson B.   Inverse method for off-shell continuation of the scattering amplitude in quantum mechanics - in "Applied Inverse Problems" Lecture Notes in Physics n° 85, Springer-Verlag Berlin, Heidelberg, New York 1978.

(8) Kay I.   The Inverse Scattering Problem. Research Report n° EM-74 New York University. Inst. Math. Sciences 1955.

(9) Kay I.   On the determination of the free electron distribution of an ionized gas. Research Report n° EM-141 New York University Inst. Math. Sciences 1959.

(10) Kay I.   The Inverse Scattering Problem when the reflection coefficient is a rational function. Comm. Pure and Appl. Math. 13, 371-393 (1960).

(11) Kay I. & Moses H.E.   Reflectionless transmission through dielectrics and scattering potentials. Journ. Appl. Phys. 27 1503-1508 (1956).

(12) Pechenick K.R. & Cohen J.M.   Exact solutions to the valley problem in inverse scattering. Journ. Math. Phys. 24 406-409 (1983).

(13) Reilly M.H. & Jordan A.K.   The applicability of an inverse method for reconstruction of electron-density profiles IEEE Trans on Ant. and Prop. AP-29 245-252 (1982).

(14) Moses H.E.   An example of the effect of rescaling of the
     reflection coefficient on the scattering potential for
     the one-dimensional equation Stud. Appl. Math. 60, 177-181
     (1979).

(15) Abraham P.B., De Facio B., & Moses H.E.   Two distinct local
     potentials with no bound states can have the same scattering
     operator : A non uniqueness in inverse spectral transforma-
     tions. Phys. Rev. Lett. 46, 1657-59 (1981).

(16) Brownstein K.R.   Non uniqueness of the inverse-scattering
     problem and the presence of c = 0 bound states. Phys. Rev.
     D. 25, 2704-2705 (1982).

(17) Moses H.   Example of two distinct potentials without point
     eigenvalues which have the same scattering operator with
     the reflection coefficient c(0) = -1, Phys. Rev. A 27,
     2220-2221 (1983).

(18) Sabatier P.C.   Theoretical considerations for inverse scatte-
     ring Radio Science 18, 1-18 (1983).

(19) Sabatier P.C.   Rational reflection coefficients in one dimen-
     sional inverse scattering and applications Proceedings of
     Tulsa Conference on Scattering Theory, to be published by
     S.I.A.M.

(20) See for example Riska D.O.   On the electro-dynamic inversion
     problem for vertically layered earth Commentationes Physico
     Mathematicae 53 1-24 (1982).

I.2 (SR.2)

TOPOLOGICAL APPROACH TO INVERSE SCATTERING IN REMOTE SENSING[*]

Werner Güttinger & Francis J. Wright

Institute for Information Sciences
University of Tübingen
Köstlinstr. 6, 7400 Tübingen, FRG

The topological problem underlying inverse scattering is analyzed
by determining the caustic singularities that an unknown surface
generically impresses on a sensing wavefield. Imposing the principle of structural stability (i.e., qualitative insensitivity to
slight perturbations) on the inversion process shows the following:
The dominant singularities that generically occur in recorded signals, travel-time curves, surface contour maps and Fresnel-zone
topographies can, together with the associated diffraction patterns,
be classified into a few topological normal forms described by
catastrophe polynomials. As the source-receiver positions vary,
the observed patterns change their morphologies according to universal bifurcation sets. These provide a new quasi-geometric processing methodology for surface reconstruction.

1. INTRODUCTION

This paper is designed to review the present state of our knowledge concerning the application of singularity and bifurcation
theory to the inverse scattering problem in remote sensing and to
sketch some new developments and techniques which promise to become important tools for future research. The central idea underlying this geometrical approach is the following. In order that
the reconstruction of surfaces and subsurface structures from
backscattered waves be physically feasible, i.e., repeatable, the
scattering process underlying remote sensing has to be structurally
stable, i.e., qualitatively insensitive to slight perturbations of
the sensing wave system. Otherwise today's experiment would not
reproduce yesterday's result. Imposing this principle of structural

[*] Research supported in part by the Stiftung Volkswagenwerk, FRG

65

W.-M. Boerner et al. (eds.), Inverse Methods in Electromagnetic Imaging - Part 1, 65–76.
© 1985 by D. Reidel Publishing Company.

stability [19] on the inversion process has the following consequences [10]. First, it permits us to classify the geometric singularities, that an unknown surface or structure generically
impresses on a sensing wavefield, into a few universal topological
normal forms described by catastrophe polynomials. These geometrical singularities produce the dominant analytic singularities and
typical configurations that can generically occur in recorded signals, traveltime curves, contour maps, and in the associated diffraction patterns. As the source-receiver positions vary, the
patterns change their morphologies according to typical universal
bifurcation sets (caustics). These, and the resulting universal
power laws for the frequency dependent diffraction amplitudes,
permit a reconstruction of 3D profiles in a genuine zero-offset
survey. Second, the topological singularity and bifurcation concepts deriving from structural stability provide a unifying framework for all sensing techniques currently in use. The resulting
conceptually simple processing methodology yields directly the
desired end result of interpretation and comes much closer to the
interpreter's intuitive qualitative approach than wave-equation
based methods [17]. Moreover, the topological singularities provide an explanation for the similarity and universality of the
high-intensity patterns encountered in surface sensing [5], [6],
[13], seismology [16], ocean acoustics [14], electromagnetic
sensing [4], [7], ultrasound tomography, phonon spectroscopy [2],
and so forth. The same singularities also govern the spontaneous
formation of qualitatively similar spatio-temporal structures in
systems of various genesis [11], [12], [18], [19].

2. WAVEFIELD SINGULARITIES

Suppose a point source at x_0 in space emits a spherical wave pulse
that propagates through a layered medium made up of partially
reflecting surfaces. The scattered wave $\psi(x_0,t)$ that is received
back at the source -- the echo -- exhibits as a function of time
t, a number of strong peaks which vary with the source position
x_0. In this echogram (a seismogram in a geological survey) caustic
structures are discernible, e.g., those shown in Figure 1 where
the intensity $|\psi|^2$ of the echo is plotted as a function of the
(vertical) time, or depth, and of the (horizontal) source position.
Known as bright spots in seismology, these caustic events and the
diffraction patterns around them are the singularities which the
medium impresses on the sensing wavefield. They are at the root
of the interpreter's intuitive geometric, i.e., qualitative,
approach to remote sensing. Indeed, an interpreter readily points
out that the layered medium which produces the echogram of Fig. 1
is the one shown in Fig. 2.

In discussing the effects of caustics we confine ourselves
to a single reflecting surface. Since source and receiver are at

Fig. 1   Fig. 2

the same place, only the geometrical optics specular reflection
points of the surface, whose distance vectors to the source are
normal to the surface, contribute to the echo. Then, as an observer
at x_0 moves along a line on an observation surface Σ above the
surface, the reflecting point moves in the same direction for a
surface with low curvature. But when the line is above the lowest
curvature center of a concave part of the surface S, the reflecting
point suddenly moves in the opposite direction. This reversal in
direction occurs when x_0 on Σ crosses the surface's evolute E (the
set of loci of the surface's curvature centers). The evolute is
the envelope of the surface normals where neighboring rays, normal
to S, touch and focusing occurs (Fig. 3(a)). When the source-receiver -- the shotpoint for short -- crosses E, then the number of

rays going through x_0 suddenly
changes by two and so does the
number of specularly reflecting
points. This implies that the
number of arrival peaks in the
echo also changes by two. Therefore, the reversals in the propagation direction of the specularly reflecting points, as
seen from x_0, are recorded as
cusps in the echo profile
(Fig. 3(b)). Indeed, while
there is but one echo when x_0
is at A, there are three when
x_0 is at B. Fig. 3(b) is revealed in Fig. 1.

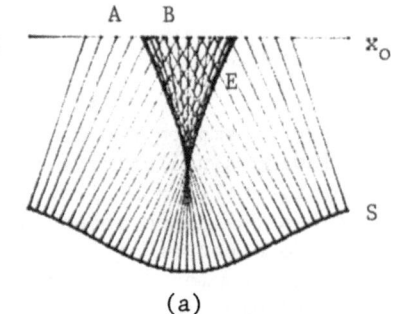

(a)

In three-dimensional
space the observation surface Σ intersects the evolute
sheet generically in smooth

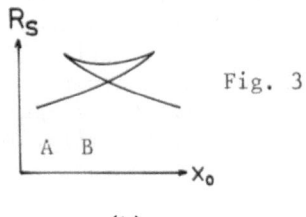

Fig. 3

(b)

curves L, called fold (A_2) lines, and in isolated cusp (A_3) points
$\hat{C}$ (Fig. 4). As the height of Σ varies, the cusp points evolve into
lines called ribs. If the rib is curved upwards (or downwards) and
Σ is nearly tangent to a point on the rib, then the observation
surface intersects the evolute in two beak-to-beak cusp singularities (or in a lip). This is shown in Fig. 4 for the winged evo-
lute sheets E of a saddle-surface S. Let $R=R(\underline{x}_0,\underline{x})=|\underline{x}_0-\underline{x}|$ be the
distance from $\underline{x}_0$ to a point $\underline{x}$ on S and $T=2R/c$ the two-way travel-
time (with speed c). The signals that are received back at the
source after reflections by the specular points $\underline{x}=\underline{x}_s(\underline{x}_0)$ on S have
travel-times $T_s=T(\underline{x}_0,\underline{x}_s(\underline{x}_0))$ or distances $R_s=R(\underline{x}_0,\underline{x}_s(\underline{x}_0))$. These
form a multisheeted 3-dimensional hypersurface in 4-dimensional
$(T_s,\underline{x}_0)$-space. For example in case of a plane observation surface
Σ, with $\underline{x}_0=(x_0,y_0)$, the saddle-surface S of Fig. 4 gives rise to
the travel-time surface on top of Fig. 5 with two dimensional
sections shown below.

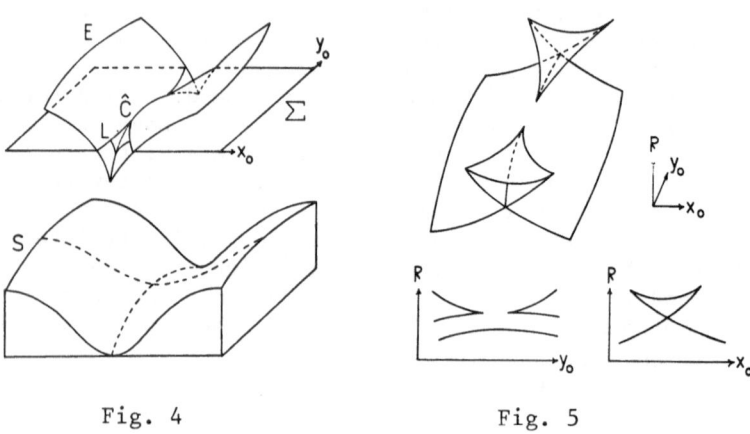

Fig. 4   Fig. 5

A rib is smooth except for isolated points where either of
the following singularities can occur: (i) the rib itself has a
cusp point, called a swallowtail point (A_4) because near it the
evolute has the shape shown in Fig. 8. (ii) a rib may touch a
fold or two other ribs; in this case two evolute sheets have a
contact in a common curvature center and are either hyperbolic
or elliptic umbilics (D_4^+ or D_4^-) shown in Figs. 9 and 10, respectively, and the surface has an umbilical point. Since the observed
singularities in echo recordings derive from the focal surfaces
which constitute evolutes, and since still more complex caustic
morphologies will arise if a surface possesses edges [1], [9], or
if source and receiver are different [10], the question arises
how to classify the singularities a surface impresses on a sensing
wavefield or ray family. To understand this, we need catastrophe
theory.

3. STRUCTURAL STABILITY OF REMOTE SENSING

Let the point $\underline{x}$ on the smooth surface S be parametrized by surface coordinates, $\underline{x}=\underline{x}(x,y)$. The distance vector $\underline{R}_s=\underline{x}_o-\underline{x}_s$ from a specular reflection point $\underline{x}=\underline{x}_s(\underline{x}_o)$ to the source is normal to S, $\underline{R}_s=R_s(\underline{x}_o)\underline{n}(\underline{x}_s)$ where $\overline{R}_s=|\underline{x}_o-\underline{x}_s|$ (cf., Fig. 6 for the basic scattering geometry). The distance R has an extremum R_s at $\underline{x}_s$: $\underline{t}\cdot\nabla_x R=0$ for vectors $\underline{t}$ tangent to S, or

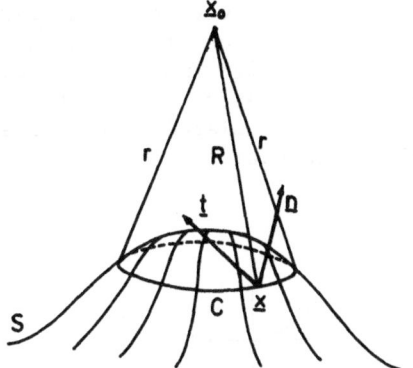

$$\nabla R = 0 \qquad (3.1)$$

with $R=R(\underline{x}_o,\underline{x}(x,y))$ and $\nabla=(\partial/\partial x,\partial/\partial y)$. By definition, a solution $\underline{x}=\underline{x}_s(\underline{x}_o)$ of (3.1) is a stationary point, or a "topological singularity" of R. The type of singularity R possesses at $\underline{x}_s$ depends on the Hessian determinant

$$H(R)=\begin{vmatrix} R_{xx} & R_{xy} \\ R_{xy} & R_{yy} \end{vmatrix} \qquad (3.2)$$

Fig. 6

evaluated at $\underline{x}_s$. If $H\neq 0$, $\underline{x}_s$ is said to be nondegenerate. In this case, R can be represented by a stable Morse quadratic form. If $H=0$, $\underline{x}_s=\underline{x}_{sc}$ is called a degenerate specular point. In that case $\underline{x}_o=\underline{x}_c$ lies in a curvature center $\underline{x}_c$ on the surface's evolute E. The equation for the evolute follows by eliminating x and y from (3.1) and $H=0$, leaving one equation for $\underline{x}_o=\underline{x}_c$ alone.

At this point we require that the scattering process underlying remote sensing be structurally stable, i.e., that the observed images preserve their quality under slight perturbations of the system caused, e.g., by slight deformations of the surface or by small variations of the source. Then Thom's theorems [19], [3] assert that -- in an appropriate curvilinear surface coordinate system denoted again by (x,y) and in a curvilinear source coordinate system $\underline{x}_o-\underline{x}_c=(u,v,w)$ -- the structurally stable distance function R can take on but five "catastrophe" polynomial normal forms near a degenerate specular point $\underline{x}_{sc}$ inside a smooth part of S, viz.,

$$R = R_o+P(x,y;\underline{x}_o)+Dy^2/2 \qquad (3.3)$$

where $R_o=|\underline{x}_c-\underline{x}_{sc}|$ and $D=\kappa_1-\kappa_2$ where κ_i are the surface's principal curvatures in $\underline{x}_{sc}$. If $D\neq 0$, P is one of the cuspoid polynomials A_2 (fold): $P=x^3+ux$; A_3(cusp): $P=x^4+ux^2+vx$; A_4 (swallowtail):

$P=x^5+ux^3+vx^2+wx$. If $D=0$, $P=xy^2\pm x^3+wy^2+uy+vx$ represents the hyperbolic (D_4^+) resp. elliptic (D_4^-) umbilic polynomials. The bifurcation set of R is just the evolute E which, therefore, is classified into the five types mentioned above. They are shown in Figs. 7 (A_3), 8 (A_4), 9(D_4^+) and 10 (D_4^-). As a consequence, there are but five structurally stable and generic travel-time singularities observable in an echogram. All others are concatenations of these. If source and receiver are at different positions, there are 14 genuine singularities [10], but the above five describe completely the observed high intensity focal surfaces. Their structural stability can be attributed to the fact that because of the high intensity accumulated in the caustics these are insensitive to small perturbations.

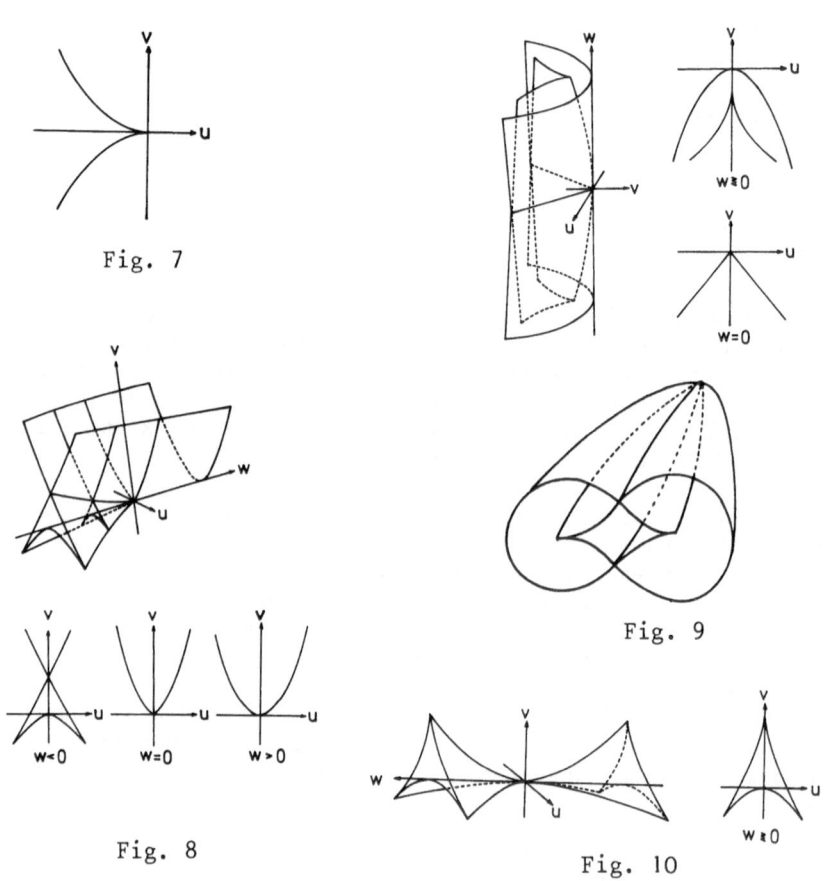

Fig. 7

Fig. 8

Fig. 9

Fig. 10

For A_3, Equ. (3.1) yields with (3.3) the equation $4x^3+2ux^2+v=0$, i.e., the "overhanging cliff" of Fig. 11 where $\rho=x_o=-v$, $X=x$ and $u=h$ are identified with the source position, surface point and height h of the observation line. Projecting the cliff's edges onto the (x_O,h)-plane gives a cusp which is precizely the evolute E of Fig. 3. A geometric interpretation of Fig. 11 in another context is given in Sec. 7.

Fig. 11

4. TRAVEL-TIMES, CONTOUR GEOMETRY AND SURFACE RECONSTRUCTION

Structural stability implies that, as the source position varies, the travel-times change their morphologies according to universal bifurcation sets. The travel-times recorded in the echo as a function of the source position are obtained by eliminating x and y from the equations (3.1) and (3.3). One example is shown in Fig. 5. A bifurcation sequence of travel-times coming from an umbilical surface point of elliptic type when $x_O=u$ varies on a line above the curvature center, is shown in Fig. 12 where $\rho_s=R_s-R_O$. The spherical wavefronts of constant radius r=R, centered on x_O, cut the surface in a series of contours $C(r,x_O)$ (cf., Fig. 6). An observer moving with x_O sees the Fresnel-zone contour topography changing according to universal bifurcation sets. A typical topographic change is shown in Fig. 13 for an umbilical surface point of hyperbolic type. The contours follow by setting $R-R_O=r=$const in (3.3).

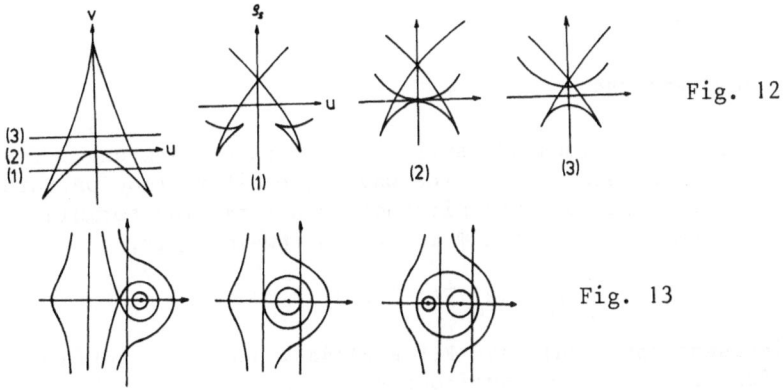

Fig. 12

Fig. 13

As the source-receiver $\underline{x}_0$ varies above the surface, each point of S eventually becomes specular. In virtue of the specularity condition (3.1), the surface normal at $\underline{x}_s$ is given by $\underline{n}(\underline{x}_s(\underline{x}_0)) = \nabla_{\underline{x}_0} R_s(\underline{x}_0)$. Then we obtain from $\underline{x}_0 - \underline{x}_s = R_s \underline{n}$ the equation for the surface

$$\underline{x} = \underline{x}_0 - R_s(\underline{x}_0) \nabla_{\underline{x}_0} R_s(\underline{x}_0) \tag{4.1}$$

Since $|\underline{n}| = 1$, only two components of $\nabla_{\underline{x}_0} R_s$ are needed and it suffices therefore, to vary $\underline{x}_0$ on a surface Σ, say, on a plane. Inferring R_s from the arrivals or peaks the echo ψ possesses at times $t = T_s = 2R_s/c$, the surface profile can be reconstructed directly from (4.1). For an input signal $F(t)$ with basic source pulse frequency ω_0 (cf., Sec. 5), the echogram near a degenerate singularity $\underline{x}_s = \underline{x}_{sc}$ has the form

$$\psi(t, \underline{x}_c) \propto \omega_0^\alpha \{ AF(t - T_{sc}) + B\bar{F}(t - T_{sc}) \} \tag{4.2}$$

where $\alpha = \{1/6, 1/4, 3/10, 1/3\}$ for the singularities $\{A_2, A_3, A_4, D_4^\pm\}$ and $\bar{F}$ is the Hilbert transform of F. Varying ω_0 and measuring the rate of change of the received power permits one to identify and to classify the structural details of the reflecting surface near travel-time reversals. Typical travel-time recordings for a surface with a fault and two domes are shown on top of Fig. 14 (D. Lang [10]) for a source-receiver moving along horizontal survey lines above the lines 1, 6, 14 and 20. In Fig. 14a, b and c the height of the observation plane is 1.5 times, 2 times and 5 times the height of the domes, respectively. Travel-times are plotted downwards. In the travel-time recording of Fig. 14a one sees four swallowtail points coming from the fault and corresponding to a rib lying entirely in the observation plane. In Fig. 14b the typical swallowtail bifurcation sets have developed into cusp lines. In addition, the slopes of the domes give rise to new swallowtail points. Fig. 14c shows the most complex travel-time recordings pertaining to the surface in which the swallowtails have fully developed.

5. DIFFRACTION PATTERNS

In classifying travel-times and surface contours the language of ray theory has been used so far. The wave-type diffraction patterns around the caustics follow from Kirchhoff's diffraction formula [10] for the backscattered wave ψ in the shortwave limit,

$$\psi(t, \underline{x}_0) = (1/8\pi^2 c) \int_0^\infty dr G(r, \underline{x}_0) \dot{F}(t - 2r/c) \tag{5.1}$$

with the incident spherical wave $F(t - R/c)/4\pi R$. Here, the surface structure function G -- the wavefront sweep velocity or scatter-

(a)   (b)

(c)

Fig. 14

ing matrix -- is given by the integral

$$G(r,\underline{x}_o)=\int_S dSR^{-3}(\underline{n}\cdot\underline{R})\delta(r-R)=r^{-3}\int_{C'}ds\sqrt{|g|}\,(\underline{n}\cdot\underline{R})/|\nabla R| \qquad (5.2)$$

where δ is Dirac's function and the contour C' is the projection of C (Fig. 6) onto the (x,y)-plane. From (5.2) the main contributions to G (and ψ) are seen to come from those values of r for which the integrand is infinite, i.e., from the geometrical optics specular reflection points given by $\nabla R=0$, Equ. (3.1). The Fourier transform of G is

$$\tilde{G}(k,\underline{x}_o) = \int_S dS(\partial R^{-1}/\partial n)\exp(ikR) \qquad (5.3)$$

where $k=\omega/c$. Hence,
$\tilde{\psi}(\omega,\underline{x}_o)=-i\omega\tilde{G}(2\omega/c,\underline{x}_o)\tilde{F}(\omega)/8\pi^2 c.$
The generic diffraction patterns associated with the travel-time singularities follow by substituting (3.3) into (5.3). Asymptotic evaluation of (5.3) gives high-intensity diffraction patterns of Airy, Pearcey and higher-order type [10]. The ω_o-dependence in (4.2) follows from (5.3) and (3.3) by scaling. A typical diffraction pattern around an elliptic umbilic caustic is shown in Fig. 15 (cf., also Fig. 1).

Fig. 15

If the surface possesses edges and faults [1], the evolutes are amputated by the surface's shadow boundaries (Fig. 16). The effects produced by such discontinuities can be classified by six "constraint" catastrophe polynomials [10], [20]. For example, a fault (Fig. 16) is determined by $P=x^3+vx^2+ux$ ($x>0$) and the diffraction pattern following from (5.3) is of combined Fresnel-Airy type: The intensity near the travel-time reversal decreases much faster than the one produced by a slope. Figs. 17 (a) and (b) show the travel-time curves when $\underline{x}_o$ varies on a line above and below the point B in Fig. 16, respectively. This effect is also present in Fig. 1.

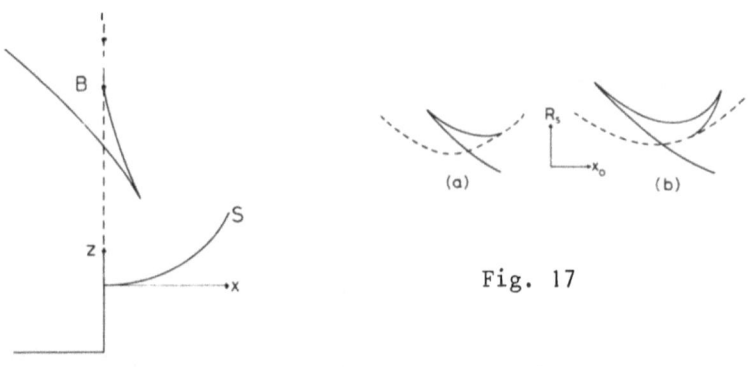

Fig. 17

Fig. 16

7. CONCLUSIONS

The above classification of singularities in wavefields and ray families scattered back by a surface can be generalized to deal with laterally inhomogeneous media by using Maslov's representation [8], [10]. Incorporating these topological concepts into S-matrix techniques of inversion offers new and practicable tools for all the sensing processes presently in use in a wide variety of physical and technical systems, ranging from ocean acoustics [14] to phonon spectroscopy [2]. An important thing to note about the universal singularities and bifurcations we have described is that they also determine the spontaneous formation of qualitatively similar spatio-temporal patterns in systems of various genesis exhibiting similar critical behavior. Comparison of, e.g., Fig. 11 with familiar phase transition diagrams, or identifying the singularities observed in flow fields [15] with the above travel-time bifurcations, strongly indicates that structure formation and structure recognition are governed by the same topological principles [11]. The reason for these analogies is that the geometry underlying pattern formation and pattern recognition is subjected to the general principle of structural stability. As a consequence, structure formation and recognition are just the two sides of the same topological coin. This fact raises fascinating questions that promise a great challenge for future research.

REFERENCES

[1] Achenbach, J.D. and Norris, A.: 1981, J. Acoust. Soc. Am. 70, pp. 165-171.

[2] Armbruster, D., Dangelmayr, G., and Güttinger, W.: Nonlinear Phonon Focusing, Proc. 4th Int. Conference on Phonon Scattering in Condensed Matter (Stuttgart 1983). Armbruster, D. and Dangelmayr, G.: Topological Singularities and Phonon Focusing, Z. Phys. B (1983), in print.

[3] Arnold, V.: 1978, Russ. Math. Surveys 33, pp. 99-116.

[4] Baltes, H.P. (ed.): Inverse Scattering Problems in Optics (Springer 1980)

[5] Berry, M.V.: 1972, J. Phys. A 5, pp. 272-291.

[6] Bleistein, N. and Cohen, J.K.: A survey of recent progress on inverse scattering problems, US Naval Res. Rep. MS-R-7806 (1978).

[7] Boerner, W.M., Polarization utilization in electromagnetic inverse scattering, in Baltes, Ref. [4].

[8]   Chapman, C.H. and Drummond, R.: 1982, Bull. Seism. Soc. Am.
      72, pp. 227-317.

[9]   Dangelmayr, G.: Singularities in quasigeometrical imaging,
      this volume.

[10] Dangelmayr, G. and Güttinger, W.: 1982, Geophys. J. R.
      Astr. Soc. 71, pp. 79-126; Güttinger, W. and Lang, D.:
      in preparation (1983).

[11] Güttinger, W. and Eikemeier, H. (Eds.): Structural Stabi-
      lity in Physics (Springer 1979); Güttinger, W.: Geometrical
      principles underlying synergetics, in preparation (1983).

[12] Haken, H.: Synergetics (Springer 1980)

[13] Hilterman, F.J.: 1975, Geophysics 35, pp. 1020-1037 and
      Geophysics 40, pp. 745-762 (1975).

[14] Keller, J.B. and Papadakis, J.S. (Eds.): Wave propagation
      and underwater acoustics (Springer 1977).

[15] Nye, J.F.: Structural Stability in evolving flow fields,
      in Güttinger, W. and Eikemeier, H., Ref. 11 .

[16] Ristow, D.: Three-dimensional finite-difference migration,
      Ph.D. thesis, Utrecht 1980.

[17] Schneider, W.A.: 1978, Geophysics 43, pp. 49-76.

[18] Stewart, I.: 1981, Physica 2D, pp. 245-305.

[19] Thom, R.: Structural Stability and Morphogenesis
      (Benjamin 1975)

[20] Wright, F.J., Güttinger, W. and Dangelmayr, G.:
      in preparation (1983).

I.3 (IS.3)

A SOLUTION OF THE TIME-DEPENDENT INVERSE SOURCE PROBLEM FOR
THREE-DIMENSIONAL ELECTROMAGNETIC WAVE PROPAGATION

Harry E. Moses

University of Lowell, Center for Atmospheric Research

The usual form of the inverse source problem for electromagnetic
theory is given in a time-independent version in which the
radiation pattern is prescribed and the sources are required to
be contained in a sphere of prescribed radius. The time-dependent inverse source problem, which we discuss in the present
paper, has currents which are switched on for a prescribed interval of time and the final time-dependent field is prescribed.
One seeks the currents which will give rise to the fields. Thus
the time-dependent inverse source problem seeks to provide sources
for pulsed fields and is complementary to the time-independent
problem. Our approach makes use of eigenfunctions of the curl
operator, whose properties are reviewed. It also represents an
extension of our earlier work. The currents which give rise to
the prescribed fields are not unique but conditions are given
which make them so.

1.0 INTRODUCTION. STATEMENT OF THE PROBLEM

Let us consider Maxwell's equations with time-dependent
sources. If, for simplicity, we take units in which the velocity
of light c is unity, Maxwell's equations are

$$\nabla \times \underset{\sim}{H}(\underset{\sim}{x};t) = 4\pi \underset{\sim}{j}(\underset{\sim}{x};t) + \frac{\partial}{\partial t} \underset{\sim}{E}(\underset{\sim}{x},t),$$

$$\nabla \times \underset{\sim}{E}(\underset{\sim}{x};t) = - \frac{\partial}{\partial t} \underset{\sim}{H}(\underset{\sim}{x};t),$$

$$\nabla \cdot \underset{\sim}{E}(\underset{\sim}{x};t) = 4\pi \rho (\underset{\sim}{x};t)$$

$$\nabla \cdot \underset{\sim}{H}(\underset{\sim}{x};t) = 0. \tag{1}$$

W.-M. Boerner et al. (eds.), Inverse Methods in Electromagnetic Imaging – Part 1, 77–85.
© *1985 by D. Reidel Publishing Company.*

For our purpose the direct time-dependent source problem for pulsed electromagnetic waves is to prescribe the current $j(\underset{\sim}{x};t)$ and density $\rho(\underset{\sim}{x};t)$ as functions of $\underset{\sim}{x}$ and t such that they vanish outside a finite time interval. Before the sources are turned on we prescribe, for the simplicity of discussion, that the field be zero. We want to find the field after the currents and sources are turned on and then off. This problem is a classical one for Maxwell's equations. After the currents and sources are turned off, the solution of Maxwell's equations represents a pulse of the electromagnetic field.

In the inverse source problem, as we treat it for the purpose of the present discussion, we prescribe the time interval within which the sources and currents are on. We also prescribe the pulse of radiation which we want to have after we have turned the source off.

Usually the inverse source problem is described in a time-independent version. One takes Maxwell's equations of Eq. (1) and separates out the time variable by taking a Fourier transformation with respect to time. The time-independent inverse problem is to find sources and currents within a finite domain in space which gives rise to a prescribed radiation pattern. Results are summarized in Ref. 1. The time-independent approach to the inverse source problem was first given by the present author in Ref. 2. The present paper represents an extension of Ref. 2. It is seen that the time-dependent problem is complementary to the time-independent problem.

The time-dependent problem has many applications. For example for use in radar systems or for ionosondes which probe the ionosphere one wishes to shape and direct pulses of radiation. What current distributions in the antennas can be used for this purpose? A second application is decoying. One can make a source which simulates the emitted electromagnetic signature of an aircraft, for example, and thus can be used to confuse detectors used by an enemy. A closely related problem is the source detection problem which is concerned with determining the position, size and shape of a source of electromagnetic radiation. The inverse source problem, as we shall formulate it, is valid in regions where the space is homogeneous, even when the homogeneous space is enclosed by or abuts an inhomogeneity. For example, one can use our method - or its extensions - to find the currents within a waveguide which will give rise to prescribed pulsed fields within the waveguide before the pulse hits the boundary of the guide. Such a shaping of the initial radiation is useful in design problems.

2.0 EIGENFUNCTIONS OF THE CURL OPERATOR AND GENERALIZATION OF THE HELMHOLTZ THEOREM

In our original treatment of the inverse source problem for Maxwell's equations (Ref. 2), we recast Maxwell's equations into

the form of Dirac's equation and expanded the fields in terms of
the eigenfunctions of the Dirac-like Hamiltonian. In this way
the transverse components of the fields and the currents which
gave rise to them were separated from the longitudinal field and
its sources. In the present paper we shall use eigenfunctions
of the curl operator to carry out this separation and find a
simple differential equation for the time amplitudes. The eigenfunctions of the curl operator were introduced by us in Ref. 3
to provide a rotationally invariant generalization of the
Helmholtz decomposition theorem in the infinite domain. [It
might be mentioned that a still further generalization which
gives analogues to the Helmholtz theorem for tensor fields is
discussed in Ref. 4.] The eigenfunctions were intended to simplify equations in which the curl operator appears, such as
Maxwell's equations and the equations of fluid mechanics.

Let us consider an arbitrary unit vector $\underset{\sim}{\eta}$ and a discrete
variable $\lambda = \pm 1, 0$. For each value of $\underset{\sim}{\eta}$ we define a vector
$\underset{\sim}{Q}_\lambda(\underset{\sim}{\eta})$ by

$$\underset{\sim}{Q}_0(\underset{\sim}{\eta}) = - \underset{\sim}{\eta},$$

$$\underset{\sim}{Q}_\lambda(\underset{\sim}{\eta}) = - \frac{\lambda}{\sqrt{2}} \left[\frac{n_1(n_1 + i\lambda n_2)}{1+n_3} - 1, \; \frac{n_2(n_1 + i\lambda n_2)}{1+n_3} - i\lambda, \; n_1 + i\lambda n_2 \right],$$

$$(\lambda = \pm 1). \tag{2}$$

The vectors $\underset{\sim}{Q}_\lambda(\underset{\sim}{\eta})$ satisfy the following orthogonality and
completeness relations:

$$\underset{\sim}{Q}_\lambda^*(\underset{\sim}{\eta}) \cdot \underset{\sim}{Q}_\mu(\underset{\sim}{\eta}) = \delta_{\lambda\mu}, \; \sum_\lambda Q_{i\lambda}^*(\underset{\sim}{\eta}) Q_{j\lambda}(\underset{\sim}{\eta}) = \delta_{ij}. \tag{3}$$

Also

$$\underset{\sim}{\eta} \cdot \underset{\sim}{Q}_\lambda(\underset{\sim}{\eta}) = - \delta_{\lambda 0}, \; \underset{\sim}{\eta} \times \underset{\sim}{Q}_\lambda(\underset{\sim}{\eta}) = - i\lambda \underset{\sim}{Q}_\lambda(\underset{\sim}{\eta}). \tag{4}$$

The eigenfunctions of the curl operator $\underset{\sim}{\chi}(x|\underset{\sim}{p},\lambda)$ are defined by

$$\underset{\sim}{\chi}(x|\underset{\sim}{p},\lambda) = \frac{1}{(2\pi)^{3/2}} e^{i\underset{\sim}{p} \cdot x} \underset{\sim}{Q}_\lambda \left(\frac{\underset{\sim}{p}}{p}\right), \; p = |\underset{\sim}{p}|. \tag{5}$$

The following orthogonality and completeness relations are
satisfied:

$$\int \underset{\sim}{\chi}^*(x|\underset{\sim}{p},\lambda) \cdot \underset{\sim}{\chi}(x|\underset{\sim}{p}',\lambda') \, dx = \delta(\underset{\sim}{p} - \underset{\sim}{p}') \delta_{\lambda,\lambda'},$$

$$\sum_\lambda \int \chi_i^*(\underset{\sim}{x} \, \underset{\sim}{p},\lambda) \chi_j(\underset{\sim}{x}' \, \underset{\sim}{p},\lambda) \, d\underset{\sim}{p} = \delta(\underset{\sim}{x} - \underset{\sim}{x}') \delta_{ij}. \tag{6}$$

Moreover, the eigenfunctions are eigenfunctions of the curl
operator with eigenvalue λp, since

$$\nabla \times \underset{\sim}{\chi}(x|\underset{\sim}{p},\lambda) = \lambda p \underset{\sim}{\chi}(x|\underset{\sim}{p},\lambda). \tag{7}$$

Also

$$\nabla \cdot \underset{\sim}{\chi}(\underset{\sim}{x}|\underset{\sim}{p},\lambda) = - i\underset{\sim}{p} \frac{e^{i\underset{\sim}{p} \cdot \underset{\sim}{x}}}{(2\pi)^{3/2}} \delta_{\lambda,0}. \tag{8}$$

We can now generalize the Helmholtz theorem.

Any vector $\underset{\sim}{v}(\underset{\sim}{x})$ can be written in the form

$$\underset{\sim}{v}(\underset{\sim}{x}) = \sum_{\lambda} \underset{\sim}{v}_{\lambda}(\underset{\sim}{x}), \text{ where } \underset{\sim}{v}_{\lambda}(\underset{\sim}{x}) = \int \underset{\sim}{\chi}(\underset{\sim}{x}|\underset{\sim}{p},\lambda) \, V(\underset{\sim}{p},\lambda) \, d\underset{\sim}{p}, \tag{9}$$

where

$$V(\underset{\sim}{p},\lambda) = \int \underset{\sim}{\chi}^{*}(\underset{\sim}{x}|\underset{\sim}{p},\lambda) \cdot \underset{\sim}{v}(\underset{\sim}{x}) \, d\underset{\sim}{x}. \tag{10}$$

Moreover,

$$\nabla \cdot \underset{\sim}{v}_{\lambda}(\underset{\sim}{x}) = 0 \text{ for } \lambda = \pm1, \quad \nabla \times \underset{\sim}{v}_{0}(\underset{\sim}{x}) = 0. \tag{11}$$

We have thereby sharpened the Helmholtz theorem by showing that there are __two__, rather than one, transverse components of a vector. As shown in Ref. 3 the decomposition is rotationally invariant.

The second part of the Helmholtz theorem states that the rotational part of a vector can be expressed as the curl of a vector and the irrotational part as the gradient of a vector. We now discuss our sharpening of this part of the theorem. We can introduce vector potentials $\underset{\sim}{A}_{\lambda}(\underset{\sim}{x})$ defined by

$$\nabla \times \underset{\sim}{A}_{\lambda}(\underset{\sim}{x}) = \underset{\sim}{v}_{\lambda}(\underset{\sim}{x}), \lambda = \pm1, \tag{12}$$

and a scalar potential $V(s)$ defined by

$$\nabla V(\underset{\sim}{x}) = \underset{\sim}{v}_{0}(\underset{\sim}{x}). \tag{12a}$$

We shall give the general form of the vector potentials $\underset{\sim}{A}_{\lambda}(\underset{\sim}{x})$ and the scalar potential $V(\underset{\sim}{x})$ which we regard as solutions of Eqs. (12) and (12a) respectively. To find the vector potentials we write the vector potentials in the form of an expansion in terms of the eigenfunctions of the curl operator, as in Eq. (9)

$$\underset{\sim}{A}_{\lambda}(\underset{\sim}{x}) = \sum_{\lambda'} \int \underset{\sim}{\chi}(\underset{\sim}{x}|\underset{\sim}{p},\lambda') B(\underset{\sim}{p},\lambda') \, d\underset{\sim}{p}. \tag{13}$$

On substituting into Eq. (12), and using the linear independence of the eigenfunctions, we obtain

$$B(\underset{\sim}{p},\lambda) = \frac{\lambda V(\underset{\sim}{p},\lambda)}{p}, \quad B(\underset{\sim}{p},-\lambda) = 0, \quad B(\underset{\sim}{p},0) \text{ arbitrary.} \tag{14}$$

Thus the vector potentials $\underset{\sim}{A}_{\lambda}(\underset{\sim}{x})$ can be written as

$$\underset{\sim}{A}_{\lambda}(\underset{\sim}{x}) = \int \underset{\sim}{\chi}(\underset{\sim}{x}|\underset{\sim}{p},\lambda) \, B(\underset{\sim}{p},\lambda) \, d\underset{\sim}{p} + \underset{\sim}{W}(\underset{\sim}{x}),$$

where

$$W(x) = \int \chi(x|p,0) \, B(p,0) \, dp. \tag{15}$$

Thus each of the two vector potentials (one for $\lambda = +1$ and one for $\lambda = -1$) consists of the sum of an essential, minimal part which depends uniquely on $v_\lambda(x)$ and an arbitrary vector $W(x)$ which has the form

$$W(x) = \nabla Q(x), \quad Q(x) = \frac{i}{(2\pi)^{3/2}} \int e^{ip\cdot x} \frac{B(p,0)}{p} \, dp. \tag{15a}$$

Thus $W(x)$ is the gradient of an arbitrary function $Q(x)$.

The function $W(x)$ is the gauge. Our procedure has isolated the essential parts of the vector potential and has removed the mystery of the gauge.

We now want to find the function $V(x)$ from $v_0(x)$. From Eq. (9) and the form of $\chi(x|p,0)$ we have

$$V(x) = -\frac{i}{(2\pi)^{3/2}} \int e^{ip\cdot x} \frac{V(p,0)}{p} \, dp. \tag{16}$$

The reality property is important. For $v(x)$ to be real

$$V(p,\lambda) = -\frac{p_1 - i\lambda p_2}{p_1 + i\lambda p_2} \, V^*(-p,\lambda). \tag{17}$$

3.0   APPLICATION OF CURL EIGENFUNCTIONS TO MAXWELL'S EQUATIONS

We rewrite Maxwell's equations in the Bateman form. Thus we introduce the complex vector.

$$\psi(x;t) = E(x;t) - iH(x;t). \tag{18}$$

Maxwell's Equations Eq. (1) become

$$\nabla \times \psi(x;t) = -i \frac{\partial}{\partial t} \psi(x;t) - 4\pi i j(x;t),$$

$$\nabla \cdot \psi(x;t) = 4\pi\rho(x;t). \tag{19}$$

The wave function $\psi(x;t)$ is generally complex. However, the current $j(x;t)$ and charge density $\rho(x;t)$ are real.

We use the eigenfunctions of curl operator and write

$$\psi(x;t) = \sum_\lambda \int \chi(x|p,\lambda) \, \psi(p,\lambda;t) \, dp, \tag{20}$$

$$4\pi j(x;t) = \sum_\lambda \int \chi(x|p,\lambda) \, \gamma(p,\lambda;t) \, dp, \tag{21}$$

$$4\pi\rho(x;t) = \frac{1}{(2\pi)^{3/2}} \int e^{ip\cdot x} \, r(p;t) \, dp. \tag{22}$$

On substituting into Eq. (19) and using the properties of the eigenfunctions Eqs. (7) and (8)

$$\psi(\underset{\sim}{p},0;t) = \frac{i}{p} r(\underset{\sim}{p};t),$$

$$-\frac{\partial}{\partial t} \psi(\underset{\sim}{p},0;t) = \gamma(\underset{\sim}{p},0;t), \tag{23}$$

which relate the longitudinal components of the electromagnetic field to the charge distribution and the longitudinal component of the current. The transverse part of the electromagnetic field is related to the transverse part of the current in the following way:

$$\frac{\partial}{\partial t} \psi(\underset{\sim}{p},\lambda;t) - ip\lambda\psi(\underset{\sim}{p},\lambda;t) = - \gamma(\underset{\sim}{p},\lambda;t), \quad (\lambda = \underline{+}1). \tag{24}$$

Our use of the eigenfunctions of the curl operator has enabled us to split the equations into a longitudinal part and two uncoupled transverse parts in a rotationally invariant way.

The longitudinal field is given essentially by $\psi(p,0;t)$ does not concern us, since it is only the radiation field which is of interest. Using the first of Eq. (23) we can set $\rho(\underset{\sim}{x};t) \equiv 0$ and thus cause the longitudinal field to vanish. The second of Eq. (23) also pertains only to the longitudinal field and is essentially the equation of continuity. This equation also can be disregarded if the charge density is identically zero.

We are then left only with the two ordinary differential uncoupled equations for $\psi(p,\underline{+}1;t)$ which are readily integrated. We take the fields before the currents are turned on to be identically zero and choose the origin of the time variable to be such that the currents are turned on at $t = - T$ and turned off again at $t = T$. Then the solution of Eq. (24) is

$$\psi(\underset{\sim}{p},\lambda;t) = - e^{i\lambda pt} \int_{-T}^{t} e^{-i\lambda pt'} \gamma(\underset{\sim}{p},\lambda;t') \, dt'. \tag{23}$$

(We recollect that $\gamma(\underset{\sim}{p},\lambda;t) \equiv 0$ for $t < - T$.)

Thus as our initial and final conditions we have

$$\psi(\underset{\sim}{x};t) \equiv 0 \text{ for } t < - T,$$

$$\psi(\underset{\sim}{x};t) \equiv \psi_{+}(\underset{\sim}{x};t) \text{ for } t > T, \tag{24}$$

where

$$\psi_{+}(\underset{\sim}{x};t) = \sum_{\lambda=\underline{+}1} \int \chi(\underset{\sim}{x}|\underset{\sim}{p},\lambda) \, e^{i\lambda p(t-T)} \, \psi(\underset{\sim}{p},\lambda;T) \, d\underset{\sim}{p}. \tag{25}$$

It is clear that $\psi_{+}(\underset{\sim}{x};t)$ is a solution of Maxwell's equations without currents and thus represents the final radiation field.

Before we proceed further it is useful to give a physical interpretation to the discrete variable $\lambda = \underline{+}1$.

It is readily seen that the general solution of Maxwell's equations without sources is of the form Eq. (25) in which $\psi(p,\lambda;T)$ is replaced by $\psi(p,\lambda)$ where this latter function is an arbitrary function of $\underset{\sim}{p},\lambda$. Moreover, it is convenient to take

$T = 0$. Let us consider the case in which

$$\psi(\underset{\sim}{p},1) = \delta(p_1)\,\delta(p_2)\,\delta(p_3-k). \quad \text{Also } \psi(\underset{\sim}{p},-1) \equiv 0. \tag{26}$$

Then

$$E_1(\underset{\sim}{x};t) = H_2(\underset{\sim}{x};t) = \frac{1}{4(\pi)^{3/2}} \cos k(x_3+t),$$

$$E_2(\underset{\sim}{x};t) = -H_1(\underset{\sim}{x};t) = -\frac{1}{4(\pi)^{3/2}} \sin k(x_3+t),$$

$$E_3(\underset{\sim}{x};t) = H_3(\underset{\sim}{x};t) \equiv 0. \tag{26a}$$

It is clear that this solution of Maxwell's equations is a positively polarized circularly polarized wave moving with wave number k in the negative z-direction. Hence $\lambda = 1$ corresponds to positive circular polarization.

Likewise the choice of $\psi(p,-1) = \delta(p_1)\delta(p_2)\delta(p_3-k)$, $\psi(p,+1) \equiv 0$ gives rise to a negative circularly polarized wave of wave number k moving in the positive z-direction. Then $\lambda = -1$ means negative circular polarization, and therefore decomposition of $\psi(\underset{\sim}{x};t)$ into eigenvectors of curl operator is equivalent to expressing the electromagnetic field as a sum of circularly polarized radiation of various frequencies moving in various directions.

4.0 THE INVERSE PROBLEM

We shall now treat the inverse source problem for Maxwell's equations.

We are given $\psi_+(\underset{\sim}{x};t)$ and we are required to find a current $\underset{\sim}{j}(\underset{\sim}{x};t)$ of the form

$$\underset{\sim}{j}(\underset{\sim}{x};t) = \underset{\sim}{j}_e(\underset{\sim}{x})\,h_e(t) + \underset{\sim}{j}_o(\underset{\sim}{x})\,h_o(t) \tag{27}$$

where $h_e(t)$ is a given real even function and $h_o(t)$ is a given real odd function of t. We shall show how real transverse vectors $\underset{\sim}{j}_{e,o}(\underset{\sim}{x})$ can be obtained uniquely from the complex vector $\psi_+(\underset{\sim}{x};t)$.

Knowing $\psi_+(\underset{\sim}{x};t)$ is equivalent to knowing $\psi(\underset{\sim}{p},\pm1;T)$. Let us define $F(\underset{\sim}{p},\lambda;\tilde{k})$ for $\lambda = \pm1$ by

$$F(\underset{\sim}{p},\lambda;k) = \frac{1}{\sqrt{2\pi}} \int_{-T}^{+T} e^{-ikt}\, \gamma(\underset{\sim}{p},\lambda;t)\, dt$$

$$= \frac{4\pi}{\sqrt{2\pi}} \int_{-T}^{+T} e^{-ikt}\, dt \int \underset{\sim}{\chi}^*(\underset{\sim}{x}|\underset{\sim}{p},\lambda) \cdot \underset{\sim}{j}(\underset{\sim}{x};t)\, d\underset{\sim}{x}. \tag{28}$$

Then we define

$$G(\underset{\sim}{p},\lambda) \equiv F(\underset{\sim}{p},\lambda;\lambda p) = -\frac{1}{\sqrt{2\pi}} e^{-i\lambda pT} \psi(\underset{\sim}{p},\lambda;T) \tag{29}$$

It is our intent to find $\underset{\sim}{j}_{e,o}(\underset{\sim}{x})$ of Eq. (27) from $G(\underset{\sim}{p},\lambda)$.
We also define,

$$F_e(\underset{\sim}{p},\lambda) = \frac{1}{2g_e(p)} [G(\underset{\sim}{p},\lambda) - \frac{p_1-i\lambda p_2}{p_1+i\lambda p_2} G^*(-\underset{\sim}{p},\lambda)],$$

$$F_o(\underset{\sim}{p},\lambda) = \frac{\lambda}{2g_o(p)} [G(\underset{\sim}{p},\lambda) + \frac{p_1-i\lambda p_2}{p_1+i\lambda p_2} G^*(-\underset{\sim}{p},\lambda)]. \tag{30}$$

where

$$g_{e,o}(k) = \frac{1}{\sqrt{2\pi}} \int_{-T}^{+T} h_{e,o}(t) e^{-ikt} dk. \tag{31}$$

Note that $g_e(k)$ is a <u>real</u>, even function of k and $g_o(k)$ is an <u>imaginary</u>, odd function of k.
Then

$$4\pi \underset{\sim}{j}_e(\underset{\sim}{x}) = \sum_\lambda \int \underset{\sim}{\chi}(\underset{\sim}{x}|\underset{\sim}{p},\lambda) F_e(\underset{\sim}{p},\lambda) d\underset{\sim}{p},$$

$$4\pi \underset{\sim}{j}_o(\underset{\sim}{x}) = \sum_\lambda \int \underset{\sim}{\chi}(\underset{\sim}{x}|\underset{\sim}{p},\lambda) F_o(\underset{\sim}{p},\lambda) d\underset{\sim}{p}. \tag{32}$$

The fact that

$$F_{e,o}(\underset{\sim}{p},\lambda) = -\frac{p_1-i\lambda p_2}{p_1+i\lambda p_2} F^*_{e,o}(-\underset{\sim}{p},\lambda) \tag{33}$$

assures us that $\underset{\sim}{j}_{e,o}(\underset{\sim}{x})$ are real vectors [see Eq. (17)].
Let us consider the special case

$$h_e(t) = \delta(t), \quad h_o(t) = \delta'(t). \tag{34}$$

Moreover, we let $T \to +0$ and define

$$\underset{\sim}{\psi}(\underset{\sim}{x}) = \underset{\sim}{\psi}_+(\underset{\sim}{x};0_+) \equiv \underset{\sim}{\psi}(\underset{\sim}{x};0_+). \tag{35}$$

That is, we prescribe the radiation fields immediately after the current is turned off. Of course, the given vector $\underset{\sim}{\psi}(\underset{\sim}{x})$ must be a purely transverse field. We obtain

$$4\pi \underset{\sim}{j}_e(\underset{\sim}{x}) = - \text{Re } \underset{\sim}{\psi}(\underset{\sim}{x}), \quad 4\pi \underset{\sim}{j}_o(\underset{\sim}{x}) = \frac{1}{4\pi} \text{Im } \underset{\sim}{\nabla} \times \int \frac{1}{|\underset{\sim}{x}-\underset{\sim}{x}'|} \underset{\sim}{\psi}(\underset{\sim}{x}') d\underset{\sim}{x}'. \tag{36}$$

However, since

$$\text{Re } \underset{\sim}{\psi}(\underset{\sim}{x}) = \underset{\sim}{E}(\underset{\sim}{x};0_+), \quad \text{Im } \underset{\sim}{\psi}(\underset{\sim}{x}) = -\underset{\sim}{H}(\underset{\sim}{x};0_+), \tag{37}$$

Eq. (36) becomes

$$4\pi \underset{\sim}{j}_e(\underset{\sim}{x}) = -\underset{\sim}{E}(\underset{\sim}{x};0_+), \quad 4\pi \underset{\sim}{j}_o(\underset{\sim}{x}) = -\frac{1}{4\pi} \underset{\sim}{\nabla} \times \int \frac{1}{|\underset{\sim}{x}-\underset{\sim}{x}'|} \underset{\sim}{H}(\underset{\sim}{x}';0_+) d\underset{\sim}{x}'. \tag{38}$$

From Eq. (38) it is clear that the current is a transverse vector as indeed are all currents as constructed by our inverse source method.

REFERENCES

1. B.J. Hoenders, Inverse Source Problems in Optics, ed. H.P. Baltes, Springer, New York, p. 41ff (1978).

2. H.E. Moses, Phys. Rev., 113, 1970 (1959).

3. H.E. Moses, SIAM J. App. Math., 21, 114 (1971).

4. H.E. Moses and A.F. Quesada, Arch. Rat. Mech., 50, 194 (1973).

ACKNOWLEDGEMENT

The work in this paper was done under the National Science Foundation Grant MCS-8024640.

I.4 (IS.2)

TRANSIENT METHODS IN ELECTROMAGNETIC IMAGING

K.J. Langenberg, G. Bollig, M. Fischer, D. Brück

Gesamthochschule Kassel
Fachgebiet Theoretische Elektrotechnik, FB 16
Wilhelmshoeher Allee 71
D-3500 Kassel, F.R.G.

The theoretical background of pulsed electromagnetic imaging is reviewed. We differentiate between stationary methods relying on the information contained in the broadband transient signal scattered into a prescribed spatial direction, like the Singularity Expansion Method, and synthetic aperture methods, which generally need data within a subset of K-space, where K is a spatial Fourier variable comprising both aperture, i.e. spatial, and frequency information. Different support of these variables yields different imaging algorithms, which can either be derived by explicit inversion of Kirchhoff's integral under low or high frequency assumptions or by the heuristic backward wave propagation argument. We outline the general potential as well as the defects and limitations of these procedures.

INTRODUCTION

Generally, imaging methods try to extract scatterer specific information out of the field scattered by a certain obstacle illuminated by an incident wave, which may either be chosen as an electromagnetic, acoustic or elastodynamic one according to applications in radar detection, remote sensing, nondestructive testing of materials, geophysics or medical diagnostics. Here, we concentrate on mathematically based methods referred to as inverse scattering or identification methods. In order to outline the common theoretical background of some presently used synthetic aperture algorithms, and the feasability of a proposed stationary procedure, we ignore the vector- or tensor-nature of certain above-mentioned applications and use a scalar notation throughout the paper; polarization effects and methods utilizing them as developed by

87

W.-M. Boerner et al. (eds.), Inverse Methods in Electromagnetic Imaging - Part 1, 87–110.
© *1985 by D. Reidel Publishing Company.*

Boerner et al. {1,2} are therefore excluded. We emphasize that part of our work is strongly related to that of Devaney and Porter et al. {3,4,5,6}.

Producing scattered wave fields and extracting information out of them can be achieved by several arrangements (Fig. 1), which can be separated into "stationary" (fixed source point Q

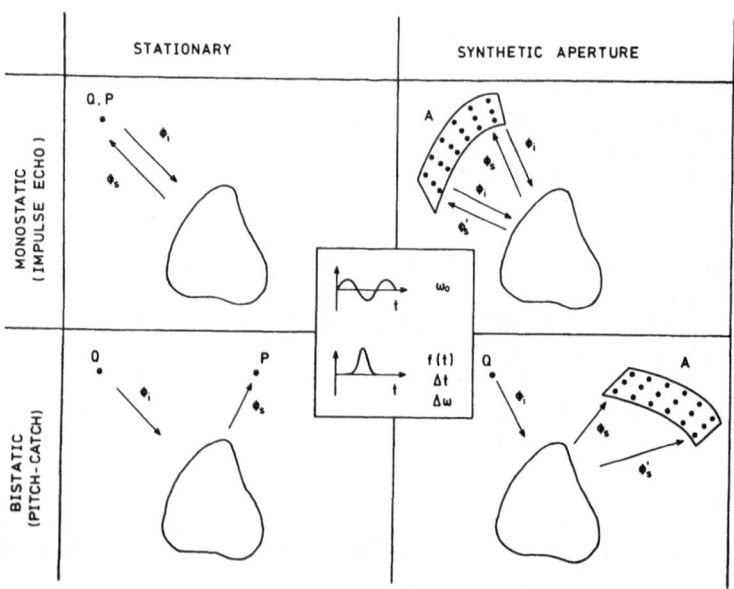

Fig. 1: Data recording arrangements (Φ_i incident field, Φ_s scattered field)

and observation point P) and "synthetic aperture" arrangements (moving observation point P within a prescribed aperture A). A further distinction can be made into "monostatic" (coinciding Q and P) and "bistatic" (spatially separated Q and P). If the application is nondestructive testing of materials the corresponding expressions are "impulse echo" and "pitch-catch". It is clear that a stationary arrangement requires a broadband incident field with a prescribed impulse structure f(t) of duration Δt or bandwidth $\Delta \omega$ in order to contain enough relevant information. In contrast, if sufficient information is collected within a synthetic aperture, a time harmonic excitation of circular frequency ω_0 might

suffice to reconstruct a scattering surface described by two variables. We will make it evident that this depends on the general
structure of the scatterer (flat or voluminous) and on resolution
requirements combined with the accessible aperture range.

Mathematical algorithms related to stationary arrangements
rely on general system identification procedures {7,8} or pattern
recognition techniques {9}. Relatively simple ideas have been
applied to nondestructive testing or underwater acoustics and
distinguish between different scattering centres estimating travel times between single distinctive scattered pulses {10,11} or
estimating their spectral counterpart in terms of resonance distances {12,13}. A further more sophisticated procedure includes
the parameters "resonance width" and is therefore based on a complex frequency plane. Originally developed by Baum {14} the mathematical background of this so-called Singularity Expansion Method
is presently not far from being settled {15,16} whereas practical
applications to experimental data are still few {17,18,19} and
lack reliability {20} on behalf of the strong ill-posedness of the
resonance extraction; Dudley {21} states that presently available
algorithms {22,23,24} match data and not resonances.

Reconstruction of the geometrical boundary of scattering objects is strongly related to the identification of the equivalent
sources induced on the scatterer by the incident field giving rise
to the scattered field via Huygens' principle. If the scattering
process is considered as a linear time-invariant filter, we know
that a time harmonic excitation $\exp(j\omega t)$ can be regarded as an
eigenfunction with the eigenvalue "frequency response" $G(\underline{\hat{r}},\underline{\hat{r}}_o,\omega)$
depending on the direction of observation $\underline{\hat{r}}$ and the direction $\underline{\hat{r}}_o$
of the incident field ("$\hat{}$" denotes unit vectors). In the far-field
G is essentially determined by the scattering amplitude or field
pattern $H(\underline{\hat{r}},\underline{\hat{r}}_o,\omega)$ via

$$G(\hat{\underline{r}},\hat{\underline{r}}_o,\omega) = H(\hat{\underline{r}},\hat{\underline{r}}_o,\omega) \; \frac{e^{jkr}}{r} \tag{1}$$

where r is the distance of the observation point to some coordinate origin, and k denotes the wave number of the circular frequency ω. As outlined in {5}, the scattered field is uniquely de-
termined by $H(\hat{\underline{r}},\hat{\underline{r}}_o,\omega)$; therefore, the scattering amplitude is the
maximum amount of information available to an observer in the far-
field. On the other hand, $H(\hat{\underline{r}},\hat{\underline{r}}_o,\omega)$ is related to the above-mentioned equivalent sources (under additional assumptions) by a spa-
tial Fourier transform, where the Fourier transform variable $\underline{K}$ is
essentially composed of the product $k\hat{\underline{r}}$. Hence, for example, a time
harmonic (bistatic) synthetic aperture experiment yields data on
an aperture subset A of the spherical surface $K = k\hat{\underline{r}}$ = const., the
so-called Ewald-sphere (Fig. 2); it is clear then that, even for
a 4π - aperture, this experiment is not sufficient to allow per-

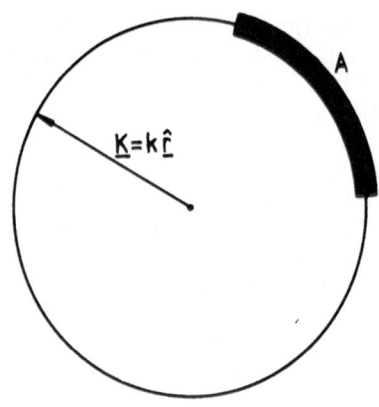

Fig. 2: The Ewald-sphere in $\underline{K}$-data space

formance of the inverse spatial Fourier transform to yield the (equivalent) sources confined to the scattering surface. In fact, the Ewald surface has to be swept throughout the $\underline{K}$-data space, i.e. broadband or transient data seem to be necessary to solve the imaging problem.

Within the visible range of electromagnetic waves our eye evolved a solution to the imaging problem during the evolutionary process in terms of "pattern matching" {25}, and we consider this as a key word in inverse scattering: our mathematical reconstruction scheme should be matched to the scattering procedure, i.e. we are looking for something like a spatial matched filter.

BROADBAND SYNTHETIC APERTURE ALGORITHMS

Basic Approaches

Contour imaging of scattering objects is mainly based on Kirchhoff's formulation of Huygens' principle (Fig. 3). Enclose the scatterer with surface S_C and volume V_C inside the surface S_M^i surrounded by another surface S_M^a, consider single and double layer sources $\partial\Phi/\partial n$ and Φ as known on $S_M^i \cup S_M^a$, then the (scalar) field Φ in the volume $V_M^a - V_M^i$ between the surfaces S_M^i and S_M^a is given by Kirchhoff's integral with the free space Green's function G

$$\Phi = \int_{S_M^a \cup S_M^i} \left(G \frac{\partial\Phi}{\partial n} - \Phi \frac{\partial G}{\partial n} \right) dS \qquad (2)$$

Fig. 3: Inverse scattering with Kirchhoff's integral

Extending S_M^a to infinity, the corresponding integral yields Φ_i the incident field produced by the source distribution Q. Therefore

$$\Phi = \Phi_i + \int_{S_M^i} (G \frac{\partial \Phi}{\partial n} - \Phi \frac{\partial G}{\partial n})\ dS \qquad (3)$$

As Bojarski already pointed out {26}, application of (2) or (3) for inverse scattering purposes, i.e. for points inside S_M^i where the scatterer lives, is useless, because the value zero and no information about the scatterer, contained in the total field Φ, is obtained. Otherwise spoken, according to (3), integration of known (measured) values of $\partial \Phi/\partial n$ and Φ on a surface S_M^i surrounding the scatterer yields nothing but the (negative) incident field inside S_M^i. Two alternative approaches seem appropriate:

1. Backward wave propagation
Induced by (3), define a quantity θ_h ("effectal field") according to

$$\theta_h = \int_{S_M^i} (G^* \frac{\partial \Phi}{\partial n} - \Phi \frac{\partial G^*}{\partial n})\ dS \qquad (4)$$

with the complex conjugate G^* of the Green's function defining incoming waves. This expression is nonzero inside V_M^i, and one can either hope that it propagates backward to the equivalent sources on the scattering surface where the data $\partial \Phi/\partial n$ and Φ originated from, or one might use it as a starting point for further processing. This backward wave propagation approach is dealt with under

the headlines "Generalized Holography" and "Exact Inverse Scattering" {26,27}.
2. Explicit inversion
Constrain the surface S_M^i to the scattering surface itself; then
the integral in (3) defines the scattered field due to the equi-
valent sources $\partial\Phi/\partial n$ and Φ on S_c, which can be measured throughout K-space. Under additional assumptions concerning the sources
(low or high frequency assumptions) and introducing the support
function of the scatterer (characteristic function) the integral
can be extended over the whole space; applying the far-field
approximation it turns out to be a Fourier integral making an explicit inversion possible. Acronymes associated with this approach
and defining (highly identical) inversion procedures are POFFIS
(Physical Optics Far-Field Inverse Scattering) {28,29} and IBA
(Inverse Born Algorithm) {5,30}. They can be both explicitly interpreted as transient methods thus rediscovering the basic idea
of SAR (Synthetic Aperture Radar) {31}.

Generalized Holography and Exact Inverse Scattering

It is worthwhile to notice that the backward wave propagation argument formulated in terms of the holographic reconstruction or
effectal field θ_h is a time harmonic approach thus reducing the
amount of data to be recorded considerably, and it comprises no
far-field approximation. The trade-off is the drawback of being
restricted to specific geometries, i.e. scatterers with planar
surfaces more or less parallel to an also planar recording surface, giving rise to the utilization of a special Green's function
being zero-valued on that surface {32}; equ. (4) then reads

$$\theta_h = - \int_{S_M^i} \Phi \, \frac{\partial G^*}{\partial n} \, dS \tag{5}$$

Denoting the kernel $\partial G^*/\partial n$ by Γ^*, the spatial Fourier transform of
(5) with respect to the measurement surface variables x and y of
a cartesian coordinate system, indicated by a tilde, yields

$$\tilde{\theta} = - \tilde{\Phi} \, \tilde{\Gamma}^* \tag{6}$$

where {33} (under the assumption kr >> 1)

$$\tilde{\Gamma} = e^{-jz\sqrt{k^2-\tilde{x}^2-\tilde{y}^2}} \tag{7}$$

can be interpreted as a spatial reconstruction filter into the
depth z from the measurement surface. On the other hand, applying
Kirchhoff's integral (3) to a planar scatterer in that depth, oriented parallel to S_M^i, and assuming a Neumann boundary condition
($\partial\Phi/\partial n = 0$ on S_c), we obtain for the (Fourier transformed) data

$$\tilde{\Phi} = -\tilde{\Phi}_s \tilde{\Gamma} \tag{8}$$

where Φ_s denote the sources, i.e. $\Phi_s = \Phi$ on S_c. Inserting (8) into (6) yields

$$\tilde{\theta}_h = \tilde{\Phi}_s \tilde{\Gamma}\tilde{\Gamma}^* \tag{9}$$

Hence, for the above special scattering arrangement, we have a transformation of the holographic reconstruction (its Fourier transform) into the Fourier transformed equivalent sources through the spatial filter $\tilde{\Gamma}\tilde{\Gamma}^*$, which, due to (7), is approximately (for kz sufficiently large) an identity transformation, as long as the spatial spectrum of the sources falls into the range $\{-k,k\}$, which is the case for (sufficiently extended) scatterers not inclined with respect to S_M^i {34}. Therefore, θ_h is actually a reconstruction of the (lateral) dimensions of such scatterers, and due to the relationship "reconstruction filter Γ^* equal to complex conjugate of scattering filter Γ", it represents a spatial matched filter with all the consequences for the noise behavior of such filters. In the field of nondestructive testing of materials this algorithm has been termed Rayleigh-Sommerfeld-reconstruction {34} and is used with good success as long as its limitations are well observed {35}. It should be noted that also axial resolution is obtained if a full 4π-aperture is available and the complete expression (4) is used (provided, $\hat{r}_o$ is not in the plane of the scatterer); the procedure has then recently been termed BIS (Basic Imaging System) {36}; numerical experiments show that this is true, in some respect, even for multidimensional scatterers (like a sphere for instance) {37}. The loss of axial resolution is always associated with a reduced aperture and the only non ill-posed remedy is the utilization of broadband or transient excitation. A further illustration and a mathematical proof of that need (apart from our introductory remarks) can be accomplished by a closer look to an extension of Generalized Holography termed Exact Inverse Scattering.

According to Fig. 4 one might consider $\tilde{\theta}_h$ as pseudo-data to fill up the Ewald-sphere for further processing. Porter {27} as well as Bojarski {26} pursued this idea deriving an integral equation of the first kind for the (equivalent) sources with θ_h as inhomogeneity. Unfortunately, due to the nonuniqueness of the solution of this equation {6}, the set of eigenfunctions is incomplete {37}, and the eigenvalues decay rapidly {36,37}, which has two consequences: θ_h is already the "best" possible solution, and, to ensure completeness of the eigenfunctions, one has to integrate over frequency confirming the above statement that transient imaging is appropriate (at least for single experiments).

Another heuristic extension to increase the resolution of the reconstruction θ_h is the use of monostatic data in (4).

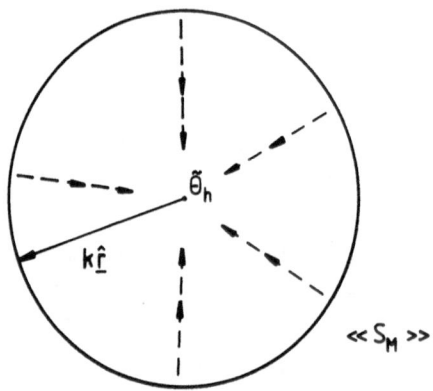

Fig. 4: Holographic reconstruction as pseudo-data in
K-space; <<S_M>> denotes the measurement surface
in K-space

Transient "inverse scattering"

For monostatic as well as bistatic excitation the surface integral
in (3) accounting for the scattered field can be converted into a
volume integral {38,39} after introducing a Dirichlet or Neumann
boundary condition (soft or rigid scatterer) on the support function (characteristic function $\gamma(r)$), assuming far-field distances,
and, after all, applying a low frequency principle of equivalent
sources of constant magnitude, and phase proportional to the incident field, on the whole surface of the scatterer. This integral
turns out to be a Fourier transform relationship making the explicit inversion of Kirchhoff's integral possible.

An alternative approach (yet yielding quite similar results) starts
with the low frequency Born approximation for a weak inhomogeneity
in an otherwise constant material {49,5}. In acoustic notation,
consider a homogeneous fluid with constant compressibility κ_0 and
density ρ_0 where a scattering body with surface S_c and spatially
varying and frequency dependent $\kappa(\underline{r},k)$ and $\rho(\underline{r},k)$ is embedded.
The total pressure field $\Phi(\underline{r},k)$ is then given by the solution of
{40}

$$(\Delta+k^2)\Phi(\underline{r},k) = \nabla \cdot \gamma_\rho(\underline{r},k)\nabla\Phi(\underline{r},k)$$

$$- k^2\gamma_\kappa(\underline{r},k)\Phi(\underline{r},k) \tag{10}$$

with

$$\gamma_\rho(\underline{r},k) = \frac{\rho(\underline{r},k) - \rho_o}{\rho_o} \quad , \quad \gamma_\kappa(\underline{r},k) = \frac{\kappa(\underline{r},k) - \kappa_o}{\kappa_o} \tag{11}$$

The right hand side of (10) can be interpreted as a source term $-q(\underline{r},k)$ giving rise to an iterative solution in terms of

$$\Phi_{n+1}(\underline{r},k) = \Phi_i(\underline{r},\underline{r}_o,k)$$

$$n=0,1,2,..$$

$$+ \int_{V_c} d^3r' \frac{e^{jk|\underline{r}-\underline{r}'|}}{4\pi|\underline{r}-\underline{r}'|} q(\underline{r}',k)\Phi_n(\underline{r}',k) \tag{12}$$

where the incident field $\Phi_i(\underline{r}',\underline{r}_o,k)$ from a source point $\underline{r}_o$ is inserted as zero-order approximation for the total field $\Phi_0(\underline{r}',k)$ under the integral, which might be sufficiently valid for a weak scatterer. Assuming a point source at $\underline{r}_o$, introducing the far-field approximation

$$\frac{e^{jk|\underline{r}-\underline{r}'|}}{4\pi|\underline{r}-\underline{r}'|} \cong \frac{e^{jk(r - \hat{\underline{r}} \cdot \underline{r}')}}{4\pi r}$$

and manipulating the differential operators in q yields for the first-order scattered field $\Phi_s = \Phi_1 - \Phi_i$

$$\Phi_s(\underline{r},k) = k^2 \frac{e^{jk(r+r_o)}}{(4\pi)^2 rr_o} \int d^3r' e^{-jk(\hat{\underline{r}}+\hat{\underline{r}}_o)\cdot\underline{r}'} \gamma(\underline{r}')p(\underline{r}',k) \tag{13}$$

with the so-called object function

$$p(\underline{r},k) = - \{\gamma_\rho(\underline{r},k) + \gamma_\kappa(\underline{r},k)\} - \frac{1}{2}k^2\Delta\gamma_\rho(\underline{r},k) \tag{14}$$

The integral (13) is a spatial Fourier transform with the transform variable $\underline{K} = k(\hat{\underline{r}} + \hat{\underline{r}}_o)$; therefore, inversion leads to an expression ("*" denoting convolution)

$$\tilde{\gamma}(\underline{K}) \; * \; \tilde{p}(\underline{K},k) = (4\pi)^2 rr_o e^{-jk(r+r_o)} \; \frac{\Phi_s(\underline{r},k)}{k^2} \qquad (15)$$

which can be experimentally determined within the volume of the "double" or limiting Ewald sphere $|\underline{K}| < 2k$. Depending on the nature and feasability of the experiment (compare Fig. 1) the subsequent alternatives can be pursued:

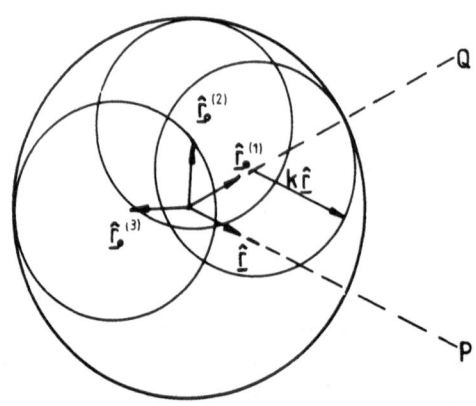

Fig. 5: Bistatic time harmonic multiple experiment

1. Bistatic time harmonic multiple experiment
The limiting Ewald sphere can be filled by a manifold of Ewald sphere surfaces with radii k and subsequent midpoints $k\hat{\underline{r}}_o^{(i)}$ (i = 1,2,3,...) (Fig. 5), indicating a new synthetic aperture bistatic time harmonic experiment for each i {5}. As outlined in {41} this means a considerable experimental effort leading to huge data storage problems. Nevertheless, for optical purposes this time harmonic approach has proved to be useful, especially since recent computational improvements have been reported {41,42} closing the gap to tomographic reconstruction.
2. Monostatic broadband single experiment (IBA, POFFIS)
As indicated by Fig. 6 monostatic broadband experiments yielding $\underline{K} = 2k\hat{\underline{r}}$ are also appropriate to fill the limiting Ewald sphere. The resulting algorithm, as extensively applied for nondestructive testing of materials, has been termed IBA (Inverse Born Algorithm) {30}; yet, the inversion requires an explicit frequency independence of the object function.

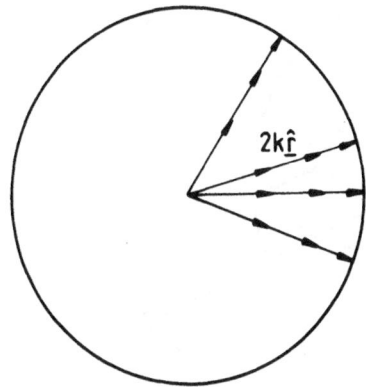

Fig. 6: Monostatic broadband single experiment
(IBA, POFFIS)

It has recently been emphasized {43} that the monostatic broadband version of (15) is very similar to the POFFIS-identity {28,
29}, even though the latter approach is a high frequency one. The
deeper reason for this coincidence is that both POFFIS and IBA
concentrate on the scattering surface, the first procedure by definition and the latter one by assuming an undisturbed incident
field inside the scatterer.
3. Bistatic broadband single experiment (FIFFIS)
According to Fig. 7, fixed $\underline{r}_o$ combined with varying k and $\hat{r}$ leads
to a subset of the data space, and the resulting algorithm has
been termed FIFFIS (Frequency Independent Far-Field Scattering)
{44}, because, once again, the object function has to be frequency
independent. As IBA, FIFFIS is derived under the low frequency
assumption, and it is not quite clear if there exists, like for
IBA and POFFIS, a formally correct high frequency counterpart,
because corresponding bistatic POFFIS identities reveal intrinsic
complications {45,46}. Nevertheless, it is somehow heuristically
justified for high frequencies by assuming a priori given sources
{47} with (frequency independent!) spatially varying magnitudes
and a phase related to the incident wave. It might then be interpreted as a broadband Fourier-type (i.e. far-field) holography,
which has been used successfully in microwave imaging {48,49}; a
broadband nearfield counterpart based on Rayleigh-Sommerfeld holography has also been reported {50}.

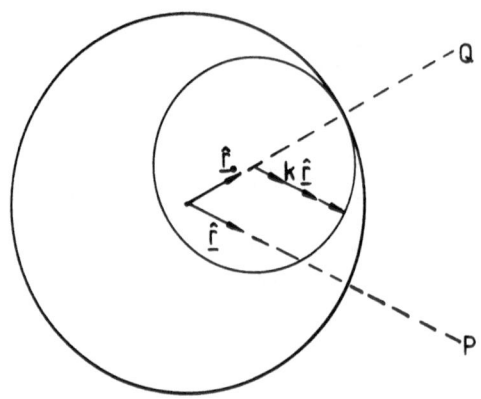

Fig. 7: Bistatic broadband single experiment (FIFFIS)

Transient interpretation of broadband "inverse scattering" (SAR)

The interpretation of IBA and POFFIS in the time domain {29,30}
reveals very nicely their intrinsic properties and identifies intuitively their possible pitfalls. Let $\underline{K} = 2k\underline{r}$ and rewrite the
Fourier transformed equ. (15) in terms of spherical coordinates
in $\underline{K}$-space $d^3\underline{K} = K^2 dKd\Omega$

$$\gamma(\underline{r}')p(\underline{r}') = \frac{32r^2}{c} \int d\Omega$$

$$\cdot \frac{1}{2\pi} \int_{-\infty}^{+\infty} d\omega \phi_s\left(\hat{\underline{r}},\frac{\omega}{c}\right) e^{-2j\frac{r}{c}\omega} e^{2j\frac{\hat{\underline{r}}\cdot\underline{r}'}{c}\omega} U(\omega) \tag{16}$$

where $U(\omega)$ denotes the unit step-function. The ω-integral is an
inverse Fourier integral for the time $t = 0$ and therefore

$$\gamma(\underline{r}')p(\underline{r}') = \frac{32r^2}{c} \int d\Omega \phi_s(\hat{\underline{r}},t)*U(t)\Big|_{t = 2\frac{r}{c} - 2\frac{\hat{\underline{r}}\cdot\underline{r}'}{c}} \tag{17}$$

with

$$U(t) = \frac{1}{2} \delta(t) + \frac{1}{2\pi} Pf \frac{1}{jt} \tag{18}$$

and $\phi_s(\hat{\underline{r}},t)$ the scattered time domain impulse response.

Hence

$$\gamma(\underline{r}')p(\underline{r}') = \frac{16r^2}{c} \int d\Omega\Phi_s(\hat{r}, 2\frac{r}{c} - 2\frac{\hat{r}\cdot r'}{c})$$

$$+ j\frac{16r^2}{c} \int d\Omega H\{\Phi_s(\hat{r}, t)\}\Big|_{t = 2\frac{r}{c} - 2\frac{\hat{r}\cdot r'}{c}}$$

(19)

where H is the Hilbert transform. We state that the broadband monostatic reconstruction under the Born assumption is composed of
a real and imaginary part, whereas the derivation of its time domain PO counterpart yields only the real part of (19) {29}. We
suspect that this is due to the fact that the POFFIS identity is
strictly valid only for *Physical Optics* data, because numerical
experiments have shown {51,52} that the additional imaginary part
of (19) accounts for the deviation of exact scattering data from
Physical Optics data emphasizing the scattering centres of the
body (edges, corners etc.) whose singularities are not accounted
for by the smooth PO equivalent sources.

The above fact of differing real life data yields a pitfall
of POFFIS and IBA even for 4π-apertures and infinite bandwith as
illustrated by Fig. 8.

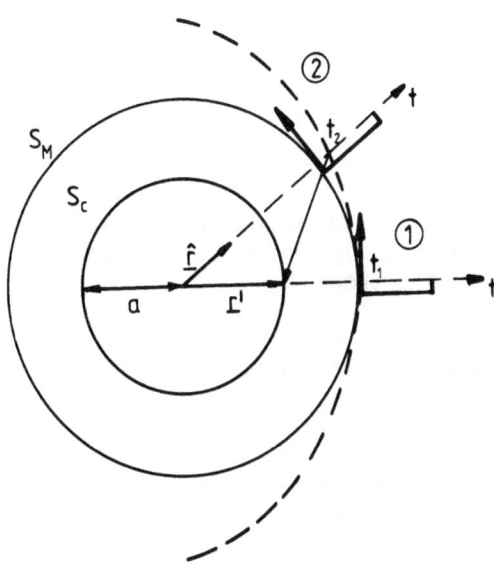

Fig. 8: Time domain illustration of IBA and POFFIS

Consider a spherical scatterer of radius a with a perfectly hard
boundary surrounded by the measurement surface S_M, and two monostatic experimental points displaying the (infinite broadband)
backscattered impulse response as function of time, which is composed of the specular reflection peak and the constant PO part
ending abruptly at the time corresponding to the shadow boundary.
According to the real part of (19), choice of a reconstruction
point $\underline{r}'$ means integration of the transient data amplitudes along
a surface indicated by the dashed line; we see that one specular
and additional PO data points are included, where the amplitude
at t_2 corresponds to the PO contribution from our reference point
thus being correctly accounted for. The integration procedure then

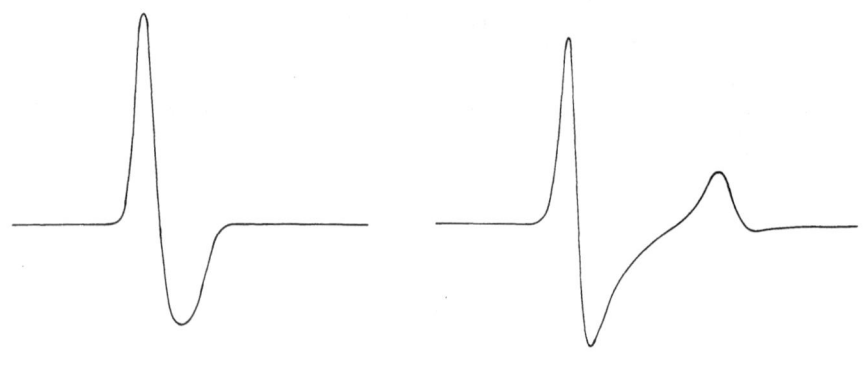

Fig. 9: Left: PO far-field impulse response of hard sphere.
        Right: Far-field impulse response of hard sphere
        (both Gaussian smoothed)

exactly yields the characteristic function of the sphere. Fig. 9
now compares the exact (Gaussian smoothed) impulse response with
the PO counterpart, and obviously the PO part of the exact response decays more rapidly than that of the PO response yielding
an indentation of the top of the characteristic function (Fig. 10).
Additionally, the creeping wave contribution to the exact impulse
response shows up as a ghost image.

    The POFFIS identity itself is independent on the boundary
condition "soft" or "hard"; nevertheless, the soft impulse response differs considerably from the hard one (Fig. 11) since its
PO part is much less pronounced compared to the specular part.

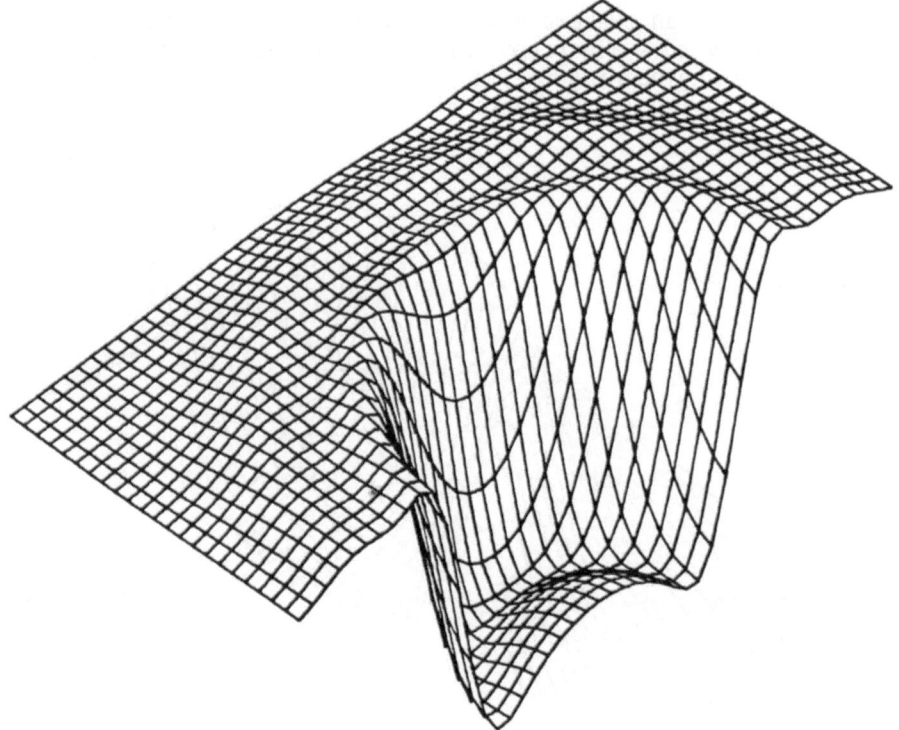

Fig. 10: POFFIS reconstruction of hard sphere

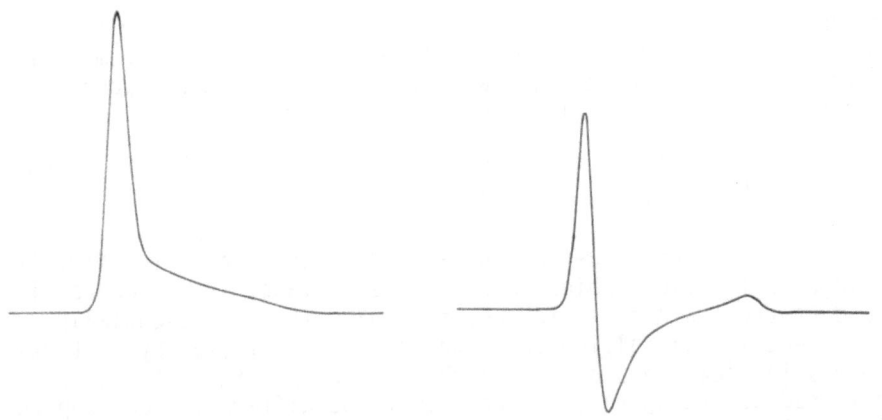

Fig. 11: Left: Far-field impulse response of soft sphere. Right: Near-field impulse response of hard sphere (both Gaussian smoothed)

This results in an overemphasis of the scattering surface when reconstructed by this algorithm (Fig. 12). Therefore, this fact un-
derlines the engineering feeling that specular reflection contributions to the impulse response give rise to a contour or surface

Fig. 12: POFFIS reconstruction of soft sphere

reconstruction when summed up properly according to their travel
times. Hence, a heuristic near-field extension of (19) seems possible in terms of (expecially when one looks at a corresponding
near-field response as given in Fig. 12).

$$\gamma_{surface} \ (\underline{r}') \stackrel{\sim}{=} \int d\Omega \phi_s(\hat{\underline{r}}, \frac{2}{c}|\underline{r} - \underline{r}'|) \tag{20}$$

Explicit surface detection can be directly accounted for, if
a singular function, which peaks on the surface, is introduced instead of the characteristic function {53}. This is essentially
equivalent in multiplying the right hand side of (15) by k, thus
suppressing low frequencies; in the time domain the data to be
collected according to (19) are then to be differentiated, suppressing again the relatively slowly varying PO portion. This suppression of low frequencies is also achieved for bandlimited excita-
tions, say amplitude modulated sine bursts or chirp signals, i.e.

real life data. We therefore prefer to use the singular function formulation for experimental purposes {43,44}.

Unprocessed reconstruction with bandlimited signals introduces unwanted image side lobes {51}; a hint for further processing is given by our "Real life data POFFIS identity" (19): in the context of analytic or radar signals the time domain data and their Hilbert transform can be combined to the computation of envelopes, correlations etc. thus improving resolution, and, if extended to the near-field version, rediscovering the synthetic aperture radar (SAR), which recently found its counterpart SAFT (Synthetic Aperture Focussing Technique) {54} for nondestructive testing of materials.

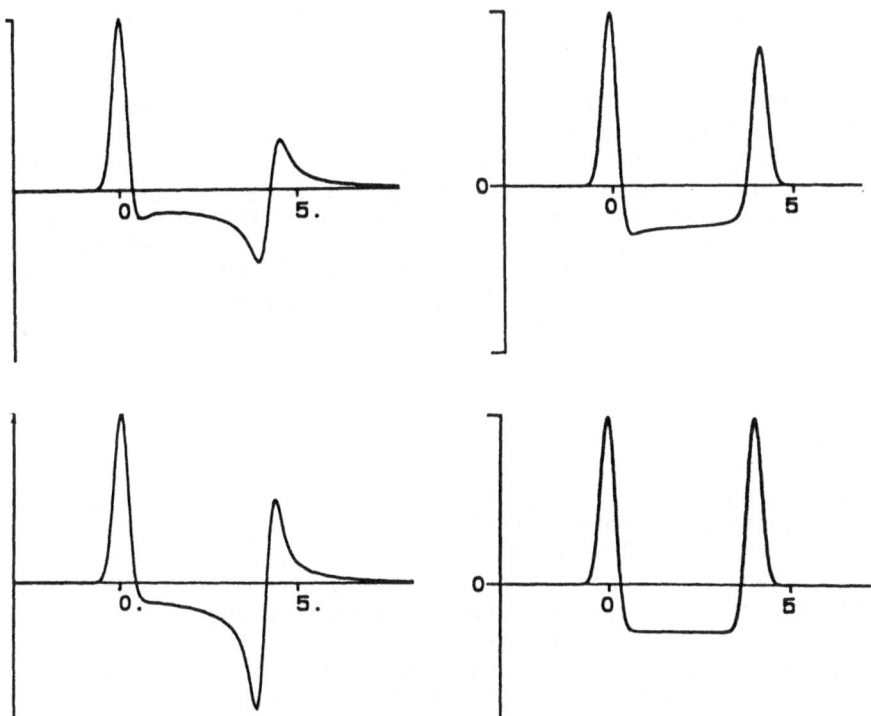

Fig. 13: Far-field backscattered Gaussian smoothed impulse response of hard elliptical scatterers for nose-on incidence (above: axis ratio 5:1, below: axis ratio 100:1).
Left: infinitely long cylinder with elliptical cross-section (two-dimensional problem).
Right: prolate spheroid (three-dimensional problem)

Certain severe drawbacks of POFFIS or IBA reconstructions are readily derived via simulations {43,44,51} and well understood through their time domain interpretation: a limited spatial aperture, for instance, misoriented with respect to a planar part of the scatterer, may lead to a lack of specular information; then, high amplitude corner or edge diffracted pulses are interpreted as such by the algorithm and placed at the proper position. When applied to nondestructive testing this missing information could be dangerous because the image of a long crack might look like a point scatterer, namely the crack tip.

When simulations are performed in testing POFFIS and IBA another problem has to be carefully considered: the time domain algorithm is essentially independent of the dimension of the scatterer but the data are not (Fig. 13); therefore, misinterpretations may arise.

We want to conclude this section in pointing out that other transient procedures essentially utilize this k^{-2}-dependence of (15) relating the so-called area function of the scatterer to its ramp response (impulse response integrated twice) for prescribed monostatic angles of incidence ("look angles") {55,56,57} instead of compensating it via the $\underline{K}$-space integration. These methods need a very large bandwidth because they rely on the PO portion of the response; on the other hand, the amount of data is reduced because three equally spaced look angles have proved to be generally sufficient.

SINGULARITY EXPANSION METHOD

If, according to Fig. 1, a stationary arrangement is used to identify a scatterer, information about its size is, for instance, contained in the arrival time of the creeping wave, which, when varying the shape from a sphere to a prolate spheroid (Fig. 9, Fig. 13), turns characteristically into the tip echo from the farther end of the spheroid. Time delay estimation methods are then appropriate to separate single scattering centres. On the other hand, timely distinguishable wave fronts yield resonances when Fourier transformed to a spectral domain and these may then serve as a tool for identification (Fig. 14). Unfortunately, both time and frequency results depend upon the look angle. An identification procedure which overcomes this problem is derived in terms of the Singularity Expansion Method. Consider for example the Mie series for the scattered far-field of a rigid sphere in the Laplace (complex frequency) domain ($k \rightarrow js/c$)

$$\Phi_s(\underline{r},s) = \frac{e^{-\frac{s}{c}r}}{\frac{s}{c}} \sum_{n=0}^{\infty} (-1)^n (2n+1) \frac{j_n'(j\frac{s}{c}a)}{h_n^{(1)'}(j\frac{s}{c}a)} P_n(cos\theta) \qquad (21)$$

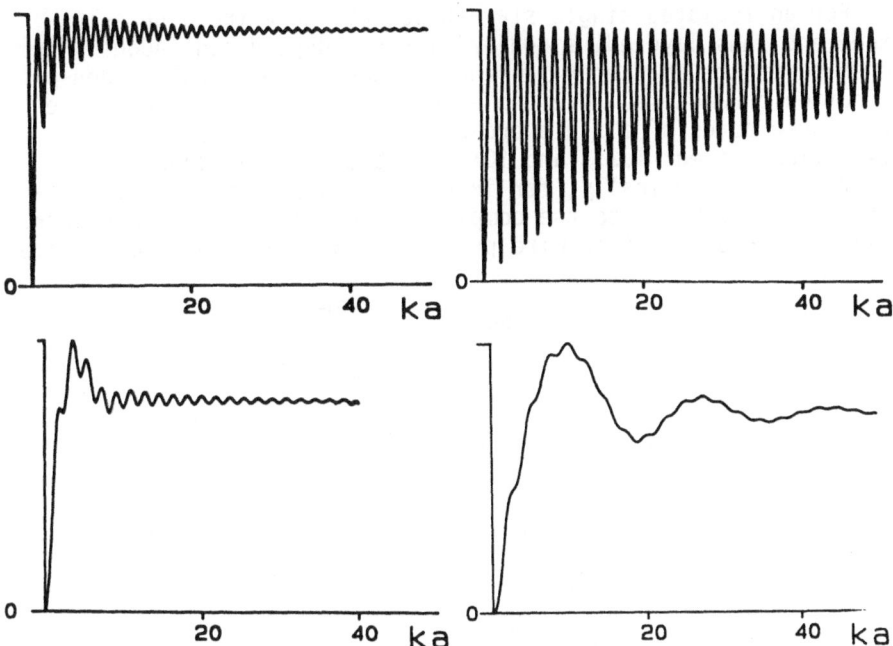

Fig. 14: Far-field backscattered frequency response (magnitude)
for nose-on incidence of hard prolate spheroid (above; left: axis
ratio 1:1, i.e. sphere; right: axis ratio 5:1) and hard oblate spheroid (below; left: axis ratio 2:1; right: axis ratio 5:1); a denotes
major axis

where Θ denotes the polar angle of a spherical coordinate system;
j_n, $h_n^{(1)}$ are spherical Bessel and Hankel functions and P_n Legendre
polynomials. A singularity representation of (21) is obtained applying the Mittag-Leffler theorem to the corresponding induced surface distribution and integrating over the surface {20} ending up
with

$$\Phi_s(\underline{r},s) = js\frac{a^2}{c}j_0'(j\frac{s}{c}a)\frac{e^{-\frac{s}{c}(r-a)}}{r} + \tag{22}$$

$$jsa \sum_{n=0}^{\infty}(-1)^n(2n+1)j_n'(j\frac{s}{c}a)P_n(cos\Theta)\frac{e^{-\frac{s}{c}(r-a)}}{r} \cdot$$

$$\sum_{l=1}^{n+1}\frac{e^{-\frac{s_{nl}}{c}a}}{\left[n(n+1)+\frac{s_{nl}^2}{c^2}a^2\right]h_n^{(1)}(j\frac{s_{nl}}{c}a)}(\frac{1}{s-s_{nl}}+\frac{1}{s_{nl}})$$

For an isolated single star-shaped obstacle the double-in-
dexed, and thus layered, singularities s_{nl} depend continuously on
the shape of the scatterer {16} and may therefore serve as identification parameters. An example is shown in Fig. 15, where the
first layer poles of a rigid prolate spheroid are displayed for
axis ratios between 1.2:1 and 10:1 as obtained via spheroidal
eigenfunctions; typically, for increasing axis ratio the singularities move further into the complex plane (in contrast to electromagnetic spheroidal "thin-wire problems" {58}) and tend to arrange

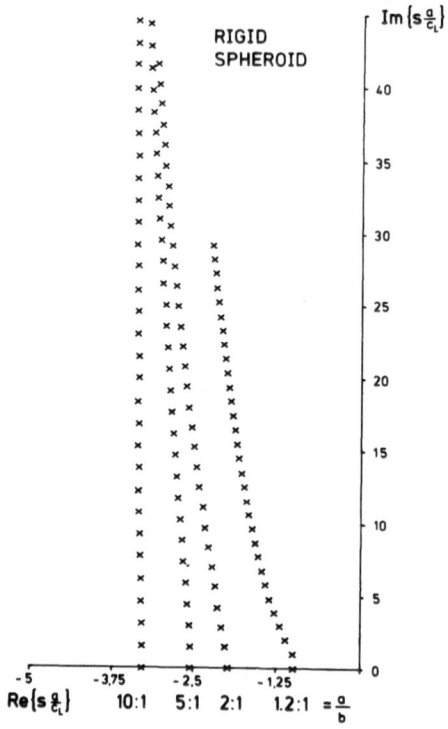

Fig. 15: First layer singularities of prolate rigid spheroid

themselves along a straight line parallel to the imaginary axis,
thus approaching the high-frequency ray theoretical result. Additional investigations of that kind using T-matrix methods are contained in {59,60}, and the "elastodynamic" spheroid is addressed
in {20}. Even though promising at this point that SEM might be
used for imaging and identification purposed, the pole extraction
from experimental data has not yet been satisfactorily solved {20,
21}. The reason for this may be illustrated the following way {61}:
consider the (simulated) identity between timely separated (by a
distance T) high-frequency wave fronts and resonances

$$\sum_{n=-\infty}^{+\infty} \delta(t - nT) = \frac{1}{T} \sum_{n=-\infty}^{+\infty} e^{jn\frac{2\pi}{T} t} \qquad (23)$$

where δ denotes the δ-distribution. Assuming 20 resonances the re-sult is shown in Fig. 16, but, if the resonances are corrupted by Gaussian noise with spectral power N according to $2\pi n(1+\sqrt{N}/n)/T$ and $N = 5 \cdot 10^{-3}$, Fig. 17 illustrates the ill-posedness of resonance extraction out of wave fronts, like creeping waves for instance.

Fig. 16: Wave fronts composed of 20 resonances

Fig. 17: Wave fronts composed of 20 noisy resonances

REFERENCES

{ 1} Boerner, W.-M.: Polarization Utilization in Electromagnetic Inverse Scattering. In: Inverse Scattering Problems in Optics. Ed.: H.P. Baltes, Springer, New York 1980

{ 2} Boerner, W.-M.: El-Arini, M.B., Chan, C.-Y., Mastoris,P.M.: IEEE Trans. AP-29 (1981)

{ 3} Porter, R.P., Devaney, A.J.: J. Opt. Soc. Am. 72 (1982) 327

{ 4} Porter, R.P., Devaney, A.J.: J. Opt. Soc. Am. 72 (1982) 1707

{ 5} Devaney, A.J.: Inverse source and scattering problems in ultrasonics. Proc. Int. Symp. IEEE Ultrasonics, San Diego/Ca 1982

{ 6} Devaney, A.J., Sherman, G.C.: IEEE Trans. AP-30 (1982) 1034

{ 7} Dudley, D.G.: Wave Motion 5 (1983)

{ 8} Ljung, L., Glover, K.: Automatica 17 (1981) 71

{ 9} Rose, J.L., Goldberg, B.G.: Basic Physics in Diagnostic Ultrasound. J. Wiley & Sons, New York 1979

{10} Tittmann, B.R.: Wave Motion 5 (1983)

{11} Carter, G.C. (Ed.): Special Issue on Time Delay Estimation. IEEE Trans. ASSP - 29 (1981)

{12} Adler, L., Fitting, D.W.: Ultrasonic Spectral Analysis for Nondestructive Evaluation. Plenum Press, New York 1980

{13} Überall, H., Moser, P.J., Murphy, J.D., Nagl, A., Igiri, G., Subrahmanyam, J.V., Gaunaurd, G.C., Brill, D., Delsanto, P.P., Alemar, J.D., Rosario, E.: Wave Motion 5 (1983)

{14} Baum, C.E.: The Singularity Expansion Method. In: Transient Electromagnetic Fields. Ed.: L.B. Felsen, Top. Appl. Phys. 10, Springer, Berlin 1976

{15} Pearson, L.W., Marin, L. (Eds.): Special Issue on the Singularity Expansion Method. Electromagnetics 1 (1981)

{16} Ramm, A.G.: J. Math. Anal. Applic. 86 (1982) 562

{17} Chan, L.C., Moffatt, D.L., Peters, L.: Proc. IEEE 67 (1979) 991

{18} Volakis, J.L., Peters, L.: IEEE Trans. AP - 31 (1983) 334

{19} Lytle, R.J.: A remote method for determining the thicknesses and constitutive properties of planar layered media. Lawrence Livermore Laboratory UCRL - 51789 (1975)

{20} Bollig, G., Langenberg, K.J.: Wave Motion 5 (1983)

{21} Dudley, D.G.: Identification and parametric modeling of transient waves. In: New Procedures in Nondestructive Testing. Ed.: P. Höller, Springer, Berlin 1983

{22} van Blaricum, M.L., Mittra, R.: IEEE Trans. AP - 23 (1975) 777

{23} Goodman, D.M., Dudley, D.G.: Transient electromagnetic identification by non-linear least squares. US National Radio Science Meeting, May 23-26, 1983, Houston/Tx

{24} Sarkar, T.K., Nebat, J., Weiner, D.D., Jain, V.K.: IEEE Trans. AP - 28 (1980) 928

{25} Lorenz, K.: Die Rückseite des Spiegels. Piper Verlag, München 1973

{26} Bojarski, N.N.: Radio Science 16 (1981) 1025

{27} Porter, R.P.: J. Opt. Soc. Am. 60 (1970) 1051
{28} Bojarski, N.N.: IEEE Trans. AP - 30 (1982) 980
{29} Bleistein, N.: J. Acoust. Soc. Am. 60 (1976) 1249
{30} Rose, J.H., Richardson, J.M.: J. Nondestr. Eval. 3 (1983)
{31} Skolnik, M.I.: Radar Handbook. McGraw Hill, New York 1970
{32} Sommerfeld, A.: Optik. Akademische Verlagsgesellschaft,
     Leipzig 1964
{33} Collier, R.J., Burckhardt, Ch.B., Lin, L.H.: Optical Hologra-
     phy. Academic Press, New York 1971
{34} Berger, M., Brück, D., Fischer, M., Langenberg, K.J., Oberst,
     J., Schmitz, V.: J. Nondestructive Evaluation 2 (1982) 85
{35} Müller, W., Schmitz, V., Schäfer, G., Langenberg, K.J.,
     Hoppstädter, K., Gräber, B.: Materialprüfung 25 (1983) 9
{36} Porter, R.P., Devaney, A.J.: J. Opt. Soc. Am. 72 (1982) 1707
{37} Fischer, M., Langenberg, K.J.: Limitations and defects of
     Exact Inverse Scattering. Submitted to Electromagnetics for
     publication (1983)
{38} Bojarski, N.N.: IEEE Trans. AP - 30 (1982) 775
{39} Bojarski, N.N.: J. Acoust. Soc. Am. 73 (1983) 733
{40} Morse, P.M., Ingard, K.U.: Theoretical Acoustics. McGraw Hill,
     New York 1968
{41} Devaney, A.J.: Inverse Source and Scattering Problems in
     Optics. In: Optics in Four Dimensions. Eds.: M.A. Machado,
     L.M. Narducci, Am. Inst. Phys., New York 1981
{42} Devaney, A.J.: Optics Letters 7 (1982) 111
{43} Langenberg, K.J., Bollig, G., Brück, D., Fischer, M.: Remarks
     on recent developments in inverse scattering theory. In: Orien-
     tation and Localization in Engineering and Biology. Ed.: D.
     Varju, Springer, Berlin 1983
{44} Langenberg, K.J., Brück, D., Fischer, M.: Inverse scattering
     algorithms. In: New Procedures in Nondestructive Testing.
     Ed.: P. Höller, Springer, Berlin 1983
{45} Rosenbaum-Raz, S.: IEEE Trans. AP - 24 (1976) 66
{46} Meckelburg, H.J.: Electronics Letters 18 (1982)
{47} Detlefsen, J.: Kleinheubacher Berichte 25 (1982) 297
{48} Chan, C.K., Farbat, N.H.: IEEE Trans. AP - 29 (1981) 312
{49} Paolini, F.J., Duffy, M.J.: IEEE Trans. AP - 31 (1983) 389
{50} Karg, R.: Archiv Elektronik Übertrag.techn. 31 (1977) 150
{51} Brück, D.: Simulationsrechnungen zur Fehlerrekonstruktion mit
     dem POFFIS-Algorithmus. Diploma Thesis, Universität des Saar-
     landes, Saarbrücken, FRG, 1982
{52} Rose, J.H.: private communication of the Ames Research Lab,
     Iowa State University, Ames/Iowa
{53} Cohen, J.K., Bleistein, N.: Wave Motion 1 (1979) 153
{54} Ganapathy, S., Wu, W.S., Schmult, B.: Ultrasonics 20 (1982)
     249
{55} Kennaugh, E.M., Moffatt, D.L.: Proc. IEEE 53 (1965) 893
{56} Moffatt, D.L., Young, J.D., Ksienski, A.A., Lin, H.C.,
     Rhoads, C.M.: IEEE Trans. AP - 29 (1981) 192

{57} Boerner, W.M., Ho, C.M., Foo, B.Y.: IEEE Trans. AP - 29 (1981) 336
{58} Marin, L.: IEEE Trans. AP - 22 (1974) 266
{59} Boström, A.: J. Math. Phys. 23 (1982) 1444
{60} Kristensson, G.: Natural frequencies of circular discs. Report of the Institute of Theoretical Physics, Chalmers University, Göteborg, Sweden, March 1983
{61} Felsen, L.B.: Progressive wave and oscillartory formulations of propagation and scattering. Plenary Session of the Joint AP-S/URSI Symposium, Houston/Tx, 1983

I.5 (IS.4)

THEORETICAL STUDY AND NUMERICAL RESOLUTION OF INVERSE PROBLEMS
VIA THE FUNCTIONAL DERIVATIVES

A. Roger

Laboratoire d'Optique Electromagnétique, ERA 597 CNRS,
Faculté des Sciences et Techniques, Centre de Saint-
Jérôme, 13397 MARSEILLE CEDEX 13, France

ABSTRACT : A rigorous method of inverse scattering is described,
which enables one to solve numerically many practical inverse
problems and throws light upon the fundamental instability of
these problems. The algorithm is described on the example of a
finite two dimensional rough surface, and several profile
reconstructions are presented.

1. INTRODUCTION

This paper presents a method which could be called "numerical
inverse diffraction". It is a rigorous method, meaning that is
it limited only by purely numerical phenomenons. Also, it is a
method of inverse diffraction, since it works directly in the
framework of functional spaces as opposed to methods of parameter
optimization which are limited to the search of a finite number
of a-priori chosen parameters. Our algorithm has permitted us to
solve successfully practical inverse problems associated with
layered media, cylinders and gratings. The method relies upon
linearization and iteration. Its implementation includes the
following two steps :

First, one must explicitly find the linear operator which
links small variations of the object profile to the corresponding
small variations of the scattering pattern ; of course it is a
first order relation. This operator is the "functional derivative"
and it may be exhibited with the help of a so called "adjoint
direct problem". Not only the functional derivative is a powerful
numerical tool, but also it provides a way of analyzing the
information contained in the data via the study of its singular

111

W.-M. Boerner et al. (eds.), Inverse Methods in Electromagnetic Imaging – Part 1, 111–120.
© *1985 by D. Reidel Publishing Company.*

values. On the other hand, the compactness of these operators
explains why the associated inverse problems are unstable and
thus require the use of regularization techniques.

Second, one must use the functional derivative to implement
an iterative algorithm capable to retrieve the characteristics of
the scatterer. Following optimization methods, one may minimize
a "cost function" with the help of conjugate gradient algorithms.
One may also construct a Newton type algorithm, often much faster
than the conjugate gradient.

The main features of these methods will be described on a
particular example, i.e. the scattering by a finite rough surface.

2. THE FINITE ROUGH SURFACE : DIRECT AND INVERSE PROBLEMS

Fig. 1A shows the physical device : x,y,z being cartesian
coordinates, we study the scattering by a perfectly conducting
plane situated at y = 0 which presents a local z-independent
deformation. This is a model for a finite rough surface S, with
equation y = f(x) (f(x) = 0 outside [a,b]).

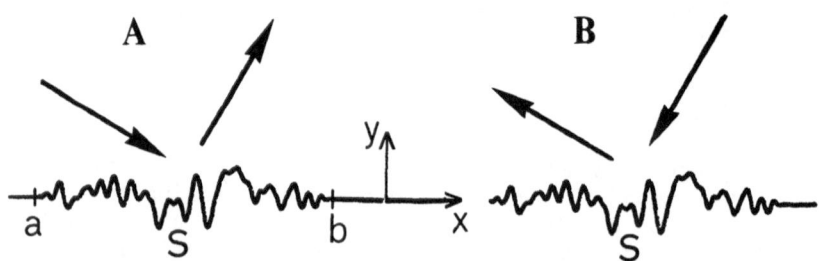

Fig. 1 : The diffraction by a finite rough surface ; fig. 1A
shows the initial direct problem, fig. 1B an associated adjoint
direct problem.

Let an electromagnetic plane wave, with wavevector $\underline{k}$ ($k = |\underline{k}|$)
parallel to the xy plane, be incident upon S with an angle of
incidence θ. The electric field $\underline{u}$ is supposed to be parallel to Oz.
Under these conditions, the component u of $\underline{u}$ (which we call "the
field" in the rest of the paper) along Oz verifies :

$$\Delta u + k^2 u = 0 \quad \text{above S} \tag{1}$$

$$u = 0 \qquad \text{on and below S}$$

$$u = \exp(i\alpha x - i\beta y) - \exp(i\alpha x + i\beta y) + \int_{-\infty}^{+\infty} B(\alpha')\exp(i\alpha' x + i\beta' y)d\alpha'$$
$$(y > \sup f(x))$$

where α, β, β' are defined by :

$$\alpha = k \sin\theta \qquad \beta = k \cos\theta$$

$$\beta' = (k^2 - \alpha'^2)^{1/2} \qquad \text{if } |\alpha'| \leqslant k$$

$$\beta' = i(\alpha'^2 - k^2)^{1/2} \qquad \text{if } |\alpha'| > k .$$

The direct problem of diffraction may be defined as : knowing $f(x)$, calculate $B(\alpha')$. It has been recently studied and solved [1], and now we have at our disposal a computer code capable to calculate the diffracted amplitude $B(\alpha')$ rigorously, at least in the limit of accuracy of the computer. The numerical resolution is achieved via the use of the Green function of the free space, which leads to an integral equation.

In the following of the paper, the following theorem will be needed : let v and w be two functions of x and y, defined and bounded for $f(x) \leqslant y \leqslant A$ (with $A > \sup f(x)$) and which verify the Helmholtz equation (1). Then, under some conditions of regularity on v and w, we have that :

$$\int_{S_0} (v \frac{\partial w}{\partial n} - w \frac{\partial v}{\partial n}) dS = -4i\pi \int_{-\infty}^{+\infty} \beta'(\alpha')[V^+(\alpha')W^-(-\alpha') -$$
$$- V^-(\alpha')W^+(-\alpha')]d\alpha' \quad (2)$$

where S_0 is any sufficiently regular surface above S, and V^+, V^-, W^+, W^- are the coefficients of the plane wave expansions of v and w (valid only for $y \geqslant A > \sup f(x)$) :

$$v(x,y) = \int_{+\infty}^{-\infty} [V^-(\alpha')\exp(i\alpha'x-i\beta'y) +$$
$$+ V^+(\alpha')\exp(i\alpha'x+i\beta'y)]d\alpha' \quad (3)$$

$$w(x,y) = \int_{+\infty}^{-\infty} [W^-(\alpha')\exp(i\alpha'x-i\beta'y) +$$
$$+ W^+(\alpha')\exp(i\alpha'x+i\beta'y)]d\alpha' \quad (4)$$

The demonstration is given in [2]. The reciprocity theorem is a straighforward consequence of this lemma : let $u^*(x,y)$ be the field corresponding to the angle of incidence θ^* ($\alpha^* = \sin\theta^*$, $\beta^* = \cos\theta^*$). The coefficients of the plane wave expansions of u and u^* similar to (3) and (4) are :

$$U^-(\alpha') = \delta(\alpha' - \alpha) \qquad U^{*-}(\alpha') = \delta(\alpha' + \alpha^*)$$

$$U^+(\alpha') \overset{def}{=} B(\alpha') - \delta(\alpha' - \alpha)$$

$$U^{*+}(\alpha') \overset{def}{=} B^*(\alpha') - \delta(\alpha' + \alpha^*).$$

Now, let us apply the theorem given above, with $v = u$ and $w = u^*$. Since u and u^* are zero on S, the left hand member of (2) is also zero. The evaluation of the right hand member gives the reciprocity theorem :

$$\beta * B(-\alpha *) = \beta B * (-\alpha).$$ (5)

3. CALCULATION OF THE FUNCTIONAL DERIVATIVE

Clearly, the connection between f and B is complicated : it is non linear and, moreover, it cannot be exhibited in closed form since B may be computed only by solving integral or differential equations. So, as a first step, we will restrict our attention to small variations δf and δB. To first order in δf, there exists a linear relation between these two quantities, and we will exhibit this relation.

Let u'(x,y) be the total field created by the incident plane wave exp i(αx−βy) in presence of the diffracting surface S' ; in order to simplify the demonstration, we suppose for a moment that S' is everywhere above S (i.e. δf(x) ⩾ 0 ∀x). Let B'(α') be the corresponding diffracted amplitude ; apply the lemma of section 2 , with v = u', w = u* and S_0 = S' ; the right hand member is :

$$-4i\pi \int_{-\infty}^{+\infty} \beta'(\alpha')[U'^{+}(\alpha')U*^{-}(-\alpha') - U'^{-}(\alpha')U*^{+}(-\alpha')]d\alpha' =$$

$$= -4i\pi(\beta * B'(-\alpha*) - \beta B*(\alpha)) = -4i\pi\beta*(B'(-\alpha*) - B(-\alpha*)) =$$

$$= -4i\pi\beta*\delta B(-\alpha*)$$

where the reciprocity relation (5) has been used. So, it is enough to evaluate the left hand member in order to obtain the expression of δB ; u' is zero on S' and we have that :

$$-4i\pi\beta*\delta B(-\alpha*) = - \int_{S'} u* \frac{\partial u'}{\partial n} dS' \quad .$$

This relation is rigorous. Now, to first order in δf :

$$\int_{S'} u* \frac{\partial u'}{\partial n'} dS' \simeq \int_{S} \frac{\partial u*}{\partial y} \cdot \partial f \cdot \frac{\partial u'}{\partial n'} dS \simeq \int_{-\infty}^{+\infty} \frac{\partial u}{\partial n} \frac{\partial u*}{\partial n} \delta f(x)dx$$

$$\delta B(-\alpha*) = \frac{1}{4i\pi\beta*} \int_{-\infty}^{+\infty} \frac{\partial u}{\partial n} \frac{\partial u*}{\partial n} \partial f(x)dx \quad .$$ (6)

For sake of simplicity, it has been supposed that δf(x) > 0, but the result is the same for any sufficiently regular δf : the demonstration may be achieved from the same lemma with a well chosen surface of integration.

The quantity $\frac{\partial u}{\partial n} \frac{\partial u*}{\partial n}$ is often called the "functional derivative" of B with respect to f, and is denoted by $\frac{\delta B}{\delta f}$. We have obtained a very compact and elegant expression for δB/δf : it involves only the normal derivatives of the fields u (initial direct problem, Fig. 1A) and u* (adjoint direct problem, Fig. 1B). This is a general feature of all the diffraction problems we have studied : the functional derivative of the diffraction pattern with respect to the profile of the scatterer may always be expressed very simply in terms of quantities related either to the initial direct problem or to an <u>adjoint direct problem</u>. When the source, in the initial problem, is a plane wave and the diffraction pattern is observed at infinity, then the adjoint problem corresponds to the same scatterer illuminated by a plane wave propagating in the direction of observation (Fig. 1B). Functional derivatives have already been calculated for 2-D and 3-D scatterers of any conductivity [3]. In the case of diffraction gratings, their expressions can be found in [4] [5].

The functional derivative of B with respect to f contains all the necessary information on the "local" relationship between B and f : in particular, it yields very simply the partial derivative of B with respect to any parameter linked with the profile. Suppose for instance that the profile f is restricted to a class of functions depending upon m parameters p_1, p_2, ... p_m and described by the known function F :

$$f(x) = F(x, p_1, p_2, \cdots p_m) \;.$$

A variation δf thus takes the form :

$$\delta f(x) = \sum_{i=1}^{m} \frac{\partial F}{\partial p_i} \, \partial p_i$$

and the corresponding variations δB is :

$$\delta B(\alpha^*) = \sum_{i=1}^{m} \delta p_i \int_{-\infty}^{+\infty} \frac{\partial u}{\partial n} \frac{\partial u^*}{\partial n} \frac{\partial F}{\partial p_i} \, dx$$

which means that :

$$\frac{\partial B}{\partial p_i} = \int_{-\infty}^{+\infty} \frac{\partial u}{\partial n} \frac{\partial u^*}{\partial n} \frac{\partial F}{\partial p_i} \, dx \;.$$

The functional derivative is simply the generalisation of the concept of gradient to the case of infinitely dimensional spaces. Eq.(6) has a straightforward simple consequence : when f = 0, $\partial u/\partial n$ and $\partial u^*/\partial n$ are analytically known since S reduces to a plane ; moreover, $B(\alpha')$ is zero when f = 0, and thus δB and δf may be confused with B and f. Eq.(6) becomes :

$$B(\alpha') = \frac{i\beta}{\pi} \int_{a}^{b} \exp[i(\alpha - \alpha')x] f(x) dx$$

which shows that the diffracted amplitude is simply the Fourier transform of the profile. <u>Of course, this relation holds only for shallow surfaces</u> !

4. CONJUGATE GRADIENT AND NEWTON METHODS

Eq.(6) is a linear relation, valid only to first order approximation in δf. However, it may be employed to construct <u>rigorous</u>
numerical algorithms capable to solve inverse scattering problems.
In order to show this fact, this subsection deals with the
following particular problem : given a diffracted amplitude $B_s(\alpha')$
$(\alpha_1 \leqslant \alpha' \leqslant \alpha_2)$, find the corresponding profile $f_s(x)$ $(a \leqslant x \leqslant b)$.

4.1. Newton-Kantorovitch algorithm

Let us start from a profile $f^{(0)}$ which, for example, may be a
rough estimate of the solution f_s. The corresponding amplitude $B^{(0)}$
may be easily calculated with the help of the computer code written
for the direct problem. We define :

$$\delta B_s^{(0)} = B_s - B^{(0)}$$

and let $\delta f^{(0)}$ be the solution of the integral equation :

$$\delta B_s^{(0)}(\alpha') = \int_a^b \frac{\partial u}{\partial n}(x) \, \frac{\partial u^*}{\partial n}(x,\alpha') \, \delta f^{(0)}(x) \, dx \qquad (7)$$

where $\partial u/\partial n$, $\partial u^*/\partial n$ are relative to the known profile $f^{(0)}$ and
thus are easily calculable with the help of the direct code.
Since the actual profile perturbation $f - f^{(0)}$ and $\delta B_s^{(0)}$ verify
approximatively eq.(6) - only approximatively because (6) is
exact to first order in $\delta f - \delta f^{(0)}$ is expected to be approximatively
equal to $f - f^{(0)}$. Thus the function $f^{(1)}$ defined as :

$$f^{(1)} = f^{(0)} + \delta f^{(0)}$$

should be a better approximation of f_s than $f^{(0)}$. Starting then
from $f^{(1)}$, the same calculation can be made and yields a new
approximation $f^{(2)}$, and so on. An iterative process is so generated,
which, when it converges, yields a numerically rigorous solution to
the inverse scattering problem.

   This algorithm is the analogue in terms of functions of the
well-known Newton Raphson algorithm often used to find the zeros
of a real function. It is called the "Newton Kantorovitch"
algorithm, because L.Y. Kantorovitch first exhibited conditions of
convergence [6]. The N.K. algorithm has been successfully applied
by the author to inverse scattering problems associated with
layered media [7], gratings [8] [9] and cylinders [10].

   It is worth noting that the resolution of the integral
equation of the first kind (7) is not a simple problem, because
the solution $\delta f^{(0)}$ is highly sensitive to rounding or experimental

errors on $\delta B_s^{(0)}$. It is necessary to make use of so-called "regularization techniques". This "instability" is highly signi-ficant and has a clear physical meaning.

4.2. Conjugate gradient algorithm

This type of algorithm is often used in Numerical Analysis. In the framework of optimization theory, the inverse problem is formulated as the minimization of a given real functional $J(f)$ called the "cost function". The choice of J expresses the desired features of the required solutions. In the particular problem defined above, one may choose :

$$J(f) = \int_{\alpha_1}^{\alpha_2} |B(\alpha') - B_s(\alpha')|^2 \, d\alpha' \tag{8}$$

the first point is to calculate the "gradient" of J, i.e. the functional derivatife $\partial J/\partial f$. The calculation is straighforward from eqs.(6) (8) :

$$\frac{\partial J}{\partial f} = 2 \, \text{Re} \, \{\int_{\alpha_1}^{\alpha_2} [\overline{B}(\alpha') - \overline{B}_s(\alpha')] \, \frac{\partial u}{\partial n}(x) \, \frac{\partial u^*}{\partial n}(-\alpha', x) \, d\alpha'\}$$

Then J may be easily minimized using a conjugate gradient method, such as the Fletcher-Powell algorithm. It is not the purpose of the present paper to describe this kind of methods precisely ; for more details see for instance [11]. The main advantage of this method is to be simpler to carry out than the N.K. algorithm, and to require less memory storage. However, its convergence is much slowier. Apparently no regularization should be needed. In fact, numerical computations have shown that the cost function (8) often leads to oscillatory reconstructed profiles, and this is not astonishing : the instability of the inverse problem is a fundamental feature and cannot be avoided so easily! It is necessary to re-introduce a kind of regularization by choosing another type of cost function, such as :

$$J(f) = \int_{\alpha_1}^{\alpha_2} |B(\alpha') - B_s(\alpha')|^2 \, d\alpha' + r \int_a^b \left(\frac{df}{dx}\right)^2 \, dx \, .$$

In this context, the physical significance of the regularizing term is not as clear as in the Newton algorithm.

5. INSTABILITY

As is now well known, the resolution of integral equations of the first kind as (7) constitutes an ill-posed problem in Hadamard's sense ; i.e. an arbitrarily small perturbation on B_s and thus on $\delta B_s^{(0)}$ may correspond to an arbitrarily large perturbation on $\delta f^{(0)}$.

Consider for instance the following variation :

$$\delta f^{(0)}(x) = A \sin \omega x \qquad a \leqslant x \leqslant b$$

it is easily seen that the corresponding variation $\delta B_s^{(0)}$ tends
towards zero when ω tends towards infinite, and thus the property
holds for sufficiently large values of ω and A. Note that in this
example, $\delta f^{(0)}$ oscillates very quickly ; in other words, high
frequency oscillations of the profile f have little influence on
the scattering pattern B, and the device acts as a low-pass filter.
This property is particularly clear for shallow surfaces : in this
case, the relation between B and f is simply a Fourier transform,
and the inverse problem reduces to the reconstruction of f knowing
its Fourier transform on a finite bandwith. For more details on
this classical problem, see for instance [12]. The restoration
of the stability may be achieved with the help of regularization
methods [13]. The problem of solving (7) is replaced by a
variational problem where physical constraints are incorporated,
and the solution is computed taking into account for instance
experimental errors on the data.

6. A FEW RESULTS

Figure 2 shows the reconstruction of a locally deformed surface.
The scattered amplitude is known for $-60° < \theta < 60°$, the angle
of incidence is $30°$, and $\lambda = 1$; line 1 is the true profile,
line 2 the initial estimation, line 3 the reconstructed profile.

    Figure 3 presents the reconstruction of the profile of a
manufactured grating with 1800 gr/mm. The efficiency curve in
the -1 order has been measured in the range $0.485~\mu < \lambda < 0.600~\mu$
and the algorithm described above has been used, in conjunction
with well-checked rigorous integral theories of gratings [14]
full line is the reconstructed profile from the efficiency in
TM polarization (magnetic field parallel to the grooves) and
the dotted one corresponds to the efficiency curve in T.E. polarization (electric field parallel to the grooves). Both of them
are in very good agreement with the profile found by the manu-
facturer with electron microscopy.

7. CONCLUSION

We have described a method capable of solving various inverse
problems numerically, which provides a theoretical explanation
of the fundamental instability of the inverse problems. This
method has proven to be successful in many problems of inverse
scattering in electromagnetics. Further developpements seem to be
straighforward, especially in the case of three dimensional

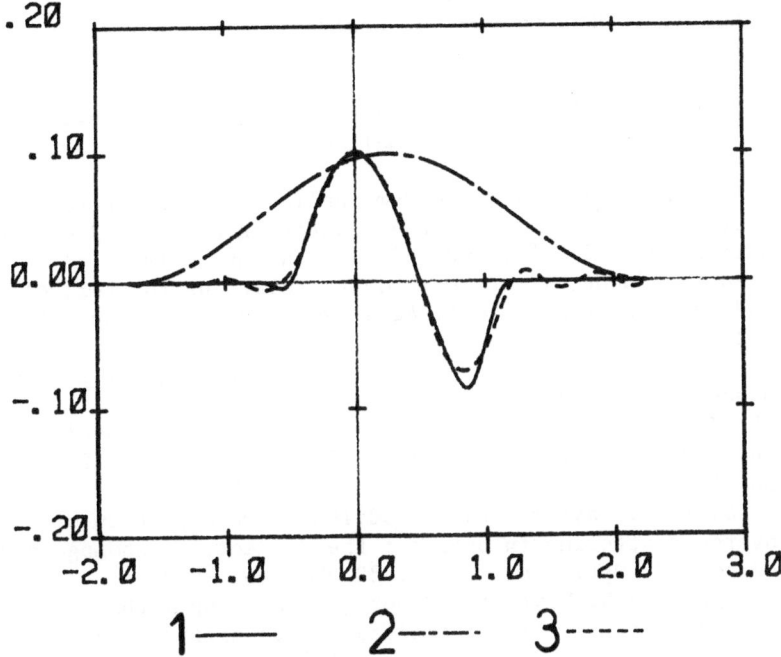

Fig. 2 : Reconstruction of the profile of a locally deformed surface from its scattered amplitude.

Fig. 3 : Reconstruction of the groove profile of a grating from actual measured values of its efficiency.

scatterers where the required functional derivatives have already been calculated [3]. So, the algorithm described here is general and convenient, and it could be of great help for the numerical solution of many inverse problems.

References

[1] D. Maystre, "Electromagnetic scattering from perfectly
     conducting rough surfaces in the resonance region", to be
     published in IEEE Trans. Antennas Propagat.
[2] D. Maystre, O. Mata Mendez and A. Roger, "A new electromagnetic
     theory for scattering from shallow rough surfaces", submitted
     to Opt. Acta.
[3] A. Roger, "Reciprocity theorem applied to the computation of
     functional derivatives of the scattering matrix", Electroma-
     gnetics, vol. 2, n° 1, pp. 69-83, Jan.-March 1982.
[4] A. Roger, "Generalised reciprocity relations for perfectly
     conducting gratings", Opt. Acta, vol. 29, n° 10, pp. 1427-1439,
     1982.
[5] A. Roger, "Generalised reciprocity for gratings of finite
     conductivity", to be published in Opt. Acta.
[6] L.Y. Kantorovitch, "On Newton's method for functional
     equations", Dokl. Akad. Nauk. SSSR, vol. 59, pp. 1237-1240,
     Jan. 1948.
[7] A. Roger, D. Maystre and M. Cadilhac, "On a problem of
     inverse scattering in optics : the dielectric inhomogeneous
     medium", J. Opt., vol. 9, pp. 83-90, Feb. 1978.
[8] A. Roger and D. Maystre, "Inverse scattering method in
     electromagnetic optics : application to diffraction gratings",
     J. Opt. Soc. Amer., vol. 70, pp. 1483-1495, dec. 1980.
[9] A. Roger and M. Breidne, "Grating profile reconstruction by
     an inverse scattering method", Opt. Commun., vol. 35,
     pp. 299-301, Dec. 1980.
[10] A. Roger, "Newton Kantorovitch Algorithm applied to an
     Electromagnetic Inverse Problem", IEEE Trans. Antennas
     Propagat., vol. 29, n° 2, pp. 232-238, March 1981.
[11] R. Fletcher and J.D. Powell, "A rapidly convergent descent
     method for minimization", The Computer Journal, vol. 6,
     pp. 163-168, 1963.
[12] M. Bertero, C. De Mol, and G. Viano, "Restoration of optical
     objects using regularization", Opt. Lett., vol. 3, pp. 51-53,
     1978.
[13] A.N. Tikhonov and V.Y. Arsenin, Solutions of Ill-Posed
     Problems. New York : Winston-Wiley, 1977.
[14] R. Petit, Ed., Electromagnetic Theory of Gratings, New York :
     Springer, 1980.

I.6 (IS.5)

DISSIPATIVE INVERSE PROBLEMS IN THE TIME DOMAIN

J.P. Corones*, M.E. Davison*, R.J. Krueger**

*Applied Mathematical Sciences, Ames Laboratory, Iowa
State University, Ames, IA 50011 USA

**Dept. of Mathematics, University of Nebraska,
Lincoln, NE 68588 USA

ABSTRACT. This work focuses on the time domain electromagnetic
inverse problem for a one dimensional inhomogeneous dissipative
medium. In practice, reflection data from a normally incident
plane wave is not sufficient for reconstructing both the permittivity and conductivity profiles. However, if one of these
profiles is known a-priori, then the other can be reconstructed.
A straightforward technique for doing this is shown. It is based
on a solution of the direct scattering problem which is derived
from a wave splitting approach coupled with an invariant imbedding
analysis of the scattering matrix. The technique is summarized
in an integro-differential equation for the kernel of the reflection operator. Numerical examples are given which study the
degradation of the reconstruction when the assumed profile differs
from the true profile.

1. PROBLEM FORMULATION

A new approach to time domain inverse problems is presented
in this paper. It is based upon a representation of the solution
of the corresponding direct scattering problem. Thus, the direct
problem will be formulated first.

Consider a one dimensional inhomogeneous dissipative medium
occupying the region $0 < z < L$ with a homogeneous nondissipative
medium outside of that region. A normally incident electromagnetic plane wave produces a field $E(z,t)$ in the medium such that

$$E_{zz} - c^{-2}(z)E_{tt} - b(z)E_t = 0 \tag{1}$$

121

W.-M. Boerner et al. (eds.), Inverse Methods in Electromagnetic Imaging - Part 1, 121–130.
© *1985 by D. Reidel Publishing Company.*

where

$$c^{-2}(z) = \varepsilon(z)\mu_0$$

$$b(z) = \sigma(z)\mu_0$$

and ε, σ, μ_0 are the permittivity, conductivity and (constant) magnetic permeability respectively. To simplify the analysis assume that c is continuous at z = 0 and z = L so that c(z) = c(0) for z < 0 and c(z) = c(L) for z > L and observe that b(z) = 0 for z < 0 or z > L. Finally, within the slab 0 < z < L assume c is c^1 and b is c^0. This model of a finite medium is used only for purposes of motivation. A semi-infinite model can also be considered, as will be obvious as the analysis proceeds.

A right moving incident plane wave in the region z < 0,

$$E^i(z,t) = f(t - z/c(0)),$$

produces reflected and transmitted fields,

$$E^r(z,t) = g(t+z/c(0)),$$

$$E^t(z,t) = h(t-z/c(L))$$

in the regions z < 0 and z > L, respectively. These fields are related by scattering operators for the slab

$$E^r(z,t) = [\tilde{R}^+(0,L)E^i(z;\cdot)](t)$$

$$E^t(z,t) = [\tilde{T}^+(0,L)E^i(z,\cdot)](t)$$

where the reflection operator is given by

$$[\tilde{R}^+(0,L)f](t) = \int_{-\infty}^{t} R^+(0,L,t-s)f(s)ds \tag{2}$$

and the transmission operator by

$$[T^+(0,L)f](t) = [\frac{c(L)}{c(0)}]^{\frac{1}{2}} \exp[-\tfrac{1}{2}\int_0^L b(s)c(s)ds]f(t-\Delta t(0,L))$$

$$+ \int_{-\infty}^{t-\Delta t(0,L)} T^+(0,L,t-s)f(s)ds, \tag{3}$$

and

$$\Delta t(0,L) = \int_0^L \frac{1}{c(s)} \, ds.$$
(4)

These expressions follow from [6] or [7]. The kernels $R^+(0,L,t)$ and $T^+(0,L,t)$ are dependent on the properties of the slab, but are independent of $E^i(z,t)$. The endpoints $z = 0$ and $z = L$ are explicitly displayed because in the analysis that follows, scattering operators for subregions of the original medium will be considered.

If a left moving plane wave in the region $z > L$ impinges on the slab, then a second pair of scattering operators, R^-, $\tilde{T}^-$ is used to determine the right going reflected and left going transmitted waves. Notice that because Eq. (1) is linear, it follows by superposition that the operators $R^\pm$, $T^\pm$ can be used to describe the outcome of any scattering experiment performed on the slab.

The definitions of the scattering operators presumes a concept of right and left moving waves. This gives no difficulty because the medium is homogeneous outside of the slab. Since the notion of a scattering operator is to be extended to subregions of the original slab, it is also necessary to extend the idea of right and left going waves to the case in which the properties of the medium are nonconstant. This is done via a splitting method.

2. SPLITTING METHOD FOR A HOMOGENEOUS MEDIUM

In order to motivate the general notion of a splitting, consider a homogeneous, nondissipative medium. The field satisfies the wave equation

$$E_{zz} - c_0^{-2} E_{tt} = 0$$
(5)

where c_0 is constant. Thus, E can be split into a sum of right and left moving waves,

$$E(z,t) = F(z-c_0 t) + G(z+c_0 t)$$
$$= E^+(z,t) + E^-(z,t).$$
(6)

Now consider the problem of rewriting Eq. (5) as a system

involving $E^{\pm}$. This is straightforward, but a general mechanism for doing it will be displayed. First, rewrite Eq. (5) as

$$\frac{\partial}{\partial z} \begin{pmatrix} E \\ E_z \end{pmatrix} = \begin{pmatrix} 0 & 1 \\ (\partial_t/c_0)^2 & 0 \end{pmatrix} \begin{pmatrix} E \\ E_z \end{pmatrix} \equiv D \begin{pmatrix} E \\ E_z \end{pmatrix} \quad (7)$$

and notice that Eq. (6) implies

$$\begin{pmatrix} E^+ \\ E^- \end{pmatrix} = \tfrac{1}{2} \begin{pmatrix} 1 & -c_0\partial_t^{-1} \\ 1 & c_0\partial_t^{-1} \end{pmatrix} \begin{pmatrix} E \\ E_z \end{pmatrix} \equiv T \begin{pmatrix} E \\ E_z \end{pmatrix} \quad (8)$$

Combining (7) and (8) it follows that

$$\begin{pmatrix} E^+ \\ E^- \end{pmatrix} = [-T\frac{\partial T}{\partial z}^{-1} + TDT^{-1}] \begin{pmatrix} E^+ \\ E^- \end{pmatrix} .$$

Since T^{-1} is independent of z, this becomes

$$\begin{pmatrix} E^+ \\ E^- \end{pmatrix} = \begin{pmatrix} -\partial_t/c_0 & 0 \\ 0 & \partial_t/c_0 \end{pmatrix} \begin{pmatrix} E^+ \\ E^- \end{pmatrix}$$

Notice that in this case, T diagonalizes D.

3.   SPLITTING METHOD FOR A NONHOMOGENEOUS DISSIPATIVE MEDIUM

Now consider Eq. (1) instead of Eq. (5). This can be written as

$$\begin{pmatrix} E \\ E_z \end{pmatrix} = \begin{pmatrix} 0 & 1 \\ (\partial_t/c)^2 + b\partial_t & 0 \end{pmatrix} \begin{pmatrix} E \\ E_z \end{pmatrix} \equiv D \begin{pmatrix} E \\ E_z \end{pmatrix} . \quad (9)$$

Using Eq. (8) as a guide, choose a splitting of $E(z,t)$ according to

$$\begin{pmatrix} E^+ \\ E^- \end{pmatrix} = \frac{1}{2} \begin{pmatrix} 1 & -c(z)\partial_t^{-1} \\ 1 & c(z)\partial_t^{-1} \end{pmatrix} \begin{pmatrix} E \\ E_z \end{pmatrix} \equiv T \begin{pmatrix} E \\ E_z \end{pmatrix} \quad (10)$$

Notice that this reduces to (8) in the region $z < 0$. Combining (9) and (10) it follows that

$$
\begin{pmatrix} E^+ \\ E^- \end{pmatrix} = \begin{pmatrix} -(\partial_t/c)-\tfrac{1}{2}(bc-c'/c) & -\tfrac{1}{2}(bc+c'/c) \\ \tfrac{1}{2}(bc-c'/c) & (\partial_t/c)+\tfrac{1}{2}(bc+c'/c) \end{pmatrix} \begin{pmatrix} E^+ \\ E^- \end{pmatrix}
$$

$$
= \begin{pmatrix} \tilde{\alpha}(z) & \tilde{\beta}(z) \\ \tilde{\gamma}(z) & \tilde{\delta}(z) \end{pmatrix} \begin{pmatrix} E^+ \\ E^- \end{pmatrix} \tag{11}
$$

4. SCATTERING MATRIX EQUATIONS

For arbitrary x and y consider the portion of the medium in the interval $x \le z \le y$, where $0 \le x, y \le L$. Assume that outside of this subinterval the medium is replaced with a nondissipative, homogeneous medium with $c(z) = c(x)$ for $z < x$, $c(z) = c(y)$ for $z > y$. Waves incident on this subinterval are described by $E^+(x,t)$ and $E^-(y,t)$ and the resulting scattered waves by $E^-(x,t)$ and $E^+(y,t)$. Hence, the scattering matrix $S(x,y)$ for the region $x \le z \le y$ takes the form

$$
\begin{pmatrix} E^+(y,\cdot) \\ E^-(x,\cdot) \end{pmatrix} = \begin{pmatrix} \tilde{T}^+(x,y) & \tilde{R}^-(x,y) \\ \tilde{R}^+(x,y) & \tilde{T}^-(x,y) \end{pmatrix} \begin{pmatrix} E^+(x,\cdot) \\ E^-(y,\cdot) \end{pmatrix}
$$

$$
= \tilde{S}(x,y) \begin{pmatrix} E^+(x,\cdot) \\ E^-(y,\cdot) \end{pmatrix} \tag{12}
$$

The elements of $\tilde{S}(x,y)$ are the scattering operators for the region $x \le z \le y$. They are defined as in (2) and (3) with x,y replacing 0,L.

A formal procedure for deriving differential equations for the scattering operators is now given. Evaluate (12) at t and differentiate with respect to x, and use (11) evaluated at x to obtain

$$
\frac{\partial \tilde{S}(x,y)}{\partial x} = - \begin{pmatrix} \tilde{T}^+(x,y) & 0 \\ \tilde{R}^+(x,y) & I \end{pmatrix} \begin{pmatrix} \tilde{\alpha}(x) & \tilde{\beta}(x) \\ -\tilde{\gamma}(x) & -\tilde{\delta}(x) \end{pmatrix} \begin{pmatrix} \tilde{I} & \tilde{0} \\ \tilde{R}^+(x,y) & \tilde{T}^-(x,y) \end{pmatrix}
$$

In particular, the equation for $\tilde{R}^+(x,y)$ is

$$
\frac{\partial \tilde{R}^+(x,y)}{\partial x} = \tilde{\gamma}(x) + \tilde{\delta}(x)R^+ - \tilde{R}^+\tilde{\alpha}(x) - \tilde{R}^+\tilde{\beta}(x)R^+ \tag{13}
$$

An alternate method for obtaining (13) is given in [5]. It uses

an invariant imbedding [1] approach to determine the behavior of
the scattering operators by appending a portion of the medium of
width Δx to the left hand edge of the region $x \leq z \leq y$ and computing the change in the scattering matrix; i.e.,

$$\lim_{\Delta x \to 0} \; - \frac{1}{\Delta x} \; [\tilde{S}(x-\Delta x,y) - \tilde{S}(x,y)].$$

This approach conveys the idea that the scattering matrix $\tilde{S}(0,L)$
for the physical medium is the *-product [8] of scattering matrices
for thin, contiguous subregions which can be thought of as comprising the original medium.

Equation (13) can be rewritten in terms of the kernel,
$R^+(x,y,t)$, of the convolution operator $R^+(x,y)$. This yields

$$\frac{\partial R^+(x,y,t)}{\partial x} - \frac{2}{c} \frac{\partial R^+(x,y,t)}{\partial t} = bcR^+ + \tfrac{1}{2}(bc+c'/c)\int_0^t R^+(x,y,t-s)R^+(x,y,s)ds,$$

$$0 < x < y, \quad t > 0 \tag{14}$$

where b,c and c' are evaluated at $z=x$. From this same calculation, it also follows that

$$R^+(x,y,0^+) = -\tfrac{1}{4}(bc^2-c')\Big|_{z=x} \tag{15}$$

$$R^+(y,y,t) = 0, \quad t > 0 \tag{16}$$

Equations for the remaining scattering kernels, $R^-(x,y,t)$,
$T^\pm(x,y,t)$, can be found in a similar way. They are not pertinent
to the present discussion and consequently are not written here.
The interested reader can readily obtain them from [2] or [5].

Notice that Eq. (14) and (15) are not dependent on the
finite thickness of the medium. Consequently, these same
equations apply to a semi-infinite ($L=\infty$) medium. It follows
from causality that

$$R^+(0,\infty,t) = R^+(0,y,t) \text{ for } 0 < t < 2\Delta t(0,y).$$

5. SOLUTION PROCEDURE

Both the direct and inverse scattering problem can be solved
via Eq. (14) - (16). Figure 1 illustrates the space-time region
in which these equations are to be considered.

In the direct problem, c(z) and b(z) are given for
$0 < z < L$ and $R^+(0,L,t)$ is to be found for $t > 0$. Thus, set
$y = L$ in Eq. (14) - (16). The integro-differential equation (14)

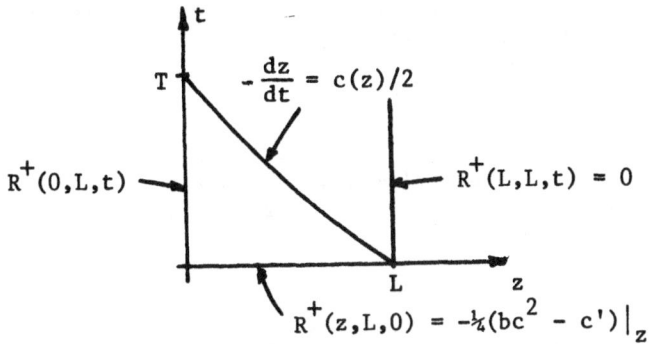

Fig. 1.   Initial and boundary values for $R^+(z,L,t)$.

then propagates the initial data (15) and the boundary data (16) to the line z = 0, thus solving the direct problem.

In the inverse problem $R^+(0,L,t)$ is known for $0 < t < T$ where T is one round trip travel time through the medium. In the situation considered here, either $c(z)$ or $b(z)$ is also assumed to be known, and the problem is to determine the remaining function. To do this, Eq. (14) is used to propagate the boundary data $R^+(0,L,t)$ onto the line $t = 0$ at which point the initial condition (15) and known function $b(z)$ (or $c(z)$) are used to determine $c(z)$ (or $b(z)$).

Details regarding the numerical implementation of these procedures are given in [3].

6.   EXAMPLES

The following examples were generated using the inversion procedure described above. In these examples, it is assumed that the relative permittivity ε_r or conductivity σ is only known approximately. Using an approximation σ_A of the true σ in the inversion procedure results in a reconstructed permittivity ε_r^* which differs from the true permittivity ε_r, with the accuracy of the reconstruction generally decreasing as the accuracy of σ_A decreases. Roughly speaking, a trade-off is occurring between permittivity gradients and energy dissipation.

If σ_A is set equal to the true σ in the inversion algorithm, then ε_r^* is found to be indistinguishable from the true ε_r in the examples that follow. In fact, the plots of $\varepsilon_{r,TRUE}$ shown below were obtained by setting $\sigma_A = \sigma_{TRUE}$ and carrying out the inversion (and conversly for the plot of σ_{TRUE}).

Some preliminary studies on the influence of noisy data in the inversion process are given in [4].

Example 1: Consider a medium of length 10 m. with

$$\varepsilon_r(z) = \begin{cases} 50 & , \ z < .25 \\ 75 + 25 \ \sin(2(z-.5)), & .25 < z < .75 \\ 100 & , \ .z > .75, \end{cases}$$

and σ (in mho/m) given by

$$\sigma(z) = \begin{cases} .02 & , \ z < .25 \\ .02 - .039 \ (z-.25) & , \ .25 < z < .75 \\ .0005 & , \ .z > .75. \end{cases}$$

Figure 2 shows ε_r^* for various approximations σ_A of σ. (The data $R^+(0,L,t)$ is the same for all cases.) Notice that over- and underestimating σ in the thin region of relatively high conductivity has considerable influence on ε_r^*. This in turn affects the apparent depth of the medium.

Example 2: Again assume L = 10 m. Let ε_r be given by

$$\varepsilon_r(z) = 50 + kz(10-z), \quad 0 < z < 10 \tag{17}$$

with k=6 and let σ be constant, $\sigma \equiv 0.01$ mho/m. Figure 3 shows ε_r^* for several σ_A's. Observe that even a 2% error in the approximation of σ can produce significant errors in the permittivity profile.

Using the same scattering data, reconstructed conductivity profiles are shown in Fig. 4. The assumed permittivity profiles are given by Eq. (17) for various values of k.

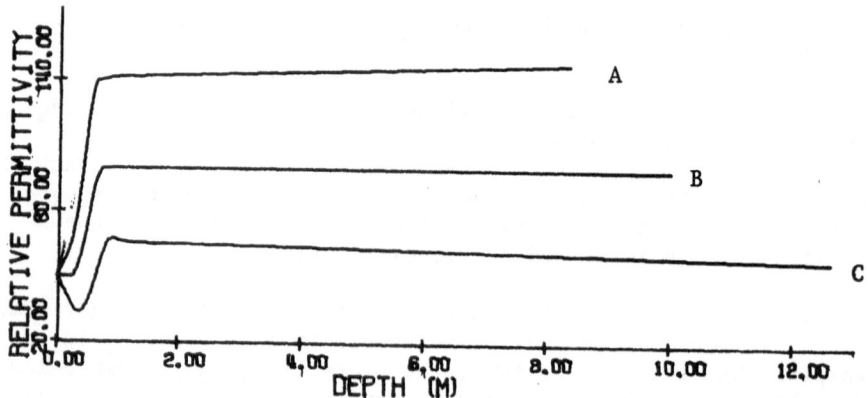

Fig. 2. Reconstructed permittivity profiles. A. σ_A = 0.01 in the thin region. B. $\sigma_A = \sigma_{TRUE}$ C. σ_A = 0.03 in the thin region.

Fig. 3. Reconstructed permittivity profiles for example 2.
A. σ_A = 0.009   B. σ_A = 0.0098   C. $\sigma_A = \sigma_{TRUE}$ = 0.01
D. σ_A = 0.0102.

Fig. 4. Reconstructed conductivity profiles for example 2.
A. k = 5.5   B. k = k_{TRUE} C. k = 10.0.

ACKNOWLEDGEMENTS

This work was done at the Ames Laboratory of the U.S.D.O.E. and the U. of Nebraska-Lincoln, and was supported in part by the Office of Naval Research Contract N0014-83-K-0038.

REFERENCES

1. Bellman, R. and Wing, G.M., "An introduction to invariant imbedding," Wiley 1975.

2. Corones, J. and Krueger, R., Obtaining scattering kernels using invariant imbedding, J. Math. Anal. Appl., Vol. 95, pp. 393-415. 1983.

3. Corones, J., Davison, M. and Krueger, R., "Effects of dissipation in one-dimensional inverse problems", Inverse Optics, A.J. Devaney, Editor, Proc. SPIE 413, pp. 107-114. 1983.

4. Corones, J., Davison, M. and Krueger, R., Direct and inverse scattering in the time domain via invariant imbedding equations, J. Acoustical Soc. Am., to appear.

5. Davison, M., A general approach to splitting and invariant imbedding for linear wave equations. Ames Laboratory Preprint.

6. Davison, M., Bremmer series for scattering operators associated with linear wave equations. Ames Laboratory Preprint.

7. Krueger, R. J., An inverse problem for a dissipative hyperbolic equation with discontinuous coefficients, Quarterly of Applied Mathematics, Vol. 34, pp. 129-147. 1976.

8. Redheffer, R., "On the relation of transmission-line theory to scattering and transfer," J. Math. and Phys. Vol 41, pp. 1-41. 1962.

I.7 (IM.3)

APPLICATIONS OF ALMOST-PERIODIC FUNCTIONS
TO INVERSE SCATTERING THEORY

A.K. Jordan*, S. Ahn* and D.L. Jaggard**
*Naval Research Laboratory
Washington, DC 20375

**University of Pennsylvania
Philadelphia, PA 19104

ABSTRACT. Inverse scattering theory is used to reconstruct
dielectric profiles that are almost-periodic functions. If
the scattering data can be represented as a rational
function of the wave number, then it is possible to obtain
solutions to the Gelfand-Levitan integral equation of
inverse scattering theory that will be, in general,
almost-periodic functions. The dielectric profile function
will also be an almost-periodic function, for which the
number and strengths of the almost-periodic "tones" can be
related to the pole-zero configuration of the scattering
data. An example is considered where the complex reflection
coefficient r(k) is represented as a rational function of
the wave number k with four poles and two zeros.

1.   INTRODUCTION

Inverse scattering theory includes the many different
analytical methods that can be used to interpret
electromagnetic and optical wave scattering data. Despite
their diversity of applications, these methods are related
by their basic theoretical structures. The mathematical
methods used are typically rigorous and "exact" for simple
and idealized physical models; these theories are useful for
providing basic understanding of physical phenomena. It is
possible to obtain knowledge about a class of complicated
systems by extending an "exact" inversion method to include
systems that can be represented by almost-periodic
functions. These functions have been used in direct
scattering theories for inhomogeneous dielectric media,
integrated optical multilayers, and heterogeneous dielectric media,
that exhibit stochastic properties.

131

W.-M. Boerner et al. (eds.), Inverse Methods in Electromagnetic Imaging – Part 1, 131–140.
© 1985 by D. Reidel Publishing Company.

As Sabatier [1,2] notes, the word "exact" refers to the fact that the inversion methods to be used are, in principle, not approximations. A one-dimensional physical model is used here, so that the inverse problem is properly termed "profile inversion"; this idealization will allow us to focus on the mathematical structure of the inverse scattering problem and its solution.

The direct scattering problem can be studied formally as a transformation or mapping M between a set C of theoretical parameters and a set E of experimental variables [1,2]. The inverse scattering problem can be studied in terms of the mapping M^{-1} between E and C; this study usually begins by considering a subset E' and its mapping into a subset C'.

In the physical model of Fig. 1, we have assumed that the plane-polarized electromagnetic field is normally incident on a semi-infinite inhomogeneous region whose permittivity $\varepsilon(k,x)$ is assumed to obey the dispersion relation

$$\varepsilon(k,x) = \begin{cases} \varepsilon_0 (1 - \frac{1}{k^2} q(x)), x \geq 0 \\ \varepsilon_0 \end{cases}$$

where ε_0 is the permittivity of free space.
Here the subset E' consists of the reflection coefficient r(k) which is an analytic function of the wavenumber k; the subset C' consists of q(x), the profile function.

The inverse problem is to determine q(x) from r(k); the mapping M^{-1} is the "exact" method developed by the Soviet mathematicians Gel'fand, Levitan and Marchenko, [8,9] circa 1950, for use in quantum mechanical problems. Subsequent work extended the availability of these results and expanded the theory to include the electromagnetic problem. Of particular interest is the treatment by Kay [10] in which several examples display the usefulness of the Gel'fand-Levitan-Marchenko theory. Much of the initial work [10,11,12] emphasized exact, closed-form solutions to the problem when the scattered data were represented by rational functions. Numerical and approximate [13,15] solutions provided additional understanding of this problem.

2. EXTENSION OF C' TO INCLUDE ALMOST-PERIODIC FUNCTIONS

We now explore the feasibility of extending the subset C' to include "almost periodic functions". For our purpose, we will consider almost periodic functions as defined by Bohr [3] and Besicovitch [4], that is, as functions which possess a discrete or line spectrum. This implies that such a function f(x) can be written in a generalized Fourier series

$$f(x) = \sum_n a_n e^{ik_n x}, \tag{1}$$

where k_n is the wavenumber or spatial frequency of the nth harmonic and a_n is its amplitude. (Often we will refer to k_n as the tone frequency and a_n as the tone strength.) This equation can be inverted to produce the harmonic amplitudes

$$a_n = \lim_{X \to \infty} \int_{-X}^{X} f(x) e^{-ik_n x} dx \tag{2}$$

These functions have been applied to electromagnetic wave propagation in inhomogeneous regions by Jaggard [5,6,7], where the resemblance to random wave propagation is apparent. Furthermore, almost-periodic functions form a large class of examples of entire functions of exponential type [14]. Clearly, (1) represents a periodic function if only one tone is present or if all of the tone frequencies are commensurable. In this case (1) and (2) become the usual expressions for the Fourier series and Fourier inversion formula, respectively. Although almost periodic functions do not possess a period, as noted above, they do possess an almost period or translation number τ such that

$$|f(x + \tau) - f(x)| < \sigma , \tag{3}$$

where $\tau = \tau(\sigma)$. This relation provides an alternative definition to the more restrictive one expressed by eq. (1). Another property of interest is the Parseval relation expressed as

$$\lim_{X \to \infty} \frac{1}{2X} \int_{-X}^{X} |f(x)|^2 dx = \sum_n |a_n|^2 . \tag{4}$$

These and other characteristics can be shown from the defining equations (1) and (2) and can be found in the literature [3,4].

If the potential function $q(x)$ is almost periodic, an approximate theory can be developed [4,5,6] for the reflection coefficient. For example, the three-tone case described by

$$q(x) = \begin{cases} -\dfrac{\kappa^2 \eta}{4}\left[\dfrac{m}{2}\cos\{(\kappa+\Delta)x\} + \cos\{\kappa x\} + \dfrac{m}{2}\cos\{(\kappa-\Delta)x\}\right], \\ \qquad\qquad\qquad\qquad\qquad\qquad\qquad\qquad 0 \le x \le \ell \\ \\ 0 , \; x < 0 , \; x > \ell \end{cases} \tag{5}$$

This function represents a modulated periodic structure with carrier wavenumber κ and side-band wavenumbers $\kappa \pm \Delta$; clearly Δ is the tone spacing. Here η is the relative amplitude of the carrier and m is the modulation index. By assuming an appropriate form for the electric field amplitude,

$$u(k,x) = F_o(x)e^{i\frac{\kappa}{2}x} + \underline{B}(x)e^{-i(\frac{\kappa}{2}-\Delta)x} + B_o(x)e^{-i\frac{\kappa}{2}x} + B_+(x)e^{-i(\frac{\kappa}{2}+\Delta)x} \tag{6}$$

in the vicinity of the primary Bragg resonance defined by

$$k = \kappa/2 \tag{7}$$

it is possible to obtain a Riccati differential equation for the reflection matrix $\underline{\underline{R}}$:

$$\underline{\underline{R}}' = i\underline{\chi} + i\delta_{\underline{\underline{F}}} \cdot \underline{\underline{R}} + i\underline{\underline{R}} \cdot \delta_{\underline{\underline{B}}} + i(\underline{\underline{R}}\underline{\chi})^\dagger\underline{\underline{R}}, \tag{8}$$

where the dagger indicates Hermitean transpose and the initial condition is

$$\underline{\underline{R}}(0) = [\ 0\ \ 0\ \ 0\ ]. \tag{9}$$

This formulation is amenable to numerical solution and can be used for any number of tones. A result for the three tone case corresponding to (5) is shown in Fig. 2. Here the scalar reflection is given by

$$\tilde{R} = \sqrt{\underline{\underline{R}}\ \underline{\underline{R}}^\dagger} \tag{10}$$

and is plotted for a single value of normalized coupling

$$\chi\ell = \sqrt{(\chi_o\ell)^2 + (\chi_+ + \chi_-)^2\ell^2} = 2 \tag{11}$$

as a function of normalized frequency $\tilde{f} = 2\Delta\kappa/\chi$.   Note

particularly the large peak at the primary Bragg resonance defined by eq. (7), or $\tilde{f} = 0$, and the secondary Bragg resonances at

$$k = (\kappa \pm \Delta)/2 \ . \tag{12}$$

The tone spacings are $\Delta = 0, 2\chi_c, 4\chi_c, 6\chi_c, 8\chi_c$

(top to bottom) where χ_c is some critical coupling and the loss is $L\ell$ with m = 1.

2.1 Example: Widely Spaced Tones

For extremely widely spaced tones ($\Delta \gg \kappa$) in which the tones act independently, the results of scalar coupled mode theory can be used in the vicinity of each Bragg resonance. For example, if an N tone medium is defined by

$$q(x) = \begin{cases} -\frac{1}{4}\sum_{n=1}^{N} n_i \kappa_i^2 \cos\kappa_i x, & 0 \le x \le \ell \\ \\ 0, & x < 0, \ x > \ell \end{cases} \tag{13}$$

where $|\kappa_i - \kappa_j| << \kappa_i$ or κ_j; $i,j = 1,2,\ldots N$,
then the scalar reflection $\tilde{R}_i$ in the vicinity of $k = \kappa_i/2$ is given
by first order coupled mode theory as

$$\tilde{R}_i \approx \frac{i\kappa_i \ell}{(\chi_i^2 - \delta_i^2)\ell \coth \sqrt{(\chi_i^2 - \delta_i^2)}\ell - i\delta_i \ell} , \tag{14}$$

where

$$\chi_i = n_i \kappa_i / 8 \tag{15}$$

$$\delta_i = k - \kappa_i / 2 . \tag{16}$$

The peak value of these primary resonances is

$$\left|\tilde{R}_i\right|_{max}^{pri} = \tanh(\chi_i \ell) , \tag{17}$$

which can be inverted to get all the $\chi_i \ell$.   The placement of
the peak gives the values of the κ_i.

The criterion for the applicability of this method is that the
tones are separated sufficiently so that they act independently.
This occurs when the bandgaps due to each periodicity are spaced in
frequency so that they do not coalesce. This condition is
described by the relative balance of the periodicity amplitudes
with the wavenumbers as expressed by

$$\left|n_i\right|^2 << 16 \left|\kappa_i - \kappa_j\right|; i,j = 1,2,\ldots N, i \neq j . \tag{18}$$

2.2 Example: Closely Spaced Tones

If the tones are closely spaced so that condition (18) does not
hold, then the simplified expression (14) for the reflection
coefficient cannot be used. The traditional approach of iterating
between the direct problem, via the matrix Riccati equation, and
the experimental data leads to formidable algebra and a problem of
physical interpretation. Alternatively, the inverse problem for
closely spaced tones can be approached by observing that the graph
of the scalar reflection $\tilde{R}$ for reflection from an almost-periodic
medium with three closely spaced tones (Fig. 2, second plot, $\Delta = 2\chi_c$
$L\ell = 0.4$) can be approximated by a reflection coefficient that is
a rational function of the wave number k,

$$r(k) = r_o \frac{(k-\mu_1)(k-\mu_2)\ldots}{(k-k_1)(k-k_2)\ldots} , \qquad (19)$$

where k_1, k_2,...are the poles and μ_1, μ_2 ,...are the zeros of $r(k)$ in the complex k-plane and r_o is a normalization constant.

Scattering data that can be represented by rational reflection coefficients belong to a class of inverse problems solvable by "exact" methods that demonstrate the mathematical structure of inverse scattering theory.

3. "EXACT" METHODS FOR INVERSE PROBLEMS

A physical description of the inverse scattering procedure [10] is obtained by using the time-dependent differential equation

$$\frac{\partial^2}{\partial x^2} U(x,t) - \frac{\partial^2}{\partial t^2} U(x,t) = q(x) U(x,t). \qquad (20)$$

The incident plane-wave impulse, $U_{inc} = \delta(x - t)$, produces a reflected transient

$$R(x+t) = \frac{1}{2\pi} \int_{-\infty}^{\infty} r(k) e^{-ik(x+t)} dk - i \sum_{\nu} r_\nu e^{-ik_\nu(x+t)} , \qquad (21)$$

where the r_ν in the second term are the residues at the poles, k_ν if any, of $r(k)$ on the positive imaginary k-axis. It is possible to relate the wave amplitude, $U(x,t)$, in the inhomogeneous region, $x \geq 0$, with the wave amplitude in the free space region, by the transformation

$$U(x,t) = U_o(x,t) + \int_{-\infty}^{x} K(x,z) U(z,t) dz, \qquad (22)$$

where

$$U_o(x,t) = \delta(x-t) + R(x+t). \qquad (23)$$

Substituting the expression for the field in the free-space, $U_o(x,t)$, gives the integral equation

$$R(x+t) + K(x,t) + \int_{-t}^{x} K(x,z) R(z+t) dz = 0, \qquad (24)$$

which is Kay's version of the Gel'fand-Levitan-Marchenko integral equation and is to be solved for the function $K(x,t)$. Conditions on $K(x,t)$ are found by substituting the expression (22) for $U(x,t)$

in the differential equation (20); the function K(x,t) will satisfy
the same differential equation as U(x,t), subject to the conditions

$$K(x, -x) = 0 \tag{25}$$

and

$$\frac{d}{dx} K(x,x) = \frac{1}{2} q(x), \; x \geq 0. \tag{26}$$

From condition (26) we see that the inverse scattering problem is
solved for q(x) if we can solve the integral equation (24).
Several exact and approximate methods of solving (24) have been
reviewed by Jordan [12]; the most important for obtaining
closed-form solutions useful for engineering applications is that
of Kay [10], where r(k) is a rational function of the wavenumber k,
as represented by eq. (17). Examples of these solutions have been
given by Jordan et al. [11,12] for several pole-zero
configurations.

The band-pass characteristics of the almost-periodic reflection
coefficient in the vicinity of the central Bragg resonance suggests
that poles on the unit circle ("Butterworth poles") could be used
as a first approximation. The number of minimum of $|r(k)|^2$ will be
determined by the number of zeros while the characteristics of the
peaks are determined by the pole locations. The energy-conserva-
tion condition

$$|r(k)|^2 \leq 1 \tag{27}$$

determines the "allowed regions" for the zeros and poles of r(k) in
the k-plane.

We note that the R(x) defined by eq. (22) will be, in general,
an almost-periodic function as defined by Bohr in equation (3),
when r(k) is a rational function of k. Thus that K(x,y) in the
Gelfand-Levitan integral equation (24), and hence q(x) will also be
an almost-periodic function, due to the multiplication theorem for
almost-periodic functions [3].

This method for solving the integral eq. (22) will be
demonstrated by using our example for closely-spaced tones. The
scalar reflection coefficient can be approximated by a rational
reflection coefficient

$$\tilde{R} \simeq r(k) = r_B(k) \cdot r_\delta(k), \tag{28}$$

where

$$r_B(k) = \frac{-k_1 k_2 k_3 k_4}{(k-k_1)(k-k_2)(k-k_3)(k-k_4)},$$

and

$$r_\delta(k) = \frac{(k-\mu_1)(k-\mu_2)}{\mu_1\mu_2}$$

The pole-zero configuration is chosen so that

$$k_2 = -\bar{k}_1 = c_1 - ic_2,$$
$$k_4 = -\bar{k}_3 = c_3 - ic_4,$$
$$\mu_2 = -\bar{\mu}_1 = c_3 - i\delta c_4, 0 < \delta \le 1.$$

The fourth-order Butterworth poles [11] are given by

$$c_1 = \cos(3\pi/8), \quad c_2 = \sin(3\pi/8),$$
$$c_3 = \cos(\pi/8), \quad c_4 = \sin(\pi/8).$$

The reflection coefficient $|r(k)|$ is shown in Fig. 3. The solution of the Gelfand-Levitan integral equation proceeds as outlined in [11,12]; it is interesting to note that the characteristic function $R(x)$, eq. (21), is related to the corresponding $R_B(x)$, for the Butterworth approximation with no zeros, by

$$R(x) = R_B(x) + \frac{1}{c_3^2 + \delta^2 c_4^2}\left[R''_B(x) - 2\delta\, c_4 R'_B(x)\right] \tag{29}$$

The function $K(x,y)$ has the form

$$K(x,y) = \sum_{n=1}^{4}\left[C_n(x)e^{if_n y} + D_n(x)e^{-if_n y}\right], \tag{30}$$

where the f_n are functions of $R(x)|_{x=0}$ and its derivatives and C_n, D_n are found from the boundary conditions on $K(x,y)$ and its derivatives. The graph of $q(x)$ found from eq. (26) is shown in Fig. 4.

4. DISCUSSION

It has been demonstrated that when the scattering data are represented as a rational function of the wave number, then it is possible to obtain solutions to the Gelfand-Levitan integral equation of inverse scattering theory that are almost-periodic functions. The profile of the equivalent dielectric constant for inhomogeneous media can be reconstructed from the scattering data. The use of zeros in addition to poles in the analytic representation of the data provides additional degrees of freedom for the reconstruction procedure. This method may be useful to analyze heterogeneous systems, such as polydisperse distributions of dielectric particles.

Fig. 1. Idealized physical model.

Fig. 3. Reflection coefficient $|r(k)|$ for example of eq. (28).

Fig. 2. Reflection coefficients $\tilde{R}$ for three-tone medium (from {7}).
——— $L\ell = 0$
---- $L\ell = 0.2$
——— $L\ell = 0.4$

Fig. 4. Profile function recon-structed from Fig. 3.

5.   ACKNOWLEDGEMENTS

The authors acknowledge the support of the University of Pennsylvania through the Committee on Faculty Grants and Awards and of the Office of Naval Research under the 6.1 Core Research Program.

6.   REFERENCES

1.   Sabatier, P.C., "Theoretical Considerations for Inverse Scattering", Radio Science 18, pp. 1-18 (1983)
2.   Sabatier, P.C., "Introduction to Applied Inverse Problems", in Applied Inverse Problems", ed. by P.C. Sabatier, Springer-Verlag, Lecture Notes in Physics No. 85 (1978).
3.   Bohr, H., Almost Periodic Functions, H. Cohn and F. Steinhardt, Trans., Chelsea, New York (1974).
4.   Besicovitch, A., Almost Periodic Functions, Cambridge Univ., London (1932).
5.   Michelson, A. and D.L. Jaggard, "Electromagnetic Wave Propagation in Almost Periodic Media", IEEE Trans. Ant. and Prop. AP-27, pp. 34-40 (1979).
6.   Jaggard, D.L. and A. Michelson, "The Reflection of Electromagnetic Waves from Almost Periodic Structures", Appl. Phys. 19, 405-412 (1979).
7.   Jaggard, D.L., and G.T. Warhola, "Characteristics of Modulated Periodic Media: I-Passive Structures", Radio Sci. 16, pp. 467-474 (1981); "II-Active Structures", pp. 475-480 (1981).
8.   Gel'fand, I.M. and B.M. Levitan, "On the Determination of a Differential Equation by Its Spectral Function", Am. Math.Soc. Trans. 1, pp. 253-304 (1955).
9.   Marchenko, V.A., "Reconstruction of the Potential Energy from the Phase of Scattered Waves", Dokl, Akad. Nauk. SSR 104, pp. 695-698 (1955).
10.   Kay, I., "The Inverse Scattering Problem When the Reflection Coefficient is a Rational Function", Comm. Pure Appl. Math. 13, pp, 371-393 (1960).
11.   Jordan, A.K. and H.N. Kritikos, "An Application of One-Dimensional Inverse Scattering Theory for Inhomogeneous Regions", IEEE Trans. Ant. and Prop. AP-21, pp. 209-211 (1973)
12.   Jordan, A.K. and S. Ahn, "Inverse Scattering Theory and Profile Reconstruction", IEE Proc. (London) 126, pp. 945-950 (1979) or "Inverse Scattering Theory; Exact and Approximate Solutions", in Mathematical Methods ana Applications of Scattering Theory, ed. by J. DeSanto, A. Saenz, W. Zachary, Springer-Verlag (1980).
13.   Kritikos, H.N., D.L. Jagard and D.B. Ge, "Numerical Reconstructions of Smooth Dielectric Profiles", Proc. IEEE 70, pp. 295-297 (1982).
14.   Ross, G., M.A. Fiddy, and M. Nieto-Vesperinas, "The Inverse Scattering Problems in Structural Determinations", in Inverse Scattering Problems in Optics, ed. by H.P. Baltes, Springer-Verlag, 1980.

I.8 (TS.3)

ONE DIMENSIONAL TIME-DOMAIN INVERSE SCATTERING APPLICABLE TO
ELECTROMAGNETIC IMAGING

HANS BLOK and ANTON G. TIJHUIS

Delft University of Technology
Department of Electrical Engineering
P.O. Box 5031, 2600 GA Delft
The Netherlands

Abstract

Various inverse-scattering methods applied to the electromagnetic
imaging of the permittivity or the conductivity profile of an
inaccessible layered medium are reviewed. Special emphasis is
placed upon the time-domain analysis. The inaccessible region is
illuminated by a pulsed plane wave at normal incidence. The problem of estimating the profile from this set of data is formulated
in terms of a source-type integral equation for the electric-
-field strength in the layered medium, e.g. a slab. In the direct-scattering problem, space-time discretization leads to a
step-by-step updating procedure; alternatively the Singularity
Expansion Method is applied. For the solution of the inverse-
-scattering problem various methods will be discussed: a direct
method, an iterative method and methods based on the iterative
minimization of an appropriately chosen integrated square error.

1. INTRODUCTION

In this paper, we consider the scattering by a one-dimensional,
inhomogeneous, lossy dielectric slab (Fig. 1.) of thickness d,
embedded in vacuum. Normally incident on the slab is a given,
pulsed, E-polarized electromagnetic wave E^i of finite duration.
The case of oblique incidence is a straightforward extension. The
electric field strength can be written as $\underline{E}(\underline{r},t) = E(z,t)\underline{i}_y$ and
the magnetic field strength as $\underline{H}(\underline{r},t) = H(\overline{z,t})\underline{i}_x$. The investigation into this configuration is subdivided into two parts: (i)
solution of the direct-scattering problem, i.e. the computation
of the total electromagnetic field for given constitutive para-

141

W.-M. Boerner et al. (eds.), Inverse Methods in Electromagnetic Imaging - Part 1, 141–156.
© *1985 by D. Reidel Publishing Company.*

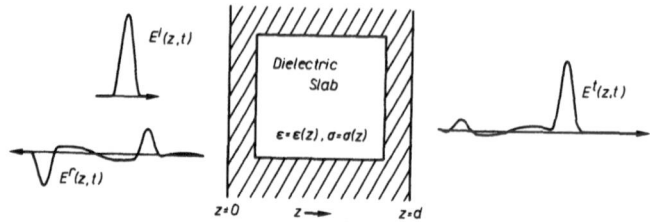

Fig. 1. Reflected and transmitted field corresponding to
a pulse incident on an inhomogeneous, lossy
dielectric slab.

meters; (ii) solution of the inverse-scattering problem, i.e. the
identification of the slab from known values of the incident and
the scattered electromagnetic field. In Fig. 2 we show an outline
of the various methods that we have applied to solve these problems. Hereafter, the time-domain case, as well as the application
of the Singularity Expansion Method (S.E.M.), will be discussed
briefly. Whenever possible, we will also show some illustrative
numerical results. The first part of this paper is devoted to the
direct-scattering problem. In Section 2, we show how, for the
time-domain case, the total field can be computed by marching-on-
-in-time [1, 2]. When the reflected field is known, e.g. in an
inverse-scattering solu-
tion problem, an alterna-
tive scheme is proposed.
There the total field can
be computed by marching-
-on-in-space. In section
3, we describe a different
way of solving the time-
-domain problem. In this
so-called Singularity
Expansion Method [3, 4, 5,
6, 7] the field quantities
are expressed in terms of
the natural modes of the
slab, each corresponding
to a complex natural fre-
quency. These natural
modes are dependent on the
slab characteristics only.
To this end, the time-
-domain electric-field
strength is written as a
Laplace contour integral.
If the contour is deformed
properly in the complex
plane, only residual con-

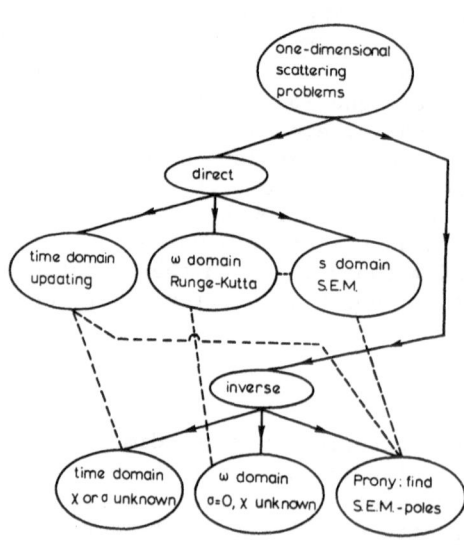

Fig. 2. Outline of the methods that
were applied to solve the
one-dimensional direct or
inverse-scattering problem.

tributions need to be taken into account. At each point in space, the total electric field can then from an initial instant be written as a series of damped exponentials with complex arguments being exactly the natural frequencies.

The second part of the paper is devoted to the inverse-scattering problem. In Section 4, we discuss the reconstruction of either the permittivity or the conductivity profile from the known re- flected field at the front of the slab. This problem can be posed in various ways: (i) the classical time-domain approach where the scattered data are used to construct the kernel and forcing term for a *Gelfand-Levitan type of integral equation*. The solution of this integral equation is then used to generate the permittivity and conductivity profiles [8, 9, 10]. This approach will not be discussed here. For an extensive review the reader is referred to [11].

(ii) An alternative approach in the time-domain where the direct- and inverse-scattering problem is formulated in terms of a *source-type integral equation*. This integral equation is subse- quently discretized in the space-time domain. For the *direct- -scattering problem* this leads to an infinite system of algebraic equations from which the total field can be solved. By virtue of the choice of equally large steps in space and time, these equa- tions can be solved in a *step-by-step updating procedure* (march- ing on in time or, when the reflected field is known, marching on in space). For the solution of the inverse-scattering problem various methods will be discussed. (a) The *direct approach* which involves a step-by-step determination of the unknown profile by accuratedly tracing the wave front of the field [12, 13, 14].

(b) The *iterative approach* which basically consists of alternate- ly solving equations for an approximate direct-scattering problem and an approximate inverse-scattering problem yielding successive approximations for the electric field in the slab and the unknown constitutive coefficient [15, 16].

(c) The *optimization approach* where the conductivity or suscepti- bility profiles are reconstructed by iterative minimization of an appropriatedly chosen integrated square error. In this review two different optimization schemes will be discussed: the *separate- optimization scheme*, where the square error depends on both the approximated profile and the approximated field and the *direct- -optimization scheme*, where the square error is introduced as a function of the unknown profile only [17, 18].

Finally, in Section 5, we describe an alternative way of *identi- fying* the layered structure, namely the decomposition of the field into a set of damped exponentials with a Prony-type algo- rithm. According to the SEM-theory, the corresponding natural frequencies depend on the scatterer geometry only and can there- fore be used for identification purposes [19].

2. DIRECT SCATTERING: TIME-DOMAIN UPDATING AND SIDE DATING

Consider the direct-scattering problem for the case of an incident field $E^i(z,t)$ with arbitrary time-dependence. With the time-
-domain Green's function technique, the following integral-equa-
tion for $E(z,t)$ can be derived

$$E(z,t) = E^i(z,t)-(Z_0/2)\int_0^d [\sigma(z')+\varepsilon_0\chi(z')\partial_t]$$
$$E(z't-|z-z'|/c_0)dz' \quad \text{when } 0<z<d \text{ and } 0<t<\infty. \tag{1}$$

The reflected field in front of the slab can be written as
$E^r(z,t)=R(t+z/c_0)=E(z,t)-E^i(z,t)$ when $-\infty<z<0$. When this reflected
field is known, e.g. in an inverse-scattering problem, an alternative integral equation can be derived

$$E(z,t) = E^i(z,t)+R(t+z/c_0)-(Z_0/2)\int_0^z [\sigma(z')+\varepsilon_0\chi(z')\partial_t]$$
$$[E(z',t-(z-z')/c_0)-E(z',t+(z-z')/c_0)]dz'. \tag{2}$$

Note that in the integral on the right-hand side of (1) besides
the unknown field $E(z,t)$ at instant t, only field values at pre-

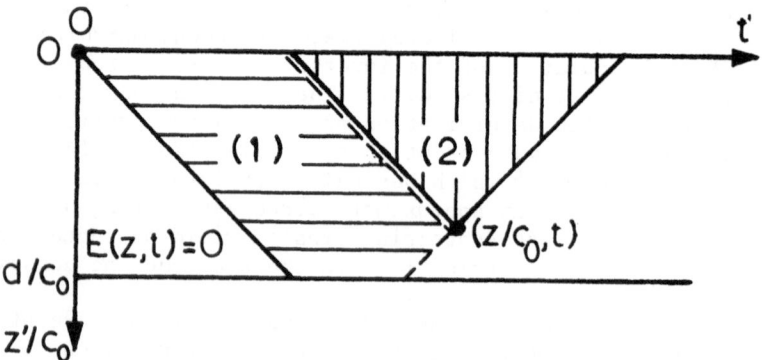

Fig. 3. Space-time points relevant in equations (1) and (2).

vious instants occur, while the corresponding integral in (2)
contains only field values at previous space points. The space-
-time points that are relevant in eqs. (1) and (2) are shown in
Fig. 3. The reflected field at the front of the slab is given by

$$R(t) = -(Z_0/2)\int_0^d [\sigma(z')+\varepsilon_0\chi(z')\partial_t]E(z',t-z'/c_0)dz'. \tag{3}$$

In order to implement the integral equations (1) and (2) numerically, they are discretized in the space-time domain. When the
time step Δt and the space step Δz satisfy $\Delta z=c_0\Delta t=\Delta=d/N$, this
discretization results for eq. (1) into

$$E(1,m) = E^i(1,m) - \tfrac{1}{4} \sum_{n=0}^{N} \alpha_n \chi^{(n)}$$

$$[\tfrac{3}{2}E(n,m')-2E(n,m'-2)+\tfrac{1}{2}E(n,m'-4)] \tag{4}$$

$$- \frac{1}{2N} \sum_{n=0}^{N} \alpha_n \sigma^{(n)} E(n,m'),$$

where $m'=m-|1-n|$, $1=0,1,2,\ldots,N$ and $m=0,1,2,\ldots,\infty$. In (4) we have
$E(1,m) \triangleq E(1\Delta z, m\Delta t)$, $E^i(1,m) \triangleq E^i(1\Delta z, m\Delta t)$, $\chi(1) \triangleq \chi(1\Delta z)$, $\sigma(1) \triangleq \sigma(1\Delta z)$,
$\alpha_n=1$ for $0<1<N$, while $\alpha_0=\alpha_N=\tfrac{1}{2}$. In this discretization, the spatial integral is approximated by a repeated trapezoidal rule and the time derivative by a three-point backward interpolation formula with a time-interval equal to the double time step. For given E^i, χ and σ, (4) represents an infinite system of algebraic equations from which the total electric field E can be solved. This special discretization allows a step-by-step updating procedure. The choice of the double time step in (4) allows the use of convenient recurrence relations similar to the ones derived in [2]. The increase in error in each iteration step is $O(\Delta^2)$. That means that for each finite time interval the maximum error is $O(N\Delta^2)=O(\Delta)$. This implies that stability can always be achieved by increasing N, i.e. by choosing Δ small enough.
In the case of eq. (2), this discretization allows a similar step-by-step side-dating procedure. The resulting schemes are often referred to as *marching on in time* and *marching on in space*, respectively. For some representative numerical results we refer to Fig. 1, Fig. 4 and Fig. 5.

SPACE -TIME FIELD DIAGRAM

Fig. 4. Three-dimensional plot of field in a homogeneous slab for 4 runs of a pulse in the slab ($\chi=1.25$, $\sigma=0$).

Fig. 5. Incident, reflected and transmitted field in the case of
(a) a lossless, homogeneous slab ($\chi=1.25$, $\sigma=0$) and (b) a
lossless slab with a parabolic susceptibility profile
($\chi=2-4(z/d-0.5)^2$, $\sigma=0$). Dashed lines: $E^i(0,t)$, solid
lines: $-E^r(0,t)$, dotted lines: $E^t(d,t)$. The time variable
is c_0t/d.

3. DIRECT SCATTERING, SINGULARITY EXPANSION METHOD

If the incident field is a pulse of finite duration T, the electric field can also be determined by expressing the field quantities in terms of a (one-sided) Laplace transform

$$\{E,H\}(z,t)\leftrightarrow\{\widetilde{E},\widetilde{H}\}(z,s) = \int_{0^-}^{\infty} \{E,H\}(z,t)\exp(-st)dt. \qquad (5)$$

For the incident electric field we obtain

$$\widetilde{E}^i(z,s) = \exp(-sz/c_0)F_T(s) = \exp(-sz/c_0)\int_{0^-}^{T} E^i(0,t)\exp(-st)dt.$$

Using the one-dimensional Green's function $G(z;z')=$
$(c_0/2s)\exp(-s|z-z'|/c_0)$, one derives the integral equation in
operator form as

$$L(s)\widetilde{E}(z,s) = \widetilde{E}^i(z,s), \qquad (6)$$

in which

$$L(s)\widetilde{E}(z,s) = \int_{-\infty}^{\infty} \{\delta(z-z')+s^2C(z,s)G(z;z')/c_0^2\}\widetilde{E}(z',s)\, dz', \qquad (7)$$

where $C(z,s) \overset{\Delta}{=} \varepsilon_r(z) - 1 + \sigma(z)/\varepsilon_0 s$ denotes the *contrast function* of the slab. From (6), we formally have $\tilde{E}(z,s) = L^{-1}(s)\tilde{E}^i(z,s)$. The time-domain response is then found by the inverse Laplace integral

$$E(z,t) = (1/2\pi i) \int_{\beta-i\infty}^{\beta+i\infty} \tilde{E}(z,s)\exp(st) \, ds, \qquad (8)$$

where in the Bromwich contour C_B, $\mathrm{Re}(\beta) > 0$.

In the evaluation of (8), we encounter simple poles and entire functions. Then the following Mittag-Leffler expansion is valid

$$\tilde{E}(z,s) = \sum_m e_m(z)/(s-s_m) + \tilde{E}^e(z,s), \qquad (9)$$

where $\tilde{E}^e(z,s)$ is an entire function and $e_m(z) \overset{\Delta}{=} \mathrm{res}\ \tilde{E}(z,s)\big|_{s=s_m}$. The complex natural frequencies s_m occur in pairs. With each m corresponds a complex natural mode, defined by

$$L(s_m)E_m(z) = 0. \qquad (10)$$

Following now the SEM procedure as outlined in [3, 4, 5, 6, 7] we can express $e_m(z)$ as

$$e_m(z) = D_m F_T(s_m)E_m(z), \qquad (11)$$

in which the coupling coefficient D_m depends on the slab configuration only and defines the strength of the relevant natural mode. The residual contribution of the poles to the time-domain response is then given by

$$E_{poles}(z,t) = \sum_m D_m F_T(s_m)E_m(z)\exp(s_m t). \qquad (12)$$

The natural frequencies and the corresponding natural modes are found as follows. Outside the slab we redefine the electric field as

$$E(z,s) = \begin{cases} \exp(-sz/c_0)/t(s) + r(s)\exp(sz/c_0)/t(s) \\ \qquad\qquad\qquad\quad \text{when } -\infty < z < 0, \\ \\ \exp(-s(z-d)/c_0) \quad \text{when } d < z < \infty. \end{cases} \qquad (13)$$

Since the boundary conditions require the continuity of $E(z,s)$ and $H(z,s)$ at both ends of the slab, the unknown coefficients $r(s)/(t(s)$ and $1/t(s)$ in (13) can be computed by direct numerical integration for the Laplace-transformed, source-free electromagnetic field equations for E and H (a Runge-Kutta type method). The natural frequencies $\{s_m\}$ are then found by searching the zero's of $1/t(s)$ with Muller's method. The natural modes $E_m(z) \overset{\Delta}{=} \tilde{E}(z,s_m)$ are obtained in this proces, while the coupling coefficients can be determined by numerically performing a multi-

ple integration of a product containing factors $C(z,s_m)E_m(z)$.
In the total time-domain response, the entire function in (9)
represents the contribution of the closing contour either in the
left half or in the right half of the s-plane. In order to analyze the role of the entire function an asymptotic solution is
constructed using the WKB method. Applying Jordan's lemma, we
then find that we can avoid computing contributions along the
closing contour by taking into account only that part of the incident pulse that has emerged into the slab and accelerating the
convergence using the obtained asymptotic solution. In this way,
the well known problem of the uniqueness of the SEM-representa-
tion has been resolved. For a more detailed description of this
procedure, the reader is referred to [7].
In the Figures 6-8, we consider the scattering by a slab with a
parabolic permittivity profile and a constant conductivity. In
Figs. 6b and 6c, the magnitudes of the corresponding coupling
coefficients are plotted. The field distributions of some of the
natural modes with $\sigma=0$ are given in Fig. 7. The incident, reflec-

Fig. 6. Configuration-dependent parameters plotted in the complex
        s-plane for a dielectric slab with a parabolic permitti-
        vity profile $\varepsilon_r(z)=6.25-16(z/d-0.5)^2$ and constant conduc-
        tivity $Z_0\sigma d=0$ and 2.
        (a): Complex natural frequencies s_m with $|m|\le 10$.
        (b,c): Magnitudes of the corresponding coupling coeffi-
              cients D_m.

ted and transmitted fields for a sine-squared pulse and a sinusoidal pulse train incident on a lossless, parabolic permittivity
profile are shown in Fig. 8.

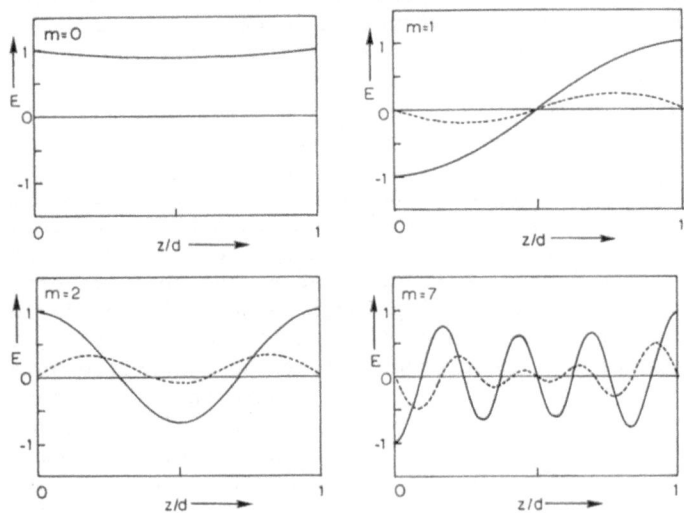

Fig. 7. Field distributions of the natural modes with m=0, 1, 2
and 7 for the lossless slab specified in Fig. 6.
(———— = real part, ---- = imaginary part).

Fig. 8. (a,b): Incident, reflected and transmitted field for the
lossless slab specified in Fig. 6 and for the incident
pulses (a) $F(t)=\sin^2(\pi t/T)\mathrm{rect}(t-T/2;T)$;
(b) $F(t)=\sin(5\pi t/T)\mathrm{rect}(t-T/2;T)$, with $c_0 T/d=1.5$.

4. INVERSE SCATTERING: TIME DOMAIN

In the inverse scattering problem, we want to determine either
the susceptibility profile $\chi(z)$ or the conductivity profile $\sigma(z)$

when the incident and reflected field at the front of the slab
are known. Since the reconstruction is completely analogous for
both constitutive parameters, we restrict ourselves to the case
of unknown susceptibility $\chi(z)$.

In the *direct approach* [12, 13, 14], a space-time discretization
of the source-type integral equation (1) is used. Calculating
step by step, in the slab, the field at the front of the pulse,
one can make an estimate of the local value of the susceptibility
or conductivity. This enables the computation of the field in the
next step. This method has the disadvantage that the front of the
pulse has to be traced very accuratedly, the relative error in
the computed field there being large, especially for dielectric
slabs. An alternative method [16], the *iterative method*, is based
on the consideration that especially those space-time points
where the electric-field
strength is relatively large
and therefore allows an accurate numerical computation,
should be taken into account.
To this end, $\chi(z)$ is assumed
to vary linearly within z-
-intervals $[kN_{cell}\Delta z,$
$(k+1)N_{cell}\Delta z]$ with N_{cell} a
positive integer. Then the
discretized version of the
integral expression for the
reflected field reduces to
a matrix equation of the
form

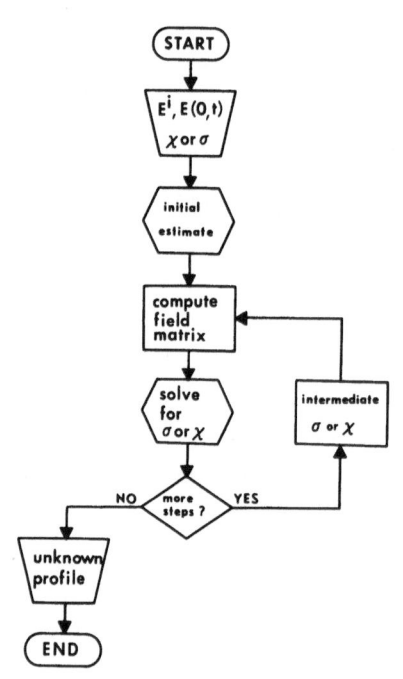

$$\sum_{k=0}^{K} E_{mk}\chi_k = R_m - V_m,$$

$$\text{with } m=0,2,4,\ldots,\infty. \tag{14}$$

In (14) we have $K \triangleq N/N_{cell}$,
$\chi_k=\chi(kN_{cell})$, while E_{mk} is
a weighted sum of the $E(1,n)$
with $|1-kN_{cell}|<N_{cell}$ and
$n=m-1$, $m-1-2$ or $m-1-4$.
R_m is given by the reflected
field, while V_m depends on
both σ and E.

Fig. 9. Simplified block diagram of
the iterative procedure for
solving the time-domain in-
-verse-scattering problem.

Equation (14) constitutes an infinite system of linear equations.
If the electric field E in the slab were known, χ_k could be determined as follows. The infinite system is reduced to a finite
one by selecting those equations that cover the "first run" of
the pulse through the slab:

$$T/2 \leq m\Delta t \leq \tau + T/2 + d/c_0 \tag{15}$$

In (15), $\tau = c_0^{-1} \int_{\tau}^{d} \varepsilon^{\frac{1}{2}}(z)dz$ is the time it takes for the front of the pulse to reach the end of the slab.

The remaining system contains considerably more equations than unknowns. So the problem is overdetermined. In order to overcome this difficulty, two methods are applied to construct $K+1$ equations, from which χ_k can be exactly solved by Gauss elimination. In the first method, the total time interval is subdivided into $K+1$ subintervals, each corresponding to an equation given by the sum of all equations with $t=m\Delta t$ in the subinterval.

The second method is a linear least squares inversion, minimizing the error

$$e = \sum_{subsystem} (R_m - \sum_{k=0}^{K} E_{mk}\chi_k)^2. \tag{16}$$

Since the electric field is known at the front of the slab only, the inverse problem is solved with the iterative procedure outlined in Fig. 9. Starting with an appropriate initial value of the unknown susceptibility, we obtain successive approximations on both the electric field in the slab and the susceptibility. In the first step of this procedure, the electric field is approximated by a field that results from the incident pulse in a slab with a suitably chosen, constant susceptibility. This field is computed with the method outlined in Section 2. The thus obtained results are converted into an equation of the form (14), solution of which yields the first approximation of χ_k. In subsequent steps, this process is repeated until a stable solution is reached.

We have solved this inverse-scattering problem numerically for various permittivity and conductivity profiles, using the time--domain response of a sine-squared incident pulse. The results indicate that, although correction procedures may be required, for low contrast, convergence is achieved within five iteration steps. Also, the method turns out to be stable with respect to variation of the initial guess and addition of random noise to the incident and reflected field. As an illustration, numerical results of the reconstruction of a susceptibility profile are shown in Figs. 10 and 11. For more details, the reader is referred to [15, 16].

Finally, we discuss an approach where the conductivity or susceptibility profiles are reconstructed by iterative minimization of an appropriatedly chosen integrated square error. Two different optimization schemes will be discussed: the *separate-optimization scheme* and the *direct optimization scheme*. First we will introduce the relevant integrated square errors. To this aim we rewrite the equations (1), (2) and (3) in the form $F_1(z,t)=0$, $F_2(z,t)=0$ and $F_3(t)=0$, respectively. Next we introduce the integrated square errors

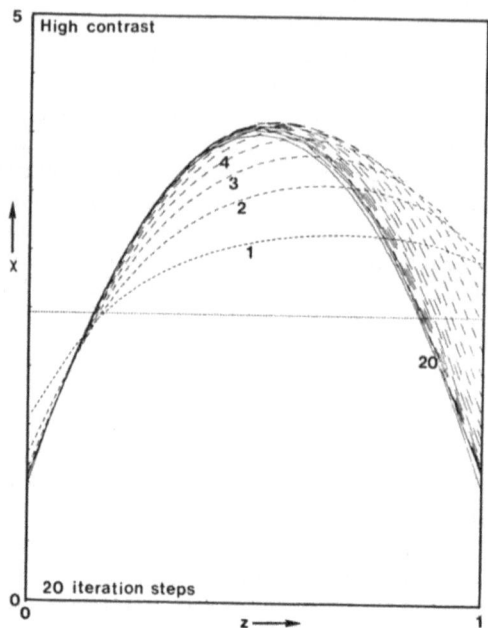

Fig. 10.   Actual susceptibility profile $\chi(z)=4-12(z/d-0.5)^2$
(solid line) compared with the results obtained after
$1,2,\ldots,20$ iteration steps for $\sigma(z)=0$, K=20, N=100,
$c_0T_p/d=0.5$ and an initial estimate computed from the
travel time (dashed lines; dash length increases with
number of the iteration step).

Fig. 11. Real discontinuous susceptibility profiles compared to
result obtained after five iteration steps for $\sigma(z)=0$,
K=40, N=400, $T_p=0.3$, and initial estimate $\chi=1.25$.

$$I_{1,2} \overset{\Delta}{=} \int_0^\infty dt \int_0^d dz \; F_{1,2}^2(z,t), \quad I_3 \overset{\Delta}{=} \int_0^\infty dt \; F_3^2(t), \tag{17}$$

where I_1 denotes the integrated square update error, I_2 the integrated square side-date error and I_3 the integrated square error
in the approximated reflection factor. In both methods the electric field strength E is discretized as indicated   in Section
2, while the susceptibility profile is, as indicated in Section
4, represented by a piecewise linear approximation.
In the *separate-optimization method* we minimize either $I(E,\chi,\sigma)$ =
$=I_1+I_2$ or $I(E,\chi,\sigma)=I_1+cI_3$, where c is some chosen constant. In
the initial step, the unknown profile has a chosen value and the
unknown field is approximated by solving $F_1(E_1,\chi,\sigma)=0$ or
$F_2(E_2,\chi,\sigma)=0$.
The profile is optimized by minimizing the integrated square
error I, while the approximation of the field is not changed.
Next, the field is improved by minimizing I without altering the
profile. This process of separatedly improving profile and field
is repeated until I has reached some minimum. Numerical experiments show that this method converges badly and only gives
reasonable results for the determination of conductivity profiles
for $\chi=0$.
In the *direct-optimization method*, the integrated square error is
introduced as a function of the unknown profile only [16, 17].
The method works for both conductivity and susceptibility profiles. We will discuss the case of a χ-profile. Here we minimize
either $I_2(E_\chi)$ or $I_3(E_\chi)$ where E_χ satisfies $F_1(E,\chi;z,t)\equiv 0 \Psi(z,t)$.
It turns out that the minimization of $I_3(E_\chi)$ leads to a more
efficient scheme. Two different minimization methods have been
applied: the quasi-Newton method and the conjugate-gradient
method [20]. In one version of the quasi-Newton method and in the
conjugate-gradient method the first derivative of the integrated
square error has to be determined analytically. It turns out that
application of the conjugate-gradient method is more efficient
than the quasi-Newton method. A numerical result is shown in Fig.
12.
As a conclusion it can be noted that the separate-optimization
scheme can only be applied to the determination of conductivity
profiles, provided that the susceptibility is equal to zero. The
direct-optimization scheme is applicable for both conductivity
and susceptibility profiles. In the numerical implementation it
turns out that the conjugate-gradient method is more efficient
than the quasi-Newton method. It is important to notice that
for the determination of the profile less parameters are needed
than for the determination of the field. This in contrast with
the approach of Lesselier [17]. Finally it is remarked that the
occurrence of noise on the measured fields may have a considerable effect on the retrieved profiles. There exist various ways to
reduce this influence.

Fig. 12. Step-index susceptibility profile with linearly varying
conductivity reconstructed by minimizing error I_3 with
the conjugate-gradient method. 3% uncorrelated noise
has been added on both incident and reflected field.
$Z_0\sigma(z)/d=z/d$, K=20, N=100, $c_0T_p/d=1$.

5. INVERSE SCATTERING: PRONY ALGORITHM

As we discussed in Section 3, the Singularity Expansion Method
shows that after the incident pulse has completely emerged into
the slab, the reflected field is given by

$$E(z,t) = \sum_{m=-\infty}^{\infty} F_T(s_m)D_mE_m(0)\exp\{s_m(t+z/c_0)\}, \text{ for } z<0. \qquad (18)$$

In (18), the constants s_m and $D_mE_m(0)$ depend on the configuration
only, while $F_T(s_m)$ is given by the incident field. Therefore, if
certain obstacle properties can be determined from the reflected
field, there is also a one-to-one correspondence between the
parameters s_m and $D_mE_m(0)$ and the unknown slab configuration.
This correspondence can then be used to identify the slab uniquely.
Recently [19], methods have been described to determine the parameters s_m and $D_mE_m(0)$ numerically. Most of these methods are
based on Prony's algorithm. This algorithm yields a characteristic equation, from which the natural frequencies s_m can be determined. The residual coefficients $D_mE_m(0)$ can subsequently be
solved from a system of linear equations. It is noted that for
the dielectric slab considered in this paper, the number of
natural modes is infinite, whereas the Prony method was devised
for a finite number of exponentials. Presently, techniques are
being studied to circumvent this difficulty.

ACKNOWLEDGEMENTS

The authors with to thank the students C.P.L. Cames van Batenburg, R. Koster, B.L. Michielsen and R.M. van der Weiden for their contributions to parts of this research.

REFERENCES

[1] Bennett, C.L. and G.F. Ross, "Time-domain electromagnetics and its applications", Proc. IEEE, vol. 66, 1978, pp. 299-318.

[2] Bolomey, J.Ch., Ch. Durix and D. Lesselier, "Time domain integral equation approach for inhomogeneous and dispersive slab problems", IEEE Trans. Antennas Propagat., vol. AP-26, 1976, pp. 1598-1615.

[3] Felsen, L.B., ed., "Transient electromagnetic fields", Springer Verlag, Berlin, 1976, Chapter 3 (C.E. Baum).

[4] Uslenghi, P.L.E., ed., "Electromagnetic scattering", Academic Press, New York, 1978, Chapter 3 (C.E. Baum).

[5] Marin, L., "Natural-mode representations of transient scattered fields", IEEE Trans. Antennas Propagat., vol. AP-21, 1973, pp. 809-818.

[6] Tijhuis, A.G. and H. Blok, "SEM approach to the transient scattering by an inhomogeneous, lossy dielectric slab", Proceedings International URSI-Symposium 1980, Munich, Aug. 26-29, 1980, 221 B.

[7] Tijhuis, A.G. and H. Blok, "SEM approach to the transient scattering by an inhomogeneous, lossy dielectric slab; Part 1: The homogeneous case. Part 2: The inhomogeneous case". To be published in Wave Motion.

[8] Weston, V.H., "On the inverse problem for a hyperbolic dispersive partial differential equation", J. Math. Phys. 13, 1972, pp. 1952-1956.

[9] Krueger, R.J., "An inverse problem for a dissipative hyperbolic equation with discontinuous coefficients", Quart. Appl. Math. 34, 1976, pp. 129-147.

[10] Krueger, R.J., "Numerical aspects of a dissipative inverse problem", IEEE Trans. Antennas Propagat., vol. AP-29, 1981, pp. 253-261.

[11] Burridge, R., "The Gelfand-Levitan, the Marchenko, and the Gopinath-Sondhi integral equations of inverse scattering theory, regarded in the context of inverse impulse-response problems", Wave Motion 2, 1980, pp. 305-323.

[12] Bolomey, J.C., C. Durix and D. Lesselier, "Determination of conductivity profiles by time-domain reflectrometry", IEEE Trans. Antennas Propagat., vol. AP-27, 1979, pp. 244-248.

[13] Lesselier, D., "Determination of index profiles by time domain reflectrometry", J. Optics, vol. 9, 1978, pp. 349-358.

[14] Bojarski, N.N.,"One-dimensional direct and inverse scattering in causal space", Wave Motion, vol. 2, 1980, pp. 115-124.

[15] Tijhuis, A.G., "Iterative determination of permittivity and conductivity profiles of a dielectric slab in the time domain", Proceedings International URSI-Symposium, 1980, Munich, Aug. 26-29, 324 C.

[16] Tijhuis, A.G., "Iterative determination of permittivity and conductivity profiles of a dielectric slab in the time domain", IEEE Trans. Antennas Propagat., vol. AP-29, 1981, pp. 239-245.

[17] Lesselier, D.,"Optimization techniques and inverse problems: reconstruction of conductivity profiles in the time domain", IEEE Trans. Antennas Propagat., vol. AP-30, 1982, pp. 59-65.

[18] Cames van Batenburg, C., "One dimensional inverse scattering in the time domain using an error criterium", Report 1983-06, Laboratory of Electromagnetic Research, Delft University of Technology.

[19] Miller, E.K., "Natural mode methods in frequency and time domain analysis", Proceedings Nato advanced study institute on theoretical methods for determining the interaction of electromagnetic waves with structures, Norwich, July 23-August 4, 1979, Chapter H7.

[20] NAG, "E04-Minimizing or maximizing a function", Fortran library manual, Mark 9, vol. 3, Numerical Algorithms Group, 1982.

I.9 (IS.2)

INVERSE ELECTROMAGNETIC SCATTERING FOR RADIALLY INHOMOGENEOUS DIELECTRIC SPHERES

C. Eftimiu

McDonnell Douglas Research Laboratories
St. Louis, Missouri 63166

The scattering of a plane electromagnetic wave by a dielectric sphere with a radially distributed index of refraction is analyzed in terms of Debye potentials. The scattering amplitude, expanded in vector spherical harmonics, is given in terms of two sets of phase shifts. The inverse problem considered consists in reconstructing the index of refraction when one of the phase shifts is known as a function of frequency.

The approach is based on the use of Liouville transformations which cast the equations for the radial parts of the Debye potentials in the form of Schroedinger-like equations with frequency independent potentials. Given the frequency dependence of one of the phase shifts, one such potential can be found by solving the quantum mechanical inverse problem by using the Marchenko formalism. The index of refraction can be reconstructed from the potential by solving a nonlinear integro-differential equation; the existence and uniqueness of its solution is discussed.

1. INTRODUCTION

The electromagnetic inverse scattering problem consists in the reconstruction of a target for electromagnetic waves from scattering data. In general, the reconstruction concerns the electromagnetic characteristics of the target (dielectric constant, magnetic permeability, conductivity) the knowledge of which includes the spatial extension of the object from a set of measurable quantities directly related to the scattered field.

No attempt will be made here to examine this problem in its total generality. Rather, in order to reduce it to manageable proportions, partial **a priori** information regarding the target will be presumed available. Specifically, it will be assumed that the target has spherical shape, the conductivity is identically zero, its magnetic permeability is everywhere that of vacuum, and its dielectric constant is a continuous, twice differentiable function of the distance from its center, only. The only unknown of the inverse scattering problem will thus be the function $\epsilon(r)$. The dielectric sphere will also be assumed to have a finite extension, implying that there exists a radius r_0 (not **a priori** known) such that $\epsilon(r) \equiv 1$ for $r \geq r_0$, whereas for $0 \leq r \leq r_0$, $\epsilon(r) \geq 1$.

W.-M. Boerner et al. (eds.), Inverse Methods in Electromagnetic Imaging - Part 1, 157–176.
© 1985 by D. Reidel Publishing Company.

The formulation of the inverse electromagnetic problem depends, of course, on the nature of the assumed set of given scattering data, and in this respect several choices are possible. To describe them, an analysis of the scattering process is necessary, and this is briefly done in the next section to make the exposition self-sufficient.

2. DIRECT ELECTROMAGNETIC SCATTERING

We begin by disposing of the complications arising from the vectorcharacter of the electromagnetic field, by reducing the electromagnetic scattering problem to two scalar problems. The procedure involves the introduction of the Debye potentials and, since it is well-known (1), will be described briefly. It follows from Maxwell's equations:

$$\nabla \times \vec{E} = -\frac{1}{c}\dot{\vec{B}} = ik\vec{H} \qquad \nabla \cdot \vec{B} = 0$$

$$\nabla \times \vec{H} = \frac{1}{c}\dot{\vec{D}} = -ik\epsilon\vec{E} \qquad \nabla \cdot \vec{D} = 0 \qquad (2.1)$$

that the total (incident + scattered) electromagnetic field (the harmonic time dependence, in which $\omega = kc$, is suppressed):

$$\vec{E} = \vec{E}^{inc} + \vec{E}^{sc} = \vec{u}_x e^{ikz} + \vec{E}^{sc}$$

$$\vec{H} = \vec{H}^{inc} + \vec{H}^{sc} = \vec{u}_y e^{ikz} + \vec{H}^{sc} \qquad (2.2)$$

satisfies the uncoupled equations:

$$\nabla \times (\nabla \times \vec{E}) = k^2\epsilon\vec{E}$$

$$\nabla \times \left(\frac{1}{\epsilon}\nabla \times \vec{H}\right) = k^2\vec{H}. \qquad (2.3)$$

Addressing the second of Eqs. (2.3) first, and taking into account the transverse character of the magnetic field, one makes the Ansatz

$$\vec{H} = \vec{M} + \vec{N}, \qquad (2.4)$$

where

$$\vec{M} = \nabla \times (n\vec{r}\chi) \qquad (2.5)$$

introduces χ, the electric multipole Debye potential, and

$$\vec{N} = k^{-1}\nabla + [\nabla + (\vec{r}\psi)] \qquad (2.6)$$

introduces ψ, the magnetic multipole Debye potential, with

$$n = \sqrt{\epsilon}. \qquad (2.7)$$

The Debye potentials satisfy the equations:

$$\nabla^2 \chi \times [k^2 n^2 - U(r)]\chi = 0$$

$$\nabla^2 \psi \times k^2 n^2 \psi = 0, \tag{2.8}$$

in which

$$U(r) \equiv n(r) \frac{d^2}{dr^2} \frac{1}{n(r)} . \tag{2.9}$$

Since the electric field can be obtained in terms of the magnetic field from the second of Maxwell's equations (2.1), the electromagnetic problem reduces to two scalar problems.

For the purpose of performing a phase-shift analysis of the scattering process, it is appropriate to introduce the vector spherical harmonics, defined as

$$\vec{\mu}_{\ell m}(\theta, \phi) = \nabla \times (\vec{r} Y_{\ell m})$$

$$\vec{\nu}_{\ell m}(\theta, \phi) = 2r \nabla Y_{\ell m}$$

$$\vec{\tau}_{\ell m}(\theta, \phi) = \ell(\ell + 1) \frac{\vec{r}}{r} Y_{\ell m}, \tag{2.10}$$

where $Y_{\ell m}$ are the standard spherical functions:

$$Y_{\ell m}(\theta, \phi) = \left[\frac{2\ell + 1}{4\pi} \frac{(\ell - |m|)!}{(\ell + |m|)!} \right]^{1/2} \cdot P_\ell^m(\cos \theta) e^{im\phi} . \tag{2.11}$$

One can check now that the second of Eqs. (2.3) **in vacuo** ($\epsilon \equiv 1$) possesses two sets of linearly independent solutions. One set, consisting of solutions regular in the origin (as it will be seen below, $\ell \geq 1$) is given by the functions

$$\vec{M}_{\ell m}^{(1)} = j_\ell(kr) \vec{\mu}_{\ell m}(\theta, \phi) \tag{2.12}$$

and

$$\vec{N}_{\ell m}^{(1)} = \frac{1}{k} \nabla \times \vec{M}^{(1)} = \frac{1}{kr} j_\ell(kr) \vec{\tau}_{\ell m} + \frac{1}{2kr} \frac{d}{dr} [r j_\ell(kr)] \vec{\nu}_{\ell m} \tag{2.13}$$

and another set, describing outgoing spherical waves, by

$$\vec{M}_{\ell m}^{(2)} = h_\ell^{(1)}(kr) \vec{\mu}_{\ell m}(\theta, \phi) \tag{2.14}$$

and

$$\vec{N}_{\ell m}^{(2)} = \frac{1}{k} \nabla \times \vec{M}^{(2)} = \frac{1}{kr} h_\ell^{(1)}(kr) \vec{\tau}_{\ell m} + \frac{1}{2kr} \frac{d}{dr} [r j_\ell(kr)] \vec{\nu}_{\ell m} \tag{2.15}$$

where j_ℓ and h_ℓ are the usual Riccati-Bessel and Riccati-Hankel functions, respectively. In terms of the first set, the incident electromagnetic field can be written as

$$\vec{E}^{inc} = \sqrt{\pi} \sum_{\ell=1}^{\infty} i^{\ell-1} \left[\frac{\ell(\ell+1)}{2\ell+1} \right]^{1/1} \left(\vec{M}_{\ell m}^{(1)} + \vec{N}_{\ell m}^{(1)} - \vec{M}_{\ell m}^{(1)} + \vec{N}_{\ell m}^{(1)} \right)$$

$$\vec{H}^{inc} = \sqrt{\pi} \sum_{\ell=1}^{\infty} i^{\ell+2} \left[\frac{\ell(\ell+1)}{2\ell+1} \right]^{1/1} \left(\vec{M}_{\ell m}^{(1)} + \vec{N}_{\ell m}^{(1)} + \vec{M}_{\ell m}^{(1)} - \vec{N}_{\ell m}^{(1)} \right) \tag{2.16}$$

The fact that only ± 1 values for m appear in the expansions (2.16) implies that $\ell \geq 1$, so that only such values for ℓ and m will appear in similar expansions for the scattered field as well. The latter will be **asymptotically** ($r \mapsto \infty$) described by the solutions from the second set:

$$\vec{E}^{sc} \underset{r \to \infty}{\sim} \sum_{\ell=1}^{\infty} \sum_{m = \pm 1} i^{\ell-1} \left[\frac{\ell(\ell + 1)}{\ell(\ell + 1)} \right]^{1/2} \left(a_{\ell m} \vec{M}_{\ell m}^{(2)} + b_{\ell m} \vec{N}_{\ell m}^{(2)} \right)$$

$$\vec{H}^{sc} \underset{r \to \infty}{\sim} \sum_{\ell=1}^{\infty} \sum_{m = \pm 1} i^{\ell+2} \left[\frac{\ell(\ell + 1)}{\ell(\ell + 1)} \right]^{1/2} \left(b_{\ell m} \vec{M}_{\ell m}^{(2)} + a_{\ell m} \vec{N}_{\ell m}^{(2)} \right) \tag{2.17}$$

An examination of symmetry properties of the fields, and of the vector spherical harmonics under the transformation $\phi \mapsto - \phi$, shows that

$$a_{\ell 1} = - a_{\ell - 1}, \qquad b_{\ell 1} = b_{\ell - 1}, \tag{2.18}$$

and hence, by introducing the phase shifts δ_ℓ and η_ℓ through the relations:

$$a_{\ell 1} = i e^{i\delta_\ell} \sin \delta_\ell, \qquad b_{\ell 1} = i e^{i\eta_\ell} \sin \eta_\ell, \tag{2.19}$$

one obtains for the asymptotic expansions of the scattered field the expressions:

$$\vec{E}^{sc} \underset{r \mapsto \infty}{\sim} \vec{a}\,(\theta, \phi) \frac{e^{ikr}}{r}$$

$$\vec{H}^{sc} \underset{r \mapsto \infty}{\sim} \frac{\vec{r}}{r} \times \vec{a}\,(\theta, \phi)\, \frac{e^{ikr}}{r}, \tag{2.20}$$

where $\vec{a}\,(\theta, \phi)$, the scattering amplitude, is

$$\vec{a}\,(\theta, \gamma) = - \frac{\sqrt{\pi}}{k} \sum_{\ell=1}^{\infty} \left[\frac{2\ell + 1}{\ell(\ell+1)} \right]^{1/2} \left[i(\vec{\mu}_{\ell 1} - \vec{\mu}_{\ell - 1}) e^{i\delta_\ell} \sin \delta_\ell \right.$$

$$\left. - \frac{1}{2} (\vec{\nu}_{\ell 1} + \vec{\nu}_{\ell - 1}) e^{i\eta_\ell} \sin \eta_\ell \right]. \tag{2.21}$$

Hence, given the orthonormality properties of the vector harmonics, one obtains for the total scattering cross-section the well-known expression:

$$\sigma = \int d\Omega |a|^2 = \frac{2\pi}{k^2} \sum_{\ell=1}^{\infty} (2\ell + 1) \left[\sin^2\delta_\ell + \sin^2\eta_\ell \right] \tag{2.22}$$

The direct electromagnetic scattering problem, consisting in finding the scattering amplitude (2.21), and hence the differential, $|\vec{a}|^2$, or total, σ, cross-sections, given the (spherically symmetric) scatterer, reduces thus to a determination of the two sets of phase-shifts, δ_ℓ and η_ℓ with $\ell = 1, 2, \ldots$, which are in general functions of the frequency ω of the incident field.

The asymptotic expressions for the total fields follow from Eqs. (2.16) and (2.17):

$$\vec{E} \underset{r \to \infty}{\sim} \frac{\sqrt{\pi}}{kr} \sum_{\ell=1}^{\infty} \left[\frac{2\ell+1}{\ell(\ell+1)} \right]^{1/2} i^{\ell-1} \Big[(\vec{\mu}_{\ell 1} - \vec{\mu}_{\ell-1}) e^{i\delta_\ell} \sin \Big[kr - \frac{\ell\pi}{2} + \delta_\ell \Big]$$

$$+ \frac{1}{2} (\vec{\nu}_{\ell 1} + \vec{\nu}_{\ell-1}) e^{i\eta_\ell} \cos \Big(kr - \frac{\ell\pi}{2} + \eta_\ell \Big) \Big]$$

$$\vec{H} \underset{r \to \infty}{\sim} \frac{\sqrt{\pi}}{ikr} \sum_{\ell=1}^{\infty} \left[\frac{2\ell+1}{\ell(\ell+1)} \right]^{1/2} i^{\ell-1} \Big[(\vec{\mu}_\ell + \vec{\mu}_{\ell-1}) e^{i\eta_\ell} \sin \Big(kr - \frac{\ell\pi}{2} + \eta_\ell \Big)$$

$$+ \frac{1}{2} (\vec{\nu}_{\ell 1} + \vec{\nu}_{\ell-1}) e^{i\delta_\ell} \cos \Big(kr - \frac{\ell\pi}{2} + \delta_\ell \Big) \Big] \tag{2.23}$$

The total fields having the above asymptotic expressions, and satisfying Eqs. (2.3) everywhere, can be written in the form:

$$\vec{E} = \frac{\sqrt{\pi}}{kn^2} \sum_{\ell=1}^{\infty} \left[\frac{2\ell+1}{\ell(\ell+1)} \right]^{1/2} i^{\ell-1} \left(\vec{M}_{\ell 1}^{(4)} + \vec{N}_{\ell 1}^{(4)} - \vec{M}_{\ell-1}^{(4)} + \vec{N}_{\ell-1}^{(4)} \right)$$

$$\vec{H} = \frac{\sqrt{\pi}}{ik} \sum_{\ell=1}^{\infty} \left[\frac{2\ell+1}{\ell(\ell+1)} \right]^{1/2} i^{\ell-1} \left(\vec{M}_{\ell 1}^{(3)} + \vec{N}_{\ell 1}^{(3)} + \vec{M}_{\ell-1}^{(3)} - \vec{N}_{\ell-1}^{(3)} \right) \tag{2.24}$$

where

$$\vec{M}_{\ell m}^{(3)} = F_\ell(r) \vec{\mu}_{\ell m}$$

$$\vec{N}_{\ell m}^{(3)} = \frac{1}{kr} G_\ell(r) \vec{\tau}_{\ell m} + \frac{1}{2kr} \frac{d}{dr} \Big[rG_\ell(r) \Big] \vec{\nu}_{\ell m} \tag{2.25}$$

and

$$\vec{M}_{\ell m}^{(4)} = \frac{1}{k} \nabla \times \vec{N}_{\ell m}^{(3)} = n^2 G_\ell(r) \vec{\mu}_{\ell m}$$

$$\vec{N}_{\ell m}^{(4)} = \frac{1}{k} \nabla \times \vec{M}^{(3)} = \frac{1}{kr} F_\ell(r) \vec{\tau}_{\ell m} + \frac{1}{2kr} \frac{d}{dr} \Big[rF_\ell(r) \Big] \vec{\nu}_{\ell m} \tag{2.26}$$

with the radial functions F_ℓ and G_ℓ **regular** (in the origin) solutions of the equations

$$\frac{d^2}{dr^2} \frac{rF_\ell(r)}{n(r)} + \left[k^2 n^2(r) - \frac{\ell(\ell+1)}{r^2} - U(r) \right] \frac{rF_\ell(r)}{n(r)} = 0$$

$$\frac{d^2}{dr^2} rG_\ell(r) + \left[k^2 n^2(r) - \frac{\ell(\ell+1)}{r^2} \right] rG_\ell(r) = 0, \tag{2.27}$$

and behaving asymptotically as:

$$rF_\ell(r) \underset{r \to \infty}{\sim} e^{i\eta_\ell} \sin \Big(kr - \frac{\ell\pi}{2} + \eta_\ell \Big)$$

$$rG_\ell(r) \underset{r \to \infty}{\sim} e^{i\delta_\ell} \cos \Big(kr - \frac{\ell\pi}{2} + \delta_\ell \Big) . \tag{2.28}$$

We will refer to the total fields defined by the asymptotic conditions (2.28) as the **physical** fields. In addition to these solutions of Eqs. (2.3), we shall also construct auxiliary solutions playing an important role in the following section. One such set of solutions, called regular, are defined as solutions of the Eqs. (2.27):

$$\frac{d^2}{dr^2} \frac{u_\ell(r)}{n(r)} + \left[k^2 n^2(r) - \frac{\ell(\ell+1)}{r^2} - U(r) \right] \frac{u_\ell(r)}{n(r)} = 0$$

$$\frac{d^2}{dr^2} v_\ell(r) + \left[k^2 n^2(r) - \frac{\ell(\ell+1)}{r^2} \right] v_\ell(r) = 0 \qquad (2.29)$$

through their behavior at the origin

$$u_\ell(r), \; v_\ell(r) \underset{r \to 0}{\sim} \frac{r^{\ell+1}}{(2\ell+1)!!} \qquad (2.30)$$

which, up to a factor, is also the behavior at the origin of the function $r j_\ell(kr)$, a solution of either of the Eqs. (2.29) with $n \equiv 1$.

Another set of solutions, called **Jost solutions**, are defined as solutions of Eqs. (2.29):

$$\frac{d^2}{dr^2} \frac{U_\ell(k, r)}{n(r)} + \left[k^2 n^2(r) - \frac{\ell(\ell+1)}{r^2} - U(r) \right] \frac{U_\ell(k, r)}{n(r)} = 0$$

$$\frac{d^2}{dr^2} V_\ell(k, r) + \left[k^2 n^2(r) - \frac{\ell(\ell+1)}{r^2} \right] V_\ell(k, r) = 0 \qquad (2.31)$$

with the asymptotic behavior:

$$U_\ell(k, r) \underset{r \to \infty}{\sim} e^{i(kr + \ell\pi/2)}$$

$$V_\ell(k, r) \underset{r \to \infty}{\sim} e^{i(kr + [\ell-1]\pi/2)} \qquad (2.32)$$

which, up to constant factors, is also the asymptotic behavior of the function $r h_\ell^{(1)}(kr)$ another solution of the equations (2.31) with $n \equiv 1$. It is important to note that, if $k \neq 0$, the solution $U_\ell(-k, r)$, is independent of $U_\ell(k, r)$, as $V_\ell(-k, r)$ is independent of $V_\ell(k, r)$ since their Wronskians are non-vanishing:

$$W\left[\frac{U_\ell(k, r)}{n(r)}, \frac{U_\ell(-k, r)}{n(r)} \right] = \frac{1}{n^2} W\left[U_\ell(k, r), U_\ell(-k, r) \right] = 2ik \, (-1)^{\ell+1}$$

$$W\left[V_\ell(k, r), V_\ell(-k, r) \right] = 2ik \, (-1)^\ell \qquad (2.33)$$

However, it follows from (2.31) and (2.32) that

$$U_\ell(-k, r) = (-1)^\ell \, U_\ell^*(k, r)$$

$$V_\ell(-k, r) = (-1)^{\ell-1} \, V_\ell^*(k, r) . \qquad (2.34)$$

Consequently, the regular solutions, which are real, even functions of k, can be written in the form of the linear combinations:

$$u_\ell(r) = \frac{i}{2k^{\ell+1}} \left[U_\ell(k) \, U_\ell(-k, r) - (-1)^\ell U_\ell(-k) \, U_\ell(k, r) \right]$$

$$v_\ell(r) = \frac{i}{2k^{\ell+1}} \left[V_\ell(k) \, V_\ell(-k, r) - (-1)^\ell V_\ell(-k) \, V_\ell(k, r) \right] \tag{2.35}$$

in which the coefficients $U_\ell(k)$ and $V_\ell(k)$ are, by definition, the so-called Jost functions. From (2.35) and (2.33) it also follows that the Jost functions can be given explicitly as:

$$U_\ell(k) = (-k)^\ell W \left[\frac{U_\ell(k, r)}{n(r)}, \frac{u_\ell(r)}{n(r)} \right]$$

$$V_\ell(k) = -(-k)^\ell W \left[V_\ell(k, r), v_\ell(r) \right] \tag{2.36}$$

In view of Eqs. (2.34) and the reality of the regular solutions, Eqs. (2.36) imply that

$$U_\ell(-k) = U_\ell^*(k)$$

$$V_\ell(-k) = -V_\ell^*(k). \tag{2.37}$$

From Eqs. (2.35) one can also obtain the asymptotic expressions of the regular solutions. These expressions in fact must be, up to constant factors, those of the physical solutions, as given by Eqs. (2.28) (the physical solutions are also regular and thus, up to multiplicative constants, equal to the regular solutions). To reproduce the behavior (2.28) from Eqs. (2.35), one must assume that

$$U_\ell(k) = |U_\ell(k)| e^{-i\eta_\ell}$$

$$V_\ell(k) = |V_\ell(k)| e^{-i\delta_\ell} \tag{2.38}$$

One obtains thus the asymptotic expressions

$$u_\ell(r) \underset{r \to \infty}{\sim} \frac{|U_\ell(k)|}{k^{\ell+1}} \sin\left(kr - \frac{\ell\pi}{2} + \eta_\ell \right)$$

$$v_\ell(r) \underset{r \to \infty}{\sim} \frac{|V_\ell(k)|}{k^\ell} \cos\left(kr - \frac{\ell\pi}{2} + \delta_\ell \right) \tag{2.39}$$

which establish the connection between the physical phase-shifts and the arguments of the complex Jost functions.

While listing the various sets of solutions of the Eqs. (2.27) it is also of interest to mention the lack of a special type of solutions (generally present in Quantum Mechanics), namely bound state solutions, defined as real, regular, square-integrable solutions corresponding to **negative** values of k^2 (purely imaginary k, with positive imaginary part). The absence of such solutions in the problem under consideration could be argued on physical, intuitive grounds, but a direct mathematical proof can be readily exhibited. Indeed considering first the second of the equations (2.27), assume that $w_\ell(r)$ is a square integrable solution of the equation:

$$\frac{d^2}{dr^2} w_\ell(r) + \left[k^2 n^2(r) - \frac{\ell(\ell + 1)}{r^2} \right] w_\ell(r) = 0. \tag{2.40}$$

Multiplying this equation by $w_\ell(r)$ and integrating the first term by parts, one obtains

$$\int_0^\infty \left[\left(\frac{dw_\ell}{dr} \right)^2 + \frac{\ell(\ell + 1)}{r^2} w_\ell^2 \right] dr = k^2 \int_0^\infty n^2 w_\ell^2 dr. \tag{2.41}$$

The integrated part is discarded because, to be square integrable, the solution w must vanish at infinity, while in the origin it vanishes like $r^{\ell+1}$. It is clear now that Eq. (2.41) cannot be satisfied with negative values of k^2.

The same conclusion can be drawn for the first of Eqs. (2.27), but not as directly as for the second equation. In this case, it is convenient to first apply to the equation

$$\frac{d^2}{dr^2} w_\ell + \left[k^2 n^2 - \frac{\ell(\ell + 1)}{r^2} - n \frac{d^2}{dr^2} \frac{1}{n} \right] w_\ell = 0 \tag{2.42}$$

a Liouville transformation, consisting of the transformation of variable:

$$r \mapsto s: s = \int_0^r n^2(t) dt \tag{2.43}$$

and simultaneously the transformation of function:

$$w_\ell \mapsto z_\ell: z_\ell(s) = m(s) w_\ell(r) \tag{2.44}$$

where $m(s) = n(r(s))$. Then

$$\frac{d^2 w_\ell(r)}{dr^2} = m^2(s) \left[m(s) \frac{d^2 z_\ell(s)}{ds^2} - z_\ell(s) \frac{d^2 m(s)}{ds^2} \right]$$

and

$$\frac{d^2}{dr^2} \frac{1}{n(r)} = - m^2(s) \frac{d^2 m(s)}{ds^2}$$

so that Eq. (2.42) becomes

$$\frac{d^2 z_\ell(s)}{ds^2} + \left[\frac{k^2}{m^2(s)} - \frac{\ell(\ell + 1)}{m^4(s) r^2(s)} \right] z_\ell(s) = 0 \tag{2.45}$$

Because the transformation (2.43) is monotonic (indeed linear for $r > r_0$) and $s \sim r$ in the origin, the argument presented above can be applied to Eq. (2.43), with the general conclusion that no bound states are present in the context of our scattering problem.

3. INVERSE QUANTUM MECHANICAL SCATTERING

In the inverse electromagnetic scattering problem we are finally ready to begin discussing, the unknown is the function n(r). The input, obviously, must involve the vector scattering amplitude, which is given by the expression (2.21) in terms of two infinite sequences of phase-shifts, each of which is a function of k. The problem of retrieving the scattering amplitude from measurements, while important, will not be considered here. We assume that whatever information regarding the phase-shifts η_ℓ (k) and δ_ℓ (k) needed to solve the inverse scattering problem is available.

The general approach adopted is that the inverse electromagnetic problem can be solved by reducing it to the quantum mechanical inverse scattering problem. The motivation for this approach[2] is based on the observation that either of the Eqs. (2.27) can be given the form of the reduced Schroedinger equation for radial functions

$$\frac{d^2}{dr^2} \psi_\ell(r) + \left(k^2 - \frac{\ell(\ell + 1)}{r^2} \right) \psi_\ell(r) = V(r) \psi_\ell(r) \tag{3.1}$$

where, for the first of Eq. (2.27)

$$V(r) = k^2(1 - n^2(r)) + n(r) \frac{d^2}{dr^2} \frac{1}{n(r)} , \tag{3.1a}$$

while for the second,

$$V(r) = k^2(1 - n^2(r)) \tag{3.1b}$$

As far as the inverse scattering problem at fixed k is concerned, this is indeed all that is required for the reduction, since the techniques available for the reconstruction of a potential V(r) from one set of phase shifts values are directly applicable to the problem at hand. Moreover, since the potential to be reconstructed is known to have finite range, the reconstruction is theoretically unique. From a practical viewpoint, there still are questions relating to the use of only a finite number of phase shifts and to numerical stability which deserve more attention.

The inverse electromagnetic problem of the reconstruction of the function n(r) from **one** phase-shift, assumed known as a function of k, also reduces to the corresponding quantum mechanical inverse problem, but not as directly as the problem at fixed k. The main reason for the complications arising in this case resides in the k dependence of the potential appearing in the Schroedinger equation (3.1). However, this dependence can be readily disposed of by means of a Liouville transformation. Namely, if it is assumed that δ_ℓ (k) is the phase shift available for the reconstruction problem, then we must investigate Eq. (3.1) with the potential (3.1b). In this case, the Liouville transformation of the equation for the physical radial solution (the asymptotic behavior of which is given by the second of Eqs. (2.28)), defined as:

$$r \mapsto \rho: \qquad \rho(r) = \int_0^r n(s)ds$$

$$\psi_\ell(r) \mapsto \overline{\psi}_\ell(\rho): \ \overline{\psi}_\ell(\rho) = \sqrt{\nu(\rho)} \ u_\ell(r(\rho)) \tag{3.2}$$

where $v(\rho) = n(r(\rho))$, yields the Schroedinger-like equation

$$\frac{d^2}{d\rho^2}\bar{\psi}_\ell(\rho) + \left[k^2 - \frac{\ell(\ell+1)}{R^2(\rho)}\right]\bar{\psi}_\ell(\rho) = \bar{V}(\rho)\,\bar{\psi}_\ell(\rho) \tag{3.3}$$

with a potential independent of k:

$$\bar{V}(\rho) = [v(\rho)]^{-1/2}\frac{d^2}{d\rho^2}\,v(\rho)^{1/2} \tag{3.4}$$

and a "centrifugal potential" in which

$$R(\rho) = v(\rho)\,r(\rho), \tag{3.5}$$

a function of ρ that can only vanish in the origin:

$$R(\rho) \underset{\rho \to 0}{\sim} \rho \tag{3.6}$$

and which for $\rho \geq \rho_0$, where

$$\rho_0 = \int_0^{r_0} n(s)ds, \tag{3.7}$$

becomes linear:

$$R(\rho) = \rho + r_0 - \rho_0 \cdot (\rho \geq \rho_0) \tag{3.8}$$

Eq. (3.3) can also be written as a Schroedinger equation:

$$\frac{d^2}{d\rho^2}\bar{\psi}_\ell(\rho) + \left[k^2 - \frac{\ell(\ell+1)}{\rho^2}\right]\bar{\psi}_\ell(\rho) = \bar{V}_\ell(\rho)\,\bar{\psi}_\ell(\rho) \tag{3.9}$$

with the ℓ-dependent potential:

$$\bar{V}_\ell(\rho) = \bar{V}(\rho) - \ell(\ell+1)\left[\frac{1}{\rho^2} - \frac{1}{R^2(\rho)}\right]. \tag{3.10}$$

Because of our assumptions on n(r) and the definition (3.5),

$$\lim_{\rho \to 0}\rho\,\bar{V}_\ell(\rho) = -\ell(\ell+1)\frac{n'(o)}{n(o)}, \tag{3.11}$$

while asymptotically,

$$\lim_{\rho \to \infty}\rho^3\bar{V}_\ell(\rho) = 2\ell(\ell+1)(\rho_0 - r_0) \tag{3.12}$$

and hence both the integrals

$$\int_0^\infty \rho |\bar{V}_\ell(\rho)| d\rho \quad \text{and} \quad \int_0^{\rho_0} \rho^2 |\bar{V}_\ell(\rho)| d\rho \quad \text{exist.} \tag{3.13}$$

It follows then that a regular solution $\bar{v}_\ell(\rho)$ of the equation (3.9), obeying the boundary condition

$$\bar{v}_\ell(\rho) \underset{\rho \to 0}{\sim} \frac{\rho^{\ell+1}}{(2\ell+1)!!}, \tag{3.14}$$

exists and can be obtained by iteration as a solution of the Volterra integral equation

$$\bar{v}_\ell(\rho) = k^{-\ell} \rho j_\ell(k\rho) + \int_0^\rho g_\ell(k, \rho, \rho') \bar{V}_\ell(\rho') \bar{v}_\ell(\rho') d\rho'. \tag{3.15}$$

where

$$g_\ell(k, \rho, \rho') = \frac{i(-1)^\ell}{2} k\rho\rho' \left[h_\ell^{(1)}(k\rho) h_\ell^{(1)}(-k\rho') - h_\ell^{(1)}(k\rho') h_\ell^{(1)}(-k\rho) \right] \tag{3.16}$$

In our case, Eq. (3.15) only needs to be solved for $0 \le \rho \le \rho_0$, since for $\rho \ge \rho_0$, $\bar{v}_\ell(\rho)$ is a linear combination of the functions $rj_\ell(kr)$ and $ry_\ell(kr)$ where $r = \rho - \rho_0 + r_0$, the value of $R(\rho)$ or of $r(\rho)$ for such ρ. It is also known that, for fixed ρ, $\bar{v}_\ell(\rho)$ is an entire function of k.

Similarly, we can define a Jost solution of Eq. (3.9) through the asymptotic condition

$$\bar{V}_\ell(k, \rho) \underset{\rho \to \infty}{\sim} e^{i(k\rho + [\ell-1]\pi/2)} \tag{3.17}$$

Actually, $\bar{V}_\ell(k, \rho)$ is known explicitly for $\rho \ge \rho_0$ to be the function

$$\bar{V}_\ell^{II}(k, \rho) = e^{i\pi\ell} kr \, h_\ell^{(1)}(kr) \, e^{ik(\rho_0 - r_0)} \tag{3.18}$$

where again r stands for $\rho - \rho_0 + r_0$. The expression $\bar{V}_\ell^I(k, \rho)$ of $\bar{V}_\ell(k, \rho)$ for $0 < \rho \le \rho_0$ is then a solution of the Volterra equation:

$$\bar{V}_\ell^I(k,\rho) = \bar{V}_\ell^{II}(k, \rho_0) + w_\ell(k\rho) - w_\ell(k\rho_0)$$

$$- \int_\rho^{\rho_0} g_\ell(k, \rho, \rho') \bar{V}_\ell(\rho') \bar{V}_\ell^I(k, \rho') d\rho', \tag{3.19}$$

in which

$$w_\ell(k\rho) = e^{i\pi\ell} k\rho \, h_\ell^{(1)}(k\rho). \tag{3.20}$$

It can be shown now that, under the conditions (3.13), $\bar{V}_\ell^I(k, \rho)$ and thus $\bar{V}_\ell(k, \rho)$ can be obtained by iterating Eq. (3.19) and that, for fixed $\rho > 0$, $k^\ell \bar{V}_\ell(k, \rho)$ thus constructed is an entire function of k of exponential type, vanishing at infinity in the upper complex k-plane. Another Jost solution, $\bar{V}_\ell(-k, \rho)$, can also be introduced, **mutatis mutanda,** and finally the Jost functions $\bar{V}_\ell(k)$ and $\bar{V}_\ell(-k)$ can be defined through the relation

$$\bar{v}_\ell(\rho) = \frac{i}{2k^{\ell+1}} [\bar{V}_\ell(k) \bar{V}_\ell(-k, \rho) - (-1)^\ell \bar{V}_\ell(-k) \bar{V}_\ell(k, \rho)] \tag{3.21}$$

or, explicitly, as

$$\bar{V}_\ell(k) = - (-k)^\ell W [\bar{V}_\ell(k, \rho), \bar{v}_\ell(\rho)] \tag{3.22}$$

From Eqs. (3.15), (3.17) and (3.21) with $\rho \mapsto \infty$ it follows that

$$\bar{V}_\ell(k) = i - k^{\ell+1} \int_0^\infty \rho' h_\ell^{(1)}(k\rho') \bar{V}_\ell(\rho') \bar{v}_\ell(\rho')d\rho' \tag{3.23}$$

and that

$$\bar{V}_\ell(-k) = i + (-k)^{\ell+1} \int_0^\infty \rho' h_\ell^{(1)*}(k\rho') \bar{V}_\ell(\rho') \bar{v}_\ell(\rho')d\rho' = -\bar{V}_\ell^{(1)*}(k) \tag{3.24}$$

Hence

$$\bar{V}_\ell(k) \underset{|k| \mapsto \infty}{\sim} i. \tag{3.25}$$

If then one sets

$$\bar{V}_\ell(k) = |\bar{V}_\ell(k)| e^{-i\bar{\delta}_\ell + i\pi/2}, \ \bar{\delta}_\ell \text{ real}, \tag{3.26}$$

it follows that

$$e^{2i\bar{\delta}_\ell} \equiv \frac{\bar{V}_\ell(-k)}{\bar{V}_\ell(k)} \underset{|k| \mapsto \infty}{\sim} 1 \tag{3.27}$$

i.e., that

$$\bar{\delta}_\ell(k) \underset{|k| \mapsto \infty}{\sim} 0. \qquad (\text{mod } 2\pi) \tag{3.28}$$

The meaning of $\bar{\delta}_\ell(k)$ emerges from an examination of the asymptotic form of the regular solution, $\bar{v}_\ell(\rho)$. Indeed, from Eqs. (3.17), (3.21) and (3.26)

$$\bar{v}_\ell(\rho) \underset{\rho \mapsto \infty}{\sim} \frac{|\bar{V}_\ell(k)|}{k^{\ell+1}} \sin(k\rho - \ell\pi/2 + \bar{\delta}_\ell)$$

$$= \frac{\bar{V}_\ell(k)}{k^{\ell+1}} e^{i\bar{\delta}_\ell} \sin(k\rho - \ell\pi/2 + \bar{\delta}_\ell)$$

$$= \frac{\bar{V}_\ell(k)}{2k^{\ell+1}} i^{\ell+1} [e^{-ik\rho} - (-1)^\ell e^{2i\bar{\delta}_\ell} e^{ik\rho}], \tag{3.29}$$

so that $\bar{\delta}_\ell$ are the phase-shifts of a quantum mechanical scattering by the potential $\bar{V}_\ell(\rho)$ described by a "physical" solution with the asymptotic behavior

$$\bar{\psi}_\ell(\rho) \underset{\rho \mapsto \infty}{\sim} e^{i\bar{\delta}_\ell} \sin(k\rho - \ell\pi/2 + \bar{\delta}_\ell). \tag{3.30}$$

The behavior (3.28) of $\bar{\delta}_\ell$ (k) as k $\mapsto$ ∞ is "physically" reasonable since, as the energy increases, the effect of the potential $\bar{V}_\ell$ (ρ), itself independent of energy, should correspondingly fade away.

However, this quantum mechanical process does not occur as such, so that it is "physical" only to the extent that it can be related to the truly physical process we are investigating, namely the scattering of electromagnetic waves by a sphere of index of refraction n(r). The phase-shift $\bar{\delta}_\ell$ is not one of the real phase-shifts δ_ℓ appearing in Eq. (2.28). Nevertheless, these phase-shifts are interrelated, since ψ_ℓ (ρ) and ψ_ℓ (r) = G_ℓ (r) are interrelated through the Liouville transformation defined by Eqs. (3.2). It follows thus that

$$\delta_\ell(k) = \bar{\delta}_\ell(k) + k(\rho_0\text{-}r_0) - \frac{\pi}{2} \tag{3.31}$$

which implies that δ_ℓ (k) increase linearly with k, a reasonable conclusion in view of the fact that the potential (3.1b) also increases with k (quadratically).

The relation (3.31) shows now that if δ_ℓ (k) for some fixed ℓ is known as a function of k, then $\bar{\delta}_\ell$ (k) is also known since, taking into account Eq. (3.28),

$$\bar{\delta}_\ell(k) = \delta_\ell(k) - k \lim_{s \mapsto \infty} s^{-1}\delta_\ell(s) + \frac{\pi}{2} \ . \tag{3.32}$$

The electromagnetic inverse scattering problem of finding the index of refraction n(r), given a phase-shift is thus reduced to the quantum- mechanical inverse scattering problem of finding the potential $\bar{V}_\ell$ (ρ), given the phase-shift $\bar{\delta}_\ell$ (k). The latter has a well-established solution we will now briefly describe.

One begins by noting that, since both w_ℓ (kρ) and kρ j_ℓ (kρ) satisfy Eq. (3.9) with an identically zero right-hand side,

$$\int_\rho^\infty w_\ell(k\rho')j_\ell(k'\rho')d\rho' = \frac{w_\ell(k\rho) \dfrac{d}{d\rho} [k'\rho j_\ell(k'\rho)] - k'\rho j_\ell(k'\rho) \dfrac{d}{d\rho} w_\ell(k\rho)}{k'^2 - (k + i\epsilon)^2}$$

$$\tag{3.33}$$

Then one considers the integral

$$\mathcal{F}_\ell(k, \rho) = \frac{(-1)^\ell}{\pi} \int_C dk' \bar{V}_\ell(k', \rho) \int_\rho^\infty d\rho' w_\ell(k\rho') k'\rho' j_\ell(k'\rho'), \tag{3.34}$$

where the closed contour of integration, C, consists of the real axis and a semi-circle of indefinitely large radius in the upper half-plane. The integral (3.34) is evaluated in two different ways.

First, because the integrand is an analytic function in the upper half-plane, except for a pole at k + iϵ, the theorem of residues can be employed. Since at this pole the numerator in (3.33) becomes a Wronskian, one obtains immediately the result:

$$\mathcal{F}_\ell(k, \rho) = \bar{V}_\ell(k, \rho) . \tag{3.35}$$

Second, the contributions from the real axis and semi-circular portions of C are separated, viz.,

$$\mathcal{F}_\ell(k, \rho) = \mathcal{F}_\ell^{(1)}(k, \rho) + \mathcal{F}_\ell^{(2)}(k, \rho) \tag{3.36}$$

and these contributions are separately evaluated. Because

$$k\rho j_\ell(k\rho) = -\frac{1}{2}[w_\ell(-k\rho) - (-1)^\ell w_\ell(k\rho)], \tag{3.37}$$

the contribution of the real axis portion can be written as

$$\mathcal{F}_\ell^{(1)}(k, \rho) = \frac{(-1)^{\ell+1}}{2\pi} \int_{-\infty}^{\infty} dk' [\bar{V}_\ell(-k, \rho) - (-1)^\ell \, \bar{V}_\ell(k', \rho)]$$

$$\times \int_\rho^\infty d\rho' \, w_\ell(k\rho') w_\ell(k'\rho') . \tag{3.38}$$

Using the relation between the Jost solution and the physical solution

$$\bar{\psi}_\ell(k, \rho) = \frac{k^{\ell+1}}{\bar{V}_\ell(k)} \, \bar{v}_\ell(\rho) = \frac{i}{2}[\bar{V}_\ell(-k, \rho) - (-1)^\ell e^{2i\bar{\delta}_\ell(k)} \, \bar{V}_\ell(k, \rho)], \tag{3.39}$$

one obtains

$$\mathcal{F}_\ell^{(1)}(k, \rho) = \frac{-1}{2\pi} \int_{-\infty}^{\infty} dk' \, [e^{2i\bar{\delta}_\ell(k')} - 1]\bar{V}_\ell(k', \rho) \int_\rho^\infty \cdot d\rho' \, w_\ell(k\rho') w_\ell(k'\rho')$$

$$+ \frac{(-1)^{\ell+1}}{\pi i} \int_{-\infty}^{\infty} dk' \bar{\psi}_\ell(k', \rho) \int_\rho^\infty d\rho' \, w_\ell(k\rho') \, w_\ell(k'\rho') . \tag{3.40}$$

Because the integrand of the term containing the physical solution $\bar{\psi}_\ell(k, \rho)$ is an entire function of k' (as we have seen, there are no bound states in this problem), this integral equals its negative evaluated along the semi-circle portion of C. Thus it can be combined with $\mathcal{F}_\ell^{(2)}$ and evaluated by using the asymptotic forms of the solutions for large k in $\text{Im} k \geq 0$. The combined result is $w_\ell(k\rho)$. Therefore the integral equation

$$\bar{\psi}_\ell(k, \rho) = w_\ell(k\rho) - \frac{1}{2\pi} \int_{-\infty}^{\infty} dk' \, [e^{2i\bar{\delta}_\ell(k')} - 1]\bar{V}_\ell(k', \rho)$$

$$\times \int_\rho^\infty d\rho' w_\ell(k\rho') w_\ell(k'\rho'), \tag{3.41}$$

is obtained which, in principle, solves the inverse scattering quantum mechanical problem: given $\bar{\delta}_\ell(k)$, one solves Eq. (3.41) for the Jost solution $\bar{V}_\ell(k, \rho)$ which, when substituted in Eq. (3.1), allows for immediate identification of the potential $\bar{V}_\ell(\rho)$.

However, Eq. (3.41) has the unpleasant feature of being singular, a circumstance which, when possible, is preferably avoided. In the present case, it can be avoided by recasting the solution according to the so-called Marchenko formulation. By introducing the function

$$A_\ell(\rho, \sigma) = \frac{(-1)^\ell}{\pi} \int_{-\infty}^{\infty} [\bar{\psi}_\ell(k, \rho) - w_\ell(k, \rho)] k\sigma j_\ell(k\sigma) \, dk \qquad (3.42)$$

and taking into account the bounds:

$$|\bar{V}_\ell(k, \rho) - w_\ell(k, \rho)| < C \left(\frac{1 + |k|\rho}{|k|\rho}\right)^\ell e^{-\rho \text{Im} k}$$

$$|k\sigma j_\ell(k\sigma)| < e^{\sigma \text{Im} k}, \qquad (3.43)$$

one recognizes that

$$A_\ell(\rho, \sigma) = 0 \qquad \text{for } \rho > \sigma . \qquad (3.44)$$

The integral representation

$$\bar{V}_\ell(k, \rho) = w_\ell(k\rho) + \int_\rho^\infty A_\ell(\rho, \sigma) w_\ell(k\sigma) \, d\sigma \qquad (3.45)$$

is obtained by taking the Hankel transform of Eq. (3.42). Substitution of this representation in Eq. (3.41) yields the Marchenko equation:

$$A_\ell(\rho, \sigma) = B_\ell(\rho, \sigma) + \int_\rho^\infty d\tau \, A_\ell(\rho, \tau) B_\ell(\tau, \sigma) , \qquad (3.46)$$

where

$$B_\ell(\rho, \sigma) = \frac{1}{2\pi} \int_{-\infty}^{\infty} dk [e^{2i\delta_\ell(k)} - 1] w_\ell(k\rho) w_\ell(k\sigma) . \qquad (3.47)$$

Finally, one establishes the direct connection between $A_\ell(\rho, \sigma)$ and the potential $\bar{V}_\ell(\rho)$:

$$\bar{V}_\ell(\rho) = -2 \frac{d}{d\rho} A_\ell(\rho, \rho) . \qquad (3.48)$$

Eqs. (3.46), (3.47) and (3.48) provide the solution to the quantum mechanical inverse scattering problem: given $\delta_\ell(k)$, one evaluates the function $B_\ell(\rho, \sigma)$ by calculating the integral (3.47). Substitution of the result in Eq. (3.46) allows integration of the Marchenko equation, the solution of which provides the potential $\bar{V}_\ell(\rho)$.

4. RECONSTRUCTION OF n(r)

Once the quantum mechanical inversion is carried out and the potential $\bar{V}_\ell(\rho)$ is found, we may return to the electromagnetic inverse problem. The task now is, given $\bar{V}_\ell(\rho)$, find $\nu(\rho)$. In other words, we must solve Eq. (3.10) which, with (3.4), reads

$$[\nu(\rho)]^{-1/2} \frac{d^2}{d\rho^2} [\nu(\rho)]^{1/2} - \ell(\ell + 1) [\rho^{-2} - R^{-2}(\rho)] = \bar{V}_\ell(\rho) \tag{4.1}$$

where, from Eqs. (3.5) and (3.2),

$$R(\rho) = \nu(\rho) \int_0^\rho \frac{dt}{\nu(t)} \tag{4.2}$$

Eq. (4.1), as an equation for $\nu(\rho)$, has a formidable aspect: strongly nonlinear, integro-differential, singular and, its solution should also be ℓ-independent. Realistically speaking, such an equation could be expected to have a solution, unique and ℓ-independent, only if a "miracle" happened.

However, the nature of the miracle one expects to occur in this case is closely related to the miracle discussed by Newton in his solution of the three dimensional quantum mechanical inverse scattering problem. The situation can be described as follows.

If one starts with a given, arbitrary function $\delta_\ell(k)$ there is no reason to expect the reconstructed potential $\bar{V}_\ell(\rho)$ to be such that Eq. (4.1) will have a solution. Only if $\delta_\ell(k)$ is a phase shift resulting from the actual scattering by a target with the assumed general properties of n(r) is it reasonable to attempt to reconstruct the underlying function n(r). It is not reasonable to attempt to **construct** a function n(r) for each $\delta_\ell(k)$ one can think of, but it does make sense to **reconstruct** the function n(r) which has yielded a given $\delta_\ell(k)$.

The implication of this attitude is that a solution of Eq. (4.1) should be sought only for that class $\mathcal{C}_\ell$ of functions $\bar{V}_\ell(\rho)$ that can be obtained from functions $\delta_\ell(k)$ representing physically performable scattering experiments on objects with the general properties outlined above. In particular, functions $\bar{V}_\ell(\rho)$ from this class must be such that, given $\delta_\ell(k)$ (and thus the limit $\rho_o - r_o = \lim_{s \to \infty} s^{-1} \delta_\ell(s)$), there exists a distance ρ_o such that, for $\rho \geq \rho_o$,

$$\bar{V}_\ell(\rho) = \ell(\ell + 1) [\rho^{-2} - (\rho - \rho_o + r_o)^{-2}] . \tag{4.3}$$

Hence, as $\bar{V}_\ell(\rho)$ is given, the distance ρ_o beyond which $\nu(\rho)$ is known to be identically unity is readily identifed. Moreover, the functions in the class $\mathcal{C}_\ell$ must all be such that

$$\bar{V}_\ell(\rho) + \ell(\ell + 1) (\rho^{-2} - R^{-2}(\rho)) \text{ is bounded for all } \rho \geq 0, \tag{4.4}$$

including the origin $\rho = 0$, which implies that only such solutions of Eq. (4.1) should be sought for which $R(\rho) > 0$ has the representation

$$R(\rho) = \begin{cases} \rho - \dfrac{[s\,V_\ell(s)]\,s = 0}{2\ell(\ell+1)}\,\rho^2 + \rho^3 S(\rho) & (0 \le \rho \le \rho_0) \\[2em] \rho - \rho_0 + r_0 & (\rho \ge \rho_0) \end{cases} \qquad (4.5)$$

where $S(\rho)$ is a continuous, twice differentiable function, bounded for all $\rho \epsilon$ [0, ρ_0]. The fact that such information is available for the function $R(\rho)$ suggests that it might be more convenient to consider this function rather than $\nu(\rho)$ as the unknown function in Eq. (4.1). This change is easily done because Eq. (3.50) can be immediately solved for $\nu(\rho)$ yielding

$$\nu(\rho) = \exp \int_\rho^{\rho_0} \frac{1 - R'(t)}{R(t)\,.}\,dt \qquad (4.6)$$

Substitution of (4.6) into Eq. (4.1) yields a second order nonlinear differential equation for R which, in turn, can be written as a system of two first order nonlinear differential equations. Namely, if we introduce the function

$$P(\rho) = \frac{1 - R'(\rho)}{2R(\rho)}\,, \qquad (0 \le \rho \le \rho_0) \qquad (4.7)$$

the following system results:

$$R' = 1 - 2\,PR$$

$$P' = P^2 - \ell(\ell+1)\,(\rho^{-2} - R^{-2}) - \bar{V}_\ell(\rho) \qquad (4.8)$$

to be solved for $P(\rho)$ and $R(\rho)$ for $0 \le \rho \le \rho_0$ with the boundary conditions $P(\rho_0) = 0$, $R(\rho_0) = r_0$, say, where r_0 may be assumed known, since both $\rho_0 - r_0$ and ρ_0 and have already been identified. The first of the Eqs. (4.8) can then be given the integral form:

$$R(\rho) = r_0 \exp\left(2\int_\rho^{\rho_0} P(s)ds\right) - \int_\rho^{\rho_0} du \exp\left(2\int_\rho^{u} P(s)ds\right)\,. \qquad (4.9)$$

Similarly, the second of Eqs. (4.8) can be given the integral form:

$$P(\rho) = -\int_\rho^{\rho_0} ds\, P^2(s) + \int_\rho^{\rho_0} [\ell(\ell+1)\,[s^{-2} - R^{-2}(s)] + \bar{V}_\ell(s)]ds \qquad (4.10)$$

The original equation (4.1) is thus reduced to a system of two coupled (non-linear) Volterra equations, (4.9) and (4.10). To prove the existence and uniqueness of a solution (R, P) of this system, all we need to show, in principle, is that there exists a domain D(s,R, P) for the variables s, R and P in which the integrands of Eqs. (4.9) and (4.10) are continuous as functions of s, and satisfy Lipschitz conditions as functions of R or P (which, of course, would be implied if the integrands had in D continuous partial derivatives of first order with respect to R and P). This can be easily seen to be the case, provided one excludes the origin $\rho = 0$.

In order to analyze the situation in the origin we must examine carefully the second integrand in Eq. (4.10). Since $R(\rho)$ must have the representation (4.5), the quantity

$$\alpha = \frac{[s\,V_\ell(s)]_{s\,=\,0}}{2\ell(\ell + 1)} \tag{4.11}$$

must exist and be ℓ-independent. This implies that the class $\mathcal{C}_\ell$ of potentials $\bar{V}_\ell\,(\rho)$ corresponding to underlying functions $n(r)$ consists of functions of the form

$$\bar{V}_\ell(\rho) = \begin{cases} \dfrac{2\ell(\ell + 1)\,\alpha}{\rho} \; + \; \bar{\bar{V}}_\ell(\rho) & (0 \le \rho \le \rho_o) \\[4mm] \ell(\ell + 1)\,[(\rho - \rho_o + r_o)^{-2} - \rho^{-2}] & (\rho \ge \rho_o) \end{cases} \tag{4.12}$$

where $\bar{\bar{V}}_\ell\,(\rho)$ is integrable in $[0, \rho_o]$.

Likewise, from (4.5) and (4.7) it follows that $P(\rho)$ must have the representation:

$$P(\rho) = \alpha + \rho\,T(\rho) \tag{4.13}$$

where $T(\rho)$ is continuous in $[0, \rho_o]$, so that

$$P(0) = \alpha \tag{4.14}$$

Using (4.14) and $R(0) = 0$ as boundary conditions, Eqs. (4.8) can be given the integral form

$$R(\rho) = \int_o^\rho du \exp\left(-2 \int_u^\rho P(s)\,ds\right) \tag{4.15}$$

$$P(\rho) = \alpha + \int_o^\rho P^2(s)\,ds - \int_o^\rho [\ell(\ell + 1)\,[s^{-2} - R^{-2}(s)] + \bar{V}_\ell(s)]ds.$$

The first of these equations coincides with Eq. (4.9) provided

$$r_o = \int_o^{\rho_o} du \exp\left(-2 \int_u^{\rho_o} P(s)ds\right) = \int_o^{\rho_o} \frac{du}{v(u)}$$

which is actually the definition of ρ_o. This equation also shows that $R(\rho)$ vanishes only at the origin (linearly), and that for all $\rho\epsilon(0, \rho_o]$,

$$\rho^{-1}\,R(\rho) = 1 - \alpha\rho + \rho^2\,S(\rho) \rangle\,0\,. \tag{4.16}$$

Taking into account the representations (4.5), (4.12) and (4.13), Eq. (4.15) becomes a system of equations for $S(\rho)$ and $T(\rho)$:

$$S(\rho) = 2\rho^{-3} \int_0^\rho s^2 ds \, [\alpha^2 + \alpha s \, T(s) - T(s)] \exp\left[-2\alpha(\rho - s) - 2 \int_s^\rho t \, T(t) dt\right]$$

$$T(\rho) = \rho^{-1} \int_0^\rho \left\{ [\alpha + s \, T(s)]^2 + \ell(\ell + 1)\left(\frac{[\alpha - s \, S(s)]^2}{[1 - \alpha s + s^2 S(s)]^2} \right.\right.$$

$$\left. + 2 \, \frac{\alpha^2 + (1 - \alpha s) S(s)}{1 - \alpha s + s^2 S(s)} \right) - \bar{\bar{V}}_\ell(s) \right\} ds \tag{4.17}$$

The values at the origin of the functions $S(\rho)$ and $T(\rho)$ (and of their derivatives) are easily seen to be given in terms of the value of $\bar{\bar{V}}_\ell(\rho)$ (and its derivatives) at the origin:

$$S(o) = \frac{2}{3} \, \frac{\ell(\ell + 1)(4\alpha^2 - 7) + 3 \, \bar{\bar{V}}_\ell(o)}{3 + 4 \ell(\ell + 1)}$$

$$T(o) = \frac{3\alpha^2 + 7\ell(\ell + 1) - 3 \, \bar{\bar{V}}_\ell(o)}{3 + 4\ell(\ell + 1)} \tag{4.18}$$

so that their existence requires corresponding existence conditions on the behavior of $\bar{\bar{V}}_\ell(\rho)$ at the origin. In any event, it is apparent now that the integrand in the second of Eq. (4.17) is no longer singular. The arguments leading now to a proof of existence and uniqueness of the solution are rather standard and will not be detailed here.

Clearly, once $R(\rho)$ is known, $\nu(\rho)$ follows via Eq. (4.6). Hence, the inverse Liouville transformation

$$r(\rho) = \int_0^\rho \frac{d\tau}{\nu(\tau)} \tag{4.19}$$

yields the function $r(\rho)$, the inverse of which, upon substitution in $\nu(\rho)$, finally produces $n(r)$.

To conclude, let us just mention that if the input of the electromagnetic scattering problem were an η_ℓ phase shift rather than a δ_ℓ, the reduced Schroedinger Eq. (3.1) with the potential (3.1a) also becomes a Schroedinger-like equation under the Liouville transformation (3.2), except that in this case, instead of the k-independent potential $\bar{V}(\rho)$ of Eq. (3.4) one obtains the potential

$$\bar{V}(\rho) = [\nu(\rho)]^{1/2} \frac{d^2}{d\rho^2} [\nu(\rho)]^{-1/2}, \tag{4.20}$$

also k-independent. The procedure outlined above can then be followed without essential modifications.

ACKNOWLEDGEMENTS

This work was performed under the McDonnell Douglas Independent Research and Development program.

REFERENCES

1. See e.g.: H. C. van de Hulst, *Light Scattering by Small Particles,* John Wiley and Sons, Inc., New York; R. Newton, *Scattering Theory of Waves and Particles,* McGraw-Hill Book Co., New York 1966; M. Kerker, *The Scattering of Light and other Electromagnetic Radiation,* Academic Press, New York 1969.

2. See e.g.: C. Eftimiu, *Direct and Inverse Scattering by a Sphere of Variable Index of Refraction,* J. Math. Phys. Vol 23, pp. 2140-2146, 1982, and the references given therein.

I.10 (SP.6)

APPLICATIONS OF THE ABEL TRANSFORM IN REMOTE SENSING AND
RELATED PROBLEMS

H. Süss

Deutsche Forschungs- und Versuchsanstalt
für Luft- und Raumfahrt
Institut für Hochfrequenztechnik
8031 Oberpfaffenhofen, FRG

ABSTRACT

The Abel transform often applied in remote sensing on account of
its analytical solution is used to determine density profiles from
integrated line measurements if cylindrical or spherical symmetry
is present. The purpose of this paper is to describe and to dispose
the common physical aspects for three different fields of application in remote sensing with electromagnetic or elastic waves:
atmospheric sounding, plasma- and geophysics. In a further application example - the sounding of the solar corona during the Helios
occultation experiment - a combined analytical-numerical method is
introduced and fully described. The physical and numerical applicability is demonstrated by inversion of simulated and measured
electron content data sets.

1. INTRODUCTION TO THE MATHEMATICAL AND PHYSICAL BACKGROUND

In 1826 Niels Henrik Abel solved in one of his earliest works [1]
the tautochrone problem, which arises in mathematical physics and
was posed by Huygens: A material point moving under the influence
of gravity along a curve $s(y)$ in a vertical xy-plane takes the time
$t(h = y_0)$ to move from the vertical height h to a fixed point O
on the curve. The question is to find a curve $s(y)$ such that
$T = T(h)$ is a given function. As usual using the mechanical energy
conservation law

potential energy: $mg (h - y)$
kinetic energy: $1/2 \ mv^2$

(m: mass, g: acceleration of gravity) and the common expression for

177

W.-M. Boerner et al. (eds.), Inverse Methods in Electromagnetic Imaging – Part 1, 177–201.
© *1985 by D. Reidel Publishing Company.*

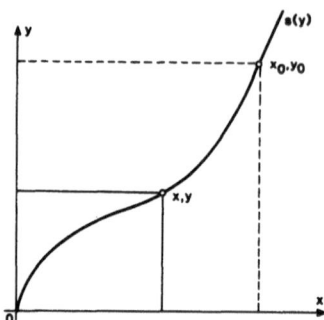

Figure 1. Representation of the coordinate frame for
a material point moving along the curve s(y).

velocity v = ds(y)/dt the expression

$$dt = - \frac{ds(y)}{\sqrt{2g(h - y)}}$$

follows. Substitution of the relation ds/dy = s'(y) and integration
from y = 0 to y = h leads to the following integral

$$T(h) = \int_0^h \frac{u(y)}{\sqrt{h - y}} \, dy, \qquad (1)$$

where $T(h) = (2g)^{-1/2} t(h)$ and s'(y) = u(y). The problem is to
find the function u(y) when T(h) is given in such a way that equation (1) holds. The difficulty arises that T(h) depends not only
on a single value but on all values of the unknown function u(y).
The above equation is known as the Abelian equation.

The arguments for detailed derivation of this problem which is relatively easy to survey and often described concern the analogies
between classical mechanics, geometrical optics and wave mechanics.
The link is given by the formal correspondence between the
Schrödinger equation, the iconal and the Hamilton-Jacobian equation
(see [2]). They lead immediately from classical mechanics to other
ranges of applications of the Abel transform e.g. to remote sensing
with electromagnetic or seismic waves. Simplifying assumptions in
mechanics namely the model of the material point, the frictionless
gliding along the curve and the presence of the gravitation potential (only depending on the radial distance from the earth's center)
are also necessary in the other scopes. In the field of remote
sensing - the main point of this paper - it will be assumed that
Fermat's principle is valid. This means that diffraction and interference effects can be neglected. Furthermore Fermat's principle
includes that the propagation medium is reciprocal and the charac-

teristical scale length is much greater than the vacuum wavelength of the sounding wave.

Equation (1) originates by a chain of simplifying assumptions of the generalized Abel transform. Now this course will be followed in the opposite direction. Equation (1) belongs to a class of integral equations of the more general type

$$T(h) = \int_{0}^{h} \frac{s(y)}{(h - y)^{\lambda}} \, dy, \qquad (2)$$

where λ takes a value between 0 and 1 (for the case $\lambda = 1/2$ the equations (1) and (2) get identity). The singular integral equation is one of the Volterra type of first the kind. For equation (2) exists the analytical solution

$$s(h) = \frac{\sin \pi \lambda}{\pi} \int_{0}^{h} \frac{T'(y)}{(h - y)^{1-\lambda}} \, dh. \qquad (3)$$

For $\lambda = 1/2$ the solution of the special case of equation (1) can easily be derived.

As a further generalization of equation (2) there exists the generalized Abel equation

$$T(h) = u(h) \int_{0}^{h} \frac{s(y)}{(h-y)^{\lambda}} \, dy + v(h) \int_{h}^{b} \frac{s(y)}{(h-y)^{\lambda}} \, dy, \qquad (4)$$

where the given functions $u(h)$ and $v(h)$ are defined on the interval $[a, b]$ with derivatives satisfying Hölder's condition. A description of their extensive solutions which rarely arise in elastic theory, is given for review in [3, 4, 5].

Only for complete termination of the pursued course should the mathematical treatement of systems of generalized Abelian integral equations be mentioned here. Whereas mathematical solutions can be found in [6, 7], practical application in remote sensing is outside the author's scope.

After this short outline of the mathematical background the basics of the physical background will follow for an illustrative understanding of the application in remote sensing. The fundamental physical ideas are limited for demonstration only to a two dimensional medium (an extension to three dimensions works in the same way). Generally the problem can be formulated as (see Fig. 2): The unknown, inhomogenuous two-dimensional density distribution $f(x,y)$

- x and y are the coordinates of the reference frame - is detected
by the measurement of the value of a particular line integral along
the integration path l

$$g(p,\Theta) = \int_l f(x,y)\ dl. \tag{5}$$

For the fixed parameter Θ and by variation of the ray or impact
parameter p the measurement quantity $g(p,\Theta)$ gives one projection
onto a straight line parallel to p. Variation of the parameter Θ
leads for practical applications to a finite set of projections,
which will be used for the (density-)reconstruction of $f(x,y)$.
The function $g(p,\Theta)$ is known as the Radon transform of $f(x,y)$. The
solution of this inversion problem arising nowadays in computerized
tomography [8], radio astronomy [9] and radar target shape estimation [10] was given by J. Radon [11].

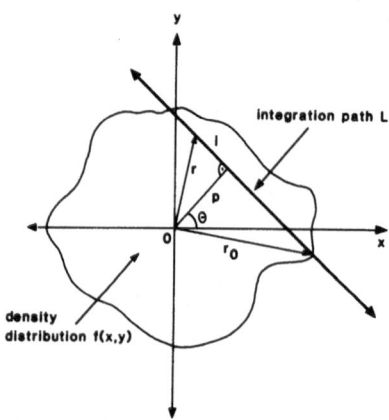

Figure 2. Geometrical representation of the line L
(represented in dependence on the parameters p and Θ)
passing the two-dimensional density distribution $f(x,y)$
to be probed indirectly.

In a special case which leads to the Abel transform the two-dimensional density distribution $f(x,y)$ takes on circular symmetry; therefore in equation (5) the function $g(p,\Theta)$ reduces to $g(p)$. Transformation of the line element dl into a cylindrical coordinate frame
(using $r^2 = p^2 + l^2$ for the derivation of dl/dr) leads to the integral equation (for the case without ray bending)

$$g(p) = 2 \int_p^{r_o} \frac{rf(r)}{\sqrt{r^2 - p^2}}\ dr. \tag{6}$$

Performing the substitution of variables $h = r_0^2 - p^2$ and $y = r_0^2 - r^2$ the equation (6) is identical to the Abel equation (1). Only in this special case for rotational symmetry all projections are identical with one another by varying the angle Θ (see Fig. 2).

As the last point of this introduction it will be hinted to some useful relationships - mainly used in radio astronomy - between the Abel, Fourier and Hankel transforms [9]. Fig. 3 demonstrates that the Fourier transform of the Abel transform is a Hankel transform and that Abel transform in the function domain corresponds to the inverse Abel transform in the frequency domain. Furthermore in [9] is given a table with various and often used functions which are related by these transforms.

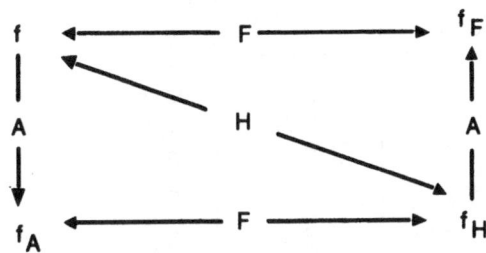

Figure 3. Schematic representation of the relationships between the Abel (A), Fourier (F) and Hankel (H) transforms [9]. f_A, f_F and f_H are the corresponding transforms of f.

2. APPLICATIONS IN VARIOUS FIELDS OF REMOTE SENSING

After some hints of possible applications in the introduction, three important areas of application of the Abel transform will be described in more detail in this section, namely the application in geo- and plasmaphysics and in atmospheric sounding. For an easy survey the meaning of the physical parameters of the integral equation are summarized in table 1.

The Abel transform found its first application in seismology in 1907, introduced by Herglotz and Wiechert [12]. It was used to determine from the measurement of travel times the velocity distribution of compressional (P) and transverse (S) waves which propagate inside the earth's mantle with different velocities. If an earthquake or an artificial explosion takes place at a point on or beneath the earth's surface the influence will travel to a distant point B (see Fig. 4). The travel time between A and B is related to the vertical distribution of the propagation speed which increases with depth. As a result the downward travelling ray paths are bent . Each ray is

field of application	integral equation	interpretation of the kernel K(p,r)	measured quantity T(p)	unknown quantity F(r)	upper integration boundary r_o
geophysics	$$T(p) = \int_{r_0}^{p} K(p,r)\,dr = 2\int_{r_0}^{p} \frac{rF(r)\,dr}{\sqrt{\gamma^2 - p^2}}$$	interaction of elastic waves with matter	travel time	density distribution inside the earth	earth's surface
plasma physics		interaction of electromagnetic waves with gaseous plasmas	integrated line intensity	distribution of particles in the plasma	plasma boundary
atmospheric sounding			ray bendig angle, phase or time delay	distribution of electrons in the atmosphere	effective atmospheric boundary

Table 1. Synoptical table of applications of the Abel integral equation in remote sensing (γ denotes the product of the refractive index of the propagation medium times the radial distance from the coordinate center).

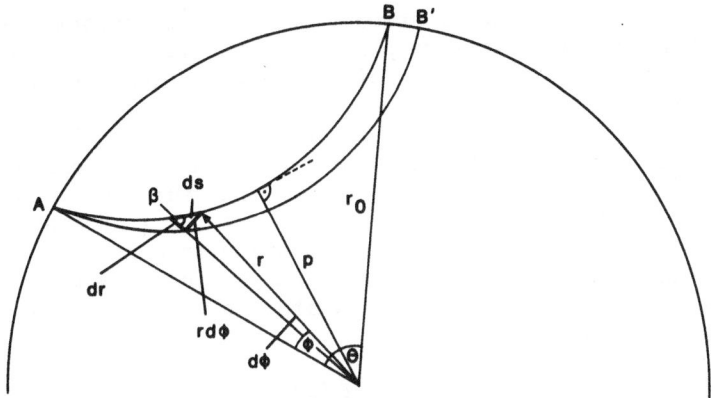

Figure 4. Geometrical notations used for the seismic problem.

characterized by the ray parameter p which is constant along the whole ray and for which the following relation holds (see Fig. 4)

$$p = \frac{r \sin \beta}{v} = \frac{r}{v} \frac{rd\phi}{ds} \tag{7}$$

(known as Bouger's rule).
Using a further relation also derived from Fig. 4

$$(ds)^2 = (dr)^2 + r^2 (d\phi)^2 \tag{8}$$

and the ray equation for the travel time

$$T(p) = \int_s \frac{1}{v} ds \tag{9}$$

$(v = v(r)$: wave velocity of the elastic medium) after elimination of $d\phi$ and straightforward computation the integral equation

$$T(p) = 2 \int_p^{r_0} \frac{\eta^2}{\sqrt{\eta^2 - p^2}} \frac{dr}{r} \tag{10}$$

(where $\eta = r/v(r)$) can be obtained.

Analytical solutions of this kind of Abel integral equations can be found in [13, 14] where a lot of results are described too. The limits of this method are given by discontinuous changes of the density e.g. on the border between the earth's core and mantle where reflections of the waves occur.

The second area of application of the Abel transform deals with
the diagnostics of gaseous plasmas in the laboratory and in space,
whereby the physical origin of the measured quantity is the radiation of the plasma. For diagnostics of laboratory plasmas the
measurement configuration is shown in Fig. 5. It will be assumed
that the plasma is optically thin and that the emission and absorption coefficient only depends on the radial coordinate. Consequently the ray path doesn't have a bent but it forms a straight
line along which the side-on observation is performed. The plasma
tube can be divided into concentrical cylinders with different
radii. For every concentrical cylinder with index k the relation
of radiation balance holds

$$dI_k(r) = \varepsilon_k(r)\ dx - \alpha_k(r)\ I_k(r)\ dx$$
$$= i_k(r)\ dx$$

(11)

(I_k, ε_k, α_k: intensity, emission and absorption coefficients of
the k-th shell: dx infinitesimal length element). The measurement
quantity, the integrated line intensity along the direction of observation can be expressed as

$$I(p) = \int_{-x}^{x} i(r)\ dx.$$

(12)

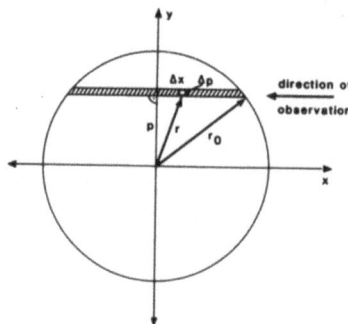

Figure 5. Sketch of the
geometrical configuration
for probing gaseous plasmas
in cylindrical tubes.

Transforming this equation into cylindrical coordinates by using
the relation $r^2 = x^2 + p^2$ for the calculation of dx the corresponding equation is yielded and nearly identical to equation (6). For
this kind of remote probing of laboratory plasmas, which is widely
used, a large number of numerical results are described in [15, 16].

Another related and currently used method is the determination of
the electron density distribution of the inner solar corona (called

K-corona which extends to about 3 R_Θ (solar radii)) by measuring
the polarization-brightness product of the white scattered light
during a natural or artificial eclipse [17].The data were collected
with a ground-based K-coronameter which records the product of polarization

$$P = \frac{I_T - I_R}{I_T + I_R}$$

(I_T, I_R: tangentially or radially emitted intensities) and the
brightness which is the denominator of P.

The visibility of the K-corona can be derived from the Thomson
scattering cross section which is related to the emitted power
(scattered per steradian and per unit area) by

$$\sigma(r) = N(r) \cdot R_\Theta \cdot \frac{8\pi}{3} \left(\frac{e^2}{mc^2}\right)^2$$

(e,m: charge and mass of a free electron; c: velocity of light;
$N(r)$ radial dependent electron density distribution). Whereas in

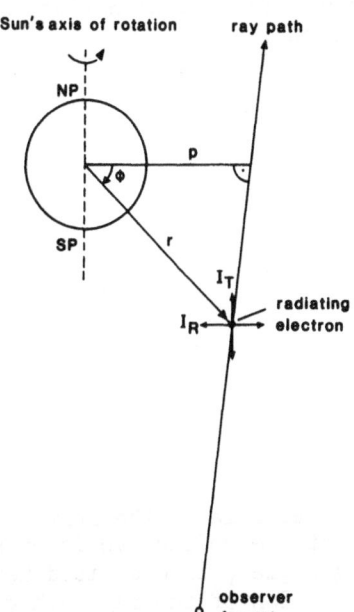

Figure 6. Sketch of the geometrical configuration for indirect probing the K-corona (NP,
SP: north-, southpole of the
sun).

equation (12) I(p) has to be replaced by the measured quantity pB,
the unkown distribution i(r) is replaced by $\sigma(r)$ or $N(r)$. The connection to the Abel transform follows in the same way as in the
previous example. Analytical results of this method and numerical

computations can be found in [18, 19].

The last part of this chapter deals with the applications for studying planetary atmospheres by radio occultation methods. The orbit
of a typical occultation experiment (e.g. the occultations of the
Helios B spacecraft) is shown in Fig. 7 demonstrating remote probing
of the solar corona. The principal observables of a radio occultation experiment are the amplitude (mainly used to determine absorption profiles) and the frequency of the signal. This last mentioned
measured quantity forms the base for deriving the radial distribution of the refractive index. After elimination of observed frequency
contributions resulting from spacecraft motion, rotation of the
earth etc. the remaining frequency residual is the time rate of the
change of the total phase caused by delay and refractive bending
of the propagation medium.

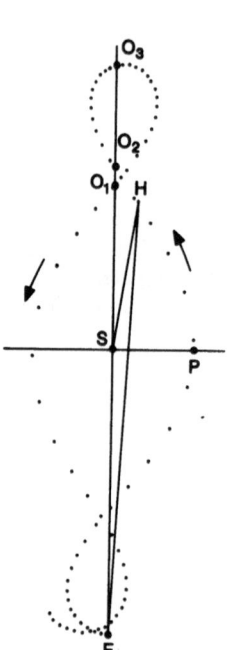

Figure 7. Geometry of the Helios B
occultation experiment presented in
a coordinate frame whose origin lies
at the center of the sun and whose
axis sun-earth remains fixed; 0_1, 0_2
and 0_3 situations of occultations
(E: earth, S: sun, H: Helios, P:
perihelion).

Since the phase path plays a fundamental role in the next chapter,
the relationships for determination of the refractive index profile
from the observed refractive bending angle ψ is described here as
an alternative example. The representation of the ray path geometry
in Fig. 8 can be used to get the relations between the angle of
incidence β, the bending angle ψ and the coordinates r and ϕ. They
are linked by the following equations

$$\tan \beta = \frac{r d\phi}{dr}, \quad \text{and} \quad d\psi = d\beta + d\phi, \tag{13}$$

whereby the last relation results from the fact that the sum of $\psi, -\beta$ and ϕ is equal to $\pi/2$. The expression for $d\beta$, which can be

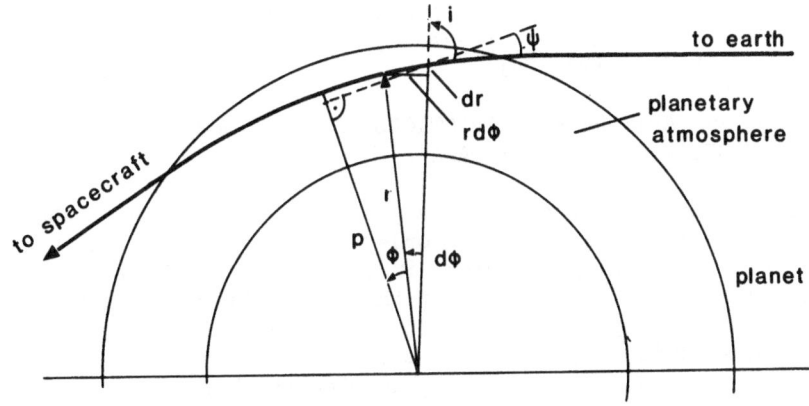

Figure 8. Refraction of a ray path in planetary atmosphere.

obtained by straightforward computation after differentiation of the first relation of equation (13), is given by

$$d\beta = \frac{- p \left(n + r \frac{dn}{dr}\right)}{nr \sqrt{n^2 r^2 - p^2}} \, dr.$$

An expression for $d\beta$ is deduced by the relation $\tan\beta = \sin\beta \cdot (1 - \sin^2\beta)^{-1/2}$ and by Bouger's rule (see equation (7)). For $d\phi$ the following holds

$$d\phi = \frac{p \, dr}{r \sqrt{n^2 r^2 - p^2}} \, .$$

Substitution of the expressions for $d\beta$ and $d\phi$ in the second relation of equation (12) leads to the integral equation for the bending angle

$$\psi(p) = -2 \int_p^{r_o} \frac{p \, dn/dr}{n \sqrt{n^2 r^2 - p^2}} \, dr \tag{14}$$

where r_o marks the effective boundary of the atmosphere (if no one exists: $r_o \to \infty$). These theories are well outlined in [20, 21] where also a lot of numerical exampels and results can be found inclusive of an exhaustive error analysis [22].

As a further possibility of probing atmosphere, which will only be hinted at here, is the radio probing the earth's atmosphere by a system of several **occultation** satellites. The experimental configuration as well as the relation between interesting meteorological parameters like density, pressure, temperature and water content are described in [23, 24].

3. PROBLEMS ARISING FROM PRACTICAL APPLICATIONS; INVERSION OF TIME DELAY OCCULTATION DATA

In this chapter a practical example of the inversion of Abel's equation will be performed in order to show in detail arising difficulties concerning numerical instabilities of the inversion algorithm and modelling problems of the propagation medium [25]. To avoid numerical problems an attempt was made to find a compromise for the solution by evaluating a combined analytical-numerical method; to reduce the modelling problem the experiment was started at a time where the best approximations of spherical symmetry of the propagation medium exist.

The physical background of this inversion problem is given by the experimental indirect probing of the solar corona measuring the time delay of radio signals(carrier frequency 2.115 GHz uplink/ 2.296 GHz downlink) during the occultation of the Helios B spacecraft [26]. The measured quantity is the difference between the phase- and group-delay time which is proportional to the electron content, the number of electrons integrated along the propagation path per unit area [27]. The main advantages of the employed measurement technique - developed at the Jet Propulsion Laboratory and called Differenced Range Versus Integrated Doppler (DRVID) - includes that all frequency independent errors (e.g. frequency drifts, inaccuracies in locus and velocity of the satellite) are left out whereas all frequency dependent errors (e.g. delay times in electronic systems on ground and on spacecraft, electron content of the terrestrial ionosphere) have to be taken into consideration *). As shown in [25] the total electron content error takes a value of $1.3 \cdot 10^{16}$ $1/m^2$ (corresponding delay time 14.84 ns) compared with characteristic electron content variations of $9 \cdot 10^{17}$ to $2.7 \cdot 10^{19}$ $1/m^2$ (variations of delay time from 100 ns to 3 μs).

As briefly mentioned above for a clear interpretation of the electron content measurements the following properties of the propagation medium concerning the launch time and the orbit of the spacecraft have to be regarded:

---

*) The DRVID technique records the change of the electron content with respect to some unknown reference value at the initial time of a measurement set.

i)   The spacecraft was launched so that during the period of data
     acquisition the solar activity had a minimum. During that time
     the density distribution of the solar corona nearly takes
     spherical symmetry above all relative to the ecliptic plane.

ii) The orbit lies nearly in the ecliptic plane, so that cylindri-
     cal symmetry is a good approximation.

iii) During the solar minimum structures of the charge density
     distribution keep for more than one solar rotation (period
     27 days).

Further simplifying assumptions concerning the modelling of the
propagation medium and the derivation of the integral equation
are necessary:

iv) The propagation medium is fully ionized and quasineutral.

v)   The magnetic field in the medium is so small that it can be
     neglected.

vi) The medium is so tenuous that the functional relation between
     signal frequency and refractive index can be linearized.

Then it is valid to write for the group and phase path (see also
Fig. 9)

$$\rho_g(p) = \int_o^{L(t)} \frac{1}{n}\, ds, \qquad \rho_{ph}(p) = \int_o^{L(t)} n\, ds$$

($L(t)$: length of the signal path as a function of time). Using the
linearization of the dispersion law, which relates the index of
refraction with the electron density N by

$$n^2 = 1 - \frac{e^2}{\pi\, m} \frac{N}{f^2}$$

(signal frequncy $f \approx 2.2$ GHz, solar plasma frequency $f_p \approx 0.9$ MHz
e.g. at $6\, R_\Theta$; this means that $f_p/f \approx 0.4 \cdot 10^{-3}$) and subtracting
the phase and group path, for the electron content holds

$$I(p) = \frac{\pi\, m}{e^2} f^2 \left(\rho_{ph}(p) - \rho_g(p)\right) = \int_o^{L(t)} N(r,\phi,t)\, ds. \qquad (15)$$

The electron density in the solar plasma depends on the radial distance r, the azimuth angle ϕ and on time t. In order to get an
analytical solution of equation (15) the following last additional
assumptions of the coronal model have to be made:

vii) The medium remains stationary for the period of measurements.

viii) The electron density only depends on the radial direction.

After transformation of equation (15) into cylindrical coordinates
the following generalized Abel equation (see eq. (4)) holds:

$$I(p) = \int_p^{r_0} \frac{rN(r)}{\sqrt{r^2 - p^2}} \, dr + \int_p^{r_1} \frac{rN(r)}{\sqrt{r^2 - p^2}} \, dr. \tag{16}$$

Figure 9. Occultation geometry
during data acquisition.

The expense for the solution of this equation can be diminuished
by assuming that r_1 is approximately equal to r_0. Estimation of
the electron content variations in the region around the earth of
about 10^{14} $1/m^2$ justifies this approximation. In equation (15) the
electron content $I(p'$ can be written as twice the first integral and
looks like equation (6). This integral equation has the following
analytical solution

$$N(r) = -\frac{1}{\pi} \frac{1}{r} \frac{d}{dr} \int_r^{r_0} \frac{p \, I(p)}{\sqrt{p^2 - r^2}} \, dp \tag{17}$$

when these assumptions are fulfilled:

- I(p) has to be continuous in the integral $[p, r_o]$

- I(p) must be zero at $p = r_o$ (includes the convergence of the integral)

- The derivation of I(p) may have a finite number of discontinuities.

Choosing a suitable solution, a combined analytical–numerical algorithm turned out to be numerically the most stable. Whereas a purely analytical solution is impossible because the measured quantity is a set of discrete values and not a continuous function the disadvantages of a purely numerical solution are:

- discretization errors caused by the transition from continuous to discrete formulation

- rounding errors caused by the finite number of places of the computer

- break-off errors caused by linearization (e.g. integration)

- measurement errors or noise with which every measured value is contaminated

- errors caused by amplification of noise by numerical differentiation in equation (18).

In order to avoid these error sources the inversion including the differentiation is performed analytically and the remaining part is done numerically. This method is outlined in the following part.

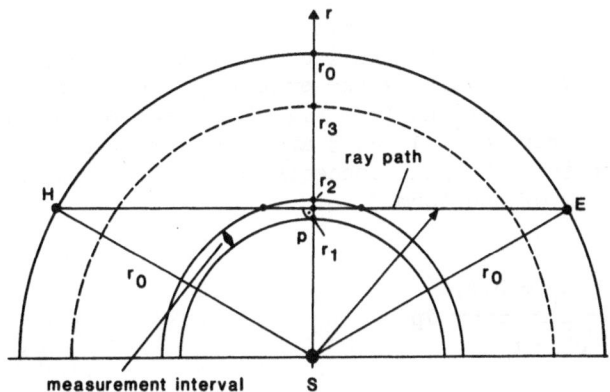

Figure 10. Occultation geometry at spherical symmetry of the density distribution.

For an analytic determination of the electron density $N(r)$ the
total ray path (see Fig. 10) is divided into three sections $[r_1,r_2]$,
$[r_2,r_3]$ and $[r_3,r_0]$ which lead to a corresponding splitting of
equation (17)

$$N(r) = - \frac{1}{\pi} \frac{1}{r} \frac{d}{dr} \qquad\qquad\qquad (18)$$

$$\left\{ \int_{r}^{r_2} \frac{p\, I_1(p)}{\sqrt{p^2 - r^2}}\, dp + \int_{r_2}^{r_3} \frac{p\, I_2(p)}{\sqrt{p^2 - r^2}}\, dp + \int_{r_3}^{r_0} \frac{p\, I_3(p)}{\sqrt{p^2 - r^2}}\, dp \right\}$$

shell I   shell II   shell III

measured electron   electron content model
content values

where each shell corresponds to one section of the ray path in the
order mentioned above. For the principal and most important point
the intervall $[r_1, r_2]$ is divided into M-1 concentrical shells
as shown in Fig. 11 (M: number of measured values). In order to
make the analytical evaluation of the integral for shell I possible,
in each partial interval $[r'_i, r'_{i+1}]$ the electron content is approximated by a straight line of the form

$$I_i(p) = a_i + b_i p \qquad\qquad \text{with} \qquad i = 1, 2, 3, \ldots, M$$

where the coefficients b_i are determined by

$$b_i = (I(r'_i) - I(r'_{i-1}))/(r'_i - r'_{i+1}).$$

(As shown in [25], a determination of the coefficients a_i is not
necessary, because they neutralize by limit calculation between
two neighboured intervals by the evaluation from table 2 to
equation (19)). As derived from Fig. 11, where for example the
courses of the ray paths 1, 2, 3 are shown explicitly, the entrance
of a ray path into the next concentrical shell causes a further
additional term of the form

$$\int_{r}^{r'_i} \frac{p(a_i + b_i p)}{\sqrt{p^2 - r^2}}\, dp$$

interval	expressions for density
$r_1' < r < r_2'$	$N_{11}(r) = \dfrac{-1}{\pi}\dfrac{1}{r}\dfrac{d}{dr}\left\{\displaystyle\int_r^{r_2'}\dfrac{p(a_1+b_1\,p)}{\sqrt{p^2-r^2}}\,dp + \int_{r_1'}^{r_3}\dfrac{p(A_2+C_2/p)}{\sqrt{p^2-r^2}}\,dp\right\}$
$r_2' < r < r_3'$	$N_{12}(r) = \dfrac{-1}{\pi}\dfrac{1}{r}\dfrac{d}{dr}\left\{\displaystyle\int_r^{r_3'}\dfrac{p(a_2+b_2\,p)}{\sqrt{p^2-r^2}}\,dp + \int_{r_2'}^{r_1'}\dfrac{p(a_1+b_1\,p)}{\sqrt{p^2-r^2}}\,dp + \int_{r_1'}^{r_3}\dfrac{p(A_2+C_2/p)}{\sqrt{p^2-r^2}}\,dp\right\}$
$r_3' < r < r_4'$	$N_{13}(r) = -\dfrac{1}{\pi}\dfrac{1}{r}\dfrac{d}{dr}\left\{\displaystyle\int_r^{r_4'}\dfrac{p(a_3+b_3\,p)}{\sqrt{p^2-r^2}}\,dp + \int_{r_3'}^{r_2'}\dfrac{p(a_2+b_2\,p)}{\sqrt{p^2-r^2}}\,dp + \int_{r_2'}^{r_1'}\dfrac{p(a_1+b_1\,p)}{\sqrt{p^2-r^2}}\,dp + \int_{r_1'}^{r_3}\dfrac{p(A_2+C_2/p)}{\sqrt{p^2-r^2}}\,dp\right\}$
$r_{M-1} < r < r_M$	$N_{1M}(r) = -\dfrac{1}{\pi}\dfrac{1}{r}\dfrac{d}{dr}\left\{\displaystyle\int_r^{r_M'}\dfrac{p(a_{M-1}+b_{M-1}\,p)}{\sqrt{p^2-r^2}}\,dp + \cdots + \int_{r_{i+1}'}^{r_i'}\dfrac{p(a_i+b_i\,p)}{\sqrt{p^2-r^2}}\,dp + \cdots + \int_{r_1'}^{r_3}\dfrac{p(A_2+C_2/p)}{\sqrt{p^2-r^2}}\,dp\right\}$
$r_j' < r < r_{j+1}'$	$N_{1j}(r) = -\dfrac{1}{\pi}\dfrac{1}{r}\dfrac{d}{dr}\left\{\displaystyle\int_r^{r_{j+1}'}\dfrac{p(a_j+b_j\,p)}{\sqrt{p^2-r^2}}\,dp + \sum_{i=2}^{j}\int_{r_i'}^{r_{i-1}'}\dfrac{p(a_{i-1}+b_{i-1}\,p)}{\sqrt{p^2-r^2}}\,dp + \int_{r_1'}^{r_3}\dfrac{p(A_2+C_2/p)}{\sqrt{p^2-r^2}}\,dp\right\}$ valid for $j \geq 2$

Table 2. Table of integrals for derivation of the recursive electron density formula

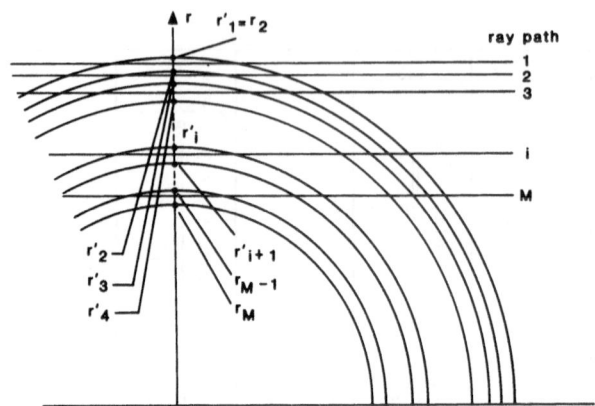

Figure 11. Model of concentrical shells of the measurement interval $[r_1, r_2]$ with passing ray paths.

to the preceding terms. The density expressions are summarized from
the first until to the (M-1)-th shell in table 2. For shell II and
III the following remarks have to be added (see [25]): first, esti-
mations of the absolute values of the integral of shell III have
shown, that it can be neglected compared with the integrals of
shell I and II; second, the electron content of shell II is approximated by $I_0(p) = A_2 + C_2/p$ (expansion of equation (A2) into a Taylor
series of second order) where the constants A_2 and C_2 are determined by continuity conditions at the boundaries. By straightforward
computation the following recursive formula for the electron density
distribution can be obtained from table 2

$$
N(r_j) = \begin{cases}
\dfrac{1}{\pi}\dfrac{A_2 + C_2\, r_3/r_1^2}{\sqrt{r_3^2 - r_1^2}} & \text{(19)}\quad \text{valid for } j = 1 \\[2em]
\dfrac{1}{\pi}\left[\dfrac{a_1 + b_1\, r_1'}{\sqrt{r_1'^2 - r_j^2}} + \dfrac{A_2 + C_2\, r_3/r_j'^2}{\sqrt{r_3'^2 - r_j^2}} - \dfrac{A_2 + C_2\, r_1'/r_j'^2}{\sqrt{r_1'^2 - r_j'^2}}\right. \\[2em]
\left. + b_1\ \text{arcosh}\ r_1'/r_j + \displaystyle\sum_{i=2}^{j-1} (b_i - b_{i-1})\,\text{arcosh}\ r_{j-1}'/r_j\right] \\[1.5em]
\hspace{6cm}\text{valid for } j \geq 2.
\end{cases}
$$

In order to examine the inversion error of equation (19) the following steps of computations were carried out:

$$N_A(r) \xrightarrow[\text{integration}]{\text{analytical}} I_A(p) \xrightarrow[\text{algorithm}]{\text{inversion}} N(r)$$

This is a suited procedure for determination of the inversion error because the calculations of the single steps can be performed analytically using for $N_A(r)$ and $I_A(p)$ the relations (A1) and (A2) (see appendix). The advantage is that break-off and discrimination errors are avoided and only rounding and approximation errors are taken into account. From the computational results - shown in Fig. 12 - can be seen that the absolute error defined as $(N_A(r) - N(r))/N_A(r)$ takes a value of 1.8% and that the slopes differ by 1.2%. Furthermore can be observed the increasing numerical stability with increasing number of simulated (also valid for measured) electron content values.

In order to demonstrate the influence of noise to the inversion algorithm, the same simulated electron content data set was contaminated by pseudo noise with a variance of 0.04% (corresponds to an electron content variance of $3.7 \cdot 10^{16}$ $1/m^2$ which is related to an absolute value of 10^{20} $1/m^2$; model calculations with formula (A2)). The result of the inversion is shown in Fig. 13. To get a nu-

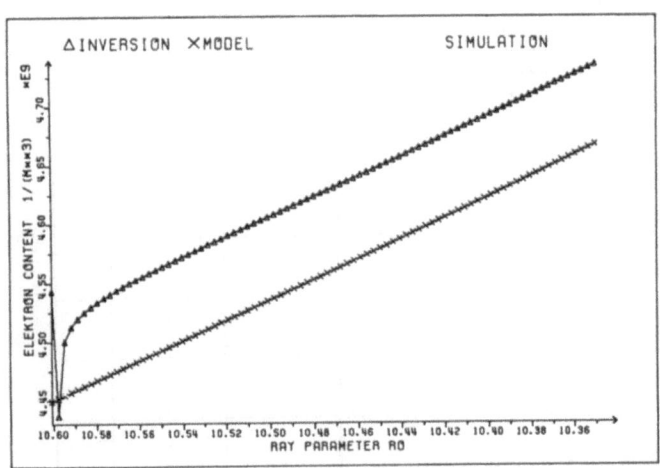

Figure 12: Illustration of the error of the combined analytical-numerical inversion algorithm.

merical quantity of the error magnification the ratios $\sigma_I/\bar{I}$ and $\sigma_N/\bar{N}$ are compared, where σ_I, σ_N and $\bar{I}$ and $\bar{N}$ are respectively

Figure 13. Illustration of the error magnification of a simulated electron content data set contaminated by (pseudo) noise.

the variances and means of the electron content and density. The quotient ρ_1 defined as

$$\rho_1 = \frac{\sigma_I / \bar{I}}{\sigma_N / \bar{N}}$$

Figure 14. Illustration of the error magnification as a function of averaged electron content values.

has the value of about 100. Averaging over measured electron content values and calculating the corresponding factors to ρ_1, called ρ_2, ρ_3, ..., lead to a decrease of the error amplification. The result of successive averaging is demonstrated in Fig. 14 and shows a continuous decrease of ρ with an asymptotical approach to $\rho = 20$. The averaging process effects that spatial and temporal resolution decrease whereas the statistical resolution will increase because of noise supression. By the asymptotical character of the function ρ it follows (shown in Fig. 14) that the straight line shown in Fig. 13 cannot be reconstructed from noisy measurements.

From the large number of measured electron content data sets of the Helios occultation experiment one characteristical series is choosen in order to demonstrate the applicability of the developed inversion method. This electron content data set, acquired at an impact parameter between 10.6 and 10.35 R_Θ, shows during the entry phase of the occultation as indicated by the least-squares approximation, the accepted slope corresponding to the undisturbed coronal

Figure 15. Measured electron content data set at a ray parameter of about 10.5 R_Θ.

plasma (see (A1) and (A2)). Furthermore the measured series proves that close to the sun immense plasma disturbances of solar origin pass the ray path. The result of the application of the inversion algorithm is presented in Fig. 16. The noisy character of the electron density distribution is obvious but on the other hand the oscillatory character as indicated in the measurements is significantly transformed into the density distribution. This shows that the physical information content gained by measurements will be

Figure 16. Electron density distribution calculated with
the inversion algorithm of equation (19).

preserved. Furthermore has to be added that the density fluctuations of $3 \cdot 10^9$ 1/m³ can be correlated with in situ-electron density measurements of the Helios B spacecraft [25, 28].

5. CONCLUSION

This paper demonstrates a lot of different examples in remote sensing and the applicability of the Abel transform which is used when
transport phenomena can be described with the ray theory in a rotational symmetric propagation medium. For the application in inhomogenuous media many constraints concerning experimental design as
well as modelling have to be considered in order to get a solution
whose results are physically realistic after a chain of simplifying
assumptions. The developed combined analytical-numerical inversion
algorithm has a relatively high stability and is well suited for the
inversion of electron content measurements as well as for the inversion of other types of integrated line intensities.

APPENDIX

Stationary Model of Coronal Electron Density Distribution

The stationary, rotational symmetric electron density distribution
of the solar corona in the ecliptic can be described by

$$N(r) = \frac{A}{r^6} + \frac{B}{r^2 + \epsilon} \qquad (A1)$$

where r is the dimensionless heliocentric distance in the units of solar radii (N given in electrons per cubicmeter). The coronal parameters A, B and ε determined by experimental data [29] as spatial and temporal averages have the following values and variances: $A = 1.0 \cdot 10^{14}$ $1/m^3$, $\sigma_A = 0.2 \cdot 10^{11}$; $B = 0.5 \cdot 10^{12}$ $1/m^3$, $\sigma_B = 0.2 \cdot 10^{14}$; ε: $0 < \varepsilon < 0.3$ (depends on the phase of the solar cycle; solar minimum: $\varepsilon \approx 0$).

Since the contribution of the first term of eq. (A1) to the electron density is effective below 4 R_Θ, for this inversion problem the density is approximated by the second term. Using only the quadratic term of $N(r)$ to calculate an electron content model using the simplification of equation (16) the following expression is yielded

$$I(p) = \frac{2 \cdot B \cdot R_o}{p} \arccos \frac{p}{r_o} .$$

6. REFERENCES

[1] Abel, N.H., Résolution d'un Problème de Mécanique, J. Reine Angew. Math. 1, pp. 153 – 157, 1826.

[2] Goldstein, H., Klassische Mechanik, Frankfurt, 1963.

[3] Sakalyuk, K.D., Abel's generalized equation, Doklady Akad. Nauk. SSSR 131, pp. 332 – 335, 1960.

[4] Samko, S.G., The generalized Abel equation and fractional integration operator, Differential equations 4, pp. 157–166, 1968.

[5] Wolfersdorf, L.V., Zur Lösung der verallgemeinerten Abelschen Integralgleichung mit konstanten Koeffizienten, Z. Angew. Math. Mech. 49, pp. 759 – 761, 1969.

[6] Lowengrub, M., Walton, J.R., Systems of generalized Abel equations, SIAM J. Math. Anal. 10, pp. 794 – 807, 1979.

[7] Walton, J.R., Systems of generalized Abel integral equations with applications to simultaneous dual relations, SIAM J. Math. Anal. 10, pp. 808 – 822, 1979.

[8] Herman, G.T., Rowland, S.W., Three Methods for Reconstructing Objects from X Rays: A Comparative Study, Computer Graphics and Image Processing 2, pp. 151 – 178, 1973.

[9] Bracewell, R.N., Strip integration in radio astronomy Aust. J. Phys 9, pp. 198 – 217, 1956.

[10] Das, Y., Boerner, W-M., On radar target shape estimation using algorithms for reconstructing from projections, IEEE Trans. Ant. and Prop. AP-26,pp. 274 - 279, 1978.

[11] Radon, J., Über die Bestimmung von Funktionen durch ihre Integralwerte längs gewisser Mannigfaltigkeiten, Ber. Verh. Sächs. Akad. Wiss. Leipzig, Math.Phys. Kl. 69, pp. 262 - 277, 1917.

[12] Wiechert, E., Geiger, L., Phys. Z. 11, pp. 294 - 311, 1910.

[13] Bullen, K.E., An Introduction to the Theory of Seismology, Cambridge University Press, New York, 1963.

[14] Kennett, B.L.N., Ray theoretical inverse methods in geophysics, in Applied Inverse Problems, ed. by P.C. Sabatier, Lecture Notes in Physics, Vol 85 (Springer, Berlin, Heidelberg, New York 1978).

[15] Bockasten, K., Transformation of observed radiances into radial distribution of the emission of a plasma, J. Opt. Soc. Am. 51, pp. 943 - 947, 1961.

[16] Cremers, C.J., Birkebak, R.C., Application of the Abel integral equation to spectrographic data, Appl. Opt. 5 pp. 1057 - 1064, 1966.

[17] Altschuler, M.D., Reconstruction of the global-scale three-dimensional solar corona, in Image Reconstruction from Projections, ed. by G.T. Herman, Topics in Applied Physics, Vol 32 (Springer, Berlin, Heidelberg, New York, 1979).

[18] Bohlin, J.D., Garrison, L.M., Numerical calculation of Thomson scattering from inhomogeneous models of the corona, and application to streamers of the 1970 and 1972 eclipses, Solar Physics 38,pp. 165 - 179, 1974.

[19] Hansen, R.T., Garcia, C.J., Hansen, S.F., Loomis, H.G., Brightness variations of the white light corona during the years 1964-67, Solar Physics 7, pp. 417 - 433, 1969.

[20] Phinney, R.A., Anderson, D.L., On the radio occultation method for studying planetary atmospheres, J. Geophys. Res 75 pp. 1819-1927, 1968.

[21] Kliore, A.J., Current methods of radio occultation data inversion, in Mathematics of Profile Inversion, ed. by L. Colin, NASA Tech. Mem. X-62, 150, 1972.

[22] Stewart, R.W., Hogan, J.S., Error analysis in the Mariner 6 and 7 occultation experiments, in Mathematics of Profile Inversion, ed. by L. Colin, NASA Tech. Mem. X-62, 150, 1972.

[23] Fjeldbo, G., Kliore, A.J., Eshleman, V.R., The neutral atmosphere of Venus as studied with the Mariner V radio occultation experiments, Astron. J. 76, pp. 123 - 140, 1971.

[24] Lusignan, B., Modrell, G., Morrison, A., Pomalaza, J, Ungar, S.G. Sensing the earth's atmosphere with occultation satellites, Proc. IEEE 57, pp. 458 - 467, 1969.

[25] Süß, H., Plasma-Fernerkundung durch Inversion von Messungen des Elektroneninhalts bei den Helios-Raumsonden (Okkultationsexperiment), Dissertation, Bochum, 1981.

[26] Edenhofer, P., Esposito, P.B., Hansen, R.T., Hansen, S.F., Lüneburg, E., Martin, W.L., Zygielbaum, A.I., Time delay occultation data of the Helios Spacecrafts and preliminary analysis for probing the solar corona, J. Geophys. 42, pp. 673 - 698, 1977.

[27] Mac Doran, F., A first-principles derivation of the Differenced Range Versus Integrated Doppler (DRVID) charged-particle calibration method, JPL Space Progr. Sum. 37-62.II, pp. 28 - 34, 1972.

[28] Rosenbauer, H., Schwenn, R., Marsch, E., Meyer, B., Miggenrieder, H., Montgomery, M.D., Mühlhäuser, K.H., Pilipp, W., Voges, W., Zink, S.M., A survey on initial results of the Helios plasma experiment, J. Geophys. 42, pp. 561-580, 1977.

[29] Esposito, P.B., Edenhofer, P., Lüneburg, E., Solar corona electron density distribution, J. Geophys. Res. 85, pp. 3414 - 3418, 1980.

I.11 (IM.1*)

SYMMETRY TECHNIQUES FOR DIRECT AND INVERSE SCATTERING

Paul F. Wacker

prepared in part at K. U. Leuven, BELGIUM
new address: 3037 Montrose Ave., Apt. 4
La Crescenta, CA 91214, USA

Abstract--Rigorous but practical procedures are presented for determining both overall and constituent symmetries of finite and infinite scatterers, guided mode junctions, and guided-spatial transducers (e.g., antennas). The systems may be lossy, anisotropic,
and/or inhomogeneous. The procedures are based on the facts that a)
all the familiar modal indices are symmetry indices, like even or
odd, b) a linear operator may be expressed as a sum of operators,
each designated by similar indices, and c) the symmetry of an operator (e.g., which EM multipoles are present in a scatterer) is indicated by its conversion of symmetry (modal) indices, just as a multiplier must be odd if it yields an even product from an odd multiplicand. Related mathematics reduces the number of measurements required to determine the pattern of a scatterer, radiator, or receiver of known symmetry. Many examples are given, with mathematical
definitions, theorems, and references.

1. INTRODUCTION

(So that a general understanding can be obtained without wading
through the complicated mathematics, the more complicated mathematical details and definitions are given in Section 5 and the notes of
section 6. Further background is given in (26) and the references
and will be given in a forthcoming paper; for more practical applications, see the latter two. Small differences between symbols
should not be ignored; e.g., presence or absence of underlining,
boldface or lightface, etc.) In this paper we provide practical
techniques for determining both overall and constituent symmetries
of both finite and infinite scatterers, of guided-mode junctions,
and of guided-spatial transducers (e.g., antennas) from measurements

W.-M. Boerner et al. (eds.), Inverse Methods in Electromagnetic Imaging - Part 1, 203–222.
© 1985 by D. Reidel Publishing Company.

made at a distance. In addition to being of interest in their own
right, knowledge of symmetries should speed inverse scattering analysis; further, related techniques reduce the number of measurements
and computations required to determine scattering, radiation, or
receiving patterns.

The procedures are based upon rigorous mathematics used in theoretical physics, particularly elementary particle theory. Specifically,
the modern theory of group[1] representations[2] is used in combination
with the invariances of natural law and the material properties of
the propagation medium with respect to translation, rotation, reflection, and/or inversion of the coordinate system and possibly
space-time transformations; we call this symmetry analysis (SA).
Although little used in engineering and applied mathematics, this
certainly constitutes one of the most powerful mathematical tools
used in physical science (26). Although the proofs of many of the
theorems require a sophisticated background indeed, the results of
many do not and their power makes for global understanding, leading
to simple direct proofs of practical problems, with rigor and sharp-
ness not often seen in the engineering literature. Further, these
techniques lead to rapidly converging modal sets, reducing the measurement and/or computational effort required for comprehensive
information (see Section 3.3).

Because the general procedures are so basic, they apply equally well
to many physical systems, including electromagnetic, acoustic, heat
flow, and mixed, but, for simplicity, we confine most of our attention to the curl-curl and scalar Helmholtz equations. However,
since a single differential equation may apply to several dozen
physical systems (17),(16, pp. 265, 343-346), simple changes of variables can translate even the detailed discussions; see Note 8 for
sonar and geophysical prospecting with both longitudinal and transverse waves. Both the propagation medium and the scatterer may be
lossy and, subject to the optional assumptions, the scatterer may be
anisotropic, inhomogeneous, and time varying. The presentation is
approximation free, apart from assumed mathematical linearity and
optional assumed invariances. Many examples are given, as well as
mathematical definitions, theorems, and references.

The procedures are also based upon the facts that a) all the familiar modal indices are symmetry indices, like even or odd, b) a linear operator may be expressed as a sum of operators, each designated
by similar indices, and c) the natural constraints upon the symmetry
of a "product." For example, the propagation constants, TM,TE
character, frequency, orders of Bessel functions, sin-cos distinction, both indices of associated Legendre functions, the azimuthal
index m, and the multipole indices are all symmetry indices. Also
for example, a multiplier must be odd if it yields an even product
from an odd multiplicand. The multiplier may also be an operator,
such as the scattering operator; thus, in the preceding example, the

multiplier may be d^n/dx^n if n is odd, but not if n is even. The power of these techniques is illustrated by two examples: For acoustics with the scalar Helmholtz equation, a _single_ ingoing mode may be used to determine the presence of each and every scalar multipole in the scatterer; the multipole is present _if_ _and_ _only_ _if_ the corresponding scattered mode is present. If the material properties of a capacitor (in a circuit or in space) are known to be symmetric with respect to a plane of symmetry and the _single_ ingoing mode is even with respect to the plane, presence of _any_ scattered mode odd with respect to the plane _proves_ that the capacitor is differentially charged.

For near field determination of radiation, receiving, and scattering _patterns_, we use SA techniques, especially the orthogonalities, to determine the modal expansion _coefficients_ and the far field patterns (25),(24),(27). In _this_ paper, we emphasize techniques related to the _presence_ or _absence_ of modes; specifically we use a) a more detailed form of the mathematics which provides the selection rules of quantum mechanics and spectroscopy to reduce both the measurement and computational effort in determining the patterns of scatterers of known symmetries (25) and b) related techniques, used to determine the symmetries of molecules and the _internal_ symmetries of crystals, to determine the symmetries of a macroscopic scatterer in inverse scattering (25).

2. MODES AND THEIR SYMMETRIES

2.1. Functions defined by symmetries

With an infinite number of symmetry operations, there are an infinite number of **symmetry types**[5] (_each_ like even or odd) and an infinite number of constraints (_each_ like $f(-x)=-f(+x)$ for an odd function) defining _each_ symmetry type by specifying its transformation under each and every symmetry operation of the group.[1] (Technically, we say that a function is of a given symmetry type if it belongs to a given row[6] of a given irreducible[4,6] unitary[5] representation[2] of a given group.) The following groups and the functions they define are pertinent to our problems: The (continuum of) rotations about an axis, i.e., the two dimensional rotation (orthogonal) group C_∞, defines the functions $\exp(im\phi)$ and yields Fourier series analysis (9, pp. 322-325). The translations T_1 along the real line yield the functions $\exp(-i\omega t)$ and $\exp(ik_x x)$, the Fourier transform, and the double-sided Laplace transform (9, pp. 486-489). The addition of reflections in planes containing the axis $(C_{\infty v})$ or in planes perpendicular to the line (T_{1h}) yields the corresponding sines and cosines. The three dimensional rotation (orthogonal) group O_3^+ yields $P_n^m(\cos \theta) \exp(im\phi)$, the definition of the cross product (3), and the theories of spherical harmonics and angular momentum (21, pp. 130-168) (O_3 is not to be confused with the point groups O and O_h (12)).

The two dimensional Euclidean (translation-rotation) group E_2^+ yields $J_m(k_R R) \exp(im\phi)$, the analogous expressions with cylindrical Neumann and Hankel functions (21, pp. 189-214), and Hankel (Bessel function) transforms. Moreover, three separate theorems, each applying to a class of symmetry groups (2, pp. 166-176, 420-472), lead to the mathematical completeness, linear independence, and orthogonalities of the preceding sets of functions. (Lack of linear independence may cause oscillation or lack of convergence in iterative procedures.) The three dimensional Euclidean (translation-rotation) group E_3^+ yields $P_n^m(\cos \theta) \exp(im\phi) j_n(k_r r)$ and the analogous expressions with spherical Neumann and Hankel functions (21, pp. 215-233).

In general, there are three linearly independent three-vector functions for each of the preceding symmetry types. However, addition of gauge invariance (related to the absence of fixed charges and the divergence condition) requires that the field be solenoidal, reducing the number of independent polarizations in the EM cases to two. The addition of spatial inversion (and so reflections) to O_3^+ and E_3^+ yields the full 3D rotation and 3D Euclidean groups, O_3^+ and E_3^+, respectively, which separate the TM and TE modes into different symmetry types and so define them (19). E_3^+ and E_3^+ define the scalar and EM multipoles, respectively, while the corresponding O groups define them apart from the radial dependence. The circular cylindrical ($E_2 X T_z$) and plane rectangular ($T_x X T_y$) modes and algebras are limits of the spherical on the pole and equator, respectively, as the radius goes to infinity (21); this explains the TM,TE separation for these cases. (The symbol X indicates the direct product of two groups; for our purposes, these are groups which occur simultaneously but involve independent sets of coordinates.) The fast Fourier transform arises from the group C_M of rotations by $2\pi q/M$ (q=0 to M-1) and separates the (discrete) terms of an exponential Fourier series into M types with an infinite number in each type; as such, the FFT is an approximation free symmetry decomposition; the approximations arise in replacing the continuum of terms in a transform by a Fourier series and assuming that there is only one significant term in each symmetry type. A fourth theorem leads to the orthogonalitity and (collective) completeness of the M symmetry types, as well as of even and odd functions. The preceding are merely pertinent examples; actually SA defines most of the named mathematical functions and gives their detailed properties, including some previously unknown (21),(15),(23). Further, it may be considered to be part of a generalization of Fourier series or transform analysis, with an expansion being a symmetry decomposition (2, pp. 420-472). We base our analysis upon the simple combination of the full Euclidean group E_3^+ with the temporal translation-reflection group T_{th} and their subgroups (for most macroscopic problems, the space-time transformations of the Lorentz group of special relativity add only pointless complexity). For finite scatterers, we use subgroup $O_3^+ X T_{th}$ and its subgroups, where O_3^+ is the full 3D rotation group. (The groups of crystallography are

discrete subgroups of E_3^+ subject to the law of rational indices,
with the 32 point groups being such subgroups of O_3^+ (14, pp. 20–
60); we apply neither constraint and so, for example, accept groups
such as C_∞ and C_7.) For simplicity in treating plane rectangular
and circular cylindrical modes in homogeneous isotropic media, we
impose $k_x^2 + k_y^2 + k_z^2 = k^2$ and $k_R^2 + k_z^2 = k^2$, respectively, as auxiliary constraints rather than basing these constraints upon the E_3^+ group.

2.2. Symmetry indices of functions

We partition the modal indices $N = n;m;p$ into n, m, and p where those
in n and m designate the irreducible[4,6] representation (major symmetry type) and its row[6] (minor, i.e., subsidiary, symmetry type),
respectively, of the given group. A function is said to "belong
to" a given row of a given irreducible representation (of a given
group); together they define our "symmetry type."[6] For example,
for $m \neq 0$, the $|m|$th representation of $C_{\infty v}$ has two rows and the
partner[6] functions $\sin m\phi$ and $\cos m\phi$ belong to different rows. For
any representation with more than one row, there are an infinite
number of representations which are said to be "equivalent"[3]; thus,
for the preceding example, the partner functions $\exp(+im\phi)$ and
$\exp(-im\phi)$ belong to different rows of an "equivalent" representation
(but neither the two sets of (partner) functions nor the two representations are equally suitable for expressing either linear polarization or circular polarization). Indices needed to specify the
mode but not the symmetry relative to the given group are designated
by p; our p indices are ordinarily symmetry indices of a larger
group. The operations of the given group acting upon the N mode do
not change the n or p indices (since we use linearly independent
partner[6] functions) but yield linear combinations of modes with
differing m.[6]

2.3. Eigenfunctions and eigenfunction equations

In addition to defining functions, SA yields, as eigenfunctions, the
complete solutions of many eigenvalue equations, including solutions
of the scalar Helmholtz and curl-curl equations; the latter include
the familiar ones for unbounded media, waveguides, and cavities, as
well as for unfamiliar waveguides and cavities. Further, it provides sets of such functions, each with its symmetry prescribed in
detail, which are linearly independent but mathematically complete
with respect to the eigenfunctions. We express both the incident
and scattered field in terms of these functions to reduce the required measurement and computing effort; see Section 3.3. We consider primarily equations of the form

$$\underline{O} \, f_N(x) = \lambda_N \, f_N(x), \tag{1}$$

where $\underline{O}$ is a linear operator, boldface is used for symbols which are
or may be compound (e.g., vectors), N represents the set of modal

indices, and x represents the set of independent variables, e.g.,
the spherical coordinates and/or time. The fields may be scalar,
vector, or tensor in n-space and the linear combinations may be
sums, integrals, or sums of integrals over the eigenfunctions. For
simplicity in presentation, we direct most of our discussion to the
scalar Helmholtz equation

$$\nabla^2 f_N(x) = - k^2 f_N(x) \tag{2}$$

and the curl-curl equation

$$\nabla \times \nabla \times E_N(x) = k^2 E_N(x), \tag{3}$$

i.e., most of our discussion to homogeneous isotropic media of
propagation and, in the electromagnetic case, to solenoidal fields
(no fixed charges). We regard (3) as a special case of

$$\varepsilon^{-1} \nabla \times [(\mu^{-1} \nabla \times) E_N(x)] = \omega^2 E_N(x), \tag{4}$$

where ε and μ are complex tensor functions of position for the
system as a whole; this automatically includes the boundary conditions at infinity, the walls of waveguides or cavities, etc.; step
functions of course introduce generalized functions such as the
Dirac delta "function." However, we use modes for the medium in the
absence of the scatterer and probes, permitting treatment of the
scatterer as a black box without even tentative assumptions concerning it, except for mathematical linearity. This limits the domain
of validity of our modal expansions (say to outside a sphere which
circumscribes the scatterer[9]) and causes our conclusions about overall and constituent symmetries to apply to a domain which includes
the scatterer and usually some medium of propagation. As previously
mentioned, the techniques and, with changes of variables, even the
detailed discussions apply to many systems; see Section 2.4 for
anisotropic systems, Sections 2.4 and 3.3 for lossy systems, and
Note 10 for pulsed fields. Note that equation (4) also yields the
solutions and eigenvalues of the Singularity Expansion Method.

2.4. Theorems relating symmetries and eigenfunctions

For modal symmetries, we assume that the operator $\underline{0}$ of Eq. (1) is
linear, self-adjoint in the Hermitian sense, and commutes with every
operation g of a group G (i.e., $\underline{0}g = g\underline{0}$) which is finite, Abelian,[1] or
compact[1] (e.g., O_3) and consider a set of irreducible unitary[5]
pairwise-inequivalent[3] representations. Then the set of eigenfunctions, each of which belongs to a row of a representation, is
mathematically complete with respect to the eigenfunctions; furthermore, all linear combinations of a given set of partner eigenfunctions belong to the same real eigenvalue, and functions (not
only eigenfunctions) belonging to different representations or to
different rows of the same representation are linearly independent

of each other and orthogonal in the Hermitian sense with respect to
the sum and/or weighted integral over the group operations (14, p.
108ff),(2, p. 144),(9, p. 103).[6] Except for the eigenvalues being
nonreal, the preceding theorems also apply if $\underline{O}$ is a complex constant times a self-adjoint operator or normal, i.e., commutes with
its Hermitian adjoint. <u>Usually</u>, for a <u>large</u> enough group, a) <u>all</u>
the eigenfunctions of a given eigenvalue λ_N belong to the same
representation, b) the number of linearly independent eigenfunctions
belonging to that eigenvalue is equal to the dimension (number of
rows[5]) of the representation, and c) no two eigenvalues belong to
the same representation (14, p. 109 ff.).

We illustrate the preceding with the differential equations
$i^{-n} d^n u/dz^n = \lambda u$. The operators $i^{-n} d^n/dz^n$ commute with all translations on the z axis and, if n is even, with reflection in every
plane perpendicular to the z axis. Thus, we use T_z and, if n is
even, T_{zh} and C_s (reflection in the z=0 plane). T_z yields the
eigenfunctions $\exp(ik_z z)$ (for all these equations) and shows the set
for k_z real to be complete with respect to the eigenfunctions, linearly independent, and orthogonal. For n even (but not odd), T_{zh}
yields the partner eigenfunctions $\exp(+ik_z z)$ and $\exp(-ik_z z)$ (or alternatively $\sin k_z z$ and $\cos k_z z$), shows that every linear combination of them for a given k_z has the same eigenvalue (k_z^n), that the
set for $k_z \geq 0$ is mathematically complete with respect to the eigenfunctions, and that all the functions of either set are linearly
independent and orthogonal; T_z is not large enough to do this for
even n, but it is large enough for odd n. For even n, C_s <u>alone</u>
separates the eigenfunctions into even and odd (here $\cos k_z z$ and $\sin
k_z z$) and shows that such a set is mathematically complete with re-
spect to the eigenfunctions and that the even functions are linearly
independent of and orthogonal to the odd ones; however, C_s is not
large enough to a) separate the k_z values, b) show that the sine and
cosine have the same eigenvalue, c) show that the complex exponentials are eigenfunctions, nor d) distinguish between even (or odd)
functions which are and which are not eigenfunctions.

Our procedures do <u>not</u> assume that the operator representing the
<u>scatterer</u> is self-adjoint; see Section 3.3 ff. Further, lossy media
and evanescent modes usually cause no problem (at least if the medium is homogeneous and isotropic) if attention is confined to groups
involving only two spatial coordinates and nonreal values are accepted only for the component of the propagation vector corresponding to the third coordinate, e.g., that perpendicular to a sphere,
circular cylinder, or plane. The theory then ordinarily leads to
<u>surface</u> eigenfunctions of a self-adjoint operator, from which one
may easily determine the three dimensional eigenfunctions. The
surface eigenfunctions are orthogonal in the Hermitian sense and
both sets are mathematically complete with respect to their kind of
eigenfunctions; however, symmetries involving only two coordinates
do not provide orthogonalities between ingoing and outgoing modes.

In addition to giving the eigenfunctions and eigenvalues for unbounded media, SA provides <u>complete</u> <u>exact</u> solutions inside various
waveguides and cavities. Since perfect conductors provide mirror
images which "extend" the space of the medium outside, unbounded
solutions consistent with the boundary conditions apply. In particular, SA provides the solutions for circular cylindrical guide
and both spherical and right circular cylindrical cavities. Further, it provides the solutions for waveguides and cavities bounded
by planes consistent with the law of rational indices (related to
the Bravais lattices of crystallography), namely a) for waveguides
with cross sections which are squares, rectangles, diamonds, paral-
lelograms, equilateral triangles, and regular hexagons and b), for
each of the 14 Bravais lattices of three space, the cavities bounded
by planar walls which form a unit cell. The solutions provided by
the mappings are mathematically complete <u>inside</u> all these structures
and their extensions valid outside the structures, but complete
outside only a sphere and an infinite cylinder.

3. THE SCATTERING AND PERTURBATION MATRICES AND THEIR SYMMETRY CONSTRAINTS

3.1. Eigenfunction expansions of fields

We express the fields in terms of linear combinations of modes which
are eigenfunctions for the medium of propagation in the <u>absence</u> of
both the scatterer and transmitting and receiving transducers, e.g.,
antennas. Thus,

$$f(x) = \sum_N \!\!\!\int f_N(x) \; c_N, \tag{5}$$

where the symbol $\sum\!\!\!\int$ indicates summation over discrete modal indices
in N and/or integration over indices which assume continuous values,
e.g., ω, k_x, k_y, k_z, k_R, and k_r, <u>as appropriate</u>.

3.2. Definitions of the scattering and perturbation "matrices"

We express scattering properties in terms of a "matrix" which relates
two sets of modal coefficients, each from an expansion of the form
of (5). We first define the scattering matrix, which applies to
guided-mode junctions and guided-spatial transducers, e.g., antennas, as well as for scatterers. We use the generalized scattering
"matrix" [S] (not the "power wave" or "source" scattering matrix);
it relates the coefficients of modes which are outgoing with respect
to a volume (possibly a half space) containing the scatterer to
those of modes which are ingoing; thus,

$$b = \Gamma a + \underline{R}A \tag{6a}$$

$$B = \underline{T}a + \underline{S}A \tag{6b}$$

or

$$
\begin{Bmatrix} b \\ \overline{} \\ B \end{Bmatrix} = \begin{bmatrix} \Gamma & \vdots & R \\ \cdots & + & \cdots \\ T & \vdots & S \end{bmatrix} \begin{Bmatrix} a \\ \overline{} \\ A \end{Bmatrix} ,
\tag{6c}
$$

where a and b are column submatrices of the detached coefficients of
the ingoing and outgoing <u>guided</u> modes, respectively, and A and B are
column submatrices of the detached coefficients of the ingoing and
outgoing <u>spatial</u> modes, respectively. We commonly use individual
modes with complex exponential dependence on ω, ϕ, and/or propagation constants, leading to coefficients and elements of [S] which
are complex numbers. Then Γ represents the <u>complex</u> reflection coef-
ficient(s) looking into the port(s), while $\underline{T}$, $\underline{R}$, and $\underline{S}$ are submatrices representing the transmitting, receiving, and scattering proper-
ties, respectively; it is assumed that the space is empty except for
the propagation medium and the object which [S] represents; see
(25). For a mathematically linear system, [S] expresses all the
properties which can be observed at a distance and all the input
information for an inverse scattering problem; see Note 10 for scattering of pulsed fields. $\underline{S}$ gives the properties of a scatterer, but
application of the techniques of this paper to Γ, $\underline{R}$, and $\underline{T}$ yields
similar results for guided-wave junctions, receiving transducers,
and transmitting transducers, respectively. Ordinarily, translation, rotation, and/or reflection of the coordinate system will
change [S]; in this paper we usually refer [S] and its submatrices
to a coordinate system with its origin within the scatterer. In
addition, the definition and even the form of the matrices will vary
with the ordered sets of eigenfunctions used in Eq. (5).

Each row and each column is indicated by a <u>set</u> N of modal indices
arranged in a dictionary order, i.e., each element is associated
with a pair of eigenfunctions, one ingoing and the other outgoing.
Depending upon the eigenfunctions involved, the row and column indices may be discrete, continuuous, or mixed; at least in a formal
sense, the elements on the left hand side of equation (6) are commonly given by multiple summation and/or multidimensional integra-
tion. For a perfectly-matched perfect absorber $\underline{S}=\underline{0}$, and for a volume containing only the medium of propagation $\underline{S}=\underline{1'}$. If the rows and
columns are discrete, $\underline{0}$ and $\underline{1'}$ are ordinary null and unit matrices,
the latter provided that the same ordered set of modes is used on
both sides of equation (6); however, for a continuous index, a row
of $\underline{1'}$ may be proportional to a Dirac delta "function." Following
Garbacz and Turpin (7), we also introduce a perturbation matrix
$[P]=([S]-[\underline{1'}])/2$ with $\delta=(b-a')/2$ and $\Delta=(B-A')/2$, where the primed
quantities differ from the unprimed ones only if different ordered
sets are used for the ingoing and outgoing modes. This leads to the
equation

$$\left\{ \frac{\delta}{\Delta} \right\} = [P] \left\{ \frac{a}{A} \right\} . \tag{7}$$

3.3. Symmetry and forms of the "matrices"

We now relate the detailed structures of the matrices [S], [Γ], [R̲],
[T], [S̲], and [P] to the symmetries of the scatterer, making use of
the matrices and submatrices given or implied by Eqs. (6) and (7).
The equation

$$u = \underline{P}v \tag{8}$$

may be interpreted in a number of ways. $\underline{P}$ may be taken as the perturbation operator, v the ingoing mode, and u the pertubation mode.
Alternatively, $\underline{P}$ may be taken as the matrix [P] with u and v the
column matrices of (7), or $\underline{P}$ may be taken as a submatrix of [P] with
u and v the appropriate column submatrices. Since the formal algebra is the same, $\underline{P}$ may also be taken as [S] or one of its submatrices with u and v as the appropriate column matrices or submatrices.
We discuss the perturbation and scattering matrices and their
submatrices together, using the expression "the matrix" and, for the
scattered or perturbation field, the "resultant" field.

We choose a group G which is convenient for expressing the properties of the scatterer as well as the medium (or media) of propagation, and base "the matrix" upon "symmetry adapted" functions,
i.e., linear combinations of basis functions so that each combination belongs to a row (m) of an irreducible representation (n) of G,
from a set of inequivalent[3] irreducible representations. However,
we use, as basis functions, eigenfunctions (f_N's) of the differential equation for the medium (or media) of propagation __alone__, i.e.,
in the __absence__ of the scatterer etc. Since G is ordinarily a subgroup of the group for the medium of propagation, judicious choice
of the eigenfunctions ordinarily reduces the "symmetry adaptation"
to mere grouping the eigenfunctions; see Koster's "compatibility
tables" (12, pp. 18-20) for relating the representations of the
spherical and point groups.

3.4. Scatterers known invariant under their symmetry groups

Prior to introducing the formal analysis, we introduce three simple
but practical and enlightening special cases, one for scattering in
this section and the other two for inverse scattering in the next
section. (However, the treatments of these two sections may not
yield __all__ the symmetries available, e.g., completely determine the
multipoles present in an EM (but not scalar) problem.) For scattering, we consider (in this section) scatterers, i.e., $\underline{P}$ operators,
which are __invariant__ with respect to (i.e., commute with) all the
operations of the given group G; see (8),(9),(2). Then, just as

multiplication by an even function yields a product which is even or
odd as the multiplicand, so here $\underline{P}$ will change neither n nor m,
i.e., the matrix will be diagonal in n and m, assuming no change in
the coordinate system or ordered set of modes. Further, due to our
use of partner[6] functions, the non-zero matrix elements will be
independent of m (but not the n's or p's). See Section 2.1 for
background of both this and the next section; for simplicity, the
propagation media in the following examples d), e), and f) are
assumed homogeneous and isotropic, but possibly lossy.

Invariant $\underline{P}$'s are adequate for the following examples, some of them
familiar in terms of conventional viewpoints: a) If the _material_
properties of the system (and all _externally_ _imposed_ fields and
forces to which it is subjected) are independent of time, the system
is invariant under the temporal translation-reflection group T_{th} and
the matrices will be diagonal in the frequency ω and, for ω≠0, the
row index; further, the diagonal matrix elements will be independent
of the row index. That is, if the mode v is proportional to any
linear combination of sin ωt and cos ωt or of exp(+iωt) and
exp(-iωt), then every resultant mode u (with nonzero amplitude) will
have the that same dependence, regardless of the spatial dependence; further, the matrix elements (_complex_ ratios of the final to
the initial coefficients) may vary with ω, but not with the linear
combination for a given ω. Unless otherwise stated, this symmetry
and (suppressed) exp(-iωt) time dependence are assumed in the following. b) For two contiguous half spaces, each with properties
which are invariant with respect to translation in x and y (T_xXT_y),
neither k_x nor k_y will change, either on transmission or reflection.
Further, if the material properties of the system are also even with
respect to reflection in the plane of incidence, there is no conversion between modes which are even with respect to reflection in the
plane (parallel E) and those which are odd (perpendicular E), explaining the separation in Fresnel's equations, which then follow
from the boundary conditions. Note that either or both media may be
lossy, anisotropic, and/or arbitrary functions of z, except as separately limited by the stated invariances. (Note that this applies
to acoustics in a medium whose density varies with depth, reducing
it to a one dimensional problem.) Hence, this provides generaliza-
tions of Snell's laws and Fresnel's equations. c) Application of
these techniques to Γ provides a treatment of guided-mode junctions
(11), but is not limited to ideal lossless waveguide leads; the
leads may contain lossy inhomogeneous anisotropic media and the
treatment of Sections 3.6 and 4.2 applies to much more general junctions. d) For a cylindrical scattering system invariant with respect to φ rotation and z translation ($C_\infty XT_z$) (but possibly lossy,
anisotropic, and radially inhomogeneous), the matrix will be diagonal in the azimuthal index m and k_z. Further, if the system is also
invariant with respect to reflection in all planes containing the
cylindrical axis, the elements of the matrix will be independent of
the sign of m, i.e., of the linear combination of sin mφ and cos mφ

or of $\exp(im\phi)$ and $\exp(-im\phi)$. For a cylinder of finite conductivity, there will be 2x2 blocks on the diagonal due to possible TM, TE conversion. e) For a scatterer which is invariant with respect to both spatial inversion and the three Eulerian angles (O_3^+), there is no change in m or n of $P_n^m(\cos \theta) \exp(im\phi)$, the TM, TE character, or in the propagation constant k_r of the medium. Thus, the matrix has discrete rows and columns and is diagonal in __all__ the modal indices, i.e., there is no modal conversion in the black box sense (except for phase and/or amplitude), even if the scatterer is lossy, radially inhomogeneous and __locally__ anisotropic. Further, these elements, which represent the complex ratios of coefficients, are independent of m but may vary with n, the TM, TE character, or k_r. In example d), if the cylinder is perfectly conducting, the surrounding medium is "extended" into the cylinder by the mirror image, leading to O_3^+ symmetry (with an infinite __spherical__ radius) and the matrix is diagonal in TM, TE character (see (20, pp. 524 ff.) for a conventional treatment). f) If the material properties of a transmitting antenna and feed are invariant with respect to rotation about the z axis (C_∞) and the mode in the feed has $\exp(im\phi)$ dependence from any combination of modes, the radiated field will have that dependence everywhere in space (with respect to the original axis). Further, if the system is invariant as well with respect to reflection in all planes containing the z axis ($C_{\infty v}$) and E_R of the feed has $\cos m\phi$ dependence from any combination of modes, then E_R, E_z, and H_ϕ of the radiated field will have $\cos m\phi$ dependence while E_ϕ, H_R, and H_z will have $\sin m\phi$ dependence, everywhere in space. This will be true regardless of the variation of the properties (or diameter) of the system with z or anisotropy consistent with the invariances, but TM and TE modes (here p indices) will mix. Thus, measurements of __two__ components on a __single radius vector__ determine an expansion which yields the phases and amplitudes of the __six__ EM components everywhere in the __forward__ half space bounded by an xy plane which just fails to intersect the antenna (see Note 9). Note that, in distinguishing between two airplanes or two missiles, it can usually be assumed that both of the former have a plane of symmetry and that both of the latter are essentially circularly symmetric.

3.5. Simple determinations of symmetry __constituents__ of a scatterer

For simple detection of symmetry constituents of an arbitrary mathematically linear scatterer, we consider one or more ingoing modes which are invariant under all the operations of the given group G. Then each resultant mode will have the same symmetry (n, m) as the symmetry constituent of P which gives rise to it, and, for every symmetry type in the resultant modes, P will have a constituent of that symmetry. Note that a __single__ resultant mode can show unambiguously the presence of a given symmetry constituent in P. As previously noted, if the scatterer is a capacitor whose material properties are symmetric with respect to a plane and the incident mode is symmetric with respect to the same plane, a perturbed mode antisymmetric with respect to the plane will show that the capacitor is differentially charged. Note however, that if the group is not large enough for the symmetry type (n, m) to completely determine the ingoing mode, the lack of a resultant mode of a given type does not prove that that type does not occur in P.

Another simple procedure arises if G is Abelian, since the product of two symmetry types yields only a single type. For example, if G is C_∞ and the incident mode has $e^{im\phi}$ dependence, then the operator has $e^{ik\phi}$ dependence if there is any resultant mode with $e^{i(m+k)\phi}$ dependence. This is useful for studying the fins on a circularly symmetric rocket, since the diagonal blocks of the scattering or perturbation matrix corresponding to circularly symmetric part (k=0) can be disregarded and attention confined, e.g., to the blocks corresponding to k equal to the number of (equally spaced) fins and perhaps its multiples.

3.6. General treatment of mathematically linear scatterers

The procedures of Sections 3.4 and 3.5 are very useful, but in some cases they do not permit determination of all the symmetries. In particular, a) the product of two symmetry types, e.g., sin mϕ sin kϕ, may yield a sum of types and b) for Section 3.5, there is no EM mode which is totally symmetric with respect to O_3, since E_3 here requires that $n \geqslant s=1$, i.e., the total angular momentum (n of the associated Legendre function) cannot be less than the spin momentum of the photon. We now discuss procedures for cases in which the preceding treatments are inadequate.

We express the scattering or perturbation operator as a linear combination of operators, each of a given symmetry type, i.e., each an irreducible tensor operator which belongs to a row of an irreducible unitary representation. This is possible for essentially arbitrary linear operators, self-adjoint or not (9, pp. 377-412),(2, pp. 242-276). The same set of indices which designate a mode or its symmetry designate a constituent operator or its symmetry.

The so-called Clebsch-Gordon coefficients express the permitted combinations of the symmetry types of the incident mode, the resultant mode, and the operator constituent; further, they give required relations between the coefficients of some of the permitted modes. These are tabulated for the three dimensional rotation group and the 32 point groups and may be easily determined for some other groups. With appropriate choice and ordering of the modes, the presence of a given symmetry constituent is indicated by the presence of one or more submatrices of the scattering or perturbation matrix. Further, the relative importance of symmetry constituents is roughly indicated by the relative magnitudes of elements of the submatrices. The details are given in Section 5.

4. PRACTICAL DETAILS

Conversion of the raw data from the usual plane wave representations
to the symmetrized modes used here involves at most a similarity
transformation, i.e., u, v, and $\underline{P}$ are replaced by Qu, Qv, and $Q\underline{P}Q^{-1}$,
respectively. In practice, these are matrix multiplications in
which the rows of Q are given by natural orthogonalities provided by
the theory; for spherical, circular cylindrical, and Cartesian
representations, these may be the orthogonalities with respect to
summation used in near field scanning, ordinarily the far field
forms of the ideal probe expressions (25),(24); more generally, the
symmetrizing combinations are given by Eq. (12) of Note 6. Further,
as explained Section 3.5, a single incident mode is commonly sufficient to determine the symmetry constituents of a scatterer. The
centers, axes, and planes of symmetry can often be located with
pulsed measurements, perhaps supplemented with cut and try or a
modified Newton-Raphson procedure.

5. MATHEMATICAL DETAILS: GENERAL SCATTERERS

For convenience in the use of the tables of Koster et al., we now
pattern our notation after theirs. We designate the resultant and
ingoing modes by u_{pn}^{γ} and v_{qj}^{β}, respectively, where the superscripts
indicate the representations, n and j indicate the rows of the
representations, and p and q indicate the partner sets. Similarly,
we use P_i^{α} to indicate the constituent of $\underline{P}$ corresponding to ith row
of the α representation (12, Eq. (3.18)). Thus, our n, m, and p
here correspond, respectively, to β, j, and q for the ingoing modes,
to γ, n, and p for the resultant modes, and to α, i, and - of $\underline{P}$. We
use lightface for the last eight indices to avoid confusion with the
first and third (our n and p). For a finite scatterer, the symmetry
operations of interest include rotation, reflection, and inversion,
but not translation, i.e., we are commonly concerned with the crys-
tallographic point groups. For the immediate purpose, we index on
the partner set before the row of the representation; thus, the row
and column locations of an individual submatrix are indicated by γ,p
and β,q respectively, and the numbers of rows and columns of the
submatrix are given by the numbers of rows in the γ and β representations, respectively. Should the operator $\underline{P}$ contain more than
one symmetry constituent, the following treats a constituent part of
the overall problem; the overall matrix is then the sum of the
constituent matrices.

We now express the constraints upon the scattering matrix elements
imposed by symmetry. From the theory of group representations (9),
(12),(8), the matrix elements are given by

$$\sum_{r} {}_r s_{\gamma p, \beta q}^{\alpha} \ {}_r U_{ij,n}^{\alpha\beta,\gamma*}. \tag{9}$$

That is, each submatrix of the matrix is expressed as the sum of one or more products, each of a complex number s independent of i, j, and n (the rows of the representations) and a submatrix U^* which depends only upon the irreducible representations of the group, not upon the partner sets or other aspects of the physical problem. In various contexts with slight differences in definitions, the U's are called Clebsch-Gordon, coupling, Wigner 3j, vector addition, or Racah V coefficients. These are well known and tabulated for the three dimensional rotation group, but exist for other groups and are tabulated for the 32 point groups (12),(8). Only for two of the point groups (O and T_d) are there two products; for the other point groups there is only one product. In the tables (12), the U's for the first product are indicated by $\not{a}$ and, when they occur, the U's for the second product by $\not{b}$. A large fraction of the U's have no nonzero elements. Further, if the symmetry constituent of either v or $\underline{P}$ is totally symmetric, then U^* is zero unless the symmetry type of u is identical to that of $\underline{P}$ or v, respectively; Sections 3.4 and 3.5 follow from the fact. Some of the U's are well known. Thus, those for the one-dimensional translation group follow from the relation $\exp(ik_x x) \exp(ik'_x x) \propto \exp(i[k_x+k'_x]x)$. Similarly, the expressions for the products of sines and/or cosines in terms of sums or differences give all the U's for the one dimensional translation-reflection group.

In determining the matrix for a scatterer of known symmetry, the partner set may be indexed before the row of the representation; thus, the row and column locations of an individual submatrix are located by γ, p and β, q, respectively, and the numbers of rows and columns of the submatrix are given by the numbers of rows in the γ and β representations, respectively. The determination of "the matrix" thus usually reduces to determining a complex constant (the s) for each submatrix. Note that measurements on a __far__ __field__ sphere determine the EM field outside a sphere __just__ circumscribes an antenna or scatterer; further, if the field is known to be outgoing or ingoing, only the __transverse__ components of the electric field are needed since Hermitian orthogonalities are supplied by the O_3^+ group and the group operations do not mix the transverse with the radial components.

5. NOTES

[1]A __group__ G is a set of elements {a,b,c,...} with a binary operation called __multiplication__ which associates each ordered pair of elements a,b in G with a __product__ ab also in G such that a) there exists in G a single element e, called the __identity__, such that ae=ea=a for __every__ element a in G, b) for __each__ a in G, there exists in G an __inverse__ a^{-1} such that $aa^{-1}=a^{-1}a=e$, and c) the __associative law__ (ab)c=a(bc) holds for all a, b, and c in G (14, p. 1). (The product of each element with itself is also in G.) The identity operation e and

reflection σ in the z=0 plane constitute the familiar reflection
group $C_s=\{e,\sigma\}$. A group is called **Abelian** if the effects of its
operations are independent of the order of application. A group is
said to be **compact** if the parameters designating the group operations vary over a finite range (9, p. 309). ⟨We enclose material
not needed for general understanding in angular brackets and confine
attention to locally compact, separable, unimodular topological
groups (2, p. 134 ff.).⟩

[2]A **representation** a→T_a of a group is a <u>set</u> of operators such that a)
each member of the group is associated with a <u>single</u> element of the
representation, b) each operator of the representation is associated
with <u>one or more</u> members of the group, and c) if members a, b, and c
of the group are associated with operators A, B, and C of the representation, respectively, and ab=c, then AB=C (29, pp. 69, 72 ff).
⟨We confine our attention to representations whose elements are
bounded linear operators in a separable complex Hilbert space (but
our group elements and $\underline{O}$'s may be unbounded, e.g., differential,
operators) (2, p. 134 ff.).⟩ A set of matrices, designated by D(a),
is commonly used for a linear representation; they act upon vectors
whose unit elements represent an orthonormal set of functions which
provide a basis for the space in which the operators of T act.

[3]Two representations T and T' of Hilbert spaces H and H', respectively, are <u>said</u> to be **equivalent** if there <u>exists</u> a one to one linear mapping S of H onto H' such that $T'_a=ST_aS^{-1}$ for all a of the
group G, independent of a. (H' may be identical to H.) For matrix
representations, S is a nonsingular matrix.

[4]A subspace H_1 of H is said to be **invariant** (with respect to a representation T) if, for every element a of G and every function w of
H_1, the function T_aw is also in H_1. A representation T is said to
be algebraically **irreducible** if it has no **proper** invariant subspace
H_1, i.e., no invariant subspace H_1 other than the whole space H and
the null space {0}. A <u>matrix</u> representation D is irreducible if no
transformation S <u>exists</u> such that every element in every matrix of
SDS^{-1} <u>below</u> a <u>common form</u> is zero; the form must consist of non-
overlapping contiguous square blocks on the principal diagonal (9,
p. 94 ff),(14, pp. 67-69). If the matrix representation is
unitary,[5] reducible, and of <u>finite</u> dimension (matrix order), one may
always obtain all elements zero above <u>and</u> below the blocks (2, p.
143); further, every representation of a <u>finite</u> ⟨or a compact linear
Lie⟩ group is equivalent to a unitary representation. In either
case, each set of submatrices corresponding to a given diagonal
block of the form also constitutes a representation.

[5]A unitary representation consists of unitary operators. An opera-
tor $\underline{U}$ is unitary if $(\underline{U}f,\underline{U}g)=(f,g)$ for all function pairs f,g of the
Hilbert space. A matrix is unitary if its complex conjugate trans-
pose is its inverse.

[6]In each matrix representation, there is a matrix for each group operation. We indicate an individual element of such a matrix by $D_{rc}^{(n)}(R)$ where n indicates the representation, R indicates the group element (and so the matrix), and r and c indicate the row and column, respectively. We define R as an operator which acts upon a coordinate system (leaving objects or fields fixed in space) such that

$$x' = R \, x \qquad (10)$$

where x' and x are the sets of coordinate values in the new and old coordinate systems, respectively; since R may represent a translation, it need not be representable as a matrix. Confusion exists in the literature concerning D's due to the failure to carefully define the associated R's (4),(32).

The symmetry of a function with respect to a given group is defined in terms of its transformations under each and every operation of the group. For a given irreducible representation n of a given group, a set of partner functions (29, p. 112) is defined as one which transforms according to the equation

$$_p f_m^n(x) = P_R \, _p f_m^n(x') = P_R \, _p f_m^n(Rx) = \sum_{m'} D_{m'm}^{(n)}(R) \, _p f_{m'}^n(x) \quad (11)$$

for each and every operation R of the group. Note that the operator P_R acts upon the coordinate values, not the function proper, and reverses the change of coordinate values caused by R in (10). Given one partner function of a set, all the others are completely determined including complex normalization. The representation must be completely defined, not just up to a similarity transformation STS^{-1}.

We say that a function belongs to a given symmetry type if it is equal to a constant times a partner function belonging to a given row of a given (completely defined) irreducible representation. There may be more than one linearly independent partner function belonging to a single symmetry type (e.g., of a subgroup of a group which defines eigenfunctions). If so, there is more than one set of linearly independent partner functions and the sets do not mix under the operations of the given group. We use the index p to distinguish between partner sets; it consists of the modal indices required to specify the eigenfunction but not required to specify either the representation or row of the given group. For a finite group of e elements, a function f may be decomposed into its symmetry types by the equation

$$f_m^n(x) = (r/e) \sum_R D_{mm}^{n*}(R) \, P_R f(x), \qquad (12)$$

where r is the number of rows in the representation (9, p. 113).

[7]Strictly speaking, some operators $\underline{O}$, e.g., differential operators, have no adjoint, i.e., there is no $\underline{A}$ such that $(\underline{O}u,v)=(u,\underline{A}v)$ for $\underline{all}$ functions u and v of the Hilbert space. However, there $\underline{may}$ be an $\underline{A}$ which fulfills the preceding equation for a set of functions which a) excludes some functions but b) is $\underline{dense}$ in the Hilbert space (2, pp. 641-645); in this sense the Laplacian and the curl-curl operators are "essentially" self-adjoint and the mathematical completeness of Section 2.4 is then with respect to the restricted set of functions.

[8]To illustrate how our detailed discussion carries over to other systems, we briefly treat geophysical prospecting and sonar. Assuming the displacement s is small, the longitudinal and transverse waves in an elastic medium split (16, p. 142 ff.) to yield

$$\nabla^2 \not{u} = [\rho/(\lambda+2\mu)] \; \partial^2 \not{u}/\partial t^2 \tag{13}$$

and

$$- \nabla \times \nabla \times A = (\rho/\mu) \; \partial^2 A/\partial t^2, \tag{14}$$

respectively, where ρ is the density, μ is the shear modulus, $(\lambda+2\mu/3)$ is the compression modulus, $s=\nabla\not{u}$ for the longitudinal waves, $s=\nabla\times A$ for the transverse waves, and the reciprocals of the numerical coefficients on the right hand sides are the squares of the respective velocities. For harmonic waves, the equations of course take the $\underline{forms}$ of Eqs. (2) and (3), respectively.

For sonar, assuming that the medium is incompressible and nonviscous, that the motion is irrotational and the displacements small, and that the surface and bottom discontinuities are unimportant, one obtains

$$\nabla^2 p = (\rho_0/K) \; \partial^2 p/\partial t^2, \tag{15}$$

where $K=p\rho_0/(\rho-\rho_0)$ is the bulk modulus, p is the pressure, and ρ and ρ_0 are the density and its mean, respectively. Again it reduces to the scalar Helmholtz equation (2) for harmonic waves.

[9]For the curl-curl and scalar Helmholtz equations, the domain of validity of a modal expansion is similar to that of a function of a complex variable, e.g., of a Taylor or Laurent series (1),(22). The expansion is $\underline{not}$ valid for the whole analytic domain, but for a domain free of singularities (such as a scatterer or probe) between two concentric spheres, two coaxial cylinders, or two parallel planes, for spherical, circular cylindrical, and rectangular coordinates, respectively.

[10]With any $\underline{single}$ frequency, there is a continuum of exact solutions for any $\underline{general}$ inverse scattering problem. For example, one may

consider a scatterer to be an assemblage of multipoles at a point
and the description is valid outside a sphere centered at the point
and circumscribing the scatterer (assuming the medium to be homogeneous and isotropic); further, one may expand about any point whose
sphere is within the domain in which no measurements are made.
However, if one uses pulses of finite duration, localization may be
obtained, giving (in principle) a unique exact solution (except for
what is inside a perfectly conducting capsule) (31).

6. REFERENCES

(13) and (6) have introductory chapters on finite groups. (10) applies the point groups, C_∞, etc. to the derivation of selection
rules. Theorems for compact Lie groups are similar to those for
finite groups, but weighted integration replaces summation (9),(14).
(23) and (2) are encyclopedic for application to special functions
and elementary particle theory, respectively, but the latter assumes
considerable mathematical background.

(1) Bers, L., _Theory of Pseudo-analytic Functions_. New York:
    N.Y.U. Inst. Math. Sci., 1953.
(2) Barut, A.O., and R. Raczka, _Theory of Group Representations and
    Applications_. Warsaw: Polish Scientific Publishers, 2nd ed.,
    1980.
(3) Boerner, H., _Representations of Groups with Special Considera-
    tions for the Needs of Modern Physics_. Amsterdam: North
    Holland, 1970.
(4) Bouten, M., "On the rotation operators in quantum mechanics":
    1969, Physica 42, pp. 572-580.
(5) Emmerson, J.M., _Symmetry Principles in Particle Physics_.
    Oxford: Clarendon Press, 1972.
(6) Eyring, H., _et al_., _Quantum Chemistry_. New York: Wiley, 1944.
(7) Garbacz, R.J., and R. H. Turpin, "A generalized expansion for
    radiated and scattered fields,": 1971, IEEE Trans. Antennas
    Propagat. AP-19, pp. 348-358.
(8) Griffith, J.S., _The Irreducible Tensor Method for Molecular
    Symmetry Groups_. Englewood Cliffs, N.J.: Prentice-Hall, 1962.
(9) Hamermesh, M., _Group Theory and its Application to Physical
    Problems_. Reading, MA: Addison-Wesley, 1962.
(10) Herzberg, G., _Infrared and Raman Spectra_. New York: Van
    Nostrand, 1945.
(11) Kerns, D.M., "Analysis of symmetrical waveguide junctions,":
    1951, J. Res. NBS 46, pp. 267-282.
(12) Koster, G.F., et al., _Properties of the Thirty-Two Point
    Groups_. Cambridge, MA: MIT Press, 1963.
(13) Margenau, H., and G. M. Murphy, _The Mathematics of Physics and
    Chemistry_, Princeton, NJ: Van Nostrand, 1959, Chap. 15.
(14) Miller, H., Jr., _Symmetry Groups and their Applications_, New
    York: Academic Press, 1972.
(15) ----, _Symmetry and Separation of Variables_. Reading, MA:

Addison—Wesley, 1977.

(16) Morse, P.M., and H. Feshbach, Methods of Theoretical Physics. New York: Academic Press, 1963.

(17) Olsen, H.F., Dynamical Analogies, New York: Van Nostrand, 1958.

(18) Opechowski, W., "Magnetoelectric Symmetry": 1974, Int. J. Magn. 5, no. 4, pp. 317–325.

(19) Rose, M.E., Elementary Theory of Angular Momentum. New York: Wiley, 1967, pp. 127–139.

(20) Stratton, J.A., Electromagnetic Theory. New York: McGraw—Hill, 1941, pp. 486–488.

(21) Talman, J.D., Special Functions: A Group Theoretic Approach. New York: W. A. Benjamin, 1968.

(22) Vekua, I.N., Generalized Analytic Functions. Reading, MA: Addison—Wesley, 1962.

(23) Vilenkin, N.Ja., Special Functions and the Theory of Group Representations. Providence, RI: Am. Math. Soc., 1978.

(24) Wacker, P.F., "Non—planar near—field measurements: Spherical scanning": 1975, Natl. Bur. Stds., Washington, NBSIR 75–809.

(25) ———, "Unified theory of near—field analysis and measurement: Scattering and inverse scattering": 1981, IEEE Trans. Antennas Propagat. AP–29, pp. 342–351.

(26) ———, "Symmetry analysis and eigenfunction expansions in engineering and applied mathematics": 1983, Third Intl. Conf. Ant. Propagat., Norwich, UK, pp. 113–117.

(27) ———, "Unified theory of near—field analysis and measurement: Reply to Dr. Steven B. Berger": Nov. 1983, IEEE Trans. Antennas Propagat. AP–31.

(28) Weinberg, S., "Light as a fundamental particle": 1975, Physics Today 28, pp. 32–37.

(29) Wigner, E.P., Group Theory and Its Application to the Quantum Mechanics of Atomic Spectra. New York: Academic Press, 1959.

(30) ———, "Unitary representations of the inhomogeneous Lorentz group including reflections," in Group Theoretical Concepts and Methods in Elementary Particle Physics, Feza Gürsey, ed. New York: Gordon and Breach, 1964.

(31) Wohlers, M.R., "On the uniqueness of the sources of electromagnetic fields": 1970, IEEE Trans. Ant. Propagat. AP–18, pp. 825–826.

(32) Wolf, A.A., "Rotation operators,": 1969, Am. J. Phys. 37, pp. 531–536.

I.12 (OI.5)

RECONSTRUCTION OF REFRACTIVE INDEX PROFILE OF A STRATIFIED MEDIUM

E.Vogelzang, H.A.Ferwerda and D.Yevick[*].

Department of Applied Physics, Nijenborgh 18, 9747 AG
Groningen, The Netherlands, Institute of Optical Research,
Stockholm, S-10044 Sweden*.

Abstract
In this paper we present a method for determining the permittivity
profile of a stratified medium terminated by a perfect conductor
from the (complex) reflectivity. The calculations are based on the
Gelfand-Levitan and the Marchenko equations. The bound modes of
the system are explicitly taken into account.

1. Introduction
In this contribution we shall describe how to reconstruct the refractive index profile of a stratified medium placed on top of a
perfect conductor using the complex reflection coefficients for
plane monochromatic waves, incident from different directions.
Because of the monochromaticity of the incident waves we do not
have to consider the dispersion of the refractive index.
It may be difficult to measure the phase of the reflection coefficient at optical frequencies. In that case a holographic recording of the reflected wave might be envisaged.

2. Outline of the method
A lossless dielectric layer of thickness d has the permittivity
profile $\varepsilon_2(z)$, where z is measured along the normal on the layer.
$\varepsilon_2(z)$ is assumed to be independent of the remaining coordinates x
and y. Such a medium is called a stratified medium. The medium
above the dielectric layer is supposed to be a dielectric with constant permittivity ε_1. We consider TE-polarized incoming plane
waves, i.e. waves whose electric vector only has a non-vanishing
component along the y-axis:

$$E_{y,inc} = E_o \exp[ik_1(x \sin\theta_1 + z \cos\theta_1)] , \qquad (1)$$

W.-M. Boerner et al. (eds.), Inverse Methods in Electromagnetic Imaging - Part 1, 223-230.
© 1985 by D. Reidel Publishing Company.

Figure 1

where k_1 is the wave number in the medium 1 and θ_1 is the angle of incidence (Fig.1). E_y has to satisfy the Helmholtz equation

$$[\partial^2/\partial x^2 + \partial^2/\partial z^2 + \omega^2 c^{-2}\epsilon(z)] E_y(x,z)=0 \tag{2}$$

where $\epsilon(z) = \epsilon_1$ for $z \geq d$
$\qquad\qquad = \epsilon_2(z)$ for $0 \leq z < d$
(the time-dependence $\exp(-i\omega t)$ of E_y has been suppressed). Substituting $E_y(x,z)=T(x)U(z)$ leads to

$$d^2T/dx^2+\beta^2 T=0 \quad \text{and} \quad d^2U/dz^2 + [\omega^2 c^{-2}\epsilon(z)-\beta^2] U=0, \tag{3}$$

where β is a constant over all space . As $E_{y,inc}$ satisfies (2) in $z \geq d$ we immediately find that $\beta=k_1\sin\theta_1$ $\tag{4}$
Eqn. (2) can be rewritten as

$$d^2U(z,m_1)/dz^2 + m_1^2 U(z,m_1) = [k_1^2-k^2(z)] \ U(z,m_1), \tag{5}$$

where $\quad m_1^2=k_1^2-\beta^2=k_1^2\cos^2\theta_1$
$\qquad\quad k^2(z)=\omega^2 c^{-2}\epsilon(z)$.

The m_1-dependence of $U(z,m_1)$ is now explicitly denoted. (5) has the same form as the radial Schrödinger equation in Quantum Mechanics if we formally define a potential $V(z)$ by

$$V(z) = k_1^2-k^2(z) \tag{6}$$

It is clear that the potential is only non-vanishing on the interval $0 \leq z < d$.
The basic equation (5) is henceforth written as

$$(d^2/dz^2+m_1^2)U(z,m_1)=V(z)U(z,m_1) \tag{7}$$

with the boundary condition $U(o,m_1)=0$, $\tag{8}$
because the layer is placed on top of a perfect conductor.
The problem to be solved is the calculation of $V(z)$ from the complex reflection coefficiënt.

3. A brief reminder of the Gelfand-Levitan and Marchenko-formalisms.

The problem discussed above shall be solved with the Gelfand-Levitan (GL) and Marchenko (M) formalisms [1]. We first introduce the regular solutions $\phi(z,m_1)$ of (7), which satisfy

$$\phi(o,m_1)=o \; ; \; [d\phi(z,m_1)/dz]_{z=o} = 1 . \tag{9}$$

Another, linearly independent solution of (5) is the singular solution (Jost solution) $f(z,m_1)$ specified by $\lim_{z\to+\infty} f(z,m_1)=\exp im_1 z$ (10)

It is straightforward to show that $f(z,m_1)$ and $f(z,-m_1)$ are linearly independent solutions. Consequently the regular solution $\phi(z,m_1)$ can be expressed in terms of $f(z,m_1)$ and $f(z,-m_1)$:

$$\phi(z,m_1)=i(2m_1)^{-1}[F(m_1)f(z,-m_1)-F(-m_1)f(z,m_1)] , \tag{11}$$

where the coefficient $F(m_1)$ is called the Jost-function and the symmetry property $\phi(z,m_1)=\phi(z,-m_1)$ has been used. The solution $U(z,m_1)$ of eqn. (7) is proportional to $\phi(z,m_1)$. On the other hand the asymptotic behavior of $U(z,m_1)$ can be written as

$$U(z,m) \underset{z\to+\infty}{\sim} \tfrac{1}{2}i[\exp(-im_1 z)-R(m_1)\exp(im_1 z)], \tag{12}$$

where $R(m_1)$ is the (complex) reflection coefficient at angle of incidence θ_1. Because of energy conservation $|R(m_1)|=1$ so that $R(m_1)$ can be written as
$$R(m_1)=\exp[i2\delta(m_1)]. \tag{13}$$
Combining (11) and (12) we find

$$R(m_1) = F(-m_1)[F(m_1)]^{-1} \tag{14}$$

$$U(z,m_1) = m_1[F(m_1)]^{-1}\phi(z,m_1) . \tag{15}$$

The bound modes correspond to poles of $R(m_1)$ and thus to zeroes of $F(m_1)$. It has been shown [2] that $F(m_1)$ has simple zeroes on the positive imaginary axis of the complex m_1-plane corresponding to bound modes.

The set of regular solutions, including the bound modes form a complete set:

$$\int_{-\infty}^{\infty} d\rho(E)\phi(z,E)\phi(z',E)=\delta(z-z'), \tag{16}$$

where $E=m_1^2$ ("energy") and the spectral density $\rho(E)$ is given by

$$d\rho(E) = \begin{cases} \pi^{-1}E^{\frac{1}{2}}|F(E^{\frac{1}{2}})|^{-2}dE, & E\geq o \\ \sum_j C_j\delta(E-E_j)dE , & E<o , \end{cases} \tag{17}$$

$$C_j^{-1} = \int_o^{\infty}\phi^2(z,m_{1j})dz.$$

m_{1j} is the value of m_1 for the bound mode with energy E_j. $|F(m_1)|$ can be calculated according to

$$|F(m_1)| = \prod_j |1-m_{1j}^2 \; m_1^{-2}| \quad \exp[\pi^{-1} P \int_{-\infty}^{\infty} \delta(m_1')(m_1-m_1')^{-1}dm_1'] \quad (18)$$

The constants C_j cannot be inferred from experiment and still pose a problem in this approach, a problem which can be circumvented by using a matching medium such that the combined structure has no bound modes.
We will now introduce the generalized Gelfand-Levitan equation which allows to calculate the difference of two potentials $V_2(z)-V_1(z)$, of which one is assumed to be known, say $V_1(z)$. Let the regular solutions of $V_1(z)$ be denoted by $\phi_1(z,m_1)$. The gene-ralized Gelfand-Levitan equation then reads:

$$K_{12}(z,t)+G_{12}(z,t) + \int_0^z K_{12}(z,s)G_{12}(s,t)ds=0, \quad (19)$$

where

$$G_{12}(z,t) = \int_{-\infty}^{\infty} \phi_1(z,m_1)\phi_1(t,m_1)[d\rho_2(E)-d\rho_1(E)], \quad (20)$$

$\rho_j(E)$ being the mode density of $V_j(z), j=1,2$. The potential is de-termined via

$$V_2(z) - V_1(z) = 2 \; dK_{12}(z,z)/dz \quad (21)$$

The Gelfand-Levitan theory is based upon the regular solution of (7). We can also start from the Jost-solution $f(z,m_1)$, eqn.(10), which leads to the Marchenko equation:

$$A(z,t)=A_o(z+t)+ \int_z^{\infty} A(z,s)A_o(s+t)ds \quad , \quad t>z, \quad (22)$$

where

$$A_o(t)=(2\pi)^{-1} \int_{-\infty}^{\infty} R(m_1)\exp(im_1t)dm_1 - \sum_j s_j\exp(im_{1j}t), \quad (23)$$

$$s_j^{-1} = \int_0^{\infty} f^2(z,m_{1j})dz. \quad (24)$$

$V(z)$ follows from $V(z) = -2dA(z,z)/dz$. \quad (25)

The Marchenko equation has the advantage over the Gelfand-Levitan equation of directly involving the measured quantity $R(m_1)$, while in the Gelfand-Levitan approach one first has to calculate the Jost function $F(m_1)$ which is quite involved! Eqn.(22) cannot direct-ly be solved numerically because the s_j turn out to be large. This leads to numerical instabilities when the equation is discretisized. For that reason we treat the contributions of the continuous and discrete parts of the reflection spectrum separately (corresponding to both terms on the right hand side of (23)). In the next section we shall explain how this can be done.

4. Reconstruction of a potential with bound states

In order to separate the contribution of the continuous and the discrete part of the spectrum (corresponding to both terms of (23)) we construct from the experimentally determined reflectivity $R(m_1)$ another reflectivity $R_1(m_1)$ which has no poles and consequently corresponds to a potential without bound modes. Denoting the propagation constants of the bound modes by m_{1j} (the poles of $R(m_1)$) $R_1(m_1)$ is according to (14) given by

$$R_1(m_1) = \prod_j (m_1 - m_{1j})^2 (m_1 + m_{1j})^{-2} R(m_1). \tag{26}$$

The corresponding potential is obtained by solving the Marchenko equation (22) with $s_j = 0$. This is done by discretisizing the equation and solving the resulting matrix equation. $V_1(z)$ then follows from (25). $V_2(z)$, which only has bound modes, is solved using the generalized Gelfand-Levitan equation (19) where $d\rho_2(E) = \sum_j C_j \delta(E - E_j) dE$. (19) becomes in this case

$$K(z,t) + \sum_j C_j \phi_1(z, m_{1j}) \phi_1(t, m_{1j}) +$$

$$+ \int_0^z K(z,s) \sum_j C_j \phi_1(s, m_{1j}) \phi_1(t, m_{1j}) ds = 0. \tag{27}$$

We ultimately find

$$V_2(z) = -2 \, d^2/dz^2 [\log \det \mathbf{M}(z)], \tag{28}$$

where $\mathbf{M}(z) = 1 + \int_0^z \mathbf{R}(s) ds. \tag{29}$

The matrix $\mathbf{R}(s)$ has elements

$$R_{ij}(s) = C_j \phi_1(s, m_{1i}) \phi_1(s, m_{1j}). \tag{30}$$

5. Numerical simulations

At this time of writing the main obstacle for the application of this approach to actual profiles as occurring in practical situations is the fact that the constants C_j cannot be determined experimentally. Therefore we test the method on some known profiles. For each profile the reflectivity, the bound modes and the normalization constants C_j are computed. Subsequently the profile is reconstructed from these data.

Example 1: a layer with profile $\varepsilon_2(z) = 1.0025 - 0.001z$. Thickness of the layer $d=1$, wavelength $\lambda = 0.2$ (arbitrary units). The layer has no bound modes. Fig.2 shows the result: the drawn line is the reconstructed profile, the original profile (input) is given by the dotted line. The wavelength is so small compared

Figure 2

Figure 3

Figure 4

to the width d that the edge of the distribution is quite well reproduced. For larger values of λ the quality of the reconstruction
deteriorates, as is to be expected.

Example 2: homogeneous layer $d=1, \varepsilon_2=1.01$, $\lambda=0.1$.
This layer supports 2 bound states. The thin line corresponds to
the intermediate potential $V_1(z)$ while the thick line corresponds
to the fully reconstructed potential $V(z)=V_1(z)+V_2(z)$.

Example 3: homogeneous layer $\varepsilon_2=1.1$, $\lambda=0.1$, $d=1$.
This system has 6 bound states. The reconstruction of the profile
from the bound modes only is sketched in fig.4. The larger the
number of bound modes the better the reconstruction.

6. Conclusions
The present method could lead to an alternative to the Newton-
Kantorovich method for the determination of permittivity profiles
from reflected intensities as developed by Roger [3]. In this approach it may happen that different permittivity profiles lead to
the same reflected intensities, because the bound modes are not
taken into consideration. In the numerical examples presented in
the preceding section the importance of the bound modes has been
established. At present the method defies application to practical
problems because
1) the phase of the reflected wave has to be measured which, in
   principle, might be done holographically but which could be
   very difficult in practice
2) the normalisation constants C_j cannot be determined from experiment. Prior knowledge of the width of the layer may restrict
   the possible values of C_j

3) it is difficult to find a perfect conductor at optical frequencies. In some cases the finite conductivity can be accounted for by correcting the phase of the reflected wave with the phase shift due to the conductor of finite conductivity. This correction is only reliable if multiple reflections may be ignored, which is the case when the permittivity profile above the terminating conductor changes smoothly (so that the WKB-approximation applies).

References

[1] K. Chadan and P.C. Sabatier, Inverse Problems in Quantum Scattering Theory, Springer, Berlin 1977
[2] Ref. 1, p.22
[3] A. Roger, "Newton-Kantorovich Algorithm Applied to an Electromagnetic Inverse Problem", IEEE Trans. Antennas Propagat., vol. AP-29, pp. 232-238, March 1981

I.13 (IM.4)

INVERSE DIFFRACTION PROBLEM

A. G. RAMM

Math. Dept., Kansas State Univ., Manhattan, KS 66506

Abstract: Let D be a convex body in $\mathbb{R}^3$ with a smooth connected boundary Γ. A constructive procedure for recovering Γ
from the scattering amplitude is given. The support function of
D is recovered from the known values of the scattering amplitude
at high frequencies. The support function defines explicitly the
parametric equation of Γ. The stability of the recovered Γ
with respect to small perturbations of the scattering data is
established and the corresponding estimates are given. A construction of the surface Γ from the knowledge of its two principal curvatures is given.

Contents

I.   INTRODUCTION

Let D be a convex bounded domain with a smooth boundary Γ,
$\Omega = \mathbb{R}^3 \setminus D$. Consider the problem

$$(\nabla^2 + k^2)u = 0 \quad \text{in} \quad \Omega, \, k > 0, \tag{1}$$

$$u\big|_\Gamma = 0, \tag{2}$$

W.-M. Boerner et al. (eds.), Inverse Methods in Electromagnetic Imaging - Part 1, 231–249.
© *1985 by D. Reidel Publishing Company.*

$$u = u_0 + v, \quad u_0 = \exp\{ik(v,x)\}, \quad r(\frac{\partial v}{\partial r} - ikv) \to 0, r = |x| \to \infty. \quad (3)$$

Here v is a unit vector, u_0 is the incident field, v is the scattered field,

$$v \sim \frac{\exp(ikr)}{r} f(n,v,k), \quad r \to \infty, \quad \frac{x}{r} = n. \quad (4)$$

The function $f(n,v,k)$ is called the scattering amplitude, $f(n,-n,k)$ is the backscattering amplitude. The inverse diffraction (scattering) problem consists in finding Γ from the knowledge of $f(n,v,k)$ $n,v \in S^2$, $0 < k < \infty$, where S^2 is the unit sphere. It is an overdetermined problem: f depends on 5 variables while the shape of Γ depends on 2 variables, for example it is determined by the equation $r = r(\theta,\phi)$ of the surface Γ in the spherical coordinates. Therefore, not any function $f(n,v,k)$ is a scattering amplitude. A necessary and sufficient condition for a function $f(n,v,k)$ to be a scattering amplitude is not known. The uniqueness of the solution to this problem is established long ago.

Proposition 1. If $f(n,v,k)$, $n,v \in S^2$, $0 \le a \le k \le b$, $a < b$ is the scattering amplitude for Γ_1 and for Γ_2, where Γ_1 and Γ_2 are piecewise smooth, closed, connected (not necessarily convex) surfaces, then $\Gamma_1 = \Gamma_2$. Here $[a,b]$ is an arbitrary segment.

There are several ways to prove this statement. One way, due to M. Schiffer ([1]), is general but non-constructive. The second way requires an assumption about convexity and smoothness of Γ, Gaussian curvature is assumed to be strictly positive. This way is based on the high-frequency asymptotics of $f(n,v,k)$ and can be based on the results of Fock [2], although Fock did not discuss the inverse problems. In this way one finds the Gaussian curvature of the surface from the high-frequency asymptotics of $f(n,v,k)$. It is well known [3] that Minkowski's problem of recovering a convex compact surface Γ from the knowledge of its Gaussian curvature K as a function of the unit normal vector to Γ has not more than one solution. Let us formulate the corresponding result.

Proposition 2 [3, p. 286]. Let S_1 and S_2 be two compact oriented surfaces of class C^2 in $\mathbb{R}^3$. Suppose that $f: S_1 \to S_2$ is a diffeomorphism such that at each pair of corresponding points, S_1 and S_2 have the same unit normal vector and equal Gaussian curvatures K_1 and K_2, respectively. Then f is a translation in $\mathbb{R}^3$.

If one knows that the Gaussian curvature of a closed compact surface satisfies the inequality $K \ge a^{-2}$, $a > 0$, then the diameter d of the surface satisfies the inequality $d \le \pi a$, i.e. the body D is not very large [3, p. 305]. One can calculate the area S of the surface Γ by the formula

$$S = \int_{S^2} \frac{dw}{K(w)} \, , \quad \text{where} \quad dw \text{ is the element of the area of the unit}$$

sphere S^2, $w \in S^2$, and the area of the projection of D onto the plane perpendicular to the unit vector n, by the formula $S(n) = \frac{1}{2} \int_{S^2} \frac{|n \cdot w|}{K(w)}$ dw. These formulas are well known in differential geometry and were used in inverse diffraction problem in [4]. In [5] a uniqueness theorem is proved. The theorem says that the shape of the obstacle and the potential in the exterior domain Ω can be uniquely determined from the scattering amplitude. In [6] the results of [5] are generalized, a constructive method for finding the shape of Γ from the scattering amplitude is given, and the stability of the solution to inverse diffraction problem is analyzed. Here we will follow the presentation in [6]. Some of the results of [4], [5] can be found in [7]. The role of high-frequency asymptotics in inverse scattering after work of Fock (1945-1950) [2] was discussed by many authors (see [8], [9], [10] and references in [9], [10]). The existence results in Minkowski's problem are given in [11]. Since the solution of the Minkowski's problem in [11] is non-constructive, it is of considerable practical interest to find a constructive procedure of recovering the surface Γ from the scattering amplitude. Such a procedure was given in [4] and its stability was analyzed in [7]. This procedure allows one to find the support function of D from the knowledge of $f(k,n,\nu)$ at one sufficiently large frequency k and the values of $n, \nu \in S^2$ such that the vector $1 = \frac{n - \nu}{|n - \nu|}$ runs through all of S^2 (e.g. $\nu = -n$ and $n \in S^2$, i.e. backscattering).

In the literature [12] there were attempts to solve the inverse two-dimensional diffraction problem for a perfectly conducting cylinder assuming that $f(k,n,\nu)$ is known for $0 < k < k_0$, $n, \nu \in S^1$ (S^1 is the unit sphere in $\mathbb{R}^2$), where k_0 is a small number (low frequency data). From the practical point of view this is not very interesting because one needs to know all of the Taylor coefficients of $f(k,n,\nu)$ at $k = 0$, and it is difficult to measure even the second coefficient. The knowledge of the first non-zero coefficient in the case when the obstacle is much less than the wavelength ($ka \ll 1$, where a is the diameter of the obstacle; practically $ka < 0,2$ is often sufficient) allows one to recover 3 numbers only [13]. The obstacle under these assumptions scatters as an equivalent ellipsoid with the semiaxes given by the above 3 numbers [13]. Thus, the first coefficient (corresponding to dipole radiation) gives an idea about the shape of the obstacle but does not allow one to recover the shape in detail, and it is quite difficult to measure the other coefficients (corresponding to multipole radiation).

In [13, p. 109] an inverse radiation problem is solved. This problem consists in finding the initial field at the point x_0 from the knowledge of the field scattered by a small ($ka \ll 1$) obstacle placed in the initial field at the point x_0.

If the boundary condition is of the form

$$\frac{\partial u}{\partial N} + hu\Big|_\Gamma = 0, \tag{5}$$

where h is constant (or a smooth function on Γ), $h > 0$ (or $\operatorname{Re} h \geq 0$, $\operatorname{Im} h \leq 0$, $|\operatorname{Re} h| + |\operatorname{Im} h| > 0$), then one can find the h and the shape of Γ from the knowledge of $f(k,n,\nu)$, $0 \leq a \leq k \leq b$, $b > a$, $n = n_0$, $\nu \in S^2$, where n_0 is a fixed unit vector (or $\nu = \nu_0$, $n \in S^2$, where ν_0 is a fixed unit vector).

In the physical literature ([2], [4], [9] etc.) the high frequency asymptotics of the scattering amplitude is obtained by using the Kirchhoff approximation (physical optics). This approximation and the high frequency asymptotics was rigorously justified in [14]-[16] for convex and for some non-convex obstacles.

This paper is organized as follows: in section 2 the support function and the shape of the surface are recovered from the scattering amplitude, in section 3 the stability of the solution to the inverse diffraction problem is proved, in section 4 a procedure for recovering the surface from the knowledge of its two principal curvatures K_1 and K_2 is given. This problem was formulated by the author and solved by F. Miller [17]. His results are presented in section 4, and the author thanks Professor Miller for permission to include these results in this paper. The interest in this problem comes from the fact that one can find the Gaussian curvature $K = K_1 K_2$ ([2,4,8,9]) and the difference of the curvatures $K_1 - K_2$ ([18]) from the scattering data, and therefore find K_1 and K_2.

We conclude this introduction with a proof of the uniqueness of the solution to the inverse diffraction problem. In particular, from the argument below the validity of Proposition 1 follows. Suppose there are two obstacles D_1 and D_2 with the same scattering amplitudes $f(n,\nu_0,k)$, $n \in S^2$, $a < k < b$, $b > a$, and the boundary condition on $\Gamma_1 = \partial D_1$ ($\Gamma_2 = \partial D_2$) is (5) with $h = h_1$ ($h = h_2$). We want to prove that $D_1 = D_2$ and $h_1 = h_2$. Consider $v = v_1 - v_2$, where v_j, $j = 1,2$, are determined from (1), (3), (4), (5) with $D = D_j$. By the assumption $v = o(1/r)$ as $r \to \infty$ and v satisfies (1). This implies that $v \equiv 0$ in $B_R = \{x : |x| > R\}$, where R is large enough so that the ball $\mathbb{R}^3 \setminus B_R$ contains D_1 and D_2. This follows from the known result of Rellich ([7, p. 298]). Thus $u_1 = u_2$ in B_R. Suppose first that $D_1 \cap D_2 = \emptyset$. Then $u_1|_{\Gamma_2} = u_2|_{\Gamma_1} \equiv 0$ for $a \leq k \leq b$, $b > a$, and $u_j \not\equiv 0$, $j = 1,2$, because $u_j = \exp\{ik(\nu_0,x)\} + 0(r^{-1})$ as $r \to \infty$. Therefore any number $k \in [a,b]$ is an eigenvalue of the problem

$$(\nabla^2 + k^2)u = 0 \quad \text{in } D_2, \quad \frac{\partial u}{\partial N} + h_2 u\Big|_{\Gamma_2} = 0$$

and u_1 is the corresponding eigenfunction. But this is a

contradiction since the spectrum of this problem is discrete.
If $D_1 \cap D_2 \neq \emptyset$ then consider a connected component of the set
$D = D_2 \setminus (D_1 \cap D_2)$. The boundary Γ of this domain has, say, two
parts: $\Gamma = \tilde{\Gamma}_1 \cup \tilde{\Gamma}_2$, where $\tilde{\Gamma}_j$ is a subset of Γ_j. Our argument
is the same: u_1 is the eigenfunction of the problem

$$(\nabla^2 + k^2)u = 0 \quad \text{in} \quad D, \quad \frac{\partial u}{\partial N} + h_2 u\big|_{\tilde{\Gamma}_2} = 0,$$

$$-\frac{\partial u}{\partial N} + h_1 u\big|_{\tilde{\Gamma}_1} = 0 \tag{6}$$

for any $a < k < b$, which is a contradiction since the spectrum
of the problem (6) is discrete. The sign minus in the last
equation in (6) is taken because we assume that the domain D
lies outside of D_1 and therefore the exterior normal to the
boundary $\tilde{\Gamma}_1$ of D differs by the sign from the exterior normal
N to the part $\tilde{\Gamma}_1$ of the boundary of the domain D_1. The contradiction proves that $D_1 = D_2$. Since $u_1 = u_2$ outside of
$D_1 = D_2$ the quantities $h_j = -\frac{\partial u_j}{\partial N} \cdot \frac{1}{u_j}$, $j = 1,2$, must be equal.

Note that u_j cannot vanish on the sets with positive measure
because if $u_j = 0$ on $\tilde{\Gamma}_j$, meas $\tilde{\Gamma}_j > 0$, then from (5) it follows
that $\frac{\partial u_j}{\partial N} = 0$ on $\tilde{\Gamma}_j$. If $u_j = \frac{\partial u_j}{\partial N} = 0$ on $\tilde{\Gamma}_j$, meas $\tilde{\Gamma}_j > 0$
then by the uniqueness of the solution to the Cauchy problem for
elliptic equations one concludes that $u_j \equiv 0$ in Ω_j. This is
impossible because of (3): $|u_j| \to 1$ as $r \to \infty$. This completes
the proof.

II. RECOVERING OF THE SUPPORT FUNCTION
FROM THE SCATTERING AMPLITUDE

1.   The Green's formula gives the following expression for the
scattering amplitude (for the problem (1)-(4))

$$f = -\frac{1}{4\pi} \int_\Gamma \exp\{-ik(n,s)\}h(s)ds, \quad h \equiv \frac{\partial u}{\partial N}, \tag{7}$$

where

$$h + Ah = 2\frac{\partial u_0}{\partial N}, \quad Ah = \int_\Gamma \frac{\partial}{\partial N_s} \frac{\exp\{ik|s - t|\}}{2\pi|s - t|} h(t)dt,$$

N_s is the exterior unit normal to Γ at the point s,

$$u = u_0 - \int_\Gamma ghdt, \quad g = \frac{\exp(ik|x - y|)}{4\pi|x - y|} = g(x,y,k)$$

2.   Let us fix the coordinate system with the origin outside of
D. Let us use the Kirchhoff approximation for h:

$$h = 0 \quad \text{on} \quad \Gamma_- , \quad h = 2ik(N_s,\nu)\exp\{ik(\nu,s)\} \quad \text{on} \quad \Gamma_+ , \quad (8)$$

where $\Gamma_+(\Gamma_-)$ is the illuminated (shadowed) part of Γ (i.e. $(N_s,\nu) < 0$ on Γ_+, $(N_s,\nu) > 0$ on Γ_-). Substitute (8) into (7) to obtain:

$$f = -\frac{2ik}{4\pi} \int_{\Gamma_+} \exp\{ik(\nu - n,s)\}(N_s,\nu)ds. \quad (9)$$

Define the support function of Γ to be

$$a(1) = -\max_{s\epsilon\Gamma_+} \; (1,s) \quad (\min_{s\epsilon\Gamma_+} |(1,s)|). \quad (10)$$

where 1 is a unit vector, and $(1,s) < 0$ on Γ_+.

3.   Integral (9) can be calculated by the stationary phase method.   The stationary point of the phase $(\nu - n,s) = -|n - \nu|(1,s)$, $1 \equiv \frac{n - \nu}{|n - \nu|}$ is the point $s_0 \epsilon \Gamma_+$ at which $1 = N_{s_0}$. If Γ is convex then there is only one stationary point. The following result is well-known [19]:

Lemma 1.   Let $f \epsilon C_0^\infty(\Omega)$, $S(x) \epsilon C^\infty(\Omega)$, $\Omega \subset R^n$ is a finite domain. Let $S(x)$ be a real valued function which has the only stationary (critical) point x_0 in Ω, and this point be nondegenerate (i.e. $dS(x_0) = 0$, $d^2S(x_0) \neq 0$).   Then

$$\int_\Omega f(x)\exp\{ikS(x)\}dx = (\frac{2\pi}{k})^{n/2} \exp\{i\lambda S(x_0) +$$

$$+ \frac{i\pi}{4} \operatorname{sgn} S''(x_0)\}|\det S''(x_0)|^{-1/2}[f(x_0) + O(\frac{1}{k})] \quad (11)$$

where

$$S''(x_0) = (\frac{\partial^2 S(x_0)}{\partial x_i \partial x_j}), \quad \operatorname{sgn} S''(x_0) = n_+ - n_-, \quad (12)$$

and n_+ (n_-) is the number of positive (negative) eigenvalues of the matrix $S''(x_0)$.
   Let us calculate the term $(1,s)$ around the stationary point.

$$(1,s) = (1,s_0) + (1,s - s_0) = (1,s_0) + \frac{1}{2}b_{ij}u^iu^j + o\{(u^1)^2$$

$$+ (u^2)^2\},$$

where one should sum over the repeated indices, b_{ij} is the matrix of the second differential form of the surface Γ^{ij} at the point $s_0, u^i, i = 1,2$ are the local coordinates in the tangential plane to Γ, $s_0 = (0,0)$ in these coordinates, and the third axis is directed along $1 = N_{s_0}$, the unit normal vector to Γ at the point s_0.

Therefore $S''(x_0) = -|n - \nu|b_{ij}$, $n = 2$, and formula (11) applied to (9) yields

$$f = -\frac{ik}{2\pi}\frac{2\pi}{k}\exp\{-ik|n - \nu|(1,s_0) + \frac{i\pi}{4}\,\text{sgn}\,b_{ij}(s_0)\}$$

$$|\det(-|n - \nu|b_{ij})^{-1/2}|[(1,\nu) + 0(\frac{1}{k})].\tag{13}$$

It remains to calculate $\text{sgn}\,b_{ij}(s_0)$ and $\det(-|n - \nu|b_{ij})$. One has $\det(-|n - \nu|b_{ij}) = |n - \nu|^2\det b_{ij}$. If the surface is convex then $n_+ = 2$, $n_- = 0$, $\text{sgn}\,b_{ij} = 2$. The eigenvalues of b_{ij} in the local coordinates described above are the principal curvatures K_1 and K_2 of the surface Γ at the point s_0, while $\det b_{ij} = K = K_1 K_2 > 0$ is the Gaussian curvature of Γ at s_0. Thus the leading term of the asymptotics in (13) takes the form

$$f = \frac{\exp\{-ik|n - \nu|(1,s_0)\}(1,\nu)}{|n - \nu|\{K(s_0)\}^{1/2}} =$$

$$\frac{\exp\{-2ik(n,1)(1,s_0)\}}{2\{K(s_0)\}^{1/2}}\tag{14}$$

Here the formulas $|n - \nu| = 2(n,1),(1,\nu) = -(1,n)$ were used. Finally note that

$$-(1,s_0) = a(1)\tag{15}$$

since s_0 is the stationary point at which 1 is the normal to Γ. Thus

$$f = -\frac{\exp\{2ik(n,1)a(1)\}}{2\{K(s_0)\}^{1/2}}\tag{16}$$

is the leading term of the asymptotics of f as $k \to \infty$. Note that the sign of the phase in (16) differs from the sign of the phase in [4] because in [4] the scatterer was centrally symmetric and the origin 0 was placed at the center of symmetry, so that $|(1,s_0)| = (1,s_0) = a(1)$ (compare with (15)). If Γ is convex and $0 \in D$ then $(1,s) > 0$ and $a(1)$ is defined as $a(1) = \max_{s\in\Gamma}(1,s)$. Let us formulate the result.

Theorem 1. Let Γ be a smooth convex surface. Then the leading term of the asymptotics of the scattering amplitude (7) as $k \to \infty$ is given by (16). In (16) $K(s_0)$ is the Gaussian curvature at the point $s_0 \in \Gamma$ at which $\frac{n - \nu}{|n - \nu|} = 1$ is the unit normal to Γ pointed outside of D, and $a(1)$ is the support function of Γ defined in (10).

<u>Remark 1</u>. Vector s_0 defines in the fixed coordinate system the position of the scatterer. Translation of the scatterer $s' = s + d$ by a vector d will lead to the phase shift $2ik(n,1)$. $(1,d)$ which reflects the new position of the scatterer.

For the back-scattering $1 = n$, $\nu = -n$, $(1,s_0) = -a(1) < 0$ on Γ_+, $(1,s_0) = a(1) > 0$ on Γ_-.

4. Parametric equations of the surface can be written explicitly. Let the unit vector 1 have components $(\alpha_1,\alpha_2,\alpha_3)$, $\alpha_1^2 + \alpha_2^2 + \alpha_3^2 = 1$. Then

$$x_j = -\frac{\partial a(1)}{\partial \alpha_j}, \quad j = 1,2,3, \quad 1 = \sum_{j=1}^{3} \alpha_j e_j \qquad (17)$$

is the parametric equation of Γ, the vectors e_1,e_2,e_3 are the orthonormal unit vectors defining the coordinate system with the origin 0.

To derive (17) let us write the equation of the tangent plane π to Γ orthogonal to the unit vector 1:

$$\sum_{j=1}^{3} x_j \alpha_j = -a(1) \qquad (18)$$

and consider $\alpha_1,\alpha_2,\alpha_3$ as parameters. The surface Γ is the envelope of the family of the planes (18) $\alpha_1,\alpha_2,\alpha_3$ being parameters. Thus, the equation of the surface can be obtained from the equations

$$\frac{\partial}{\partial \alpha_m} \{ \sum_{j=1}^{3} x_j \alpha_j + a(\sqrt{\alpha_1^2 + \alpha_2^2 + \alpha_3^2})\} = 0, \quad m = 1,2,3, \qquad (19)$$

and (18). Equations (19) coincide with (17) and give the parametric equation of Γ.

<u>Example 1</u>. Consider a sphere $\sum_{j=1}^{3} (x_j - b_j)^2 = R^2$. The unit normal to the sphere $\frac{1}{R} \sum_{j=1}^{3} (x_j - b_j)e_j$. The tangent plane to the sphere orthogonal to the unit vector $1 = \sum_{j=1}^{3} \alpha_j e_j$ is $\sum_{j=1}^{3} (x_j - b_j + R\alpha_j)\alpha_j = 0$. Therefore, assuming that $(b,1) < 0$ and the origin is outside of the sphere, i.e. $|b|^2 \equiv \sum_{j=1}^{3} b_j^2 > R^2$, one obtains $a(1) = -(b,1) + R$. Equations (17) are parametric equations of the sphere: $x_j = b_j - R\alpha_j$, $1 \leq j \leq 3$, $R = R(\alpha_1^2 + \alpha_2^2 + \alpha_3^2)^{1/2}$.

Remark 2. From the knowledge of the support function $a(1)$ only for $1 \in \widehat{S^2}$, where $\widehat{S^2} \subset S^2$ is a domain in S^2, one can reconstruct effectively the corresponding part $\widehat{\Gamma}$ of the surface Γ: equation (17) with $1 = (\alpha_1, \alpha_2, \alpha_3) \in \widehat{S^2}$ is the parametric equation of this part of Γ.

5. Let us chose such set Δ of the unit vectors ν, n that $1 = \frac{n - \nu}{|n - \nu|}$ runs through $\widehat{S^2} \subset S^2$. For example, one can choose $n = -\nu$ and have n run over $\widehat{S^2}$. This gives the back scattering amplitude $f(n, -n, k)$. For back scattering $(n, 1) = 1$ and (16) takes the form

$$f = -\frac{\exp\{2ik\, a(1)\}}{2\{K(s_0)\}^{1/2}} \tag{20}$$

Let us assume that $f(n, \nu, k)$ is known for $n, \nu \in \Delta$ and all sufficiently large k. Then formula (16) allows one to find $a(1)$ and the surface Γ by formulas (17). Let us formulate this as

Theorem 2. If Γ is a smooth convex surface and the scattering amplitude $f(n, \nu, k)$ is known for all n, ν, such that $\frac{n - \nu}{|n - \nu|} = 1$ runs through $\widehat{S^2} \subset S^2$, and all sufficiently large k, then the support function $a(1)$ is uniquely determined for $1 \subset \widehat{S^2}$. Therefore the position and the shape of $\widehat{\Gamma} \subset \Gamma$ are uniquely determined. Moreover, explicit equation of the part $\widehat{\Gamma}$ of the surface Γ is given by formula (17).

6. Remark 3. The Kirchhoff approximation (7) has been rigorously justified in the literature (see e.g. [14]-[16]) as the leading term in the asymptotics of h as $k \to +\infty$. Therefore the argument in this paper is rigorous. Other boundary conditions can also be treated. For example, if $\frac{\partial u}{\partial N}\big|_\Gamma = 0$ then $f(n, \nu, k) =$

$= \frac{1}{4\pi} \int_\Gamma dsu\, \frac{\partial \exp\{-ik(n,s)\}}{\partial N_s}$. The Kirchhoff approximation gives $u = 0$ on Γ_- , $u = 2u_0$ on Γ_+ . Thus $f(n, \nu, k) = -\frac{ik}{2\pi} \int_\Gamma ds$

$\exp\{ik(\nu - n, s)\}(N_s, n)$ in this approximation (compare with (9)). Further study is similar to the given above.

III. STABILITY OF THE RECOVERED SURFACE

Suppose that the scattering amplitude is measured with an error $\delta > 0$, that is f_δ is known, $|f_\delta - f| \leq \delta$. Let

$f_\delta = f(n,\nu,k_0) + \delta f$, where δf is noise, $|\delta f| \leq \delta$, $\delta = $ const > 0.
Let us assume that $\sup_{s \in \Gamma} K(s) \leq d^2$, $(n,1) \geq b > 0$, where b and
d are some constants, and that k_0 is large, so that $0(\frac{1}{k_0})$ in
(13) satisfies the inequality $0(\frac{1}{k_0}) \ll b$. Then, from (13) and
(16) one obtains

$$f_\delta(n,\nu,k_0) = -\frac{\exp\{2ik_0(n,1)a(1)\}}{2\{K(s_0)\}^{1/2}} (1 + 0(\frac{1}{bk_0}) + 0(\delta d)). \quad (21)$$

Thus

$$a_\delta(1) \equiv \frac{1}{2ik_0(n,1)} \ln \frac{-f_\delta(n,\nu,k_0)}{|f_\delta(n,\nu,k_0)|} =$$

$$a(1) + 0(\frac{1 + k_0 b\delta d}{(k_0 b)^2}) \equiv a(1) + n,$$

$$n = 0((k_0 b)^{-2} + (k_0 b)^{-1}\delta d). \quad (22)$$

It follows from (22) that the noisy data f_δ determines explicitly
the support function $a(1)$ with the error n of order given in
(22). The stable reconstruction of the surface Γ is therefore
reduced to the stable differentiation of $a(1)$ known with the
error n. This problem was solved in [20]-[22].
    The result is as follows. Assume that $|a''(\phi)| \leq M$,
$-\infty \leq \phi \leq \infty$ and a_n is given, $|a_n - a| \leq n$. Let

$$\beta(n) \equiv (\frac{2n}{M})^{1/2}, \quad \hat{a}' \equiv \frac{a_n(\phi + \beta(n)) - a_n(\phi - \beta(n))}{2\beta(n)}. \quad (23)$$

Then

$$|\hat{a}' - a'| \leq (2Mn)^{1/2} \equiv \varepsilon(n). \quad (24)$$

Therefore $\hat{a}'$ is a stable approximation of a' with the
error $\leq \varepsilon(n)$. It can be proved that among all linear and non-
linear estimates of a' the above estimate is the best possible
(in a certain sense) on the class of all data given by the two
numbers (M,n) (see e.g. [21]). Namely,

$$\inf \sup_\phi |Ta_n(\phi) - a'(\phi)| \geq (2Mn)^{1/2}$$
$$T \in A, \ |a''| \leq M,$$
$$|a_n - a| \leq n$$

where A is the set of all estimates and T is an estimate.

IV. RECOVERING OF THE SURFACE FROM THE
KNOWLEDGE OF ITS PRINCIPAL CURVATURES

1.   Let Γ be a closed compact oriented surface in $\mathbb{R}^3$ and
$\psi:\Gamma \rightarrow S^2$ be such a map that the vector $\psi(s)$ when placed at
the point s is the positive unit normal to Γ. Here $\psi(s)$ is
the vector with the initial point at the origin (the center of the
unit ball with the boundary S^2) and the end point at the $\psi(s)$.
Let us assume that ψ is a diffeomorphism of Γ onto some part
$\tilde{S}^2$ of S^2 and let $\Phi:\tilde{S}^2 \rightarrow \Gamma$ be the inverse mapping ψ^{-1}. Let
$(e_1(s),e_2(s),e_3(s))$ be a moving orthonormal frame on Γ,
$e_3(s) = \psi(s)$, $e_1(s)$ and $e_2(s)$ be the eigenvectors of $-de_3(s)$
with eigenvalues $K_1(s)$ and $K_2(s)$ respectively, where
$K_j(s)$, $j = 1,2$ are the principal curvatures of Γ at the point
s. Let us define differential forms ω_1,ω_2 by the equations
$\omega_i(e_j) = \delta_{ij}$, $1 \le i, j \le 2$. Let $de_i = \omega_{ij}e_j$. Here and below
one should sum over the repeated indices. It is known [3] that
$d\omega_{ij} = \omega_{im} \wedge \omega_{mj}$, $d\omega_1 = -\omega_2 \wedge \omega_{12}$, $d\omega_2 = \omega_1 \wedge \omega_{12}$, where $\wedge$
denotes the wedge product of the differential forms, $\omega_{13} = K_1\omega_1$,
$\omega_{23} = K_2\omega_2$, and the fundamental forms on Γ are
$I = \omega_1^2 + \omega_2^2$, $II = K_1\omega_1^2 + K_2\omega_2^2$, $III = K_1^2\omega_1^2 + K_2^2\omega_2^2$. Let us pull
back to S^2 the basic differential forms using the mapping
$\Phi = \psi^{-1}$. The pulled back forms ω_i^*, ω_{ij}^* satisfy the equations

$$d\omega_{ij}^* = \omega_{im}^* \wedge \omega_{mj}^*, \quad d\omega_1^* = -\omega_2^* \wedge \omega_{12}^*, \quad d\omega_2^* = \omega_1^* \wedge \omega_{12}^*,$$

$$\omega_{ij}^* = -\omega_{ji}^*, \quad \omega_{13}^* = K_1\omega_1^*, \quad \omega_{23}^* = K_2\omega_2^*. \tag{25}$$

Here $K_i = K_i(\theta,\phi) = K_i(\Phi(\theta,\phi))$. These equations imply that

$$dK_1 \wedge \omega_1^* = \sigma_1(K_1 - K_2)\omega_1^* \wedge \omega_2^* \tag{26}$$

$$dK_2 \wedge \omega_2^* = \sigma_2(K_1 - K_2)\omega_1^* \wedge \omega_2^* \tag{26'}$$

where σ_1 and σ_2 are defined by the equation

$$\omega_{12}^* = \sigma_1\omega_1^* + \sigma_2\omega_2^* \tag{27}$$

Assume that $K_1(\theta,\phi)$ and $K_2(\theta,\phi)$ are given. If one finds the
forms ω_i^*, then determines vectors e_i^* such that

$$\omega_i^*(e_j^*) = \delta_{ij}, \quad 1 \le i, j \le 2, \tag{28}$$

and finally uses the equation

$$d\Phi = -\frac{\omega_1^*e_1^*}{K_1} - \frac{\omega_2^*e_2^*}{K_2} \tag{29}$$

then, assuming that K_1 and K_2 are the principal curvatures of a surface, one concludes that $d\Phi$ is a closed differential. Integrating this differential one recovers the surface. In what follows the equations for the forms ω_i^*, $1 \le i \le 2$ are derived and analyzed and a constructive method for finding the surface is given under the assumption that K_1 and K_2 are the principal curvatures of a surface. Let

$$\omega_1^* = a_1 d\phi + a_2 d\theta, \quad \omega_2^* = b_1 d\phi + b_2 d\theta. \tag{30}$$

The e_j^* are the eigenvectors of the mapping $d\Phi$ with the eigenvalues $-1/K_j$. One can check, using the orthonormality of the vectors $\dfrac{e_j^*}{K_j}$ with respect to III* metric, or the fact that

$III^* = \sin^2\theta(d\phi)^2 + (d\theta)^2$ (see Appendix), that

$$b_1 = \mp a_2 \frac{K_1}{K_2} \sin\theta, \quad b_2 = \pm a_1 \frac{K_1}{K_2} \frac{1}{\sin\theta}, \tag{31}$$

$$\frac{K_1^2 a_1^2}{\sin^2\theta} + K_1^2 a_2^2 = 1 \tag{32}$$

One has

$$\omega_{13}^* = K_1\omega_1^*, \quad \omega_{23}^* = K_2\omega_2^*, \quad \omega_{12}^* = \sigma_1\omega_1^* + \sigma_2\omega_2^* \equiv \alpha d\phi + \beta d\theta, \tag{33}$$

where α and β are to be found. Furthermore

$$d\omega_{12}^* = \omega_{13}^* \wedge \omega_{32}^* = -K\omega_1^* \wedge \omega_2^* = -K(a_1 b_2 - a_2 b_1)d\phi \wedge d\theta =$$

$$-K \frac{\sin\theta}{K} d\phi \wedge d\theta = -\sin\theta \, d\phi \wedge d\theta, \tag{34}$$

where $K = K_1 K_2$. If one uses the definition of ω_i^* in terms of a_i and b_i then one finds

$$a_1\sigma_1 - \frac{K_1}{K_2} \sin\theta \, a_2\sigma_2 = \alpha,$$

$$a_2\sigma_1 + \frac{K_1}{K_2 \sin\theta} a_1\sigma_2 = \beta.$$

Solving this system for σ_1 and $-\sigma_2$ one obtains:

$$\begin{pmatrix} \sigma_1 \\ -\sigma_2 \end{pmatrix} = T \begin{pmatrix} a_1 \\ a_2 \end{pmatrix}, \quad T \equiv \begin{pmatrix} \dfrac{K_1^2 \alpha}{\sin^2\theta} & K_1^2 \beta \\ -\dfrac{K\beta}{\sin\theta} & \dfrac{K\alpha}{\sin\theta} \end{pmatrix}. \tag{35}$$

Here the equation

$$\det \begin{vmatrix} a_1 - \dfrac{K_1}{K_2}\sin\theta\, a_2 \\[2mm] a_2 \dfrac{K_1}{K_2\sin\theta}\, a_1 \end{vmatrix} = \dfrac{\sin\theta}{K}$$

was used. Using the equations (26) one obtains

$$Q \begin{pmatrix} a_1 \\ a_2 \end{pmatrix} = \frac{K_2 - K_1}{K}\sin\theta \begin{pmatrix} \sigma_1 \\ -\sigma_2 \end{pmatrix} \tag{36}$$

where

$$Q = \begin{pmatrix} -\dfrac{\partial K_1}{\partial\theta} & \dfrac{\partial K_1}{\partial\phi} \\[3mm] \dfrac{K_1}{K_2\sin\theta}\dfrac{\partial K_2}{\partial\phi} & \dfrac{K_1\sin\theta}{K_2}\dfrac{\partial K_2}{\partial\theta} \end{pmatrix}. \tag{37}$$

Therefore

$$\left(Q - \frac{K_2 - K_1}{K}(\sin\theta)\, T\,\right) \begin{pmatrix} a_1 \\ a_2 \end{pmatrix} = 0. \tag{38}$$

This equation can be written as

$$\begin{pmatrix} A & B \\ C & D \end{pmatrix} \begin{pmatrix} a_1 \\ a_2 \end{pmatrix} = 0, \qquad \begin{aligned} A &= -\frac{\partial K_1}{\partial\theta} - \frac{K_2 - K_1}{K}\frac{K_1^2\alpha}{\sin\theta} \\[2mm] B &= \frac{\partial K_1}{\partial\phi} - \frac{K_2 - K_1}{K}\sin\theta\, K_1^2\beta \end{aligned} \tag{39}$$

where A and D are linear in α, B and C are linear in β. Since a_1 and a_2 are not zeros simultaneously, one has

$$AD - BC = 0. \tag{40}$$

If $K_2 - K_1 \neq 0$ this is a quadratic equation in α, β. Equations (39) can be written as

$$Aa_1 + Ba_2 = 0, \quad Ca_1 + Da_2 = 0. \tag{41}$$

Therefore

$$A^2a_1^2 = B^2a_2^2, \quad K_1^2\sin^2\theta A^2a_1^2 = B^2\sin^2\theta\left(1 - \frac{K_1^2a_1^2}{\sin^2\theta}\right). \tag{42}$$

Assume that $B \neq 0$. Then

$$a_1 = \pm\frac{\sin\theta}{K_1}\frac{1}{\left(1 + \dfrac{A^2}{B^2}\sin^2\theta\right)^{1/2}}, \quad a_2 = -\frac{A}{B}a_1. \tag{43}$$

Therefore a_1, a_2 (and b_1, b_2) are determined by α, β.

From (34) it follows that

$$\omega_{12}^* = -\cos\theta \, d\phi + dg, \tag{44}$$

where $g(\theta,\phi)$ is a scalar function. Therefore

$$\alpha = \frac{\partial g}{\partial\phi} - \cos\theta, \ \beta = \frac{\partial g}{\partial\theta}, \tag{45}$$

so that α and β are determined by g. If one solves the quadratic equation (40) for α and write $\alpha = H(\theta,\phi,\beta)$, then (43) can be written as

$$a_j = F_j(\theta,\phi,\frac{\partial g}{\partial\theta}), \ j = 1,2. \tag{46}$$

The equation $d\omega_1^* = -\omega_2^* \wedge \omega_{12}^*$ can be written as

$$\frac{\partial a_2}{\partial\phi} - \frac{\partial a_1}{\partial\theta} = \frac{K_1}{K_2} \sin\theta \ (a_2\beta + a_1\alpha \ \frac{1}{\sin^2\theta}) = G(\theta,\phi,\frac{\partial g}{\partial\theta}) \tag{47}$$

From (46) and (47) one obtains

$$\frac{\partial F_2}{\partial\phi} + \frac{\partial F_2}{\partial\beta}\frac{\partial}{\partial\theta}(\frac{\partial g}{\partial\phi}) - \frac{\partial F_1}{\partial\theta} - \frac{\partial F_1}{\partial\beta}\frac{\partial^2 g}{\partial\theta^2} = G(\theta,\phi,\frac{\partial g}{\partial\theta}). \tag{48}$$

From (45) it follows that

$$\frac{\partial}{\partial\theta}\frac{\partial g}{\partial\phi} = -\sin\theta + \frac{\partial H}{\partial\theta}(\theta,\phi,\frac{\partial g}{\partial\theta}) + \frac{\partial H}{\partial\beta} \cdot \frac{\partial^2 g}{\partial\theta^2} \tag{49}$$

From (48) and (49) one obtains

$$(\frac{\partial F_2}{\partial\beta}\frac{\partial H}{\partial\beta} - \frac{\partial F_1}{\partial\beta})\frac{\partial^2 g}{\partial\theta^2} + \frac{\partial F_2}{\partial\phi} - \frac{\partial F_1}{\partial\theta} + (\frac{\partial H}{\partial\theta} - \sin\theta) \cdot$$

$$\frac{\partial F_2}{\partial\beta} = G \tag{50}$$

This equation can be written as

$$\frac{\partial^2 g}{\partial\theta^2} = G_1(\theta,\phi,\frac{\partial g}{\partial\theta}). \tag{51}$$

Similarly, the equation $d\omega_2^* = \omega_1^* \wedge \omega_{12}^*$ leads to

$$\frac{\partial^2 g}{\partial\theta^2} = G_2(\theta,\phi,\frac{\partial g}{\partial\theta}), \tag{52}$$

where G_2 is explicitly determined. From (51) and (52) one obtains the equation for $\frac{\partial g}{\partial\theta}$

$$G_1(\theta,\phi,\frac{\partial g}{\partial\theta}) = G_2(\theta,\phi,\frac{\partial g}{\partial\theta}). \tag{53}$$

Assuming that this equation is solvable one finds $\frac{\partial g}{\partial \theta} = \beta$, then $\alpha = H(\theta,\phi,\beta)$, then A and B from (39), then a_1 and a_2 from (43), then b_1 and b_2 from (31), then ω_1^* and ω_2^* from (30), then e_1^* and e_2^* from (28), then $d\phi$ from (29). Integrating (29) one finds the mapping ϕ which determines the parametric equation of the surface Γ (parameters are θ,ϕ). The above construction is valid under the assumption that K_1,K_2 are the principal curvatures of a surface.

It was assumed above that $K_1 \neq K_2$. If $K_1 = K_2$ then it is easily seen that $K_1 = K_2 = \text{const}$. Indeed (36) shows, that in this case

$$Q \begin{pmatrix} a_1 \\ a_2 \end{pmatrix} = 0.$$

Since a_1 and a_2 are not both zero one concludes that

$$0 = \det Q = -\left(\frac{\partial K_1}{\partial \theta}\right)^2 \sin \theta - \left(\frac{\partial K_1}{\partial \phi}\right)^2 \frac{1}{\sin \theta}. \quad \text{Thus}$$

$\frac{\partial K_1}{\partial \theta} = \frac{\partial K_1}{\partial \phi} = 0$, $K_1 = \text{const}$. If $K_1 = K_2 = \text{const}$ then the surface is spherical, its radius is K_1^{-1}.

In conclusion of this section let us formulate some open problems:

1) Characterization problem: what are the necessary and sufficient conditions on $K_1(\theta,\phi)$, $K_2(\theta,\phi)$ in order that these functions be principal curvatures of a closed surface of class c^2?

The following problem is closely connected with characterization problem:

2) If K_1 and K_2 are the principal curvatures of a surface then $d\phi$ (see (29)) is a closed differential. Is the converse true? What is the relationship between the condition that $d\phi$ is closed and the solvability conditions for equations (40) and (53)?

APPENDIX I. Derivation of formulas (31-32).

The identity $III^* = \sin^2\theta(d\phi)^2 + (d\theta)^2 \equiv K_1^2(a_1 d\phi + a_2 d\theta)^2 +$ $+ K_2^2(b_1 d\phi + b_2 d\theta)^2$ is the starting point. One obtains the system

$$K_1^2 a_1^2 + K_2^2 b_1^2 = \sin^2\theta$$

$$K_1^2 a_1 a_2 + K_2^2 b_1 b_2 = 0$$

$$K_1^2 a_2^2 + K_2^2 b_2^2 = 1$$

Let $\dfrac{K_1}{K_2} = \lambda$, $K_2 \neq 0$, $\dfrac{b_i}{\lambda} = c_i$. Then the system takes the form

$$a_1^2 + c_1^2 = \sin^2\theta, a_1 a_2 + c_1 c_2 = 0, a_2^2 + c_2^2 = 1.$$

One has $\dfrac{a_1}{c_2} = -\dfrac{c_1}{c_2}$, $c_1^2\left(1 + \dfrac{c_2^2}{a_2^2}\right) = \sin^2\theta$, $\dfrac{c_1^2}{a_2^2} = \sin^2\theta$,

$c_1 = \mp a_2 \sin\theta$, $c_2 = \pm \dfrac{a_1}{\sin\theta}$. Thus, $b_1 = \mp \dfrac{K_1 \sin\theta}{K_2} a_2$,

$b_2 = \pm \dfrac{K_1}{K_2 \sin\theta} a_1$, $\dfrac{K_1^2 a_1^2}{\sin^2\theta} + K_1^2 a_2^2 = 1$. The choice of the sign in

(31) is (- +). One could chose (+ -) without any changes in the argument.

Remark. The unit sphere is the only compact closed surface such that $III^* = I^*$, that is the pull-backs of the third and first fundamental forms are equal on this surface. Indeed, $III^* = I^*$ iff $K_1^2 \omega_1^{*2} + K_2^2 \omega_2^{*2} = \omega_1^{*2} + \omega_2^{*2}$. The last equation implies that $K_1^2 = 1$, $K_2^2 = 1$. This means that $K_1 = 1$, $K_2 = 1$, i.e. the surface is the unit sphere. The case $K_1 K_2 = -1$ does not correspond to any compact closed smooth surface as is clear from the geometrical reasons.

APPENDIX 2.

A)   Recovery of a convex surface from the sum of its two principal radii.

Let $h(\ell) = K_1^{-1} + K_2^{-1}$ be known on all of S^2. Then

$\nabla_\alpha^2 a(\ell) = h$, $\ell = (\alpha_1, \alpha_2, \alpha_3)$. Let $Y_{nm}(\theta, \phi)$, $n \geq 0$, $-n \leq m \leq n$

be the normalized spherical harmonics, $h_{nm} = \int_{S^2} h \overline{Y}_{nm} \sin \theta \, d\theta \, d\phi$.

Then $a(\ell) = \sum_{n \geq 0, n \neq 1} h_{nm}[(n-1)(n+2)]^{-1} Y_{nm}$. One proves

that $h_{1m} = 0$, $m = 0, \pm 1$. If $a(\ell)$ is found then the surface

is given by (17). If h_δ is given, $|h - h_\delta| \leq \delta$, then the

coefficients $h_{nm}(\delta)$ are given, $|h_{nm}(\delta) - h_{nm}| \leq \delta(4\pi)^{1/2}$. One

takes $h_{1m}(\delta) = 0$, $m = 0, \pm 1$. Assuming that $|h_{nm}| \leq \dfrac{A}{(n^2 + m^2)^a}$,

A, a are constants, and knowing $h_{nm}(\delta)$ one estimates $a(\ell)$

with an error of order $\delta^{2 - 1/a}$ (see [20, p.355] for details) and finds Γ stably as in section III. This is presented in [22] and the relevant references are [23]-[27]. It is not clear if this method can be extended to the case when h is given on $\widetilde{S}^2$ (and then extended in a certain way to S^2). On the other hand, the construction in section IV is entirely local.

B)   A uniqueness theorem in scattering theory [28].

Theorem. (A. G. Ramm) The knowledge of $f(n, \nu_0, k_0)$, $n \in \widetilde{S}^2$ defines a strictly convex reflecting obstacle uniquely.

Here f is defined in (4), ν_0 and k_0 are fixed, $\widetilde{S}^2$ is any solid angle (open set in S^2), reflective obstacle means that (2) holds.

C)   The author gave recently an analytical reconstruction of the three-dimensional refraction coefficient (velocity inversion in geophysics) [29].

REFERENCES

[1]   P. Lax, R. Phillips, Scattering theory. New York: Acad.
       Press, 1967.
[2]   V. Fock, Electromagnetic diffraction and propagation pro-
       blems. New York: Pergamon Press, 1965.
[3]   C. C. Hsiung, A first course in differential geometry.
       New York: Wiley, 1981.
[4]   A. G. Ramm, "Determination of the shape of the reflecting
       body from the scattering amplitude", Radiofizika, Izvestija
       vuzov, vol. 13, N5, pp. 727-731, 1970.
[5]   A. G. Ramm, "Reconstruction of the shape of an obstacle
       from the scattering amplitude", Radiotechnika i electronika,
       vol. 11, pp. 2068-2070, 1965.
[6]   A. G. Ramm, "On inverse diffraction problem", J. Math. Anal.
       Appl., (to appear).
[7]   A. G. Ramm, Theory and applications of some new classes of
       integral equations. New York: Springer-Verlag, 1980.
[8]   J. B. Keller, "Diffraction by a convex cylinder", IRE Trans.,
       vol. AP-4, N3, pp. 312-313, 1956.
[9]   R. Lewis, "Physical optics inverse diffraction", IEEE Trans.,
       vol. AP-17, pp. 308-314, 1969.
[10]   W. Boerner, C. Ho, "Analysis of physical optics far field
       inverse scattering for the limited data case using radar
       theory and polarization information", Wave motion, vol. 3,
       pp. 311-333, 1981.
[11]   L. Nirenberg, "The Weyl and Minkowski problems in differen-
       tial geometry in the large", Comm. Pure Appl. Math., vol. 6,
       pp. 337-394, 1953.
[12]   D. Colton, "The inverse electromagnetic scattering problem
       for a perfectly conducting cylinder", IEEE Trans., vol. AP-
       29, N2, pp. 364-367, 1981.
[13]   A. G. Ramm, Iterative methods for calculating static fields
       and wave scattering by small bodies. New York: Springer-
       Verlag, 1982.
[14]   M. E. Taylor, Pseudo-differential operators. Princeton:
       Princeton Univ. Press, 1981.
[15]   H. D. Alber, "Justification of geometrical optics for non-
       convex obstacles", J. Math. Anal. Appl., vol. 30, pp. 372-
       386, 1981.
[16]   V. Petkov, "High frequency assymptotics of the scattering
       amplitude for non-convex bodies", Comm. Part. Differ.
       Equations, vol. 5, pp. 293-329, 1980.
[17]   F. Miller, The construction of a surface from its principal
       curvatures. Math. Dept. Kansas State Univ., Preprint,
       Feb. 1983.
[18]   S. Chaudhuri, W. Boerner, B. Foo, A high frequency inverse
       scattering model to recover the specular point curvature
       from parametric scattering matrix data, Univ. of Illinois,
       Chicago, Preprint, 1983.

[19] M. Fedorjuk, Metod perevala, Nauka, Moscow: 1977. (Russian)
[20] A. G. Ramm,"Stable solution of some ill-posed problems,"
      Math. Methods in the appl. sci., vol. 3, pp. 336-363, 1981.
[21] A. G. Ramm,"Estimates of the derivatives of random functions",
      J. Math. Anal. Appl. (to appear).
[22] A. G. Ramm, "On numerical differentiation", Mathem.,
      Izvestija vuzov, 11, (1968), 131-135. 40 # 5130.
[23] A. G. Ramm, "Remarks about inverse diffraction problems".
[24] E. Christoffel, Uber die Restimmung der Gestalt einer
      krummen Oberflache durch lokale Messungen auf Derselben,
      J. reine u. angew. Math., 64, 193-209, (1865).
[25] A. Hurwitz, Sur Quelques Applications Geometriques des
      Series de Fourier, Ann Ecol. Norm., 19, 357-408, (1902).
[26] W. Blaschke, Vorlesungen über Differential Geometrie,
      Springer-Verlag, Berlin, 1945, Vol. 1, 204.
[27] W. Firey, The Determination of Convex Bodies from their
      Mean Radius of Curvature Function, Mathematika, 14, 1-13,
      (1967).
[28] A. G. Ramm, A uniqueness theorem in scattering theory,
      (preprint Aug. 1983).
[29] A. G. Ramm, "Inverse scattering for geophysical problems".

I.14 (IM.5)

APPLICATION OF THE CHRISTOFFEL-HURWITZ INVERSION IDENTITY TO
ELECTROMAGNETIC IMAGING

Brett Borden

Michelson Laboratory, Physics Division,
Naval Weapons Center, China Lake, California 93555

ABSTRACT.   We review the theoretical framework of an algorithm
which determines the shape of a conducting convex scatterer from
the depolarization properties of the (monostatic) scattered
signal and show it to be effective when the return signal is
from a limited range of aspects.   Numerical examples are
included.

1.   INTRODUCTION

The problem of interest in many applications of
electromagnetic inverse scattering theory is that of determining
the shape of a body from the scattered field it creates in
interacting with sampling radiation, and the shape-recovery
methods resulting from the many investigations in this field are
well reviewed [1].   Of special importance for the present
discussion are those techniques which use the vector nature of
the problem as a means of extending the type of information
available for such reconstruction algorithms [2].

Below we summarize and further develop a
polarization/depolarization approach of the author to
applications involving more realistic data sets than previously
examined. These data sets are assumed to be made up of complete
(i.e., full scattering matrix) information from perfectly
conducting smooth convex surfaces over a limited range of
aspects.

W.-M. Boerner et al. (eds.), Inverse Methods in Electromagnetic Imaging - Part 1, 251–259.
© *1985 by D. Reidel Publishing Company.*

2.   THE VECTOR INVERSE ALGORITHM

Our approach results from the observation that the principal radii of curvature local to the specular points of a perfectly conducting convex body are radar measurables [3]-[5]. Illuminating such a target with an impulse $\underline{H}_I =$ $H_u\hat{u} + H_v\hat{v} = \underline{H}\delta(t - z/c)$ will effectively isolate the specular points, and it can be shown that the leading edge of the return in the far-field approximation obeys [3],[6]

$$\underline{H}_s = \frac{1}{2\pi r}\left[\frac{\partial^2 A_z}{\partial t^2}\underline{H} + \frac{1}{2}\frac{\partial A_z}{\partial t}(K_u - K_v)(H_u\hat{u} - H_v\hat{v})\right] \quad ,$$

where $\underline{H}_s$ is the scattered magnetic field, r is the distance from the transmitter/receiver to the target, A_z is the target area projected onto the u-v plane (the so-called silhouette function), and K_u and K_v are the curvatures in the principal directions $\hat{u}$ and $\hat{v}$, respectively. This is the Kennaugh-Cosgriff formula [7],[8] with Bennett's first-order correction [6].

Denoting the Fourier transform of the m-n element of the scattering matrix by $s_{mn}(\omega)$, it is then straightforward to see that [3],[4]

$$K_u - K_v = -\frac{2\omega\,\tan(\phi_d/2)}{\left[1 + \left|\frac{s_{21}(\omega)}{s_{11}(\omega)}\right|^2 \csc^2(\phi_d/2)\right]^{\frac{1}{2}}} \equiv B \quad , \tag{1}$$

where ϕ_d is the phase difference between $s_{11}(\omega)$ and $s_{22}(\omega)$. By virtue of the high-frequency relationship [9]

$$\sigma = \frac{\pi}{K_u K_v}$$

between the differential cross section σ and the principal curvatures, we can conclude

$$R_u + R_v = \frac{B\sigma}{\pi}\sqrt{1 + \frac{4\pi}{\sigma B^2}} \quad . \tag{2}$$

($R_u = 1/K_u$ and $R_v = 1/K_v$ are the principal radii of curvature.)

The significance of this result lies in its relationship to a differential equation developed by Stoker to study the Christoffel-Hurwitz problem of classical differential geometry:

Can the surface be reconstructed from a knowledge of the
sum of its principal radii of curvature as a prescribed
function of direction of surface normal?

Let $r(u,v)$ be a parametric representation of a convex
surface S. Denote the unit normal to S at r by $\hat{n}$ =
(n_1,n_2,n_3). Then we can define the spherical image as the
mapping induced by

$$h(n_1,n_2,n_3) = \underline{r} \cdot \hat{n} \quad .$$

Requiring for any vector $\underline{X} = (X_1,X_2,X_3)$ that

$$\mathcal{H} = \underline{r} \cdot \underline{X} = |\underline{X}|h$$

defines the Minkowski support function (X_1,X_2,X_3).

In [10], Stoker obtains a differential relationship between
$\mathcal{H}$ and $\Phi = R_u + R_v$, a relationship which for the present
problem reduces to [3]

$$\nabla^2 \mathcal{H} = X^2 \nabla^2 h + 2h = -\Phi \quad , \tag{3}$$

expressing h as a function of the spherical coordinates (θ,ϕ) of
$\hat{n}$. Since $\underline{r} \cdot d\underline{X} = d\mathcal{H}$, the cartesian coordinates of S can then
be written

$$x = h \sin\theta\cos\phi + \frac{\partial h}{\partial\theta} \cos\theta\cos\phi - \frac{\partial h}{\partial\phi} \frac{\sin\phi}{\sin\theta}$$

$$y = h \sin\theta\sin\phi + \frac{\partial h}{\partial\theta} \cos\theta\cos\phi - \frac{\partial h}{\partial\phi} \frac{\cos\phi}{\sin\theta}$$

$$z = h \cos\theta - \frac{\partial h}{\partial\theta} \sin\theta \quad ,$$

and so the target shape is recovered with the solution of (3).

In radar applications, Φ rarely will be determined as a
prescribed function of aspect, and, because of this, a numerical
solution to (3) is required. To this end, we investigate the
functional

$$F = \int_V [(\nabla\mathcal{H})^2 - 2\mathcal{H}\Phi]dV - 2 \int_T \mathcal{H}\beta dT \tag{4}$$

$$= \int_D \left(\frac{\partial h}{\partial\theta}\right)^2 + \frac{1}{\sin^2\theta} \left(\frac{\partial h}{\partial\phi}\right)^2 - 2h(h + \Phi)\sin\theta d\theta d\phi - 2 \int \mathcal{H}\beta dT \quad ,$$

where D is the intersection of V with a sphere of radius 1
centered on $\underline{r} = \underline{0}$. If $h(\theta,\phi)$ is a solution to (3) over D with

boundary $T = \partial D$, then $h(\theta,\phi)$ also makes (4) stationary (β depends on the appropriate boundary conditions [11]). Given a discrete set of data $\{\Phi_i\}$ over a sparse set of aspects, we can divide D into N subdomains (one subdomain for each element of $\{\Phi_i\}$ and integrate a linearization $h_i(\theta,\phi)$ of $h(\theta,\phi)$ over each subdomain. Seeking the values $\{h(\theta_i,\phi_i)\}$ that satisfy $\delta F = 0$, we require

$$\frac{\partial F}{\partial h(\theta_i,\phi_i)} = 0 \qquad i = 1,\ldots, N \tag{5}$$

and solve the resulting set of linear algebraic equations for $\{h(\theta_i,\phi_i)\}$ by the usual methods. Of course, for aspect-limited data sets, the boundary integral would cause problems for most typical radar situations were it not for the following fortuitous circumstance.

We can deal with the boundary term in (4) by recognizing that two support functions which differ by only a linear function will have corresponding surfaces which differ by at most a translation [10]. Moreover, for the Christoffel-Hurwitz problem, such a linear function need only be specified on the boundary of S, and the resulting translation will be a rigid body motion [12]-[15]. Because of this, we need only specify the boundary values of up to an _arbitrary_ linear function.

Let $\mathcal{H}'$ be a support function differing from $\mathcal{H}$ by a linear function L,

$$\mathcal{H}' = \mathcal{H} + L \quad .$$

Suppose now that the boundary conditions corresponding to the Neumann problem are known for $\mathcal{H}$ [11],

$$\beta = \frac{\partial \mathcal{H}}{\partial \xi} = \hat{\xi} \cdot \nabla \mathcal{H}$$

($\hat{\xi}$ is the outward directed unit normal to D.) Then the boundary conditions on $\mathcal{H}'$ become

$$\beta' = \frac{\partial \mathcal{H}'}{\partial \xi} = \beta + \hat{\xi} \cdot \nabla L \quad . \tag{6}$$

Clearly, $L = 0$ corresponds to our original problem.

Since L is arbitrary, then whenever

$$\int_T \mathcal{H}'\hat{\xi}dT \neq \underline{0} \quad ,$$

we can choose L so that

$$\nabla L \cdot \int_T \mathcal{H}' \hat{\xi} dT = - \int_T \mathcal{H}' \beta dT \quad . \tag{7}$$

Because the boundary condition (6) is natural, it will be satisfied automatically by any solution to (4) [11]. Consequently, we need never specify L any further than requiring it to satisfy (7)! Doing so, we may now consider the modified problem of determining a function $\mathcal{H}'$ that makes

$$\int_V [(\nabla\mathcal{H}')^2 - 2\mathcal{H}'\Phi] dV$$

stationary over aspect-limited data sets, and ignore the fact that such a solution will differ from $\mathcal{H}$ by some linear function.

3.   NUMERICAL RESULTS

For convenience, all formulas will be developed in spherical coordinates. Dividing the domain of $h(\theta,\phi)$ into N subdomains $\mathcal{D}_i$ (with boundaries defined by constant θ and ϕ curves), we approximate over each $\mathcal{D}_i$

$$h = h_i(\theta,\phi) = h(\theta_i,\phi_i) + \frac{h(\theta_i,\phi_i+\Delta\phi_i) - h(\theta_i,\phi_i)}{\Delta\phi_i} (\phi-\phi_i)\sin\theta$$

$$+ \frac{h(\theta_i + \Delta\theta_i,\phi_i) - h(\theta_i,\phi_i)}{\Delta\theta_i} (\theta - \theta_i) \quad ,$$

where $\Delta\theta_i$ and $\Delta\phi_i$ are the differences between the θ curves and the ϕ curves bounding $\mathcal{D}_i$, respectively. Assuming $\Phi = \Phi_i$, a constant, on each $\mathcal{D}_i$, the functional of interest becomes

$$F = \sum_{i-1} \int_{\mathcal{D}_i} \left[\left(\frac{\partial h_i}{\partial\theta}\right)^2 + \frac{1}{\sin^2\theta} \left(\frac{\partial h_i}{\partial\phi}\right)^2 - 2h_i(h_i+\Phi_i) \right] \sin\theta d\phi d\phi \quad .$$

The set of equations (5) readily lends itself to numerical solution [16]. The Fortran code developed was tested on simulated data sets containing the appropriate aspect-limited curvature information for convex sample targets.

For an ellipsoid, this information takes the form [6]

$$R_u+R_v = -\{[a^2(b^2+c^2)\sin^2\theta\cos^2\phi+b^2(a^2+c^2)\sin^2\theta\sin^2\phi]$$

$$+[c^2(a^2+b^2)\cos^2\theta]\}/[a^2\sin^2\theta\cos^2\phi+b^2\sin^2\theta\sin^2\phi+c^2\cos^2\theta]^{3/2}$$

where a, b, and c are the semi-axes of the ellipsoid. Using
this relation, data representative of what would be obtained by
(2) were created, and the algorithm was tested for accuracy.
Results when a = 1, b = 1, and c = 2 over the aspects $\theta\epsilon(0,\pi/2)$
and $\phi\epsilon(0,\pi)$ are displayed in figure 1.

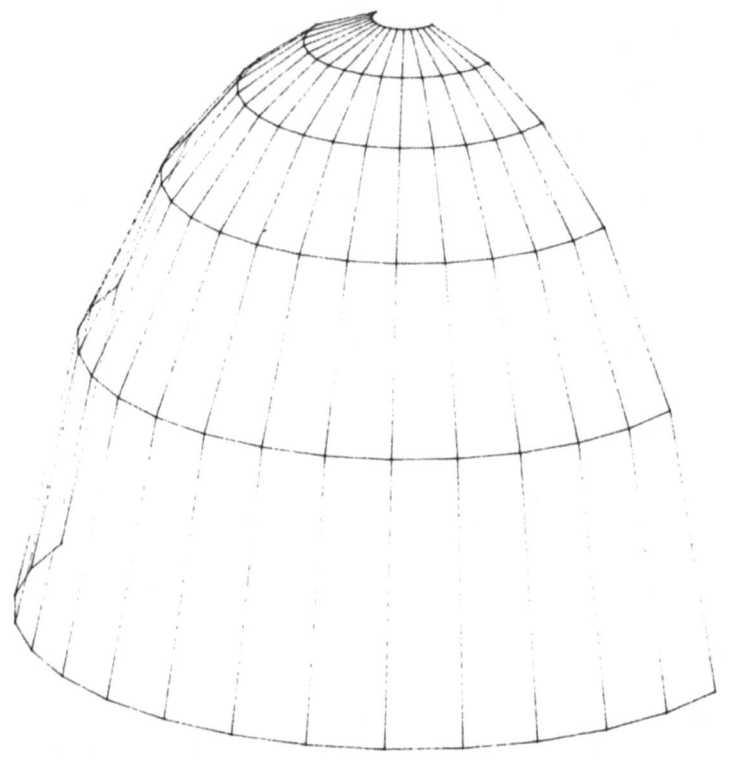

Figure 1. Reconstructed shape from synthetic data
generated for an ellipsoid with semi-axes a = 1,
b = 1, c = 2. The data were aspect-limited to the
spherical directions $\theta\epsilon(0,\pi/2)$ and $\phi\epsilon(0,\pi)$.

In a similar way, a sphere-capped paraboloid was examined
for data over $\theta\epsilon(0,\pi)$ and $\phi\epsilon(0,\pi/2)$ (see figure 2).

All calculations were performed in single precision on an
HP-1000/F mini-computer.

Figure 2. Reconstructed shape from synthetic data
generated for a sphere-capped paraboloid. The
aspect-limited data were created for the spherical
directions $\theta\varepsilon(0,\pi)$ and $\phi\varepsilon(0,\pi/2)$.

CONCLUSION

The foregoing analysis and examples have demonstrated that
the variational solution of the Stoker's differential equation
approach to inverse scattering is not limited to full-aspect

data. This means that data of the kind obtainable in radar encounters can be used to recapture the shape of convex portions of the scattering body. Numerical testing of the developed algorithms has borne this out.

Still unanswered, however, are the questions of how accurately this data can be obtained from actual radar systems and how best to obtain it. To the author's knowledge, the relation (1) has not been completely verified.

ACKNOWLEDGMENT

The author would like to express his gratitude to Dr. Charles Kenney of the Naval Weapons Center and to Dr. Alan Wolf of The University of Texas at Austin for their helpful comments and suggestions.

REFERENCES

[1] Baltes, H. P., Progress in inverse optical problems, Chapter 1 in: H. P. Baltes, Ed., "Inverse Scattering Problems in Optics, Topics in Current Physics." New York: Springer-Verlag, 1980.

[2] Boerner, W.-M., Polarization utilization in electromagnetic inverse scattering, Chapter 7 in: H. P. Baltes, Ed., "Inverse Scattering Problems in Optics, Topics in Current Physics." New York: Springer-Verlag, 1980.

[3] Borden, B., A vector inverse algorithm for electromagnetic scattering, SIAM Appl. Math., to appear.

[4] Foo, B. Y., A High Frequency Inverse Scattering Model to Recover the Specular Point Curvature from Polarimetric Scattering Data, University of Illinois at Chicago Circle Communications Laboratory Report Number EMID-CL-1982-05-21-01, 1982.

[5] Chadhuri, S. K., and Boerner, W.-M., A monostatic inverse scattering model based on polarization utilization, Appl. Phys. 11, p. 337, 1976.

[6] Bennett, C. L., Auckenthaler, A. M., Smith, R. S., and DeLorenzo, J. D., Space Time Integral Equation Approach to the Large Body Scattering Problem, Sperry Rand Research Center, Sudbury, MA, Final Report on Contract F30602-71-C-0162, RADC-CR-73-70, AD763794, May 1973.

[7] Kennaugh, E. M., and Cosgriff, R. L., The use of impulse response in electromagnetic scattering problems, IRE National Convention Record, Part I, pp. 72-77, 1958.

[8] Kennaugh, E. M., and Moffatt, D. L., Transient and impulse response approximation, IEEE Proc. 53, pp. 893-901, 1965.

[9] Silver, S., "Microwave Antenna Theory and Design." New York: McGraw-Hill, 1949.

[10] Stoker, J. J., On the uniqueness theorems for the embedding of convex surfaces in three-dimensional space, Commun. Pure Appl. Math. 3, pp. 231-257, 1950.

[11] Mikhlin, S. G., "Variational Methods in Mathematical Physics." New York: Macmillan, 1964.

[12] Hsiung, Ch.-Ch., A theorem on surfaces with a closed boundary, Math. Zeitschr. 64, pp. 41-46, 1956.

[13] Hsiung, Ch.-Ch., A uniqueness theorem for Minkowski's problem for convex surfaces with boundary, Illinois J. Math. 2, pp. 71-75, 1958.

[14] Busemann, H., Minkowski's and related problems for convex surfaces with boundaries, Michigan J. Math. 6, p. 259, 1959.

[15] Oliker, V. I., On certain elliptic differential equations on a hypersphere and their geometric applications, Indiana Univ. Math. J. 28, p. 35, 1979.

[16] Dongarra, J. J., Moler, C. B., Bunch, J. R., and Stewart, G. W., "LINPACK Users' Guide." Philadelphia, PA: SIAM, 1979.

[7] Kraeumer, R. M., and Compton, R. T., "The Analog Computer Approach to Electromagnetic Scattering Problems," IEEE National Convention Record, Part 1, pp. 15-21, 1970.

[8] Rumsey, V. H., Van Nostrand, Inc., "Ultra-wideband Antennas," IEEE Trans. Antennas and Propagation, AP-17, pp. 150-155, 1961.

[9] Elliott, R., "Antenna Theory and Design," New York: McGraw-Hill, 1981.

[10] Zucker, F. J., "Surface-wave antennas," in Antenna Engineering Handbook, R. C. Johnson, Editor, New York: McGraw-Hill, 1993.

[11] Wolff, I., "On Theoretical Methods of Antennas," Englewood, New York: Academic, 1971.

[12] Harrington, R. F., "Theory of Loaded Scatterers," IEEE Trans., AP-112, pp. 617-623, 1964.

[13] Harrington, R. F., "Time-Harmonic Electromagnetic Fields," New York: McGraw-Hill, 1961.

[14] Mautz, J. R., Harrington, R. F., and Aronson, E. A., Report, Syracuse University, Syracuse, N.Y., 1980.

[15] Mittra, R., "Computer Techniques for Electromagnetics," Elmsford, New York, 1973.

[16] Knott, E. F., "Radar Cross Section," Dedham, MA, 1985.

[17] Balanis, C. A., "Antenna Theory, Analysis and Design," New York: Harper & Row, 1982.

RECENT ADVANCES IN THE THEORY OF INVERSE SCATTERING WITH SPARSE
DATA

V. H. Weston

Department of Mathematics
Purdue University

Abstract

Multifrequency inverse scattering by an object is considered
for the case of sparse and non-redundant data, and where the
measured object is not restricted to be a small perturbation of a
known or comparison object. Previous quasi-linear techniques
suffer from the severe restriction that the measured object is a
small perturbation of a comparison object and each iterative step
represents a perturbation of the same magnitude or else the data
set must be complete. A non-linear iteration approach is
discussed here which removes the restriction that the comparison
body is close to the measured body, and the data set is sparse.
This approach has the further advantage that the regularization
process that must be used (for inverting almost singular matrices)
is much simpler since there is only one constraint (involving the
error in the matrix). In addition the method yields a condition
that will tell whether the comparison body is close to the
measured body and this can be used to terminate the iteration
process. This is important because it has been shown that for
sparse data, the solution of the exact system of equations is not
an isolated set of points (values of the quantity to be determined)
but a set of regions about these points (resolution cells). The
details are given here for the case where the scattering object
is characterized by the index of refraction $n(x)$ in the reduced
wave equation $\Delta u + k^2 n^2 u = 0$. The method has application to
other scattering problems, and applies to any spatial dimension.

W.-M. Boerner et al. (eds.), Inverse Methods in Electromagnetic Imaging - Part 1, 261–275.

1. Introduction

The multi-frequency inverse problem with sparse data for a
three dimensional object in an uncontrolled environment is
treated here. The uncontrolled environment implies that the
scattering object or medium is not simply a small perturbation
of a known object or medium, hence the usual linearizing
techniques are invalid. In addition, in an uncontrolled
environment, one would have little or no knowledge about the
scatterer other than knowledge of its position and rough estimate
of size. The data set itself is sparse and incomplete, as one
would not be able to get all the measurements that would be
desired. This could be due to physical interference or obstruc-
tion by other objects such as may occur in non-destructive
testing, or else the object itself may be an uncooperative
non-stationary target.

The main focus of this paper will be on the inverse problem
connected with the reduced wave equation

$$\Delta u + k^2 n^2(x)u = 0, \ x \in R^3 \tag{1}$$

(The method presented here can be abstracted and applied to
other scattering problems such as scattering by objects
characterized by Neumann or Dirichlet boundary conditions and
vector problems. More details of the present paper are given in
Weston (1984)).

In an uncontrolled environment associated with the reduced
wave equation, all that can be specified a priori about the
scattering object is that it is located in a prescribed bounded
region D (with piecewise smooth (C^1) boundary), and is
characterized by a real, sectionally continuous index of
refraction n(x) (n(x) is continuous everywhere except for a
finite set of piecewise smooth surfaces across which it has
finite jump discontinuities). The host medium exterior to D
is known, and for this paper will be taken to be homogeneous
with $n(x) \equiv 1$.

The data consists of a finite set of measured values of the
scattered field at different locations external to the scatterer,
and at different frequencies. (It will be assumed that the real
and imaginary parts of the scattered field are measured. The
case where only the amplitude is measured can be treated in a
similar fashion). Since the data set is sparse, the solution to
the inverse problem is not unique. In particular it has been
shown (Weston, 1984b) that the solution (of system (5)) is not
an isolated set of points (values of n(x)) but a set of regions
(resolution cells). Additional constraints have to be imposed
to get uniqueness. Analytical results for this have been

developed for case of fixed frequency scattering (Weston (1981, 1983). However the use of such constraints make the problem more complicated and less tractable for computation. Without the use of such constraints, the important thing to know is when one has an approximation that is close to the linearized region of a possible scatterer. Such conditions have been developed and are given by Eq (19).

It should be pointed out that other non-linear schemes based upon on-the-self techniques (conjugate-gradient, Newton's, etc.) have been employed for inverse scattering (Roger (1981), (1984) Coen et al (1981)). However these results are applied to the case where the data set is assumed <u>complete</u>, and no analysis has been done by these authors for the <u>incomplete</u> and sparse set of data.

2. Basic Mathematical Problem

To simplify analysis set

$$v(x) = n^2(x) - 1 \qquad (2)$$

and note that $v(x) \equiv 0$ outside D. Let v_* and v_m denote the value of v corresponding to the comparison and measured body respectively.

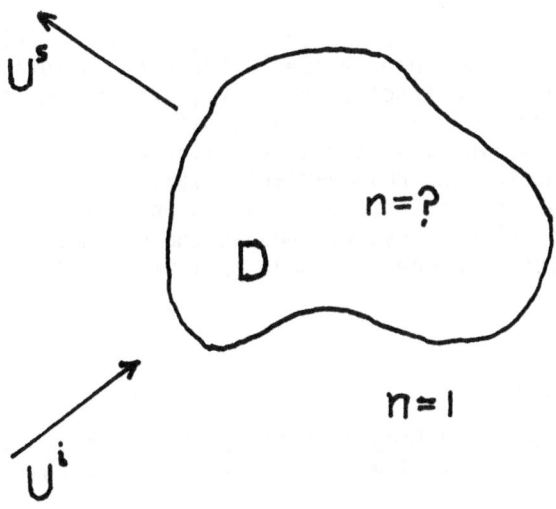

The data set will then consist of the set of N complex numbers $u_m^S(x_\ell,k_\ell)$ which is the measured value of the scattered field at a point x_ℓ outside D. The scattered field associated with each measurement is generated by a particular incident field $u^i(x,k_\ell)$ impinging upon the obstacle $(v=v_m)$ at a frequency ω_ℓ (corresponding to the wave number k_ℓ).

Let $G_*(x,y;k_\ell)$ denote the Green's function for the reduced wave equation associated with the comparison body $(n=n_*)$ and which satisfies the Radiation Condition. The total field $u(x,k_\ell)$ (where $u = u^i + u^S$) generated by the incident wave $u^i(x,k_\ell)$ impinging upon an arbitrary scatterer v (with support in D) will satisfy the integral equation

$$u(x,k_\ell) = u_*(x,k_\ell) + k_\ell^2 \int_D G_*(x,y;k_\ell)[v(y)-v_*(y)]u(y,k_\ell)dy \quad (3)$$

Here $u_*(x,k_\ell)$ is the total field produced by the same incident wave $u^i(x,k_\ell)$ impinging upon the comparison body $(v=v_*)$.

The scattered field at a point x_ℓ exterior to D is then given by

$$u^S(x_\ell,k_\ell) = u_*^S(x_\ell,k_\ell) + k_\ell^2 \int_D G_*(x_\ell,y;k_\ell)[v(y)-v_*(y)]u(y,k_\ell)dy \quad (4)$$

where the term in the integrand $u(y,k_\ell)$ is obtained by solving integral equation (3).

The inverse problem consists of finding the value of $v(x)$ such that

$$u^S(x_\ell,k_\ell) = u_m^S(x_\ell,k_\ell), \quad \ell = 1,2,\cdots N \quad (5)$$

where the left-hand side of system (5) is given by Eq.(4) and the right-hand side are measured values. Since the solution $u(y,k_\ell)$ of Eq. (3) is a function of $v-v_*$, expression (4) is a non-linear function of $v-v_*$. The basic mathematical problem is thus reduced to solving a system of non-linear complex functional equations for the real variable $v(x)$.

The usual non-linear techniques like Newton's method cannot be applied to the non-linear system because the Frechet derivative of the non-linear function does not have an inverse due to non-uniqueness.

3. Linearized Problem

If the actual measured body is sufficiently close to the comparison body so that v_m lies in the set v such that

$$\underset{\substack{k_\ell \\ \ell=1,2,\cdots N}}{\text{Max}} \quad k_\ell^4 \int\int_{D\times D} |G_*^2(x,y;k_\ell)|(v-v_*)^2 \, dxdy \ll 1 \quad (6)$$

then the solution of integral equation (3) could be approximated by

$$u(x,k_\ell) \sim u_*(x,k_\ell),$$

and system (5) in turn approximated by the linear system

$$b_\ell = k_\ell^2 \int_D G_*(x_\ell,y;k_\ell)u_*(y,k_\ell)w(y)dy , \quad \ell = 1,2\cdots N \qquad (7)$$

with

$$w = v - v_*, \qquad (8)$$

where the complex number

$$b_\ell = u_m^S(x_\ell,k_\ell) - u_*^S(x_\ell,k_\ell) \qquad (9)$$

is the difference between the measured value of the scattered field and the calculated value corresponding to v_*, at the point x_ℓ and wave number k_ℓ.

The system of N linear complex equations can be reduced to a system of $2N$ real equations. The complex quantities involved in the system are decomposed as follows

$$k_\ell^2 G_*(x_\ell,y;k_\ell)u_*(y,k_\ell) = H_\ell(y) + i\, H_{\ell+N}(y), \qquad (10)$$

$$b_\ell = B_\ell + i\, B_{\ell+N}. \qquad (11)$$

System (7) can then be reduced to the system of $2N$ real equations

$$(H_\ell,w) = B_\ell, \quad \ell = 1,2,\cdots 2N \qquad (12)$$

where the notation $(f,g) = \int_D f(x)g(x)dx$ is employed. With x_ℓ exterior to $\bar{D}$, both $G_*(x_\ell,y;k_\ell)$ and $u_*(y,k_\ell)$, and hence $H_\ell(y)$, will be continuous functions of y in D, when v_* is a sectionally continuous function.

If the measured body v_m is in the linearized region of the comparison body v_* (i.e., condition (6) holds for $v=v_m$) then the system of $2N$ real linear equations (12) provides a useful model for solution. The system as it stands, is ill-posed, with a non-unique solution, and any resulting matrix that may be used to invert the system may be ill-conditioned. Moreover, due to errors in data, one would not want an exact solution of system (12), but a solution that satisfies the equation to within the errors of the data (errors in B_ℓ). The usual techniques of regularization may be employed. In particular the technique of Backus and Gilbert (1970) is most suitable.

4. Non-Linear Process

4.1. Conceptual Approach

If one has a catalogue of comparison bodies, then linear system (12) will be solved for each value of v_* belonging to a catalogue of prescribed comparison bodies $\{v_*\}$, obtaining a solution w which is a function of v_*. Then the comparison body that is closest to the measured body is selected. If this comparison body is in the linearized region of the measured body, then the usual techniques such as Backus and Gilbert or others may be employed to get the optimum solution. There are a number of criteria for closeness that can be used. The usual approach is to employ the following quantity

$$\phi(v_*) = \sum_{\ell=1}^{N} |u_m^S(x_\ell,k_\ell) - u_*^S(x_\ell,k_\ell)|^2$$

which is the sum of the squares of the differences in the measured and calculated scattered fields. This quantity by itself will not indicate when the linearization process is valid, but is useful to eliminate possible choices of comparison bodies.

The best choice from the theoretical viewpoint is to select the comparison value v_* such that

$$\underset{v_*}{\text{Min}} \quad \underset{k_\ell}{\text{Max}} \quad k_\ell^4 \iint_{DxD} |G_*^2(x,y;k_\ell)| w^2(v_*) dxdy \tag{13}$$
$$\ell=1,2,\cdots N$$

If the minimum value is much less than unity then the linearization process employing the value of v_* yielding the minimum, would be valid. A better choice from the practical computational standpoint would be to choose v_* so that

$$\underset{v_*}{\text{Min}} \int_D w^2(v_*) dx \tag{14}$$

This latter choice provides an obvious answer of non-uniqueness of linear system (12). Condition (14) immediately infers that one would want the least squares solution of system (12). The main difficulty in using condition (14) is that it will not signify directly whether the linearization process is valid. As it stands, it can be used only for comparison. However an additional condition to go along with (14) will be given.

If either one doesn't have a catalogue of comparison bodies to begin with, or else the comparison body belonging to a catalogue is not in the linear region of the measured body, a non-linear iterative scheme will be used based upon the above concepts. The idea is to obtain a sequence $\{v_*^n\}$ of comparison bodies which minimizes the function in expression (14). Note that when $v_* = v_m$ then the absolute minimum is obtained.

4.2 Iteration Procedure

An iterative procedure based upon a descent type process applied to the non-linear functional $f(v_*)$ where

$$f(v_*) = \int_D w^2(v_*)dx \tag{15}$$

will be used. A minimizing sequence v_*^n will be sought such that

$$f(v_*^{n+1}) < f(v_*^n).$$

This will require the Taylor expansion

$$f(v_*+\delta v_*) = f(v_*) + (F(v_*),\delta v_*) + \frac{1}{2}(\mathcal{F}'(v_*)\delta v_*,\delta v_*) + \cdots$$

where $F(v_*)$ is the gradient of $f(v_*)$ and $\mathcal{F}'$ the Frechet derivative of F

$$\delta F = \mathcal{F}'\delta v = (F'(v_*),\delta v_*). \tag{16}$$

Here $F'(v_*)$ denotes the kernel of the integral operator $\mathcal{F}'$. The descent approach (Vainberg (1973)) is given by the sequence

$$v_*^{n+1} - v_*^n = -\omega_n\alpha_n F(v_*^n) \tag{17}$$

where ω_n is the relaxation factor. The real positive coefficient α_n in the descent method is obtained by minimizing the function $g(t) = f(v_*^n - tF(v_*^n))$ of the positive real variable t, or in practical applications, the quadratic approximation to $g(t)$, (Ortega and Rheinboldt (1970), Rheinboldt (1974)). The latter yields the value

$$1/\alpha_n = ((F'(v_*^n),F(v_*^n)),F(v_*^n))/(F(v_*^n),F(v_*^n)) \tag{18}$$

However for the problem on hand, F' is not a positive kernel, and a modified version of (18) has to be used by splitting F' into two parts, one a positive kernel and the other, a non-positive kernel, and retaining the contribution from the positive kernels. Explicit expressions are given for $F(v_*;x)$ and $F'(v_*;x,y)$ in Weston (1983), where in particular it is shown that the gradient has the form

$$-\frac{1}{2}F(v_*;x) = w(v_*;x) + q(v_*;x), \quad x \in R^3$$

When the comparison body v_* is close to the measured body v_m i.e., when $\|w(v_*;x)\|$ is small, then

$$-\frac{1}{2}F(v_*;x) = w(v_*;x) + 0\,(\|w\|^2)$$

and

$$\alpha (v_*) \sim \frac{1}{2} ,$$

hence the descent sequence approaches the usual linear approximation

$$v_*^{n+1} - v_*^n \sim w(v_*^n)$$

in the linear region. One consequence of this, is that a natural criterion exists to indicate when the comparison body is in the linear region of the measured body. It is given by

$$\|q\|/\|w\| \ll 1, \ ((F',F),F)/(F,F) \sim 2 \tag{19}$$

The descent process will terminate when a stationary point is reached. There are two types of stationary points, relative or absolute minimum, and saddle points, as expected. Of main interest is the absolute minimum point given by $\|w\| = 0$, yielding $B_\ell = 0$, $\ell = 1,2\cdots 2N$. There may be more than one such point due to the non-uniqueness of the problem. The descent approach can be modified to go through the saddle point. When the descent process has reached the linear region of the measured body (close to the absolute minimum of $f(v_*)$), it may be terminated and other techniques like Backus and Gilbert can then be employed to get the optimum fine structure.

4.3 Regularization

The critical juncture of the procedure is the solving of linear system (12). For the initial step (involving the initial choice of v_*) it is assumed that the data points $\{x_\ell\}$ and wave numbers $\{k_\ell\}$ are chosen so that $\{H_\ell(x)\}_{\ell=1}^{2N}$ is a linearly independent set of continuous functions over D; and any redundant set of measurements which violates this assumption will be culled out. Theoretically this only has to be assured of initially, since the descent process will tend to retain the linear independence of $\{H_\ell\}$ for each v_*^n. As a consequence the matrix $H(v_*)$ whose elements H_{ij} are given by

$$H_{ij} = \int_D H_i(x)H_j(x)dx \tag{20}$$

will be positive definite and invertible, since the quadratic form

$$\sum_{i,j=1}^{2N} c_i H_{ij} c_j = \int_D (\sum_{i=1}^{2N} c_i H_i(x))^2 dx$$

will vanish only if $\sum\limits_{i=1}^{2N} c_i H_i(x) \equiv 0$. The minimum $\mathcal{L}_2(D)$ norm

solution of system (12) is given by setting

$$w(x) = \sum_{j=1}^{2N} n_j H_j(x) \tag{21}$$

yielding the algebraic system of equations

$$\sum_{j=1}^{2N} H_{ij} n_j = B_i, \quad i = 1,2,\cdots,2N \tag{22}$$

In the numerical computation of the solution of system (22) constraints have to be imposed that involve some of the following:
(i)   Errors in H
(ii)   Errors in the data, or $\{B_j\}$
(iii)   Bounds on the solution $\vec{n}$,

The errors in the matrix elements H_{ij} are the most serious. These errors arise from two sources, numerical errors (due to computational procedures, round-off, etc.) and model error due to the approximation of the space of continuous functions over D by a finite dimensional space. With H a positive definite matrix, the errors in H will distort the component of the solution belonging to the eigenspace spanned by the eigenfunctions corresponding to the small eigenvalues. If the error in the matrix H is denoted by $E = \{E_{ij}\}$, then the numerical solution of system (22) will be required to the following accuracy

$$\|H\vec{n} - \vec{B}\|_2 \leq \|E\|_2 \|\vec{n}\|_2 \tag{23}$$

where $\|\vec{n}\|_2 = \sum\limits_{j=1}^{2N} n_j^2$.

No constraint need be placed upon the bound of the solution $\vec{n}$ for the iterative scheme developed here. This is in contrast to the schemes employed by Coen, Mei and Angelakos (1981), and Levenberg-Marquardt Algorithm (Moré (1978), Levenberg (1944), Marquardt (1963)), which require such a constraint due to a condition like (6) being imposed on their iteration process.

Errors in the data (provided they are not too large) can be ignored in the non-linear iterative scheme. They become important in the linear region where v_* is close to v_m. Generally, since no bounds are placed on $\|\vec{n}\|_2$, the error term $\|E\vec{n}\|_2$ may be larger than the error in the data.

With only the constraint (23) being considered, the system of equations can be regularized, Miller (1970) by replacing them by

$$(H + \varepsilon I)\vec{n} = \vec{B} \tag{24}$$

where $\varepsilon = \|E\|_2$.

Note the Levenberg-Marquardt iterative scheme is

characterized by a sequence of the form

$$v_\star^{n+1} - v_\star^n = w(\varepsilon_n, v_\star^n)$$

with w found from Eq(21) and a regularized form of Eq(22) with the regularization parameter ε_n chosen at each step.

4.4 Constrained Iterative Process

The descent procedure can be modified to give a solution satisfying the constraint

$$n^2(x) \geq 1 , \quad x \varepsilon D$$

This can be achieved by replacing the iterate

$$v_\star^{n+1}(x) = v_\star^n(x) - \alpha_n F(v_\star^n, x) \tag{25}$$

by $v_\star^{n+1}(x) = 0$, at all points x in D where the right hand side of Eq (25) is negative. This process is equivalent to adding a function $\psi(x)$ to the gradient F so that

$$v_\star^{n+1} = v_\star^n - \alpha_n(F(v_\star^n, x) + \psi(x)) \geq 0$$

The sequence $\{v_\star^n\}$ will still be a minimizing sequence provided that

$$- (F,\psi)/(F,F) \ll 1$$

When $- (F,\psi)/(F,F)$ is larger but still less than unity, α_n will have to be modified.

5. Computation

The non-linear process discussed here applies to any spatial dimension. At present, computations are being performed for the one-dimensional case where the domain D is given by $-1 \leq x \leq 1$. Preliminary results are given as follows.

Backscattered data corresponding to the "measured" body

$$n = 1 , \quad -1 \leq x < 0 , \quad n = 2, \quad 0 < x \leq 1$$

are generated for the wave numbers $k = \pi/4, \pi/2, 3\pi/4$ and π. Starting from the initial value $n \equiv 1.5$ for $-1 \leq x \leq 1$, the iteration procedure $(N = 4)$ employing the constraint $n \geq 1$, was used and the results for the 13th iterate are presented in Fig. 1. The values of $\int_D w^2(x)dx$ after each iteration are presented in Fig. 2.

The iteration procedure using an additional "measured" value of the backscattered field (N = 5) gave similar results but converged faster.

The effect of different initial choices of $n(x)$ was considered. The value $n \equiv 1.25$, $-1 \leq x \leq 1$, gave similar results (with slower convergence) as $n \equiv 1.5$. However the iteration procedure with the initial value $n \equiv 1.75$ for $-1 \leq x \leq 1$, converged to a different value of $n(x)$ for the case of four measurements (N = 4). The effect of additional "measurements" (N = 5,6) indicated that the value so obtained was a local minimum and not a true solution. However with the values $k = \pi/8$, $\pi/4$, $\pi/2$, $3\pi/4$, the iteration procedure converged to the correct solution when the initial value $n \equiv 1.75$ was taken.

6. Acknowledgement

The research reported herein was supported in part through O.N.R. grant N14-83-K-0038.

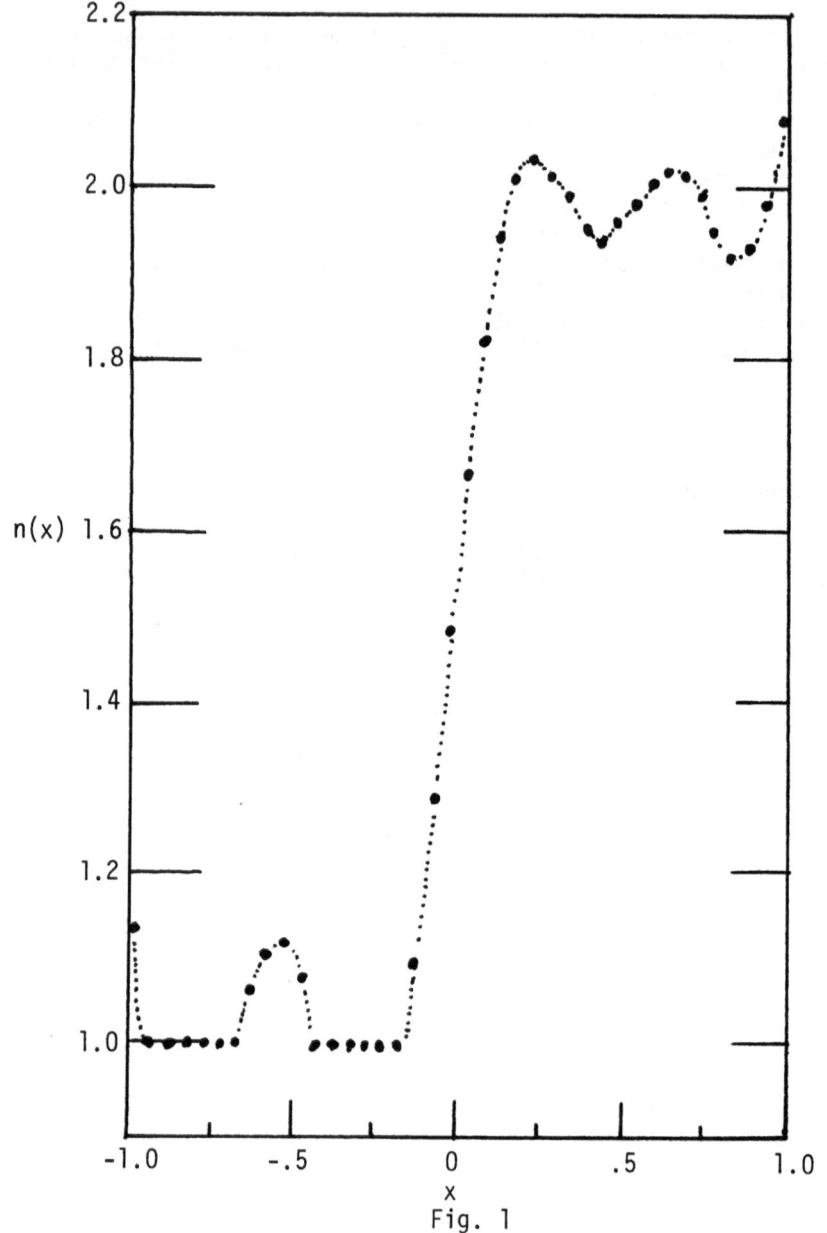

Fig. 1

Profile of the 13th iterate of n(x), corresponding to the case
where N = 4, the first iterate is n(x) ≡ 1.5, -1 ≤ x ≤ 1, and
"measured" data corresponds to n = 1, -1 ≤ x < 0, n=2, 0 < x ≤ 1.

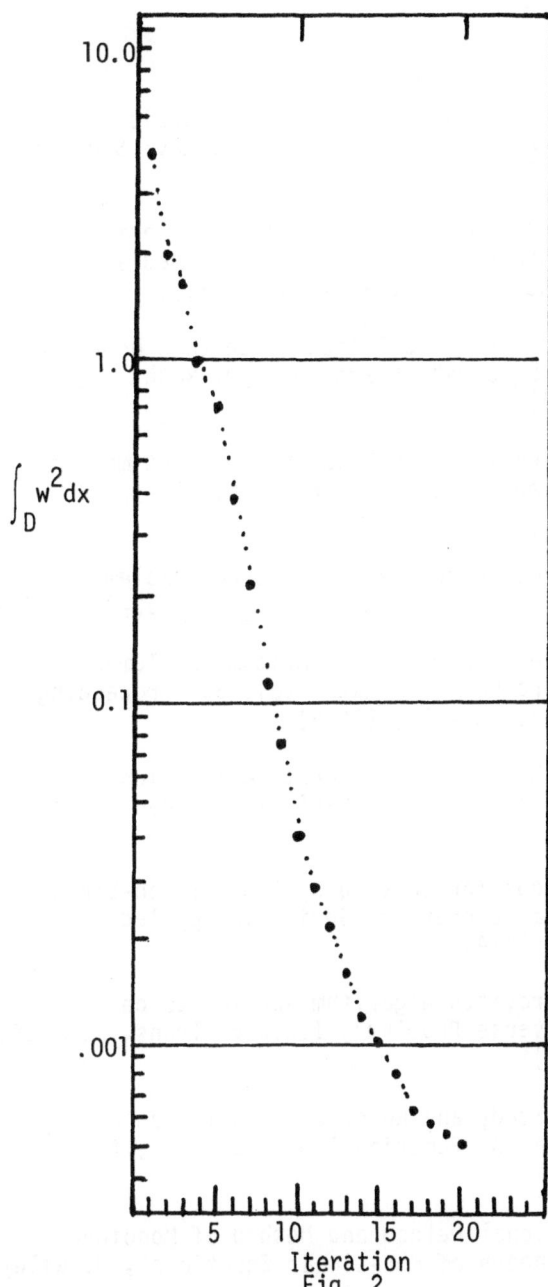

Fig. 2

Value of $\int_D w^2 dx$ after each iteration step for the case where
N = 4, initial value of $n(x)$ = 1.5, $-1 \leq x \leq 1$, and the
"measured" data corresponds to n = 1, $-1 \leq x < 0$, n = 2, $0 < x \leq 1$.

References

G. Backus and F. Gilbert, "Uniqueness in the Inversion of
    Inaccurate Gross Earth Data", Phil. Trans. Roy. Soc. Lon.
    (a), 266, pp. 123-192, (1970).

S. Coen, K. Mei, and D.J. Angelakos, "Inverse Scattering
    Technique Applied to Remote Sensing Layered Media",
    I.E.E.E. Trans. A.P. 29, pp. 298-306, (1981).

K. Levenberg, "A Method for the Solution of Certain Non-Linear
    Problems in Least Squares", Quart. J. App. Math. 2,
    pp. 164-168, (1944).

D. W. Marquardt, "An Algorithm for Least Squares Estimation of
    Non-Linear Parameters", SIAM J. App. Math. 11, pp. 431-441,
    (1963).

K. Miller, "Least Squares Methods for Ill-Posed Problems with a
    Prescribed Bound", SIAM J. Math. Anal. 1, pp. 52-74, (1970).

J. J. More, "The Levenberg-Marquardt Algorithm, Implementation
    and Theory", Lecture Notes in Math. Vol. 30, Proceedings,
    Springer Verlag, pp. 105-116, (1978).

J. M. Ortega and W. C. Rheinboldt, "Iterative Solution of
    Non-Linear Equations in Several Variables", Acad. Press,
    New York (1970).

W. C. Rheinboldt, "Methods for Solving Systems of Non-Linear
    Equations", Regional Conference Series in Applied
    Mathematics, SIAM (1974).

A. Roger, "Newton-Kantorovitch Algorithm Applied to on
    Electromagnetic Inverse Problem", I.E.E.E. Trans. A.P. 29,
    pp. 232-238, (1981).

A. Roger, "Theoretical Study and Numerical Resolution of the
    Inverse Problem Via the Functional Derivatives", this
    volume.

M. M. Vainberg, "Variational Method and Method of Monotone
    Operators in the Theory of Non-Linear Equations", J. Wiley,
    New York (1973).

V. H. Weston, "Inverse Problem for the Reduced Wave Equation
    with Fixed Incident Field, Part II", J. Math. Phys. 22,
    2523-2529, (1981).

V. H. Weston, "Inverse Problem for the Reduced Wave Equation with Fixed Incident Field, Part III", J. Math. Phys. 24, 828-833, (1983).

V. H. Weston, "Multi-Frequency Inverse Problem for the Reduced Wave Equation With Sparse Data", to be published (1984).

V. H. Weston, "The Resolution Cell for the Inverse Problem for the Reduced Wave Equation With Sparse Data", to be published (1984b).

II.2 (PT.2)

THE LIMITED ANGLE PROBLEM IN RECONSTRUCTION FROM PROJECTIONS

F. Alberto Grünbaum

Department of Mathematics
University of California, Berkeley

Abstract

The relation between "picture quality" and "amount of data" can be thoroughly analyzed both in X-ray and NMR imaging. The first case involves Euclidean geometry, the second one being three dimensional is best related to spherical geometry. In both of these cases one has the machinery of "prolate spheroidal wave functions". We consider the failure of this property for the case of hyperbolic geometry, except for Minkowsky space.

Table of Contents

---

[1] The research reported here was supported in part by National Science Foundation Grant MCS81-07086 and by the Director, Office of Basic Energy Sciences, Engineering, Mathematics, and Geosciences Division of the U.S. Department of Energy under contract DE-AC03-76SF00098.

W.-M. Boerner et al. (eds.), Inverse Methods in Electromagnetic Imaging – Part 1, 277–298.
© 1985 by D. Reidel Publishing Company.

1. Introduction

The motivation for this paper is the problem of reconstructing a real valued function f defined in R^n -- the unknown density -- from its projections $P_w f$, where the "directions" w are restricted to lie in a "limited range". In the case of the parallel beam geometry this can be put in the framework of "time and band limited functions". In some conditions this leads to a highly favorable and exceptional situation: the singular value decomposition of the corresponding "Finite Radon Transform" can be accomplished and the degree of illconditioning fully analyzed. The reason for this accident is poorly understood; we give a number of examples which include the cases of X-ray as well as NMR tomography.

After reviewing the "successful cases" we show how a most natural mathematical extension of this setup: from euclidean and spherical geometries to the hyperbolic case leads into a breakdown of "prolate spheroidal wave functions" bag of tricks. While we do not yet understand fully either the true reasons or the consequences of this mathematical fact, we present it with the hope that other people will find it useful. The last section of the paper gives indications of another piece of nineteen century mathematics lurking behind these matters.

2. Mathematical formulation

The problem of recovering, or estimating an unknown function $f(\underline{x})$, $\underline{x} \in R^n$, from

$$Pf(\underline{w},t) = \int_{(\underline{x},\underline{w})=t} f, \quad \underline{\dot{w}} \in S^n$$

can be put in the familiar form

Af = ♭ = data.

Here A is a linear map from $L^2(D)$ to $\oplus L^2(I)$, f denotes the unknown density, b stands for the collection of one dimensional projections.

The standard way to analyze a problem of this kind, although not necessarily the best way to go about solving it, is as follows: one considers the singular value decomposition of A. One brings in the vectors ψ_i, defined as the orthonormal eigenvectors of AA*, with positive real eigenvalues μ_i^2, $\mu_i > 0$. Then setting $\phi_i = A^*\psi_i/\mu_i$ one gets for the smallest f which minimizes the error Af - b the expression

$$\tilde{f} = \sum (b, \psi_i)/\mu_i \phi_i.$$

A small "singular value" μ_i spells troubles in obtaining the component of $\tilde{f}$ along ϕ_i if the data has any component along ψ_i.

We thus see that a complete analysis of the degree of ill-conditioning requires finding the triple $\{\phi_i, \psi_i, \mu_i\}$. For most problems of any interest this is the end of the road in that these quantities cannot really be found with any accuracy.

An example of the use of this "triple" is given in [1]. From some small scale numerical experiments reported there, one can see that the μ_i's satisfy the condition $0 < \mu_i < 1$ and split into two subsets: some of "close to one" and the rest are "close to zero." We saw in [1] that the effect of this split is that standard iteration methods for limited angle tomography -- without use of nonlinear constraints, like positivity -- give, instead of the desired function f, its projection into the linear span of the ϕ_i's with μ_i close to one. The analysis given in [1] was carried out in Fourier space and could not be used beyond small scale models because of numerical difficulties in obtaining the singular vectors and values.

3. X-ray and NMR imaging

To go beyond generalities one has to pick the situation very carefully or be very lucky. The simplest such case is 2 dimensional X-ray tomography where w becomes an angle and $P_w f \equiv P_\theta f$ becomes a collection of line integrals and finally

$$-\frac{\pi}{2} \le -a \le \theta \le a \le \frac{\pi}{2}$$

The next simplest case is given by 3 dimensional NMR imaging -- in one of its modalities -- where w becomes a pair of angles in R^3, $w = (\theta, \phi)$, and $P_w f$ becomes a collection of plane integrals.

B. Marr [2] pointed out that for X-ray tomography if one keeps the problem in its original form, one runs into an operator which is essentially identical to that studied by Slepian [3] in connection with "index and band limited sequences." For the case of X-ray tomography this analysis has been carried out by Davison [4,5], where the reader can find a detailed presentation.

The bottom line is that one can effectively compute (with high numerical accuracy) the required eigenvalues and eigenvectors.

The case of NMR tomography leads naturally to look for extensions of the work of Slepian, Landau and Pollak [6,7,8,9]. One considers the surface of the sphere instead of the circle and replaces the usual Fourier series by the expansion in spherical harmonics. The role of a segment $-a < \theta < a$ is now played by a spherical cap $0 < \theta < a$, ϕ arbitrary, and band limiting is taken to mean: keep only the components of f along $Y_{1m}(\theta, \phi)$ with $0 \le 1 \le L$. As it happens, this case is also exceptional in that the search for the appropriate singular vectors can be reduced to a numerically manageable problem. This, along with a host of related results, can be found in [10]. In all of these cases this reduction to a numerically feasible problem is accomplished by exhibiting a second order differential operator which has simple spectrum and commutes with the relevant operator AA*. This latter one is typically a finite convolution integral operator, and A itself is the finite Radon transform alluded to in the abstract.

The phenomenon which was so well exploited in [6,7,8,9], and to be explained below, had already appeared in mathematical form in the work of Whittaker and Ince [11,12]. It also appears, in totally independent form in the work of Mehta and Gaudin on "random matrices" see [13].

In the notation used above take A to denote the finite Fourier Transform

$$(Af)(\lambda) = \int_{-T}^{T} e^{i\lambda t} f(t) \, dt, \quad \lambda \in [-\Omega, \Omega]$$

then the integral operator given by A*A, namely

$$(kf)(t) = \int_{-T}^{T} \frac{\sin \Omega(t-s)}{t - s} f(s) \, ds \quad t \in [-T, T]$$

admits a second order differential operator commuting with it, namely

$$(Df)(t) = ((T^2 - t^2)f')' - \Omega^2 t^2 f.$$

Slepian proved that this same result holds if R^1 is replaced by R^n [9], and then in (14) -- mentioned earlier in connection with tomography -- extended this fact for functions defined on the circle or on the integers. We became interested in this problem and tried to extend its range. In [15] one finds an extension to the case of the nth roots of unity and then in [16] one finds a classification of all Toeplitz matrices which have a tridiagonal matrix, with single spectrum, in its commutator.

In [17] this property is seen to hold for the Hilbert matrix and in [24] one finds that the property (essentially) holds for the Discrete Fourier Transform matrix.

In trying to deal with the "limited angle" problem in R^3, we were led to consider the extension of Slepian's [14] results to the case of S^2, or more generally S^n. This was found to be true, as mentioned above.

Since the property holds for R^n and S^n it is only natural to expect that it should hold for hyperbolic space H^n. Indeed in [10] the result is shown to hold for H^3, see also [19] for a much more transparent proof.

And yet the result is not true for any H^n, $n \geq 2$, except when $n = 3$.

This is the topic of the next sections.

4. Harmonic analysis for hyperbolic space

Consider second order differential operators of the form

$$D \equiv \frac{1}{w(x)} \frac{d}{dx}\left(w(x) \frac{d}{dx}\right) \quad x \geq 0$$

and $w(x)$ given by

$$w(x) = (e^x - e^{-x})^p (e^{2x} - e^{-2x})^q$$

D is formally selfadjoint in $L^2(w(x) \, dx, R^+)$.

Operators of this form appear as the radial part of the Laplace Beltrami operator for noncompact symmetric spaces of rank one, with (p,q) a pair of nonnegative integers, see [20].

If we formally put $p = q = 0$, we obtain the Laplacian corresponding to the real line.

$$D = \frac{d^2}{dx^2} .$$

When $p \geq 0$, $q \geq 0$, $p + q > 0$ we are dealing with noncompact symmetric spaces arising from nonabelian groups.

The simplest examples come from $p = 0$, $q = 1$ and $p = 0$, $q = 2$ corresponding to $SO(2,1)/SO(2)$ and $SO(3,1)/SO(3)$

respectively. The first space is the upper half plane with the hyperbolic metric.

$$y^{-2}(dx^2 + dy^2)$$

-- or equivalently the unit disk with Poincaré metric. The second space is the Minkowski space of relativity theory obtained from factoring out the Lorentz group $SO(3,1)$ by the group of spatial rotations $SO(3)$.

We will consider only the cases $p \geq 0$, $q \geq 0$, $p + q > 0$. The compact and euclidean cases have been treated in [10].

Given D, let $\phi_\lambda(x)$ denote the unique solution to

$$(1) \qquad D\phi_\lambda(x) = -(\lambda^2 + \rho^2)\phi \qquad x \geq 0$$

satisfying $\phi_\lambda(0) = 1$ $\phi_\lambda'(0) = 0$. Here

$$\rho \equiv \frac{p + 2q}{2}$$

These functions take up the role of the radial (i.e. even) eigenfunctions of the Laplacian in R^1, namely cos x, and permit a harmonic (or Fourier) analysis of $L^2(w(x)\,dx)$ in terms of eigenfunctions of the operator D.

To obtain a useful analog of the Fourier analysis on R^1 one needs to find the "spectral measure" for the operator D. This gives an inversion formula and a Plancherel identity.

This is done below after some preparations.

Set

$$a \equiv a(\lambda) \equiv \frac{\rho + i\lambda}{2} \qquad b \equiv b(\lambda) \equiv \frac{\rho - i\lambda}{2} \qquad c \equiv \frac{p + q + 1}{2},$$

Then one has

$$\phi_\lambda(x) = F(a,b,c, - \sinh^2 x).$$

and setting

$$c(\lambda) = \frac{2^{2b}\Gamma(c)\Gamma(i\lambda)}{\Gamma(c-b)\Gamma(a)}$$

with F the usual hypergeometric function, one has

Theorem. For any $f \in L^2(w(x)\,dx, x \geq 0)$ and $\lambda > 0$ the integral

$$(2) \qquad \tilde{f}(\lambda) = \frac{1}{\sqrt{2\pi}} \int_0^\infty f(x)\phi_\lambda(x)w(x)\,dx$$

is well defined in the $L^2(w(x)\,dx)$ sense. Moreover $\tilde{f}(\lambda) \in L^2(|c(\lambda)|^{-2}d\lambda,\ \lambda \geq 0)$ and the map $f \rightarrow \tilde{f}$ is a unitary operator between these two L^2 spaces. The inverse map is given by

$$(3) \qquad f(x) = \frac{1}{\sqrt{2\pi}} \int_0^\infty \tilde{f}(\lambda)\ \phi_\lambda(x)\,|c(\lambda)|^{-2}\ d\lambda$$

This is a particular case of the Weyl-Titchmarsh-Kodaira theory. Notice that if one considers the function

$$\tilde{\phi}_\lambda(x) = F\left(\frac{\rho + i\lambda}{2},\ \frac{\rho - i\lambda}{2},\ \frac{p + q + 1}{2},\ -\sinh^2 x/2\right)$$

one has

$$D\tilde{\phi}_\lambda(x) = -\left(\frac{\lambda^2 + \rho^2}{4}\right)\tilde{\phi}_\lambda(x)$$

$$= -(\tilde{\lambda}^2 + \tilde{\rho}^2)\tilde{\phi}_\lambda(x)$$

with $\tilde{\lambda} = \frac{\lambda}{2}$ and $\tilde{\rho} = \frac{\rho}{2}$. In other words

$$\Psi_\lambda(x) \equiv F\left(\frac{p + 2q}{4} + i\lambda,\ \frac{p + 2q}{4} - i,\ \frac{p + q + 1}{2},\ -\sinh^2 x/2\right)$$

satisfies

$$D\Phi_\lambda(x) = -\left(\lambda^2 + \frac{p + 2q}{4}\right)^2 \Psi_\chi(x).$$

For example if $p = 0$, $q = 1$ corresponding to $SL(2,R)/SO(2)\ (= SO(2,1)/SO(2))$

$$D\Psi_\lambda(x) = -(\lambda^2 + \frac{1}{4}\,\Psi_\lambda(x)$$

while if $p = 0$, $q = 2$, corresponding to $SO(3,1)/SO(3)$

$$D\Psi_\lambda(x) = -(\lambda^2 + 1)\Psi_\lambda(x).$$

The relation between $\Psi_\lambda(x)$ and $\phi_\lambda(x)$ above is given by

$$\Psi_\lambda(x) = \phi_{2\lambda}(x/2)$$

and using Ψ_λ or ϕ_λ is a matter of convenience. We will use ϕ_λ, but some people are more familiar with Ψ_λ.

5. <u>A general formulation for "time and band limiting"</u>

Given arbitrary values $T > 0$, $\Omega > 0$ consider an operator acting on functions $f \in L^2(w(x) \, dx, \, x \geq 0)$ as follows
  1) Restrict f to $[0,T]$
  2) Expand f in terms of the functions $\phi_\lambda(x)$ as in (3) above.
  3) Restrict $\tilde{f}(\lambda)$ to $[0,\Omega]$
  4) Invert the expansion
  5) Restrict the resulting function to $[0,T]$.

In simpler terms the map in question can be defined, for $x \in [0,T]$ by means of

$$(Kf)(x) = \frac{1}{2\pi} \int_0^\Omega \left[\int_0^T f(y) \phi_\lambda(y) w(y) \, dy \right] \phi_\lambda(x) \, |c(\lambda)|^{-2} \, d\lambda$$

or exchanging the order of integration

$$(Kf)(x) = \frac{1}{2\pi} \int_0^T \left[\int_0^\Omega \phi_\lambda(x) \phi_\lambda(y) \, |c(\lambda)|^{-2} \, d\lambda \right] f(y) \, w(y) \, dy$$

This turns out to be a selfadjoint, compact operator in $L^2(w(x) \, dx, \, [0,T])$ whose eigenfunctions and eigenvalues play a crucial role in analyzing a number of practical extrapolation problems.

Even in thecase when $D = \dfrac{d^2}{dx^2}$, it is impossible to find these eigenfunctions "analytically", and one has to resort to approximate numerical methods. This sound practical but becomes impossible to do if one needs lots of eigenfunctions with any accuracy. The size of the corresponding matrix problem obtained from discretization becomes too large to handle since we are dealing with a "full" matrix.

The <u>ideal</u> solution to this problem would be to produce a second order differential operator D--with simple spectrum--which commutes with K and thus has the same eigenfunctions. While the problem still could not be analytically solved we would then have a feasible numerical problem which can be solved accurately and quickly using for instance (the last steps of) the Q-R algorithm.

In the case when

$$D = \frac{d^2}{dx^2}$$

one has

$$\tilde{D} = \frac{d}{dx}\left((T^2 - x^2)\frac{d}{dx}\right) - \Omega^2 x^2$$

This happens to be -- for totally mysterious reasons-- one of the three ordinary differential operators which appear when one separates the Helmholtz equation in "prolate spheroidal coordinates" in R^3. The proof of the relation

$$\tilde{D}K = K\tilde{D}$$

consists in writing out explicitly the kernel

(4) $K_\Omega(x,y) = \displaystyle\int_0^\Omega \phi_\lambda(x)\,\phi_\lambda(y)\,|c(\lambda)|^{-2}\,d\lambda$

which in this case becomes

$$\frac{\sin \Omega(x-y)}{x-y} + \frac{\sin \Omega(x + y)}{x + y}^\dagger$$

and checking that

(5) $(\tilde{D}_x - \tilde{D}_y)\, K\,(x,y) =$

Then with some integration by parts and using the fact that the <u>leading coefficient of</u> $\tilde{D}$ vanishes at $x = T$ one concludes that $\tilde{D}K = K\tilde{D}$.

The <u>crucial property</u> is (5), and for the rest of the paper we <u>concentrate on it</u>. Notice that the same property holds for the operator.

$$\tilde{D} = \frac{d}{dx} x^2 \frac{d}{dx} + \Omega^2 x^2$$

since this differs from $-\tilde{D}$ by the addition of a multiple of $\dfrac{d^2}{dx^2}$.

We now concentrate on the

---

†Strictly speaking this differs from the case in Slepian, Pollak, and Landau since we consider radial or even functions. This is an unimportant difference.

MAIN PROBLEM

Given D, as above, construct $K_\Omega(x,y)$ as in (4) and find functions

$$a(x,\Omega), \quad b(x,\Omega) \quad \text{with} \quad a(T,\Omega) = 0$$

such that for every Ω, the operators

(6) $\tilde{D}_x \equiv \dfrac{1}{w(x)} \dfrac{d}{dx} \left(a(x,\Omega) \, w(x) \, \dfrac{d}{dx} \right) + b(x,\Omega)$

(7) $K(x,y) = \displaystyle\int_0^\Omega \phi_\lambda(x) \phi_\lambda(y) \, |c(\lambda)|^{-2} \, d\lambda$

satisfy

(8) $(\tilde{D}_x - \tilde{D}_y) K_\Omega(x,y) = 0$

Observe that in the case of $D = \dfrac{d^2}{dx^2}$ we could choose $a(x,\Omega)$ independent of Ω, and $b(x,\Omega)$ with a simple factorized form.

This problem has a solution for higher dimensional Euclidean spaces, see [9], and for spheres and other compact examples in [10].

In [10] one finds a noncompact nonabelian example corresponding to the symmetric space SO(3,1)/SO(3) -- then so called Minkowski space -- where $a(x,\Omega)$ could still be picked independent of Ω but $b(x,\Omega)$ no longer had a factorized form.

Now we move on to consider the general case of $p,q \geq 0$ $p + q > 0$.

6.   The search for a commuting differential operator

1.   Eliminating $b(x,\Omega)$

Since the functions $a(x,\Omega)$ and $b(x,\Omega)$ are only important up to an additive constant we can assume that

$$a(0,\Omega) = b(0,\Omega) \equiv 0$$

If we use (6) and (7) to write (8) in full we get

$$(8') \quad (a(x,\Omega) - a(y,\Omega)) \int_0^\Omega - (\lambda^2 + \rho^2) \phi_\lambda(x) \phi_\lambda(y) |c(\lambda)|^{-2} \, d\lambda$$

$$+ \, a'(x,\Omega) \int_0^\Omega \phi_\lambda'(x) \, \phi_\lambda(y) |c(\lambda)|^{-2} \, d\lambda$$

$$- \, a'(y,\Omega) \int_0^\Omega \phi_\lambda(x) \, \phi_\lambda'(y) |c(\lambda)|^{-2} \, d\lambda$$

$$= \, (b(y,\Omega) - b(x,\Omega)) \int_0^\Omega \phi_\lambda(x) \phi_\lambda(y) |c(\lambda)|^{-2} \, d\lambda$$

Here a prime means a derivative with respect to x or y. We now make the "ansatz" that a solution can be found with $a(0,\Omega) = a'(0,\Omega) = 0$.

Setting $y = 0$ in (8') we obtain for $b(x,\Omega)$ an expression in terms of $\phi_\lambda(x)$, $c(\lambda)$, ρ and the unknown function $a(x,\Omega)$. Plugging this expression back in (8') we conclude that satisfying (8) is equivalent to demanding that the function

$$\Phi(x,y,\Omega) = a(x,\Omega) \left[\int_0^\Omega - (\lambda^2 + \rho^2) \phi_\lambda(x) \phi_\lambda(y) |c(\lambda)|^{-2} \right.$$

$$+ \, \frac{\displaystyle\int_0^\Omega - (\lambda^2 + \rho) \phi_\lambda(x) |c(\lambda)|^{-2} d\lambda}{\displaystyle\int_0^\Omega \phi_\lambda(c) |c(\lambda)|^{-2} \, d\lambda} \cdot \left. \int_0^\Omega \phi_\lambda(x) \phi_\lambda(y) |c(\lambda)|^2 \, d\lambda \right]$$

$$+ \, a'(x,\Omega) \left[\int_0^\Omega \phi_\lambda'(x) \phi_\lambda(y) |c(\lambda)|^{-2} \, d\lambda \right.$$

$$- \, \frac{\displaystyle\int_0^\Omega \phi_\lambda'(x) |c(\lambda)|^{-2} \, d\lambda}{\displaystyle\int_0^\Omega \phi_\lambda(x) |c(\lambda)|^{-2} \, d\lambda} \left. \int_0^\Omega \phi_\lambda(x) \phi_\lambda(y) |c(\lambda)|^{-2} \, d\lambda \right]$$

should be symmetric in the pair (x,y).

It is convenient to introduce the following notation

$$\phi_\lambda(x) = \phi_0(x) + \frac{\lambda^2}{2}\phi_2(x) + \frac{\lambda^4}{4!}\phi_4(x) + \ldots$$

$$|c(\lambda)|^{-2} = p_2\lambda^2 + p_4\lambda^4 + \ldots$$

Of course functions $\phi_{2i}(x)$ and the coefficients c_{2i} depend on the choice of parameters (p,q) which determine every thing else.

A good part of the analysis below applies in total gener- ality.   Some finer points hinge on the dependence of the functions $\phi_0(x)$, $\phi_2(x)$, $\phi_4(x)$, introduced above, on the parametrs (p,q). Recall that $c = \frac{p + q + 1}{2}$ and set

$r \equiv \frac{\rho}{2} = \frac{p + 2q}{4}$. We have, for later use,

$$\phi_0(x) = 1 - \frac{r^2x^2}{4c} - \frac{(2r^2c - 3r^4 - 6r^3 - r^2)}{96(c^2 + c)} x^4 + \ldots$$

$$\phi_2(x) = \frac{x^2}{2c} = -\frac{x^2}{2c} - \frac{2c - 6r^2 - 6r - 1}{43(c^2 + c)} x^4 - \ldots$$

$$\phi_4(x) = \frac{3}{4}\frac{x^4}{c^2 + c} + \ldots$$

These explicit relations will not be used until much later. In this sense much of what follows can be used to analyze more general situations.

The expansions given above are not hard to do by hand, but they are better carried out using a symbol manipulator. We used VAXIMA.

We also used VAXIMA for some of the manipulations described below.

2. Looking for $a(x,\Omega)$

We try to find $a(x,\Omega)$ in the form

$$a(x,\Omega) = \sum_{n \geq 0} a_{2n}(x)\Omega^{2n}$$

such that the function $\Phi(x,y,\Omega)$ defined earlier turns out to symmetric in (x,y) for every Ω.

Setting

$$\Phi(x,y,\Omega) = \sum_{i \geq 0} \Phi_{2i+1}(x,y)\Omega^{2i+1}$$

one can see that $\Phi_1(x,y) \equiv \Phi_3(x,y) \equiv \Phi_5(x,y) \equiv 0$.

The first useful relation comes from $\Phi_7(x,y)$. We get

$$\Phi_7(x,y) = \left[\frac{P_2}{175} \; a_0'(x)(\phi_2'(x) - \phi_2(x)\phi_0'(x)/\phi_0(x)) \right.$$
$$\left. - 2a_0(x)\phi_0(x)\right]\phi_2(y)$$

Notice that we get

$$\Phi_7(x,y) = \alpha^{(7)}(x)\beta^{(7)}(y)$$

which leads, from symmetry, to

$$\alpha^{(7)}(x) = A^{(7)}\beta^{(7)}(x).$$

Writing this out and dividing by $\phi_0(x)$ we get

(9)
$$\left[\frac{\phi_2(x)}{\phi_0(x)}\right]' a_0'(x) - 2a_0(x) = \frac{175}{P_2} A^{(7)} \frac{\phi_2(x)}{\phi_0(x0}$$

Going on to $\Phi_9(x,y)$ we obtain

$$\Phi_9(x,y) = \alpha_1^{(9)}(x)\beta_1^{(9)}(y) + \alpha_2^{(9)}(x)\beta_2^{(9)}(y)$$

and from the assumed symmetry we deduce

$$\alpha_1^{(9)}(x) = A_{11}^{(9)} \beta_1^{(9)}(x) + A_{12}^{(9)} \beta_2^{(9)}(x)$$

$$\alpha_2^{(9)}(x) = A_{21}^{(9)} \beta_1^{(9)}(x) + A_{22}^{(9)} \beta_2^{(9)}(x)$$

with a symmetric matrix $A_{ij}^{(9)}$.

This is an easy consequence of the symmetry of $\Phi_9(x,y)$ and the linear independence of $\beta_1^{(9)}(x)$, $\beta_2^{(9)}(x)$.   In our case

$$\beta_1^{(9)}(x) = \phi_2(x), \ \beta_2^{(9)} = \phi_4(x)$$

The last coefficient which has to be looked at with care is $\Phi_{11}(x,y)$.   One gets

$$\Phi_{11}(x,y) = \sum_{i=1}^{3} \alpha_i^{(11)}(x) \beta_i^{(11)}(y)$$

with $\beta_1^{(11)}(x) = \phi_2(x)$, $\beta_2^{(11)}(x) = \phi_4(x)$, $\beta_3^{(11)}(x) = \phi_6(x)$

and we conclude, as before, that

$$\alpha_i^{(11)}(x) = \sum_{j=1}^{3} A_{ij}^{(11)} \beta_j(x)$$

with a symmetric 3 × 3 matrix $A_{ij}^{(11)}$.

We display now two of the equations obtained in the manner indicated above.

From the equation involving $\alpha_1^{(9)}(x)$,   after dividing by $\phi_0(x)$,   we get

$$\frac{1}{1890} P_2 a_0'(x) \left(\frac{\phi_4(x)}{\phi_0(x)}\right)' - \frac{3}{1750} P_2 a_0'(x) \frac{\phi_2(x)}{\phi_0(x)} \left(\frac{\phi_2(x)}{\phi_0(x)}\right)'$$

(10)
$$+ \frac{23}{7875} P_4 a_0'(x) \left(\frac{\phi_2(x)}{\phi_0(x)}\right)' - \frac{23}{7875} P_2 a_0(x) \left(\frac{\phi_2(x)}{\phi_0(x)}\right)$$

$$- \frac{46}{7875} P_4 a_0(x) + \frac{P_2}{175} a_2'(x) \left(\frac{\phi_2(x)}{\phi_0(x)}\right)' - \frac{2}{175} P_2 a_2(x)$$

$$= A_{11}^{(9)} \frac{\phi_2(x)}{\phi_0(x)} + A_{12}^{(9)} \frac{\phi_4(x)}{\phi_0(x)}$$

From the equation involving $\alpha_2^{(11)}(x)$, after dividing by $\phi_0(x)$, we get

$$\frac{1}{19404} P_2 a_0'(x) \left[\frac{\phi_4(x)}{\phi_0(x)}\right]' - \frac{1}{6300} P_2 a_0'(x) \frac{\phi_2(x)}{\phi_0(x)} \left[\frac{\phi_2(x)}{\phi_0(x)}\right]'$$

$$+ \frac{73}{242550} P_4 a_0'(x) \left[\frac{\phi_2(x)}{\phi_0(x)}\right]' - \frac{73}{242550} P_2 a_0(x) \frac{\phi_2(x)}{\phi_0(x)}$$

(11)
$$- \frac{73}{121275} P_4 a_0(x) + \frac{P_2}{1890} a_2'(x) \left[\frac{\phi_2(x)}{\phi_0(x)}\right]' - \frac{P_2}{945} a_2(x)$$

$$= A_{21}^{(11)} \left[\frac{\phi_2(x)}{\phi_0(x)}\right] + A_{22}^{(11)} \frac{\phi_4(x)}{\phi_0(x)} \quad .$$

Here we have used the fact that $A_{23}^{(11)} = 0$. This comes from the fact that the equation for $\alpha_3^{(11)}(x)$ combined with (9) give $A_{32}^{(11)} = A_{33}^{(11)} = 0$. The symmetry of $A^{(11)}$ does the job.

Eliminating $a_2(x)$ from (10) and (11) we get

$$- \frac{10}{2079} P_2 a_0'(x) \left[\frac{\phi_4(x)}{\phi_0(x)}\right]' + \frac{40}{693} P_2 a_0(x) \frac{\phi_2(x)}{\phi_0(x)}$$

$$+ 1890 A_{22}^{(11)} \frac{\phi_4(x)}{\phi_0(x)} - 175 A_{12}^{(9)} \frac{\phi_4(x)}{\phi_0(x)}$$

$$= \frac{40}{693} P_4 a_0'(x) \left[\frac{\phi_2(x)}{\phi_0(x)}\right]' - \frac{80}{693} P_4 a_0(x)$$

$$- 1890 \; A^{(11)}_{21} \; \frac{\phi_2(x)}{\phi_0(x)} + 175 \; A^{(9)}_{11} \; \frac{\phi_2(x)}{\phi_0(x)}$$

Now we claim that the right hand side of this equation vanishes identically. Using (9) one can replace the first two terms on the right hand side of (12), and then all the right hand side of (12), by a scalar multiple of ϕ_2/ϕ_0. We claim that this scalar is zero.

Indeed if $a_0(x)$ is regular at $x = 0$ it follows from (9) that $a_0(x) \approx \mathbb{C}x^2$ for small x. This combined with

$$\frac{\phi_2(x)}{\phi_0(x)} \sim -\frac{x^2}{2c} \quad \text{and} \quad \frac{\phi_4(x)}{\phi_0(x)} \sim \frac{3}{4} \frac{x^4}{c^2 + c}$$

gives the desired result by plugging in (12).

Using this and some scaling we have replaced (12) by

$$(12') \qquad a_0'(x) \left[\frac{\phi_4(x)}{\phi_0(x)} \right]' - 12a_0(x) \frac{\phi_2(x)}{\phi_0(x)} = \tilde{k} \frac{\phi_4(x)}{\phi_0(x)}$$

From here one can proceed in many different ways.

If one is interested in the special case under discussion here one can solve for $a_0(x)$ in (9), and insert the resulting function in the left hand side of (12'). Expanding the ratio of this expression by ϕ_4/ϕ_0, setting equal to zero the coefficients of x^2 and x^4 one obtains a set of two algebraic equations in c, r with solutions

$$c = 0, \quad r = 0$$

$$c = -2, \quad r = -1 \pm i \frac{\sqrt{6}}{3}$$

$$c = -1, \quad r = -\frac{1}{2} \pm \frac{i}{2}$$

$$c = \frac{3}{2}, \quad r = 1$$

$$c = \frac{3}{2}, \quad r = \frac{1}{2}$$

$$c = 2, \quad r = 1$$

$$c = \frac{1}{2}, \quad r = 0$$

$$c = \frac{1}{2}, \quad r = \frac{1}{0}.$$

Recalling that

$$c = \frac{p + q + 1}{2}; \quad r = \frac{p + 2q}{4}; \quad p,q \geq 0; \quad p + q > 0$$

one can rule out the first three and the last two solutions.

The fourth and fifth correspond to $p = 0$, $q = 2$ and $p = 2$, $q = 0$ respectively.

The sixth solution gives $p = 2$ $q = 1$.

A look at the list in pages 30-32 of [8] shows that $p = 0$ $q = 2$ has no geometrical meaning, $p = 2$ $q = 0$ corresponds to

SO(3,1)/SO(3)

and finally, $p = 2$ $q = 1$ corresponds to

SU(1,1.)/SU(1)

Unfortunately this case has to be ruled out since the coefficient of x^6 does not vanish, and we are left with SO(3,1)/SO(3) as the only viable case.

A more general approach is to solve (9) and (12') for $a_0(x)$ and $a_0'(x)$ and obtain in this fashion an equation relating $\phi_0(x)$, $\phi_2(x)$ and $\phi_4(x)$. From the known behavior of these functions one can see again that only $(p,q) = (0,2)$ or $(2,0)$ are allowed. This method can be used in more general situations.

The conclusion is that at least for the class of second order differential operators considered here, only the case SO(3,1)/SO(3) leads to a differential operator of order two which commutes with the integral operator K_Ω in (5).

We mention that for $p = 2$, $q = 0$ the eigenfunctions are given

$$\phi_\lambda(x) = \frac{\sin \lambda x}{\lambda \sinh x}$$

and we have

$$\frac{1}{\lambda^2} \frac{d}{d\lambda}(\lambda^2 \frac{d}{d\lambda} \phi(\lambda,r)) = -r^2 \phi(\lambda,r)$$

This "differential equation in the spectral parameter" does not hold for any other example of p,q: a remarkable coincidence. Such an equation is valid every time that the Slepian-Landau-Pollak phenomenon holds, see [10,21], and is the springboard for the next and last section.

7. Differential equations in the spectral parameter

Given a second order differential operator

$$L \equiv -\frac{d^2}{dx^2} + V(x)$$

denote by $\phi(\lambda,x)$ a family of eigenfunctions of L, $L\phi = \lambda^2 \phi$. The previous consideration lead us to the following

Question. For what $V(x)$ does there exist a differential operator D_λ such that

$$D_\lambda \phi(\lambda,x) = \Theta(x) \phi(\lambda,x)$$

We have attacked this question in collaboration with H. Duistermatt [22].

We give below some few examples of this situation leaving a more complete discussion for [22].

(a) Take $V(x,t) = \dfrac{6x(x^3-2t)}{(x^3+t)}$, then if

$$\phi(\lambda,x) = \frac{(\lambda^2 x^3 + 3i\lambda x^2 - 3x + \lambda^2 t)}{\lambda^2(x^3 + t)} e^{i\lambda x}$$

we've

$$-\left[\frac{d^2}{dx^2} + V(x,t)\right] \phi(\lambda,x) = \lambda^2 \phi(\lambda,x)$$

and

$$\left(\left(\frac{d}{d\lambda} - \frac{2}{\lambda}\right)\left(\frac{d}{d\lambda} + \frac{2}{\lambda}\right)\right)^2 - 4it\frac{d\phi}{d\lambda} = (x^4 + 4tx)\phi$$

(b) Take $V(x,t,s) = \dfrac{12x^{10} + 324sx^5 + 450t^2x^4 + 300t^3x + 162s^2}{(x^6 + 5tx^3 + 9sx - 5t^2)^2}$

then for an appropriate family of eigenfunctions $\phi(\lambda,x)$ we've

$$\left[i\left(\frac{d}{d\lambda} - \frac{3}{\lambda}\right)\left(\frac{d}{d\lambda} - \frac{2}{\lambda}\right)\left(\frac{d}{d\lambda} - \frac{1}{\lambda}\right)\frac{d}{d\lambda}\left(\frac{d}{d\lambda} + \frac{1}{\lambda}\right)\left(\frac{d}{d\lambda} + \frac{2}{\lambda}\right)\left(\frac{d}{d\lambda} + \frac{3}{\lambda}\right)\right]\phi$$

$$+ \frac{63}{2} s\left(\frac{d}{d\lambda} - \frac{1}{\lambda}\right)\left(\frac{d}{d\lambda} + \frac{1}{\lambda}\right)\phi - 35it^2 \frac{d}{d\lambda} \phi$$

$$+ \frac{35}{4} t \left[\left(\left(\frac{d}{d\lambda} - \frac{2}{\lambda}\right)\left(\frac{d}{d\lambda} + \frac{2}{\lambda}\right)\right)^2 - \frac{36}{\lambda^4}\right] \phi$$

$$= (x^7 + \frac{35}{4} tx^4 - \frac{63}{2} sx^2 - 35t^2x)\phi$$

We make the observation that the function $V(x,t)$ in example (a) is a solution of the Korteweg-deVries equation. Also $V(x,t,s)$ in example (b) can be written in the form

$$e^{tX_1 + sX_2\left(\frac{12}{x^2}\right)}$$

where X_1, X_2 denote the Kdv flow and the fifth order differential operator given by $[(-L^{3/2})_+, L]$ and $[(-L^{5/2})_+, L]$ respectively, (up to scaling). See [23-28].

As one may suspect this works for the whole "hierarchy." More interesting still, not all examples are obtained this way. For much more detail about this whole development see [22,29].

References

[1]   Grünbaum, F.A. A study of Fourier space methods for
      "limited angle" image reconstruction. Numerical Functional
      Analysis and Optimization 2 (1) (1980), 31-42.

[2]   Marr, B. Private communication.

[3]   Slepian, D. Prolate spheroidal wave functions, Fourier
      analysis and uncertainty. Bell System Tech. Journal
      57, No. 5 (1978).

[4]   Davison, M.E. The ill conditioned nature of the limited
      angle tomography problem. SIAM J. Applied Math.,
      43, 2, (1983) 428-448.

[5]   Davison, M.E. A singular value decomposition for the
      Radon transform in n-dimensional euclidean space. Numer.
      Funct. Anal. and Optimiz. 3(3) (1981) 321-340.

[6]   Slepian, D., Pollak, H.P. Prolate spheroidal wave functions,
      Fourier analysis and uncertainty I. Bell System Tech.
      Journal 40, No. 1 (1961) 43-64.

[7]   Landau, H.J., Pollak, H.O. Prolate spheroidal wave
      functions, Fourier analysis and uncertainty II. Bell
      System Tech. Journal 40, No. 1 (1961) 65-84.

[8]   Landau, H.J., Pollak, H.O. Prolate spheroidal wave
      functions, Fourier analysis and uncertainty III. Bell
      System Tech. Journal 41, No. 4 (1962) 1295-1336.

[9]   Slepian, D. Prolate spheroidal wave functions, Fourier
      analysis and uncertainty IV. Bell System Tech. Journal
      43, No. 6 (1964) 3009-3058.

[10] Grünbaum, F.A., Longhi, L., Perlstadt, M. Differential
      operators commuting with finite convolution integral
      operators: some nonabelian examples. SIAM J. Applied
      Math., 42, 5, (1982) 941-955.

[11] Whittaker, E.T. Proc. London Math. Soc. (2) 14,(1915)
      260-268.

[12] Ince, E.L. 'On the connection between linear dif.
      systems and integral equations', Proc. of the Royal
      Society of Edinburgh (42),(1922) pp. 43-53.

[13] Mehta, M.L. Random matrices, Academic press, N.Y., 1967.

[14] Slepian, D. 'Prolate spheroidal wave functions, Fourier analysis and uncertainty', Bell System Tech. Journal 57, No. 5, (1978), 1371-1430.

[15] Grünbaum, F.A., 'Eigenvectors of a Toeplitz matrix: discrete version of the prolate spheroidal wave functions', SIAM J. Alg. Disc. Math. 2(2), (1981) 136-141.

[16] Grünbaum, F.A. 'Toeplitz matrices commuting with a tridiagonal matrix', Linear Algebra and its Applications 40, (1981) 25-36.

[17] Grünbaum, F.A., 'A remark on Hilbert's matrix', Linear Algebra and its Applications, 43, (1982), 119-124.

[18] Grünbaum, F.A. 'The eigenvectors of the discrete Fourier transform: a version of the Hermite functions', J. Math. Anal. Applic. 88, (1982), 355-363.

[19] Grünbaum, F.A., 'Doubly concentrated functions for three hyperbolic space', IHES preprint, Bures-sur Yvette, France (1982).

[20] G. Warner, Harmonic analysis on semisimple Lie groups, Vol. II, Springer Verlag (1972).

[21] Grünbaum, F.A. to appear, 'A new property of reproducting kernels for classical orthogonal polynomials', J. Math. Anal. Applic. 95, (1983).

[22] Duistermaat, J. and F.A. Grünbaum: in preparation, 'Differential equations in the eigenvalue parameter'.

[23] Lax, P. 'Integrals of nonlinear equations of evolutions and solitary waves', Communications on Pure and Applied Mathematics 21, (1968), 467-490.

[24] Gelfand, I. and Dikii, 'Fractional powers of operators and Hamiltonian systems', Funkts. Anal. Prilozhen, 10, 4, (1976), 13-39.

[25] H. Airault, H. McKean, and J. Moser, 'Rational and elliptic solutions of the Korteweg-deVries equation and a related many body problem', Communications in Pure and Applied Math.(30), (1977), 95-148.

[26] D. Chudnovski and G. Chudnovski, 'Pole expansions
      for nonlinear partial differential equations', Nuovo
      Cimento 40B, (1977), 339-353.

[27] M. Adler and J. Moser, 'On a class of polynomials con-
      nected with the Korteweg-deVries equation', Communica-
      tions in Mathematical Physics (61), (1978), 1-30.

[28] M. Ablowitz and H. Airault, 'Perturbations finies et
      forme particuliere de certaines solutions de
      l'equation de Korteweg-deVries. C.R. Acad. Sci. Paris
      t. 292, (1981), 279-281.

[29] Grünbaum, F.A., Band-time limiting, recursion relations
      and nonlinear evolution equations, a Chapter in "Special
      functions: group theoretical aspects and applications,
      R.A. Askey et al. (eds.), D. Reidel, Holland (1984), p. 271.

II.3 (IM.1)

DIRECT AND INVERSE HALFSPACE SCALAR DIFFRACTION:
SOME MODELS AND PROBLEMS

Giovanni Crosta

Istituto di Cibernetica dell'Università
Via Viotti 5 - I-20133 MILANO (Italy)

We consider herewith two main items: a) approximations of
Helmholtz equation and b) aperture identification. We begin with
Helmholtz equation , which is a model of the direct diffraction
process. If we approximate Rayleigh-Sommerfeld's propagator in
the spatial frequency domain we get the paraxial propagator ,
whereas if we approximate it in the spatial domain (far field) we
get both the Fresnel propagator and the Fourier optics formula.
We show the Fresnel propagator formally coincides with the
paraxial one. We investigate to which extent approximate
solutions are physically related to the exact one. Then we deal
with some consequences on the so-called "superresolution"
methods, which aim at identifying some subwavelength details in
the aperture field.We show Fourier optics is no adequate model
for superresolution. As of aperture identification, we deal with
the system theoretical approach to the problem, which consists of
minimising an adequate functional containing both an estimation
error and a regularising term. We also show how duality theory
helps in designing a minimisation algorithm based on the discrete
gradient. We apply this method to identification of a Dirichlet
boundary condition in two cases: where either complex field or
intensity data are available.The latter is closely related to the
phase reconstruction problem, for which we write the functional
to be minimised . Several open problems are also presented.

W.-M. Boerner et al. (eds.), Inverse Methods in Electromagnetic Imaging - Part 1, 299–318.
© *1985 by D. Reidel Publishing Company.*

TABLE OF CONTENTS

INTRODUCTION

The solution of any inverse problem consists of an interaction between a natural system (S) and a model (R), which on its turn is affected by the interaction process until model predictions are believed to be accurate with respect to observations performed on the natural system.

If we assume both systems , S and R , are described by maps, say σ and ϱ , which relate their respective inputs and outputs, then according to Conant and Ashby's definition (1970):

R is a model of S if there is an isomorphism between σ and ϱ. The same Authors also give an optimality criterion for system R to predict accurately the output of S when the input to both is known:

R must be a model of S.

These ideas affect our behaviour in describing natural processes related to electromagnetic phenomena and diffraction in particular, our aim being to solve what is named an inverse problem. This term means

either: the identification of the system parameters, in order to construct or improve a model,

or   the identification of an input when both the model and the output are known.

The need for accurate models of the so-called direct problems, which consist of evaluating the electromagnetic field at a point when field sources, medium properties and boundary conditions are known, has paradoxically been emphasised by the failure of some inversion methods based on oversimplified models. As an example we shall deal with approximate solutions of Helmholtz equation in the halfspace.

1.   MODELS OF DIRECT ELECTROMAGNETIC PROCESSES

The usual setting is in a classical framework, where field equations seem easy to solve , especially when the interaction between the fields and the detector is neglected. Actually, since experimental data are yielded by the detector, we expect some difficulty to arise precisely from this simplification when a data inversion procedure must be applied.

The ignorance about the photodetection process will therefore add "noise" to data , together with all other unknown processes for which we are too lazy to work out a model.

Let us assume Maxwell's equations for free fields make sense. At this point we choose

<u>either</u> to deal with the ($\underline{B},\underline{E}$) tensor and be ready to solve the coupled equations it entails, by thus keeping track of polarisation effects (see e.g. Boerner,1980,1981 and the Radar Polarimetry section of this volume),

<u>or</u> to deal with a scalar equation , we believe to represent the process adequately. We thus come to an equation in a single function w, which may be e.g. a component of $\underline{E}$.

Then we get rid of the time dependence, which we assume to be $\exp(i\omega t)$ and arrive at Helmholtz equation. If only one component of $\underline{E}$, say E_z , were non-vanishing then Helmholtz equation <u>alone</u> would be sufficient. This corresponds however to two-dimensional (x,y) problems. Halfspace diffraction from an arbitrary aperture could not be dealt with. The assumption we make is then to use Helmholtz equation for the complex scalar field w instead of three such equations together with the supplementary condition on div E.

Actually w could also stand for a component, say A_z , of the vector potential A , but <u>not</u> the scalar potential Φ , as it can be easily shown by working e.g. in the Coulomb gauge, where $\text{div}\underline{A}$= = 0.

We assume our detectors are energy sensitive. Measured

quantities are functions of $|w|^2$ both if $w = E_z$ or if $w = A_z$.

In the latter case $|E_z|^2 \approx \omega^2 |A_z|^2$, because Φ does not depend on time. All of this physically motivates energy norms in the Hilbert spaces we shall work with.

A further support towards adopting Helmholtz equation is the relative abundance of existence, uniqueness and regularity theorems for several geometries and Hilbert or Sobolev space problem formulations.

2. HELMHOLTZ EQUATION IN THE HALFSPACE

There is a well-known procedure which leads to a unique solution w of Helmholtz equation in the empty halfspace R^3_+:

$$(\text{divgrad} + k^2)\, w = 0 \qquad \text{in } R^3_+ \, ,$$

subject to Dirichlet boundary conditions (BCs) at the z = 0 plane (also named R^3_o, where position is denoted by (x_1, x_2)):

$$w\,|_{Ro}^3 = v \in U \, ,$$

and to some asymptotic conditions (<u>radiation conditions</u> or RCs for short) which also affect the radial derivative w_r :

$$|w| = 0(r^{-1}) \qquad \text{as } r \to \infty, \qquad \forall \text{ colatitude angle } \Theta \, , \, \forall \varphi$$

azimuth angle φ ,

$$|w_r - ikw| = o(r^{-1}) \qquad\qquad 0 < \Theta \leq \pi/2 - c, c > 0, \forall \varphi,$$

$$|w_r| = 0(r^{-1}) \qquad\qquad \pi/2 - c < \vartheta < \pi/2, \underline{\forall} \varphi \quad .$$

The <u>Rayleigh-Sommerfeld propagator</u>

$$g_{1n} := \partial_n g_1 = \partial_n \left[(1/r)e^{ikr} - (1/r^*)e^{ikr^*} \right]_{RO}^3$$

is introduced, where ∂_n stands for the normal derivative operator and r* is the distance between the origin and the point

x* symmetrical of x with respect to R^3_o. On its turn g_1 is the <u>fundamental solution</u> or Green's function of the problem, i.e. the <u>right and left inverse</u> (Hoermander,1963, pp 249-251) of the

differential operator $(\text{divgrad} + k^2)$ plus BCs we are considering. w is yielded by

$$w(\underline{x}_o) = (1/2\pi)g_{1n} * v \, , \tag{1}$$

where now * stands for convolution over R^3_o. <u>Existence</u> of w then is insured wherever the convolution integral makes sense. The set U of <u>boundary conditions</u> v can therefore be defined by giving e.g. a number of sufficient specifications. Let us assume the support of v is a bounded subdomain S of R_o^3, which corresponds to a finite aperture in a plane screen. Moreover we take the aperture field to behave regularly enough. One possible constraint is :

$$v \in W^{1,\infty}(S),\tag{2}$$

which implies the boundedness of v, its first partial derivatives and its energy integral, i.e. $v \in L^2(S)$. On the other hand, if the support of v is unbounded, some asymptotic conditions must hold for v, which are equivalent to radiation conditions in R_o^3. We shall however avoid this case, which to our present knowledge has nowhere been dealt with in a satisfactory way, particularly as far as the existence proof is concerned. Boundary conditions having an unbounded support can be alternatively dealt with by the limit amplitude or the limit absorption principles (Ramm, 1978). We stress that a solution w of Helmholtz equation yielded by either principle need not comply with the radiation conditions, because the asymptotic requirements on v may differ.

3.   APPROXIMATIONS OF HELMHOLTZ EQUATION

3.1 General

It is well-known that the convolution integral (1) gives rise to a distinction between homogeneous and evanescent waves . The corresponding procedure is rather straightforward and we shall avoid the details. We need however some notations:
$\underline{y} := (y_1, y_2)$ is the two-dimensional spatial frequency vector,
      i.e. a position vector in the reciprocal domain,
D_k is the disk $\{ \underline{y} \mid |\underline{y}| \leq k \}$,

$Fw := \hat{w}$ is the Fourier transform (FT) of w with respect to x_1 and x_2.

The FT of all operands in (1) is defined, in particular because v is in U of (2) . By means of tables (Jahnke, Emde, Loesch,1960) we get:

$$w = \begin{cases} F^{-1}\exp(iz\sqrt{k^2-y^2})Fv & ; |\underline{y}| \leq k \\[2em] F^{-1}\exp(-z\sqrt{y^2-k^2})Fv & ; |\underline{y}| > k . \end{cases}\tag{3}$$

Since S is bounded, the support of $\hat{v}$ is not, hence the aperture will contribute both homogeneous and evanescent waves to

w. We shall decompose $\hat{v}$ as the sum of two parts (Montgomery, 1968), according to the following diagram:

$$F : L^2(R^2) \supset W^{1,\infty}(S) \longrightarrow L^2(R^2) := B_H \oplus B_E$$

$$v \longmapsto \hat{v} := \hat{v}_H + \hat{v}_E$$

where supp $\hat{v}_H \subseteq D_k$, supp $\hat{v}_E \subseteq R^2 \setminus D_k$; therefore v_H and v_E mean the inverse FTs of $\hat{v}_H$ and $\hat{v}_E$ respectively.

The exponential operators appearing in (3) can be commuted with F , by extending the relationship between the multiplication operator $(Q\hat{~})$ in the reciprocal domain and the differentiation operator $(P\hat{~})$ in the spatial one (Jordan,1969,p 64):

$$Q\hat{~}F = FP\hat{~} .$$

This leads to the following result:

$$w = w_H + w_E = \exp\left[ikz \sqrt{1+\text{divgrad}_T/k^2}\right]. \ v_H +$$
$$\exp\left[-kz \sqrt{-1-\text{divgrad}_T/k^2}\right]. \ v_E ,$$

where divgrad_T is the laplacian with respect to x_1 and x_2.

It can now be easily understood that if the last term on the right hand is neglected, the corresponding field w_H need no longer satisfy the Helmholtz equation defined in Section 2. w_H belongs to the class of bandlimited fields , obtained by applying the Rayleigh-Sommerfeld propagator to a function v_B . The latter comes from restricting supp $\hat{v}$ of a BC v (which yields an

acceptable solution w) to the bounded domain $D_B \subset R^2$. Bandlimited fields usually fail to satisfy the radiation conditions mentioned above. A sufficient argument is given by Baltes et al. (1982), who investigate the asymptotic behaviour of w in the z direction and on the R_0^3 plane. A general proof relating to an arbitrary direction in R_0^3 is not known to us.

Yet, it makes sense to introduce the following abstract evolution problem, of which w_H is the formal solution:

$$\begin{cases} -i\partial_z \ w_H = k \sqrt{1 + \text{divgrad}_T/k^2} \ . \ w_H \\ w_H(.,z=0) = v_H \in B_H , \end{cases} \qquad (4)$$

where we have chosen v_H in B_H to stress that BCs used here need not have a bounded support.

3.2 Approximations in the reciprocal domain

The above equation is the starting point for an approximation step in the reciprocal domain , leading to the paraxial equation. If

$$\text{supp } \hat{v}_H = D_b := \left\{ \underline{y} \in R^2 \mid y^2/k^2 < b \ll 1 \right\},$$

then we perturb the infinitesimal generator in equation (4) by expanding the square root and neglecting all but the first two terms. We then get another abstract evolution problem governed by the paraxial equation (Deschamps,1981) :

$$\begin{cases} -i\, \partial_z\, w_P(.,z) = k(1 + \text{divgrad}_T)\, w_P(.,z) & (5) \\[2mm] w_P(.,z{=}0){=}v_P & (6) \\[2mm] \text{supp } v_P = D_b, \end{cases}$$

the solution of which is :

$$w_P = e^{ikz(1+\text{divgrad}_T/2k\,)} . v_P := P_P \cdot v_P. \qquad (7)$$

The paraxial propagator P_P is both right and left inverse of the differential operator defined by (5) and (6) , as it could be easily shown.
The paraxial equation be also arrived at from Helmholtz's by a slightly different procedure (Bellman & Kalaba, 1959; Corones, 1975; Fock, 1960). We now examine the domain of

P_P: it is a non dense subset of $L^2(R^2)$, hence the operator P_P can

be extended to the whole of $L^2(R^2)$ in several ways. Let P_P^e denote the extended operator. Particularly relevant to inverse diffraction is the following alternative:

$$\text{e1)} \quad P_P^{(e1)} = P_P \text{ of (7)}, \ \forall\, v \in L^2(R^2), \qquad (8)$$

$$\text{e2)} \quad P_P^{(e2)} = P_P \quad \text{if supp } \hat{v} \subseteq D_b$$

$$= 0 \quad \text{elsewhere.}$$

Only extension e2) preserves the connection between the original Helmholtz equation and the approximated one.
As an example let us consider the following boundary condition:

$$v := \cos(k_o x_1) . \cos(k_o x_2) . \ X_R(S),$$

where $k_o > k$, and $X_R(S)$ is a regularised characteristic function of S, such that v satisfies (2). If v is acted on by the Rayleigh

- Sommerfeld propagator, the energy of the corresponding $w(v)$ decays very rapidly as z increases, because most of the BC energy lies outside D_k. On the other hand if the extended propagator of (8), which is unitary, is used, the energy of the corresponding field $w_1(v)$ does not decrease. Then, at a given z $\gg$ 0 the relative estimation error :

$$\frac{\| w_1(z) - w(z) \|_{L^2(R^2)}}{\| w(z) \|_{L^2(R^2)}} \quad ; \quad z \gg 0$$

we make by taking $w_1(.)$ for $w(.)$, becomes very large.

Whenever we choose to model direct diffraction by the paraxial equation because, say, it leads to simpler computations, we must be aware of the hypotheses which led to it.

Let us now deal with another example of diffraction data inversion by the paraxial propagator. w_M is the measured field at a given plane $z=z_M$. After inverse propagation, the estimated $\hat{v}_M$ may well have a support extending outside D_b , as a consequence of the already mentioned processes we cannot model (photodetection ,apparatus behaviour etc.). The physical plausibility constraint we must enforce is then to reject the values $\{ v_M(\underline{y}) | \ |\underline{y}| \notin D_b \}$. We have traded computational simplicity for accuracy in the reconstructed aperture field.

3.3   Approximations in the spatial domain

Another way of perturbing the operator relating v to w is to neglect some terms in g_{1n} of (1), which can be written as $\partial_z((1/s)e^{iks})$, where

$$s := \sqrt{(x_1-x_{1a})^2 + (x_2-x_{2a})^2 + z^2}; \quad (x_{1a}, x_{2a}) \in R_o^3. \tag{9}$$

If we are interested in the far-field, we start with approximating $z/s \doteq 1$ and leave the phase factor unchanged. We could easily show that we eventually get a propagator which still tells homogeneous waves apart from evanescent ones, but is undefined at $|\underline{y}| = k$. Moreover it solves the following Neumann problem:

$$\begin{cases} (\text{divgrad} + k^2)w=0 \\ w_n \big|_{R_o^3} = v \ ; \ \text{supp } v \subset\subset R_o^3 \\ \text{radiation conditions.} \end{cases}$$

The consequence of a seemingly innocuous approximations yields a far-field solution which does not keep track of the type of BC we have to identify. If in addition we expand to first order the square root in (9) we arrive at the well-known <u>Fresnel optics</u> formula:

$$w_F = (-ik/2\pi z).\exp\left(ikz(1+ \frac{x_1^2 + x_2^2}{2z^2})\right) * v(x_1,x_2).$$

By means of FT tables we can show that:

$$w_F = w_P$$

defined in Section 3.2, i.e. that the solution of the paraxial equation and the Fresnel approximate the original w in the same way. The same plausibility constraints defined in the preceeding Section hold here too. To greater extent this applies to the Fourier optics formula, which results from further simplifying (9). The consequences will be analysed in the following Section.

4.   SUPERRESOLUTION, FOURIER OPTICS AND OTHER METHODS

In our case we define by <u>superresolution</u> any method aiming at estimating $\hat{v}_E$ starting from <u>measured field data</u>. $\hat{v}_E$ contains the so-called sub-wavelength details. In the realm of Fourier optics it has been often suggested that (Goodman,1968,pp 133-136) if the field w_M defined above were measured in an adequate way, then $\hat{v}_M$ could be estimated with arbitrarily high accuracy , even for $|\underline{y}| > k$. The procedure was believed to hold if a numerable set of values, the samples $\{w_M(\underline{x})\}$ were available . The link between the latter and the unknown function $\hat{v}_M$ with bounded support being due to Shannon's theorem. From the remarks of Section 3.3, nothing physically meaningful can however be achieved if supp $\hat{v}_M$ extends outside D_k. The constraint on supp $\hat{v}_M$ is a necessary condition for the Fourier optics solutions to approximate w. Then we see Shannon's theorem used in the wrong context. No wonder then if a superresolution algorithm based on Fourier optics yields some results in the values $\{\hat{v}_M(\underline{y})| \ |\underline{y}| > k\}$ the dependence of which on slight perturbations of $\{w_M(.)\}$ cannot be physically motivated. A model is being used outside its validity range.

Shannon's theorem can be easily linked to other results available for analytic functions, i.e. Plancherel-Polya's theorem stating that the FT ,now denoted by u, of a function f with bounded support is analytic. Then we can apply in principle at least, some analytic continuation method to get u outside the region where its values have been given and eventually reconstruct f everywhere. This holds in particular if u satisfies Helmholtz equation. Necessary and sufficient conditions for the latter to have analytic solutions are also well-known: Sherman (1969) defines as <u>source free</u> a scalar field u which results from superposing homogeneous waves only , i.e. such that   supp û = = supp f C D$_k$. He then shows , we quote, u "has all the properties exhibited by an entire function", to which analytic continuation applies. We see that analiticity of the scalar field u is at variance with supp û extending outside D$_k$: this is another argument against superresolution based on analytic continuation.

Any approach liable to yield anything meaningful must therefore start with a different model , which does not neglect <u>evanescent</u> waves. As an example, let us mention the algorithms based on nonuniform plane wave expansion. The convolution (1) can be carried out with respect to the wavevector angles θ and φ in the domain $0 \leq \varphi < 2\pi$, $0 \leq$ Re $\theta < \pi/2$, Im $\theta < 0$, which are linked to the spatial frequencies by the class C$^{\infty}$ diffeomorphism:

$$\begin{cases} y_1 = (k/2\pi) \; \sin\theta \; \cos\varphi \\ y_2 = (k/2\pi) \; \sin\theta \; \sin\varphi . \end{cases}$$

Baltes,Schmidt-Weinmar and coworkers (see e.g. Schmidt - Weinmar,1978) have thoroughly investigated both theoretically and numerically, the role of this expansion in superresolution algorithms and have given some numerical noise sensitivity results, which <u>should</u> be interpreted by the general stability theory (Bertero,DeMol,Viano,1980).

5.   SYSTEM THEORY AND INVERSE DIFFRACTION

5.1 Control and observation

Even in the crudest approach towards the model of a system,we need a distinction between input and output, i.e. between control and observation. Let us consider the direct problem for Helmholtz equation discussed in Section 2: since it

consists of finding w everywhere in R$^3_+$ given the BC v, it makes sense to define v the <u>input</u> or the boundary control, and w the system <u>state</u>. Therefore U is the input set and W, to which w belongs, the state space. The direct problem is then a control problem, because the <u>control map</u> B :

B : U $\longrightarrow$ W
        v $\longmapsto$ w(v)

is well-defined, thanks to radiation conditions which insure uniqueness of w(v). If we want the field at x$_o$, we perform the so-called point observation, hence we introduce the output or observation map C_o :

$$C_o : W \longrightarrow C$$
$$w \longmapsto w(\underline{x}_o).$$

The procedure which leads from v to $w(\underline{x}_o)$ can be interpreted in the system language and by introducing the adjoint system. Let us assume control and observation can be interchanged . Then we can define the following system:

$$\begin{cases} (\text{divgrad} + k^2) \ p = \delta(\underline{x} - \underline{x}_o) \ ; \ \underline{x}_o \ \text{in} \ R^3_+ \\ p\big|_{R_o^3} = 0 \\ \text{radiation conditions.} \end{cases}$$

A source term , i.e. an input, has appeared where we wanted to observe w(.) ; a homogeneous BC holds where w was controlled. We say p is the state vector adjoint to w. If we write the second Green's formula for w and p , we get precisely $w(\underline{x}_o)$ as a function of v. In this case p coincides with the already defined fundamental solution g_1 of Helmholtz equation. Green's formula is precisely the relationship which links w and p in a sesquilinear form (linear in w and antilinear in p). This extends to other control and observation maps and is a result of duality theory. Actually duality has a wider meaning either in cathegory theory (Goguen & Varela,1979), system theory (Delfour & Mitter,1972), or functional analysis and the calculus of variations in particular (Ekeland & Temam,1977;Lions,1971). We note in passing that if

the sesquilinear form is the inner product in $L^2(.)$, then the standard adjoint operators and vectors appear.

5.2 Variational approach

We have introduced the adjoint system because by solving it at the same time as the w-system we restate inverse diffraction problems as the variational approach suggests. The latter leads to the minimisation of some functional. For the general theory we refer to Lions's (1971) textbook.

As an example let us assume data, hereinafter denoted by z(.), have been measured in a domain D and that we have to find v. Then the problem becomes :

$$\text{minimise } J(v) := \int_D \left| Cw(v) - z \right|^2 dD + a \left\| v \right\|^2_{Uad}$$
$$\forall \ v \in U_{ad}$$

where :

U_{ad} is the admissible input set, to be adequately defined,

w(v) is the field corresponding to an estimated v,
C is the observation map which turns w(v) into something
    comparable to z(.),
a > 0.

The input u in U_{ad} such that :

$$J(u) = \inf_{v \in U_{ad}} J(v)$$

if it exists, is named an <u>optimal control</u> and corresponds to a solution in the least-squares sense of the identification problem.

This setting is widely established and is known by the names of some Authors such as Lions and Chavent (see e.g. Chavent, 1977), Miller (1970), Tikhonov and Arsenin (1976). The standard optimal control theory is however based on real valued of real valued variables . In our case we have a real functional physically related to error and control "energies", which depends on <u>complex</u> valued fields. Since we must minimise J(.) by a gradient method, we first verify it is continuous and differentiable , then apply some differentiation rules which may easily be implemented by a numerical scheme. Let us assume the observation map is :

$$C_1 : W \longrightarrow Y_1$$

$$w \longmapsto C_1 w := w(\underline{x} \in D),$$

where D is a three-dimensional domain.
Let us the consider the first term in the functional $J_1(v)$ defined by:

$$J_1(v) := \int_D |w - z|^2 \, dD + a \|v\|^2_{Uad} :=$$

$$= J_{11} + J_{12},$$

i.e. J_{11}. We write it as an inner product in $L^2(D)$, which makes

sense if D is bounded. We recall that if D were unbounded, we would have to work in a weighted Hilbert space such as Vogelsang 's (1975), because a theorem (Ramm,1981) states Helmholtz equation in an unbounded domain , say T, has no nonvanishing

solutions in $L^2(T)$. Physically available data refer however to bounded domains, hence our model is satisfactory. Then :

$$J_{11} = \int_D (w^* - z^*)(w - z) \, dD = \langle w - z \mid w - z \rangle_{L^2(D)},$$

where w^* is the complex conjugate of w. If we formally differentiate J_{11} with respect to w, we get:

$$\delta J_{11} = \int_D dD \; (w^*-z^*) \; \delta w + (w-z) \; \delta w^* =$$

$$= 2 \, Re \int_D dD \; (w^* - z^*) \; \delta w. \qquad (10)$$

We have no linear operator acting on δw, hence we cannot introduce the Gateaux derivative of J_{11} with respect to w. Nor it makes sense to introduce a direction independent derivative as it does for analytic functions. The way out of this difficulty is suggested by Lagrangian theory : we double the degrees of freedom and deal both with w^* and w as independent variables. Equivalently we may take the real (Re w) and imaginary (Im w) parts separately, according to a straightforward transformation, which can be made unitary:

$$\begin{bmatrix} w \\ w^* \end{bmatrix} = \begin{bmatrix} 1 & +i \\ 1 & -i \end{bmatrix} \cdot \begin{bmatrix} Re \; w \\ Im \; w \end{bmatrix}$$

Since Re w and Im w are real valued functions, we can properly define a <u>Gateaux derivative</u> of J_{11}, i.e. a bra vector :

$$J_{11} := \int_D \begin{bmatrix} grad_w J_{11} & grad_{w^*} J_{11} \end{bmatrix}^* \begin{bmatrix} w \\ w^* \end{bmatrix} \; dD.$$

From now on we shall denote the pair (w,w^*) by $\underline{w}$ and other pairs accordingly. Then

$$\delta J_{11} := < \; \underline{grad_{\underline{w}} J_{11}} \; | \; \delta \underline{w} \; > \; ,$$

which gives sense to the last term in (10). A similar procedure holds for another continuous and differentiable functional , J_{22} related to observation map C_2 :

$$C_2 : W \longrightarrow Y2$$
$$w \longmapsto C_2 w := \left| w(\underline{x} \in D) \right|^2,$$

whereby field <u>intensity</u> is given. The task is to find v, thus solving an inverse diffraction plus a phase reconstruction problem at the same time. J_2 reads:

$$J_2(v) := \int_D \left| |w|^2 - z \right|^2 dD + a \, \|v\|^2_{Uad} \; .$$

The gradient of J_2 with respect to w or v can be readily evaluated by the same method. We have e.g.

$$\delta J_{21}(\underline{w}) = 4 \ \text{Re} \int_D \left[|w|^2 - z\right] w^* \delta w$$

5.3 Functional minimisation

The functional gradient depends on w and p where p is the solution of an adjoint system. We shall deal with two examples, related both to J_1 and J_2. Given J_1 the adjoint field p_1 satisfies:

$$\begin{cases} (\text{divgrad} + k^2)p_1 = w-z \ ; \ \text{supp} \ (w-z)=D \\ p_1|_{R_O^3} = 0 \\ \text{radiation conditions.} \end{cases} \qquad (11)$$

For this to make sense we need (w-z) to be sufficiently regular, i.e. at least in $L^2(D)$. We see the state equation is error – driven. If the error were $\underline{=} 0$ then $p_1 \underline{=} 0$. Moreover the Dirichlet BC for p_1 is homogeneous.

The minimisation of a functional can be restated in an equivalent form , which is more easily translated into a practical algorithm. Reformulation in straightforward if the functional has a unique minimising element . this is the case for J_1, if D of (10) is a (bounded) three dimensional domain . It can be shown that u is unique : identification of u is related to solving in the least squares sense a set of Cauchy problems for Helmholtz equation. u i then characterised by the so-called optimality system made of :

    the original Helmholtz equation or w-system,
    the adjoint or p_1-system of (11),which needs w(v) and z(.)
       to be solved,
    the variational inequality:

$$2 \ \text{Re} \int_S dS\left[p_{1n}^*|_S \cdot (v-u)\right] + \langle \underline{\text{grad}}_v J_{12}(v)| \ (\underline{v}-\underline{u})\rangle_{Uad} \geq 0,$$

$$\forall \ v \in U_{ad} \ ,$$

where p_{1n} is the normal derivative of p_1 at S .
Minimisation of J_2 must be dealt with in a different way , because the minimising elements are not unique. Hence no optimality system in the strict sense can be introduced. Yet, if we solve:

    the w-system as above,
    the following p_2-system, defined by:

$$\begin{cases} (\text{divgrad} + k^2)p_2 = 2(w^*w-z)w^* \\ p_2|_{R_O^3} = 0 \\ \text{radiation conditions,} \end{cases}$$

and the variational inequality:

$$2 \text{ Re} \{ \int_S \rho_{2n}^* \cdot (v-u) \ dS \} + \ \langle \underline{\text{grad}}_v J_{22}(v) \,|\, (\underline{v}-\underline{u}) \rangle_{Uad} \geq 0,$$

$$\forall \ v \in B_u , \tag{12}$$

where B_u is an adequate neighbourhood of an optimal control u, we find both minimising elements lying on the boundary of U_{ad} and all stationary points of J_2 which are strictly inside U_{ad}. Hence

the above set is necessary but not sufficient to identify the u's. It gives however the functional gradient , which again will be used in an algorithm.

For the p_2-system to make sense we need the source term to

be regular enough. A sufficient condition is the same required for J_2 to be well-defined, i.e. :

$$w \in L^4_{loc}(R^3_+) ,$$

otherwise J_2 would not be continuous. Hence the need to constrain v. In this case we may choose the input space

$$U_{ad} = C^0_S(R^3_o) , \tag{13}$$

of continuous functions defined on R^3_o which have support in S.

Inverse diffraction in some other geometries are dealt with by (Crosta,1983) , together with their optimality systems. Other applications of the discrete gradient method to electromagnetic imaging have been considered by Lesselier (1981) ,Block(1983), Roger(1983) .

6.   SIMULTANEOUS APERTURE IDENTIFICATION AND PHASE
       RECONSTRUCTION

6.1 Theory

We shall study the minimisation of J_2 with some more detail: it is strictly related to the phase problem (Fiddy,1983a) , for some recent results about which we refer to Fiddy (1983b) . Our aim is to give the available results and list some open problems about existence,uniqueness and continuous dependence (regularity) of the identified control v on the data set z(.) . The problem may be either given an infinite or a finite dimensional setting , the former being mainly of use for existence and regularity estimates, the latter leading to a practical algorithm. In principle the infinite dimensional approach could deal with data having any support, ranging from a countable set of points to a three dimensional domain ($D_3 \subset\subset R_+^3$). The former case is however best analysed by discretising the equations.

Let us briefly investigate the link between existence and uniqueness of v and the dimension of the data support D. If D is a two-dimensional regular surface D_2, hereinafter named Beckert's surface, which does not divide R_+^3 into two and does not intersect R_o^3, then it can be shown, by extending Beckert's (1960) theorem, that both the traces of w and w_n on D_2 approximate in the $L^2(D_2)$ sense any function in $L^2(D_2)$ by choosing an adequate v in U_{ad} of (13).If we are given Cauchy data on D_2, existence of the corresponding v is not insured in the strict sense but in the least-squares sense. If only intensity data are available, the least-squares solution is for sure not unique. Nor things improve if we take two nearby Beckert's surfaces , because they can be joined together and yield another Beckert's surface. This statement differs from the standard uniqueness result for the phase problem ,which requires data e.g. on two planes both parallel to R_o^3: they are not Beckert's surfaces because they are unbounded. We shall therefore choose a three dimensional bounded domain D_3 as data support.

In addition to having the properties discussed in Section 5.3, J_2 is coercive, because of the second term, i.e.:

$$\lim_{\|v\|_{Uad} \to \infty} J_2(v) = \infty$$

therefore it has at least one minimising element.

J_2 is not convex,i.e. does not satisfy :

$$J_2(\Theta v_1 + (1- \Theta)v_2) \leq \Theta J_2(v_1) + (1- \Theta)J_2(v_2),$$

$$0 \leq \Theta \leq 1 , \forall v_1, v_2 \in U_{ad},$$

as it could be shown. Then we expect the minimum not to be unique, as announced in Section 5. From the above discussion, we try to reduce the multiplicity of minima by taking D_3.

The dependence of an optimal input u on $z(.)^3$ is still an open problem : it could be easily investigated if an explicit relationship between u and z were available. We have however an implicit relationship, i.e. the variational inequality (12). If it were an equation, then by extending Dini's theorem to functionals, we could get something like $\partial u/ \partial z$ and then carry out a stability estimate, but stationary points of J_2 satisfy a variational equation if they are strictly inside U_{ad}.Boundary minima satisfy the strict inequality. The separate study of inner and boundary optimal controls would then become unrewardingly complicated.

Another interesting open problem is how to combine the topological relationship a variational inequality gives about the minimising elements of J_2 with the non-local analysis of an inverse problem, recently suggested by Sabatier (1983) . This should reduce ambiguity and eventually lead to a better algorithm.

6.2 Discretisation

We denote by
$\{x_n\}$ a finite countable set of points in R_+^3, $1 \leq n \leq N$,

$\{x_{an}\}$ a finite countable set of points in R_o^3,

$v_m := v(x_{am})$ the estimated control values at R_o^3,

$w_n := w(x_n)$ the estimated field values at x_n, which depend on v

$z_n := z(x_n)$ the measured intensity values at x_n,

$J_3 := J_{31} + J_{32}$ the functional to be minimised, which
consists of (estimation error)+(quadratic form on
v_m), and is the discretised counterpart of J_2.

In order to minimise J_3 we may:
either minimise J_{31} subject to the constraint that also J_{32} be a
minimum,
or consider J_3 as a whole.
If we look for the set $\{w_n\}$ such that

$$J_{31} := \sum_{n=1}^{N} (w^*_n w_n - z_n)^2 = \text{minimum},$$

we have to solve the non-linear system of N equations

$$(w^*_n w_n - z_n) w^*_n = 0$$

in 2N unknowns. The study of the Hessian matrix does not prove useful because the ambiguity on w_n is not reduced. Then we may consider J_{32} as a constraint and follow a procedure similar to

Weston's (1980,1981, and in particular 1983,pp 830-32), who aims at identifying the refractive index starting from measured intensity data.
On the other hand if we consider the minimisation of J_3, we have better write v_m and v^*_m, which appear in the regularising

term, as a function of w_n (for simplicity we write this term as the discretised counterpart of the norm of v in $L^2(S)$; other cases could be dealt with similarly). This step involves a discretised inverse propagator , usually an ill-conditioned matrix.

Since:

$$w_n = \sum_m G_{nm} v_m,$$

where G_{nm} are the matrix elements of the discretised direct propagator (Rayleigh-Sommerfeld's ,the paraxial one or any other approximated propagator), we must choose the points $\underline{x}_n, \underline{x}_{am}$ in such a way as $G^{-1} := H$ exists. Then

$$v_m = \sum_n H_{mn} w_n$$

and J_{32} becomes:

$$\sum_m (\sum_j H^*_{mj} w_j) (\sum_n H_{mn} w_n) := \sum_{m,n} w^*_m E_{mn} w_m$$

where $\quad E := (H^T)^* H.$

Then we get the non-linear system:

$$2(w^*_n w_n - z_n) w^*_n + \sum_m w^*_m E_{nm} = 0$$

of 2N equations in 2N unknowns, to which convergence theorems for Newton's method (see e.g. Demidovitch & Maron,1979,pp 454-475) apply. A stability estimate of the local solution may now be carried out according to Weston (1983,p 763) . The minimisation of J_3 can thus take advantage of the discrete gradient method (Chavent , 1977). The weak point of the procedure consists of the critical role the $\underline{x}_n$'s play in conditioning the matrix H. A tradeoff between the accuracy of data point location and of location of the v_m's, may then follow, according to (Bertero and DeMol,1981,p 370 ff).

ACKNOWLEDGEMENTS

The author thanks prof. Gianni Degli Antoni for encouragement, prof. Wolfgang M. Boerner for accepting this contribution and prof. Alexander D. Ramm for critically reading a preliminary version of this manuscript.

REFERENCES

Beckert H,1960,"Eine bemerkenswerte Eigenschaft der Loesungen des
    Dirichletschen Prob. bei lin. ellipt. Differentialgl.",
    Math. Ann.,139,255-64
Bellman R, Kalaba R,1959 "Functional equations,wave propagation &
    invariant imbedding", J. Math. Mech. 8, 683
Bertero M, De Mol C , 1981 "The stability problem in inverse
    diffraction", IEEE AP 29 368-372
Bertero M, De Mol C, Viano G A, 1980 "The stability of inverse
    problems",in Baltes H P "Inverse scattering problems in
    optics", pp 161-214, Berlin: Springer
Blok H,1983,"One dimensional time domain inverse scattering
    applicable to electromagnetic imaging", this volume
Boerner W M ,1980,"Polarisation utilisation in electromagnetic
    inverse scattering",in Baltes H P "Inverse scattering
    problems in optics",Berlin: Springer,
Boerner W M ,1981 "Polarisation dependence in electromagnetic
    inverse problems" IEEE AP 29 262-271
Chavent G,1977 "Identification problems for distributed
    systems",New Dehli IIT Winter School lecture notes,
    Rocquencourt: INRIA
Conant R C,Ashby W R,1970,"Every good regulator of a system must
    be a model of that system",Int.J.Syst. Scie.,1,pp 89-97,
Corones J,1975 "Bremmer series that correct parabolic
    approximations", J.Math.Analys.Appl.50 361-372
Crosta G,1982 "Inverse diffraction, duality and optimal control",
    J.Phys.A 15 645-660
Crosta G,1983 "Models for diffracting aperture identification: a
    comparison between ideal and convolutional observations",
    SPIE Proc.413, paper n° 24
Delfour M D & Mitter S K,1972,"Controllability and observability
    for infinite-dimensional systems",SIAM J.C.& O. 10,329-33
Demidovitch B & Maron I,1979,"Elements de calcul numerique",
    MIR:Moscow
Deschamps G A,1981 "Gaussian beams: paraxial theory", in J K
    Skwirzynski Ed."Theoretical methods for determining the
    interaction of electromagnetic waves with structures" ,
    Rockville,MD:Sijthoff & Noordhoff
Ekeland I & Temam R,1974,"Analyse convexe et problèmes
    variationnels", Dunod,Paris
Fiddy M A,1983a,"The phase problem", SPIE Proc. 413,paper n° 26
Fiddy M A,1983b,"Object reconstruction from partial information",
    this volume
Fock V A,1960 "Electromagnetic diffraction & propag. problems",
    New York: McMillian
Goguen J A and Varela F J,1979,"Systems and distinctions:
    duality & complementarity",Int. Journ. Gen. Syst.,5,pp 31-43
Goodman J W,1968 "Introduction to Fourier optics",New York:McGraw

Hill

Hörmander L,1963 "Linear partial differential operators",Berlin:
     Springer, pp 249-51

Jahnke,Emde,Lösch,1960 "Tables of higher functions",New York:
     McGraw Hill

Jordan T F,1969,"Linear operators for quantum mechanics", Jordan:
     Duluth,MN

Lalor E,1968 "Conditions for the validity of the angular spectrum
     of plane waves",J.Opt.Soc.Amer. 58 1235-37

Lesselier D,1981 "Diagnostic des milieux inhomogénes
     unidimensionnels par échographie éléctromagnetique",Rev.du
     CETHEDEC 68 1-42

Lions J L,1971 "Optimal control of systems governed by partial
     differential equations",Berlin:Springer

Miller K,1970,"Least squares method for ill-posed problems with a
     prescribed bound",SIAM J. Math. Analys.,1,52-74

Montgomery W D,1968,"Algebraic formulation of diffraction applied
     to self imaging",J.O.S.A 58,pp 112-24

Ramm A D,1978,"Necessary and suff. conditions for the validity of
     the limit absorption principle", Math. Izv. Vizov. 5

Ramm A D,1981,"On some properties of solutions of the Helmholtz
     equation",J.Math. Phys. 22,pp 275-6

Ramsay W B,Baltes H P,Schmidt-Weinmar H G,1982 "Anomalies in
     bandlimited asymptotic fields",J.O.S.A.72,1618-29

Roger A J,1985,"Theoretical study and numerical resolution of
     inverse problems via the functional derivatives", this
     volume

Sabatier P C,1985,"Critical analysis of the mathematical methods
     used in electromagnetic inverse theories", this volume

Schmidt-Weinmar H G,1978,"Spatial resolution of subwavelength
     sources from optical far field data",in H P Baltes,Ed.
     "Inverse source problems in optics",Berlin: Springer

Sherman G C,,1969,"Diffracted wave fields expressible by plane
     wave expansions containing only homogeneous waves",
     J.O.S.A,59,pp 697-711

Tikhonov A,Arsenin V,1976 "Méthodes de résolution des problèmes
     mal-posés",Moscow:MIR

Vogelsang V,1975 "Das Ausstrahlungsproblem für elliptische
     Differentialgleichungen in Gebieten mit unbeschränktem
     Rand",Math. Zeit. 144 101-124

Weston V H,1980,"Inverse problem for the reduced wave equation
     with fixed incident field,part 1",J. Math. Phys. 21,pp
     758-64

Weston V H,1981,ditto, part 2,J. Math. Phys. 22,pp 2523-30

Weston V H,1983,ditto,part 3,J. Math. Phys. 24,828-33.

II.4 (NM.2)

STABILITY AND RESOLUTION IN ELECTROMAGNETIC INVERSE SCATTERING
- PART I

M.Bertero   C.De Mol   E.R.Pike

Univ.Genova   Univ.Bruxelles   RSRE,Malvern

ABSTRACT: A powerful technique for the solution of a number of
experimental inverse problems, described by an underlying first-
kind Fredholm equation, is presented. Such problems include, for
example, diffraction-limited EM imaging and quasi-elastic laser
light scattering. The technique requires the construction of the
"singular system" of the problem which then provides exact orthonormal bases both for the description of sampled and truncated
measured data and for the reconstructed continuous object "solution"
of the inversion. The singular-system approach may be regarded as a
theory of information which generalises in several directions the
well known classical concepts of Shannon and Nyquist.

1.   HISTORICAL INTRODUCTION

1.1. The Shannon Theory of Information

     One of the most well studied inverse problems is that of the
temporal frequency band-limited communication channel or, equivalently, the spatial frequency band-limited imaging system. The mathematical description of this problem is easily stated as follows,
where we use the latter system in one dimension for illustration
and notation. An "object", $O(x)$, $- 1 \leqslant x \leqslant + 1$, is resolved into
its Fourier components $\widetilde{O}(k)$ (by straightforward generalisations
x and k may be multidimensional)

$$\widetilde{O}(k) = \int_{-1}^{+1} O(x) \; e^{-ikx} \, dx \qquad (1.1)$$

319

W.-M. Boerner et al. (eds.), Inverse Methods in Electromagnetic Imaging - Part 1, 319–328.
© *1985 by D. Reidel Publishing Company.*

and a "geometric" image, $I(x')$, is formed by re-summing only those spatial frequencies lying below a "band-limit" and which may be taken without loss of generality in $-1 \leq x' \leq +1$

$$I(x') = \frac{1}{2\pi} \int_{-\Omega}^{+\Omega} \widetilde{O}(k) \, e^{ikx'} \, dk \quad . \tag{1.2}$$

Thus

$$I(x') = \frac{1}{2\pi} \int_{-1}^{+1} \left(\int_{-\Omega}^{+\Omega} e^{-ik(x-x')} \, dk \right) O(x) \, dx$$

$$= \int_{-1}^{+1} \frac{\sin\left[\Omega(x-x')\right]}{\pi(x-x')} O(x) \, dx \quad . \tag{1.3}$$

This has the form of a general Fredholm equation of the first kind

$$(Kf)(p) = g(p) \tag{1.4}$$

where

$$(Kf)(p) = \int_{-1}^{+1} K(p,t) \, f(t) \, dt \quad , \quad -1 \leq p \leq +1 \tag{1.5}$$

with the "kernel" $K(p,t)$ of the integral operator, K, in this case a "sinc" function.

The remarkable solution of equation (1.3) in terms of its orthonormal eigenfunctions, namely the prolate spheroidal wave functions (Slepian and Pollack, 1961), helps to understand the classical theory of information and is the source of the accepted theory of resolution in diffraction-limited imaging problems. We recall that the eigenfunctions, φ_n, are the solutions of the homogeneous equation

$$\lambda_n \, \varphi_n(x') = \int_{-1}^{+1} \frac{\sin\left[\Omega(x-x')\right]}{\pi(x-x')} \varphi_n(x) \, dx \tag{1.6}$$

$$-1 \leq x' \leq +1$$

so that the object may be reconstructed as a series

$$O(x) = \sum_{n=0}^{\infty} \frac{c_n}{\lambda_n} \; \varphi_n(x) \tag{1.7}$$

where

$$c_n = \int_{-1}^{+1} I(x') \; \varphi_n(x') \; dx' \tag{1.8}$$

The mathematical theory will be discussed in more detail in part II of this paper but we note here that the use of this solution in the presence of noise is only possible by limiting the reconstructed object to components c_n which correspond to sufficiently high eigenvalues λ_n and therefore are not to be strongly affected by the noise, that is to say, those components which appear in the image above the ambient "noise level". For the prolate spheroidal functions the eigenvalue series λ_n drops sharply from approximately unity to nearly zero at a value of n known as the Shannon number $(2\Omega/\pi)$ which means that only this number of components may be recovered almost independently of the actual noise level in any particular situation. One may thus, with this problem, speak of a number of degrees of freedom, number of "bits" of information, Nyquist sampling limit and other concepts related only to the band limit and not to the noise. These concepts are used in what we may call the classical Shannon theory of information.

1.2. Laplace Transform Inversion

First-kind Fredholm equations appear in many more physical inverse problems, of course, than that with the sinc kernel discussed above. We shall consider in this section the case where the kernel is an exponential function, giving rise to the problem of the inversion of the Laplace transform

$$g(p) = \int_0^{\infty} e^{-pt} f(t) \; dt \quad , \quad 0 \leqslant p < +\infty \quad . \tag{1.9}$$

This problem arises in diverse fields of science, our particular interest being its application to the determination of molecular size distributions from laser light scattering data (Ostrowsky et al., 1981). Many ad hoc methods of numerical inversion of eq. (1.9) have been proposed but it was not until five years ago that the problem was approached in the spirit of the information theory sketched above when McWhirter and Pike (1978) proposed the solution of (1.9) in terms of the generalised eigenfunctions, φ_ω (where ω

is now a continuous variable), given by

$$\varphi_\omega(t) = t^{-1/2-i\omega} \frac{\text{Re}\left[\Gamma(\tfrac{4}{2}+i\omega)\right]^{1/2}}{\left[\Gamma(\tfrac{1}{2}+i\omega)\right]^{1/2}} \quad , \quad \omega > 0 \qquad (1.10)$$

$$\varphi_\omega(t) = t^{-1/2+i\omega} \frac{\text{Im}\left[\Gamma(\tfrac{4}{2}+i\omega)\right]^{1/2}}{\left[\Gamma(\tfrac{1}{2}+i\omega)\right]^{1/2}} \quad , \quad \omega < 0 \qquad (1.11)$$

Furthermore the corresponding continuous spectrum of the operator was shown to have the simple form

$$\lambda_\omega = \frac{\omega}{|\omega|} \sqrt{\frac{\pi}{\cosh(\pi\omega)}} \qquad (1.12)$$

This spectrum shows an unfortunate exponential fall-off causing a severe ill-posedness and a strong limitation in the presence of noise on the band of frequencies ω which may be recovered. It must be noted that in this case, contrary to the sinc-kernel case, the resolution limit depends on the noise level. The reason is that the spectrum (1.12) no longer has a plateau before a sharp fall as observed in the previous case.

1.3. The Discrete-Discrete Inversion

In order to use the McWhirter-Pike results for numerical inversion Ostrowsky et al.(1981) found that, provided a sufficiently low number of sample points of the solution were demanded, a discrete matrix inversion based on

$$g(p_i) = \sum_{n=1}^{S} a_n e^{-p_i t_n} \qquad i = 1,\ldots,S \qquad (1.13)$$

could be used. A geometric ratio was used for the spacing of the points p_i. The somewhat surprising fact emerged that the number S of sample points which could be recovered was somewhat larger than the number of degrees of freedom predicted by the McWhirter-Pike theory.

2.   THE EFFECT OF FINITE OBJECT SUPPORT, SINGULAR VALUE ANALYSIS

Following the work of Ostrowsky et al.(1981), Bertero, Boccacci and Pike (1982) traced the increase of information capacity found by these authors to the implicit restriction of object support in the matrix inversion procedure used. The effect of a known finite support of the object in the Laplace transform was given a thorough investigation using an approach based on singular functions.

2.1. Continuous–Continuous Case

2.1.1. Laplace Transform. Bertero et al.(1982) considered the problem of the finite Laplace transform

$$(Kf)(p) = \int_a^b e^{-pt} f(t) \, dt \quad , \quad 0 \leqslant p < +\infty \qquad (2.1)$$

restricting the transformation to square integrable object functions on the interval (a,b). It is easy to show that the operator K is compact and injective and admits a singular system $\{\alpha_k; u_k, v_k\}$ $(k = 0,1,\ldots)$ given by the solutions of the coupled equations (see e.g. Miller, 1974)

$$Ku_k = \alpha_k v_k \quad , \quad K^* v_k = \alpha_k u_k \qquad (2.2)$$

where K^* is the adjoint operator given by

$$(K^* g)(t) = \int_0^\infty e^{-tp} g(p) \, dp \quad , \quad a \leqslant t \leqslant b \qquad (2.3)$$

Since both the object and ~~the~~ the image data are considered as functions of a continuous variable, we will call such a case a "Continuous–Continuous" inverse problem. In Bertero et al.(1982) it was shown that the singular value spectrum of operator (2.1) has a slower decrease rate than the corresponding spectrum for the problem where $f(t)$ is assumed to have an infinite support. As a consequence, the knowledge that $f(t)$ is a priori restricted to a finite known support allows to recover more information from the data and this in quantitatively the way observed by Ostrowsky et al.

2.1.2. Diffraction-Limited Imaging. The work of Bertero et al. described above gave rise to the suggestion that the increase in information transfer obtained in the Laplace inversion by using a known restricted support for the solution might have an analogy

in the diffraction-limited imaging problem discussed in section 1
with coherent illumination, allowing an improvement in resolution
over that defined by the Shannon number to be obtained. This
possibility was investigated and proved to be correct in a paper
by Bertero and Pike in 1982. In this paper it was shown that the
restriction of the image to a support equal to that of the geometrical image, which is implicit in the eigenvalue analysis, causes
a loss of information which can be regained by the use of the total
diffraction image. It is then required to solve a first kind
Fredholm equation as given by (1.4) where the integral operator K
is now defined as follows

$$(Kf)(p) = \int_{-1}^{+1} \frac{\sin [\Omega(p-t)]}{\Pi(p-t)} f(t) \, dt \ , -\infty < p < +\infty \quad (2.4)$$

The singular values of this operator are $\sqrt{\lambda_n}$ and clearly they
fall off more slowly than the eigenvalues λ_n defined by (1.6),
allowing more coefficients to be calculated in the reconstruction
and hence a higher resolution to be obtained. The two-dimensional
case with square and circular apertures was also considered in this
work and in a later paper (Bertero et al., 1982) the analysis was
carried out also for incoherent illumination. Furthermore it has
been remarked in these problems that resolution beyond the classical
diffraction limit (i.e. superresolution) is much easier to achieve
when the space-bandwidth product is small (i.e. the constant Ω in
equation (2.4)). A similar effect has also been demonstrated for
the problem of inverse diffraction from plane to plane, when it is
known a priori that the source is restricted to a small support
with linear dimensions comparable to the wavelength of the EM field
(Bertero et al., 1983).

2.2. Continuous-Discrete Case

    2.2.1. Laplace Transform. In the last two sections we dis-
cussed the improvement in resolution in the Laplace transform
and diffraction-limited imaging cases obtained when using a priori
knowledge of the support of the object function. The singular
systems used were constructed in bases of functions of continuous
variables. In real experiments, however, clearly the data are
measured over truncated regions of finite extent and at a finite
number of sampled points. A possible route is then to discretise
the unknown function. However, the singular value method offers
a very interesting alternative since the types of functions, u
describing the object and v describing the data, may be different.
In particular, we may regard the data as a vector set but yet
reconstruct the object in a basis of square integrable functions.
In the previous discretisation mentioned above the object was
constructed only on a set of sample points and, usually, an interpolation scheme was used between them. If we adopt the above
suggestion the interpolation step is completely avoided.

The mathematics of this situation will be discussed in part II of this paper but here we may say that the method has been put into practice and will be the subject of a publication by Bertero et al. in the near future. It has been found that computations will produce accurately the required bases in the functional object space, which, of course, will be represented in the numerical calculations by a rather large number of discretisation points, and in the vector valued image space, which will normally be represented by rather smaller dimension.

The practical application of this method is under investigation at the present time in the context of the laser light scattering problem mentioned above.

2.2.2. Diffraction-Limited Imaging. In the diffraction-limited imaging problem, just as in the Laplace problem discussed in the last subsection, the same idea of using a singular vector basis for the data and a singular function basis for the object can be applied. Successful numerical computation of such bases have been performed in Genoa and at Malvern and are currently in the process of being applied to practical problems in microscopic and other imaging. The theory looks particularly promising for application, for example, in EM imaging problems in the microwave region.

An attractive feature of the method is the speed at which the reconstruction can be effectuated, all that is required is to form a scalar product of the data set with each of the basis data vectors in turn and these scalar products are used as the coefficients, divided by the corresponding singular value, of the singular function expansion of the object. Both the basis vector set and the basis function set are, of course, stored numerically once and for all for a given problem.

3. EXTENSION TO OTHER PROBLEMS

So far in this paper we have only mentioned two cases of the kernel $K(p,t)$ of the integral equation, namely, the sinc kernel and the Laplace exponential kernel. There are, however, many other problems of a similar type and in this last section we will mention briefly some problems which are under study using the singular system technique.

3.1. Fraunhofer Diffraction

In this problem the kernel has the form $K(p,t) = J_1^2(pt)/(pt)^2$. This represents the form of the intensity scattered in the far field by a spherical particle illuminated by a coherent laser beam and the solution of the problem will be the distribution of sizes of the scattering particles. The problem is of great importance in many industrial and scientific applications and the technique of particle sizing by Fraunhofer diffraction is quite widely used.

As a first attempt to quantify the information content in a typical data set in such an experiment we have used the results of the paper of McWhirter and Pike in 1978 and found the following spectrum for the present problem (Bertero and Pike, 1983)

$$\lambda_\omega^\pm = \pm \left| \frac{2^{-\frac{5}{2}} \Gamma\left(\frac{5}{2} - i\omega\right) \Gamma\left(\frac{1}{4} + \frac{i\omega}{2}\right)}{\Gamma^2\left(\frac{7}{4} - \frac{i\omega}{2}\right) \Gamma\left(\frac{11}{4} - \frac{i\omega}{2}\right)} \right| \tag{3.1}$$

This spectrum can be used to calculate the number of resolution cells in particle size, which can be found in a certain level of experimental noise. For example we can ask how many size fractions can be resolved in the range 2 to 100 microns when the signal-to-noise ratio is 10^2. The result is that 7 or 8 fractions on a geometrical scale may be resolved in this range. Further work is required on this problem to extend the calculations to the use of singular functions and vectors, which will provide a very practical approach to the inversion in this type of experiments.

3.2. Laser Anemometry

Yet another example of this sort of inversion problem is provided by laser anemometry. In this type of experiment a laser beam scattered off a particle in uniform motion at velocity v, gives rise to a measured data set with the form of an exponentially damped cosine wave. In turbulent flows or non-uniform flows the data represent the motion of particles with a probability distribution of velocities which one needs to recover as the solution. The kernel in this case has the form

$$K(p,t) = e^{-(pt)^2/2\sigma^2} (1 + a \cos(bpt)) \tag{3.2}$$

where a,b and σ are constants. A certain amount of work with this kernel has been done using the concepts of information theory discussed above and one finds, in this case, that the fall-off of the eigenvalue or singular value spectrum is a function of the value of σ, characterising the damping of the cosine signal. In the limit of small damping a large amount of information is available from the data, while if the damping is more severe there is much less velocity information to be found. Again one can introduce a concept of resolution associated with a number of components of the solution which may be transmitted above the noise level. The pressure to push the analysis of this particular kernel to the same degree as has been done for the Laplace problem above is not so strong since in many cases the damping term may be arranged to be small. However in critical circumstances, as for example are met in very high speed supersonic flows, or where very high spatial resolution is required it will be important to extract the maximum information from data of this type and further work along the lines described above will allow this to be done.

3.3. Gaussian-Beam Illumination

In this last section we come back to the diffraction-limited imaging problem but consider the case where the object is illuminated by a coherent laser beam of infinite spatial extent but with a Gaussian amplitude profile. This is not the place to report detailed results of this calculation but it will suffice to say that singular value spectra may be calculated numerically (Bertero et al., in preparation) and these allow a discussion of such an imaging problem in terms of the other concepts of number of degrees of freedom as a function of noise and the implications of this in terms of the limits of resolution which may be achieved. Our preliminary results indicate that the singular value spectrum for this problem falls off a little more slowly than for the associated diffraction-limited imaging problem in which the illumination is assumed to have a "top hat" form (sinc kernel) of the same total power and the same width. However, the zeroes of the singular functions do not increase in density at the same rate as for the "top hat" case and therefore the overall resolution limits might be worse.

ACKNOWLEDGEMENT

C. De Mol is Research Associate with the National Fund for Scientific Research (Belgium).

REFERENCES

Bertero, M., Boccacci, P. and Pike, E.R., 1982, Proc. Roy. Soc. Lond. A383 pp. 15-29.

Bertero, M., Boccacci, P. and Pike, E.R., 1982, Optica Acta 29 pp. 1599-1611.

Bertero, M., De Mol, C., Gori, F. and Ronchi, L., 1983, Optica Acta 30 pp. 1051-1065.

Bertero, M., De Mol, C., Pike, E.R. and Walker, J., in preparation.

Bertero, M. and Pike, E.R., 1982, Optica Acta 29 pp. 727-746.

Bertero, M. and Pike, E.R., 1983, Optica Acta 30 pp. 1043-1049.

Mc Whirter, J.G. and Pike, E.R., 1978, J. Phys. A 11 pp. 1729-1745.

Miller, G.F., 1974, Numerical Solution of Integral Equations (eds L.M. Delves and J. Walsh) , p. 175, Oxford: Clarendon Press.

Ostrowsky, N., Sornette, D., Parker, P. and Pike, E.R., 1981, Optica Acta 23 pp. 1059-1070.

Slepian, D. and Pollack, H.O., 1961, Bell Syst. Tech. J. 40 pp. 43-64.

II.5 (NM.3*)

STABILITY AND RESOLUTION IN ELECTROMAGNETIC INVERSE SCATTERING - PART II

M. Bertero   C. De Mol   E.R. Pike

Univ. Genova   Univ. Bruxelles   RSRE, Malvern

ABSTRACT: Singular function expansions, whose applications to EM inverse problems have been reviewed in Part I, provide a very simple method for obtaining stable solutions and for defining resolution limits in a number of inverse problems. Their extension to cover the case of problems with discrete data is discussed and the connection between the case of continuous data and the case of discrete data is clarified. Then the general method is applied to the particular case of the finite Laplace transform inversion, a problem which is important both from the pedagogical and from the practical point of view. By means of this example we show how an appropriate choice of a small number of data points can provide almost the resolution achievable with continuous data.

1. THE MATHEMATICAL THEORY OF SINGULAR SYSTEMS

As was shown in the Part I of this paper, many inverse problems in EM imaging can be reduced to the solution of a Fredholm integral equation of the first kind

$$\int_a^b K(p,t)f(t)dt = g(p) \ , \ c \leqslant p \leqslant d. \tag{1.1}$$

When the kernel satisfies suitable conditions (for instance it is an integrable function) then the solution of eq.(1.1) is a classical example of an ill-posed problem.

329

W.-M. Boerner et al. (eds.), Inverse Methods in Electromagnetic Imaging - Part 1, 329–340.
© 1985 by D. Reidel Publishing Company.

The mathematical analysis is greatly simplified if we assume that both the "object" f and the "image" g are elements of L^2-spaces, respectively denoted by X and Y.

1.1. Eigenfunction expansions

The solution of problem (1.1) can be easily investigated when one can introduce eigenvalues and eigenfunctions of the integral operator

$$(Kf)(p) = \int_a^b K(p,t) f(t) dt, \quad c \leqslant p \leqslant d . \tag{1.2}$$

If the supports of the "object" f and of the "image" g coincide (namely, a = c and b = d) and if the kernel K(p,t) is square integrable and satisfies the condition: $K(p,t) = K^*(t,p)$ (here the star denotes complex conjugation), then K is a compact, self-adjoint operator in $L^2(a,b)$ and the set $\{\phi_k\}_{k=0}^{+\infty}$ of the eigenfunctions associated with the eigenvalues λ_k is an orthogonal basis of the orthogonal complement of the null space of K and therefore an orthogonal basis of $L^2(a,b)$ when K is injective (we will consider only this case). By expanding both f and g as series of the ϕ_k we obtain the solution of eq.(1.1) in the absence of noise

$$f(t) = \sum_{k=0}^{+\infty} \frac{g_k}{\lambda_k} \phi_k(t) \tag{1.3}$$

where

$$g_k = \int_a^b g(p) \phi_k^*(p) dp . \tag{1.4}$$

Since $\lambda_k \to 0$ when $k \to +\infty$ (this is a general property of the eigenvalues of compact operators) the series (1.3) generally is not convergent when the "image" g is affected by noise or experimental errors. Then the most simple technique for obtaining stable approximate solutions is to truncate, at a suitable value of k, the series (1.3). This technique, often called numerical filtering (Twomey, 1965), can be justified both from the point of view of regularization methods and from the point of view of Wiener filter methods (Bertero et al., 1980). If we assume, as usual, that the eigenvalues are ordered in a decreasing sequence, $\lambda_0 \geqslant \lambda_1 \geqslant \lambda_2 \geqslant \ldots$, then we take in the expansion (1.3) only the eigenvalues which are greater than a prescribed constant C and if $K_0 = \max\{k | \lambda_k \geqslant C\}$, then we get the filtered (regularized) solution

$$\tilde{f}(t) = \sum_{k=0}^{K_o} \frac{g_k}{\lambda_k} \phi_k(t) .$$ (1.5)

The quantity $1/C$ can be often interpreted as a "signal-to-noise ratio" and the number K_o as the "number of degrees of freedom" of the "image" g.

·An important property of the eigenfunctions of many integral operators encountered in experimental inverse problems is that the eigenfunction ϕ_k has exactly k zeros inside the interval $[a,b]$. Then, if the distribution of the zeros is approximately uniform one can define a <u>resolution distance</u>, associated with the solution (1.5), as follows

$$d = (b-a)/K_o$$ (1.6)

or, if the zeros of ϕ_k form approximately a geometric progression ,a <u>resolution ratio</u> given by (Bertero et al., 1982)

$$\delta^{K_o} = b/a .$$ (1.7)

In those cases where the eigenvalues drop to zero very rapidly, the number of degrees of freedom K_o is practically independent of the signal-to-noise ratio and the same property holds for the resolution distance d or the resolution ratio δ .

1.2. Singular function expansions

The previous analysis does not apply to the case where the supports of the "object" f and of the "image" g do not coincide. But if the operator K satisfies the condition

$$\int_c^d dp \int_a^b dt \left| K(p,t) \right|^2 < + \infty$$ (1.8)

(more generally, if K is compact), a powerful generalization of the eigenfunction method is provided by the singular function method (Picard, 1910). The singular system $\{\alpha_k; u_k, v_k\}_{k=0}^{+\infty}$ of the operator K is the set of the solutions of the coupled equations

$$Ku_k = \alpha_k v_k \quad , \quad K^*v_k = \alpha_k u_k$$ (1.9)

where K* is the adjoint of K

$$(K*g)(t) = \int_{c}^{d} K*(p,t)g(p)dp \quad , \quad a \leqslant t \leqslant b. \qquad (1.10)$$

The numbers α_k are called the <u>singular values</u> of K (usually they are ordered in a decreasing sequence) and the functions u_k, v_k are called the <u>singular functions</u> of K associated with α_k'. The singular functions u_k are also eigenfunctions of the operator K*K associated with the eigenvalues α_k^2 and they form an orthogonal basis of the orthogonal complement of the null space of K (of $L^2(a,b)$ if K is injective). Analogously the singular functions v_k are eigenfunctions of KK*, also associated with the eigenvalues α_k^2, and they form an orthogonal basis of the orthogonal complement of the null space of K* (of $L^2(c,d)$ if K* is injective).

In the absence of noise the solution of problem (1.1) is now given by

$$f(t) = \sum_{k=0}^{+\infty} \frac{g_k}{\alpha_k} u_k(t) \qquad (1.11)$$

where

$$g_k = \int_{c}^{d} g(p)v_k^*(p)dp . \qquad (1.12)$$

This solution has a structure very similar to that of the solution (1.3), which applies to the case of a self-adjoint operator, except for the fact that now two orthogonal systems, instead of one, are used. Again one can introduce a filtered solution as follows

$$\tilde{f}(t) = \sum_{k=0}^{K_0} \frac{g_k}{\alpha_k} u_k(t) \qquad (1.13)$$

and therefore it is clear that the concepts of "number of degrees of freedom" and of resolution distance (or resolution ratio) can be easily extended to the present case.

1.3. The case of discrete data

A very important property of the singular function methods is their flexibility which allows their extension to problems with discrete data.

Let us assume that the "image" $g(p)$ is measured only on a finite set of points, $p_1, p_2, \ldots, p_N$, so that eq(1.1) is replaced by

$$g(p_n) = \int_a^b K(p_n, t) f(t) dt \; ; \; n = 1, \ldots, N. \tag{1.14}$$

Now the data space Y is an N-dimensional vector space equipped with the scalar product

$$(g,h)_Y = \sum_{n=1}^{N} w_n \, g(p_n) h^*(p_n) \tag{1.15}$$

where the weights w_n must be introduced in order to take into account a non-uniform distribution of the data points p_n.

Let K_N be the operator, defined by eq.(1.14), which transforms an "object" of $L^2(a,b)$ into a vector of Y. The adjoint K_N^* transforms a vector of Y into a function of $L^2(a,b)$ and it is given by

$$(K^*g)(t) = \sum_{n=1}^{N} w_n \, g(p_n) K^*(p_n, t) \; . \tag{1.16}$$

Then $K_N^* K_N$ is a self-adjoint integral operator of finite rank in $L^2(a,b)$, whose kernel is given by

$$T_N(t,s) = \sum_{n=1}^{N} w_n \, K^*(p_n, t) K(p_n, s) \; , \tag{1.17}$$

while the operator $K_N K_N^*$ is a matrix $N \times N$ whose matrix elements are given by

$$t_{nm}^{(N)} = w_m \int_a^b K(p_n, t) K^*(p_m, t) dt. \tag{1.18}$$

We denote by $\left\{ \alpha_{N,k} ; u_{N,k}, v_{N,k} \right\}_{k=0}^{N-1}$ the singular system of K_N, the $u_{N,k}$ being the eigenfunctions of $K_N^* K_N$ and the $v_{N,k}$ being the eigenvectors of the matrix $K_N K_N^*$.

The singular value $\alpha_{N,k}$ provides a good approximation of the corresponding singular value α_k of the problem with continuous data, if the integral operator $K_N^* K_N$ is a good approximation of the integral operator K^*K. Indeed, let $T(t,s)$ be the kernel of K^*K:

$$T(t,s) = \int_c^d K^*(p,t) K(p,s) dp \tag{1.19}$$

and let us write

$$\epsilon_N(t,s) = T(t,s) - T_N(t,s) \tag{1.20}$$

$T_N(t,s)$ being defined in eq.(1.17). Clearly $K_N^*K_N$ is a good approximation of K^*K if $\epsilon_N(t,s)$ is small, and by the Weyl-Courant lemma (Riesz & Nagy, 1955) we have

$$\left| \alpha_{N,k}^2 - \alpha_k^2 \right| \leqslant \left(\int_a^b dt \int_a^b ds \left| \epsilon_N(t,s) \right|^2 \right)^{1/2} . \tag{1.21}$$

Therefore if K_o is the number of degrees of freedom in the problem with continuous data, we can look for a distribution of the data points which allows a good approximation of the singular values α_k up to $k = K_o$. If K_o is small, it is reasonable to argue that, using a small number of data points, conveniently placed, it is possible to obtain such a good approximation.

We conclude by remarking that the filtered solution of the problem with discrete data is

$$\tilde{f}_N(t) = \sum_{k=0}^{K_o} \frac{g_{N,k}}{\alpha_{N,k}} u_{N,k}(t) \tag{1.22}$$

where

$$g_{N,k} = \sum_{n=1}^{N} w_n \, g(p_n) v_{N,k}^*(p_n) . \tag{1.23}$$

In the last equation $v_{N,k}(p_n)$ denotes the n-th component of the singular vector $v_{N,k}$.

2. THE FINITE LAPLACE TRANSFORM INVERSION

An example of the applications of the method presented in Sect. 1 is the inversion of the finite Laplace transform which, as shown in the Part I of this paper, is important, for example, in the problem of polydispersity analysis in quasi-elastic light scattering.

Given a function $f \in L^2(a,b)$ $(0 < a < b < +\infty)$, we define the <u>finite Laplace transform</u> of f as

$$g(p) = \int_a^b e^{-pt} f(t) dt \quad , \quad 0 \leqslant p < + \infty \; . \tag{2.1}$$

The function $g(p)$ is square integrable and therefore eq.(2.1) defines an operator K from $L^2(a,b)$ into $L^2(0,+\infty)$. It is easy to verify that the kernel $K(p,t) = \exp(-pt)$ satisfies the condition (1.8) and therefore K is compact.

2.1. Laplace inversion with continuous data

As a consequence of the scaling properties of the Laplace transform the singular values of the operator K depend on the upper bound, b, and on the lower bound, a, of the support of f through the ratio

$$\gamma = b/a \quad . \tag{2.2}$$

Therefore we find convenient to take as a support of f the interval $[1,\gamma]$ and to define the operator K as follows

$$(Kf)(p) = \int_1^\gamma e^{-pt} f(t) dt \quad . \tag{2.3}$$

The singular values α_k and the singular functions $u_k(t)$ are given by the solutions of the eigenvalue problem

$$\int_1^\gamma T(t+s) u_k(s) ds = \alpha_k^2 u_k(t) \tag{2.4}$$

where

$$T(t) = \frac{1}{t} \; . \tag{2.5}$$

Then the singular functions $v_k(p)$ are given by the finite Laplace transforms of the singular functions $u_k(t)$:

$$v_k(p) = \frac{1}{\alpha_k} \int_1^\gamma e^{-pt} u_k(t) dt. \tag{2.6}$$

All the eigenvalues α_k^2 have multiplicity one (Bertero et al., 1982) and therefore they can be ordered in a strictly decreasing sequence. They tend to zero exponentially fast as follows from general results on the asymptotic behaviour of the eigenvalues of integral operators (Hille & Tamarkin, 1931). As a consequence, for moderate values of γ, very few singular values are significant. For instance, in the case $\gamma = 5$, we have (Bertero et al., 1982):

$$\alpha_0 = 8.751 \times 10^{-1}, \; \alpha_1 = 1.935 \times 10^{-1}, \; \alpha_2 = 3.827 \times 10^{-2} \qquad (2.7)$$

$$\alpha_3 = 7.343 \times 10^{-3}, \; \alpha_4 = 1.435 \times 10^{-3}, \; \alpha_5 = 2.765 \times 10^{-4}, \ldots$$

Accurate computations of the singular values α_k and of the singular functions u_k can be performed along these lines (Bertero et al., 1983): i) approximation of the kernel $T(t+s)$ by means of a tensor product of splines (Hämmerlin & Schumaker, 1980); the approximating kernel $\tilde{T}(t,s)$ is of finite rank and its eigenvalues coincide with the eigenvalues of a non-symmetric matrix Q - ii) the largest eigenvalues of Q are computed by a combination of the power method and of the deflation method (Ralston, 1965); this procedure works very efficiently when the eigenvalues are well separated as in our case - iii) once the eigenvalues have been computed, the corresponding eigenvectors of Q are determined by means of the inverse power method (Wilkinson, 1965) - iv) finally the eigenfunctions $u_k(t)$ are computed as linear combinations of splines, the coefficients being provided by the components of the eigenvectors of Q.

It was confirmed by these computations, that the singular functions $u_k(t)$ have the property mentioned at the end of Sect.1.1, namely $u_k(t)$ has exactly k zeros in the interval $[1,\gamma]$. Furthermore the distribution of the zeros is approximately geometric and therefore one can define a resolution ratio δ as in eq.(1.7).

2.2. Laplace inversion with discrete data

The analysis of Sect.1.3 can be immediately applied to the problem of the finite Laplace transform inversion and, due to the simplicity of the kernel, some computations can be done analytically. For instance the matrix $t_{nm}^{(N)}$ of eq.(1.18) takes the very simple form

$$t_{nm}^{(N)} = \frac{w_m}{p_n + p_m} \left\{ \exp\left[-(p_n + p_m)\right] - \exp\left[-\gamma(p_n + p_m)\right] \right\}. \qquad (2.8)$$

In the case of equidistant data points, $p_n = (n-1)\sigma, (n=1,\ldots,N)$, we can take $w_n = \sigma$ so that the matrix (2.8) becomes

$$t_{nm}^{(N)} = \frac{1}{n+m-2}\left\{\exp\left[-\sigma(n+m-2)\right] - \exp\left[-\sigma\gamma(n+m-2)\right]\right\} \qquad (2.9)$$

(remark that $t_{11}^{(N)} = \sigma(\gamma-1)$). The largest eigenvalues and the corresponding eigenvectors of this matrix can be easily compu-ted by means of a combination of the power method and of the Hotelling deflation method (Ralston, 1965), since the matrix is symmetric. When we have the eigenvalues $\alpha_{N,k}^2$ and the corre-sponding singular vectors $v_{N,k}$, the singular functions $u_{N,k}(t)$ can be obtained by means of the relation

$$u_{N,k}(t) = \frac{\sigma}{\alpha_{N,k}} \sum_{n=1}^{N} v_{N,k}(p_n)e^{-p_n t} , \qquad (2.10)$$

where $v_{N,k}(p_n)$ denotes the n-th component of $v_{N,k}$.

The singular values $\alpha_{N,k}$ can provide a good approximation of the singular values α_k of the problem with continuous data. The kernel of eq.(1.20) is given by $\epsilon_N(t,s) = \epsilon_N(t+s)$ where

$$\epsilon_N(t) = \frac{1}{t} - \sigma\frac{1 - \exp(-N\sigma t)}{1 - \exp(-\sigma t)} \qquad (2.11)$$

and this function tends to zero uniformly over the interval $2 \leq t \leq 2\gamma$ when $\sigma \to 0$ and $N \to +\infty$. Therefore from eq.(1.21) it is expected that $\alpha_{N,k} \sim \alpha_k$ if one takes a sufficiently large number of data points at a sufficiently small distance σ. The following singular values

$$\alpha_{N,0}=9.075 \times 10^{-1}, \alpha_{N,1}=1.983 \times 10^{-1}, \alpha_{N,2}=3,555 \times 10^{-2} \qquad (2.12)$$

have been obtained in the case $\gamma = 5$ using $N = 64$ and $\sigma = 0.03$ (Bertero et. al., 1983). If one compares with the singular values of the continuous case, eq.(2.7), one finds an average error of about 4%. This is the best approximation which can be obtained using 64 data points.

3. THE EFFECT OF GEOMETRIC SAMPLING OF DATA

In the case of Laplace inversion the use of equidistant points is not the best one. Such a choice is only convenient for deconvolution problems, i.e. for the solution of problems which have simple transformation properties with respect to the tran-slation group. In the case of Laplace transform inversion the underlying group is the dilation group and for this reason a geometric distribution of data points seems to be more natural, at least from the theoretical point of view.

Therefore we consider now the following data-point distribution

$$p_n = p_1 \triangle^{n-1} \quad ; \quad n = 1, \ldots, N \qquad (3.1)$$

where p_1 and $\triangle$ can be considered as free parameters. As concerns the choice of the weights w_n, the following result can be proved (Bertero et al., 1983): if the p_n are as in eq.(3.1) and if

$$w_n = (\ln \triangle) \, p_n \qquad (3.2)$$

then, given ε, it is possible to choose $\triangle$ and N so that, for any $t \in [2, 2\gamma]$

$$\left| \epsilon_N(t) \right| = \left| \frac{1}{t} - \sum_{n=1}^{N} w_n e^{-p_n t} \right| \leqslant \varepsilon . \qquad (3.3)$$

This result, combined with eqs.(1.17),(1.20), suggests that we introduce the weights (3.2) into the scalar product (1.15).

As concerns the computation of the eigenvalues and eigenvectors of the matrix (2.8) we remark that it is not symmetric in the usual sense ($t_{nm}^{(N)} \neq t_{mn}^{(N)}$) since it is self-adjoint with respect to the weighted scalar product (1.15). Then, for computational convenience, we transform the matrix (2.8) into a symmetric one having the same eigenvalues

$$t_{nm}^{(N)} = \frac{\sqrt{w_n w_m}}{p_n + p_m} \left\{ \exp\left[-(p_n + p_m) \right] - \exp\left[-\gamma (p_n + p_m) \right] \right\} . \qquad (3.4)$$

We also remark that if $\bar{v}_{N,k}$ is an eigenvector of this matrix, the corresponding eigenvector of the matrix (2.8) is given by

$$v_{N,k}(p_n) = w_n^{-1/2} \, \bar{v}_{N,k}(p_n) . \qquad (3.5)$$

Therefore singular values and singular vectors can be computed again by a combination of the power method and of the Hotelling deflation method.

The fundamental property of geometric sampling of data is that, using a very small number of data points, it is possible to obtain extraordinarily good approximations of the singular values of the problem with continuous data (and also of the singular functions). For instance, again for the case $\gamma = 5$, using 5 data points, $\Delta = 5.5$ and $p_1 = 1.366 \times 10^{-3}$ we have obtained the following singular values (Bertero et al., 1983)

$$\alpha_{N,0} = 8.742 \times 10^{-1}, \; \alpha_{N,1} = 1.913 \times 10^{-1}, \; \alpha_{N,2} = 3.827 \times 10^{-2}. \qquad (3.6)$$

If we compare with the singular values of the problem with continuous data, we find an average error of about 1/2% . Also the singular functions $u_{N,k}(t)$ (k=0,1,2) provide an excellent approximation of the singular functions $u_k(t)$ (k=0,1,2) corresponding to the problem with continuous data.

Now, if we notice that, when the "signal-to-noise ratio" is equal to 100, the "number of degrees of freedom" in the case $\gamma=5$ is just 3, we can conclude that using 5 data points optimally placed, the filtered solution (1.22) practically coincides with the filtered solution (1.13), the main error deriving from the different noise contribution to the "image" components g_k and $g_{N,k}$. Therefore 5 data points allow almost the same resolution as continuous data.

We believe that the previous result can be extended to any severely ill-posed problem (namely a problem where the "number of degrees of freedom" is small): a number of data points, optimally placed, approximately equal to the "number of degrees of freedom" is practically equivalent to continuous data.

R E F E R E N C E S

Bertero, M., De Mol, C. and Viano G.A., 1980, Inverse Scattering Problems in Optics, ed. H.P. Baltes, Topics in Current Physics,Vol.20 (Berlin: Springer), pp.161-214.

Bertero, M., Boccacci, P. and Pike, E.R., 1982, Proc. R. Soc. Lond. A383, pp. 15-29.

Bertero, M., Brianzi, P. and Pike, E.R., 1983 (in preparation).

Hämmerlin, G. and Schumaker, L.L., 1980, Numer. Math. 34, pp.125-141.

Hille, E. and Tamarkin, J.D., 1931, Acta Math. 57, pp.1-76.

Picard, E., 1910, R.C. Mat. Palermo 29, pp. 615-619.

Ralston, A., 1965, A First Course in Numerical Analysis (New York: McGraw-Hill)

Riesz, F. and Sz. Nagy, B., 1955, Functional Analysis, (New York: Ungar).

Twomey, S., 1965, Jour. Franklin Inst. 279, pp.95-109.

Wilkinson, J.H., 1965, The Algebraic Eigenvalue Problem (Oxford: Claredon Press).

II.6 (OI.3)

OBJECT RECONSTRUCTION FROM PARTIAL INFORMATION

M.A. Fiddy

Physics Department, Queen Elizabeth College,
University of London, Campden Hill Road,
Kensington, London W8 7AH, UK.

ABSTRACT. The first Born and Rytov approximations are briefly described and compared. If they provide a suitable model, limited
data on the Fourier transform of the object can be calculated.
Optimum use of these data through the incorporation of prior knowledge is described. The problem when only intensity data are available is also discussed.

1. INTRODUCTION

    There are two approaches to determining the internal structure
of objects from scattered-field data. The first relies on a geo-
metrical-optics description which requires the assumption that the
wavelength used is considerably smaller than the structural detail
of interest. This assumption is adequate at X-ray wavelengths, for
example, when used in computerised tomography. When the wavelength
is not relatively small, scattering and diffraction effects become
important and object information, e.g. a scattering potential, is
approximately reconstructed by some method of inverse scattering.
These methods usually rely on the assumption that the scattering
is weak, and have become known as backpropagation algorithms for
diffraction tomography [e.g. 1]. Inversion formulae based upon the
first Born or Rytov approximations [2,3,4] are essentially Fourier-
reconstruction techniques and the questions of uniqueness and stability of the estimate for the object obtained in this way have received much attention [e.g. 5]. We briefly describe here the nature
of these approximations (§ 2), the inversion problem when limited
complex Fourier data are available (§ 3) and the phase retrieval
problem when only limited intensity data are available.

W.-M. Boerner et al. (eds.), Inverse Methods in Electromagnetic Imaging - Part 1, 341–349.
© *1985 by D. Reidel Publishing Company.*

2. THE BORN AND RYTOV APPROXIMATIONS

A solution to the wave equation is difficult to calculate if a small perturbation approximation is not taken. Consider the scaler wave equation

$$[\nabla^2 + k^2(1 + \delta n)^2] U = 0 \tag{1}$$

where k is the average wave-number and δn the fluctuation of, say, the refractive index about the mean. It is this function, δn, the function describing the object, that we wish to reconstruct. It is usual [e.g. 6] to choose a series solution for U in terms of an ordering parameter ε having either of the forms

$$U = \sum_{j=0}^{\infty} \varepsilon^j U_j = \exp(\sum_{j=0}^{\infty} \varepsilon^j \Psi_j) \tag{2}$$

Converting equation (1) into an integral equation and using the Liouville-Neumann method of successive substitutions gives [7]

$$U = U_o + U_s$$

$$= U_o + U_{s1} + U_{s2} + \ldots$$

$$= U_o + k^2 \int_V (2 \delta n + \delta n^2) G \, U_o \, dr + k^2 \int_V (2 \delta n + \delta n^2) G \, U_{s1} \, dr + \ldots \tag{3}$$

where U_o is the incident field, U_s the scattered field and V the scatterer volume, and the Green function G is defined by

$$[\nabla^2 + k^2] G = -4\pi \delta (r) \tag{4}$$

If we write δn as $\varepsilon \delta n$ and U_{sj} as $\varepsilon^j U_j$ we find that

$$U_1 = k^2 \int_V 2 \delta n \, G \, U_o \, dr$$

$$U_2 = k^2 \int_V \delta n^2 \, G \, U_o \, dr + k^2 \int_V 2 \delta n \int_V 2 \delta n \, G \, U_o \, dr \, G \, dr \text{ etc. (5)}$$

upon equating equal powers of ε; U_1 rather than U_{s1} is frequently taken to be the first Born approximation to U_s. From equation (2) it is straightforward to show that [8]

$$U_o = e^{\Psi_o} \qquad\qquad \Psi_o = \ln U_o$$

$$U_1 = U_o \Psi_1 \qquad\qquad \Psi_1 = U_1/U_o$$

$$U_2 = U_o(\Psi_1^2/2 + \Psi_2) \qquad \Psi_2 = U_2/U_o - \tfrac{1}{2}(U_1/U_o)^2 \tag{6}$$

and, obviously, a sufficient condition for the convergence of the series in equation 2 is that successive terms get smaller sufficiently quickly.

If one represents the scattered field U, with unwrapped phase, by $U = \exp(\Psi)$, where Ψ is a complex phase, then equation (1) gives

$$\nabla^2 \Psi + |\nabla \Psi|^2 + k^2 (1 + \delta n)^2 = 0 \tag{7}$$

which has the solution [e.g. 6,9]

$$\Psi = \Psi_o + \Psi_s$$

$$= \Psi_o + \frac{k^2}{U_o} \int_V (2 \delta n + \delta n^2) G \, U_o \, dr + \frac{1}{U_o} \int_V |\nabla \Psi_s|^2 G \, U_o \, dr \tag{8}$$

The assumption that $k^2 (2 \delta n + \delta n^2) \gg |\nabla \Psi_s|^2$ is the Rytov approximation which can be seen to be equivalent to taking, as an approximation to Ψ, the expression $\Psi_o + \Psi_1$ (from equation (6) if the term in δn^2 is neglected). The complex phase Ψ within the Rytov approximation is simply related, therefore, to the first Born approximation to U_s. Formally, the neglect of the δn^2 term gives a Rytov solution which is equivalent to the first order perturbation solution, as was pointed out by Fried [9]. While this mathematical equivalence has been emphasised (in particular by Devaney [1]) the validity of the two approximations for U_s depends upon the terms neglected. If one considers an exponentiated Born series as in equation (2), one might deduce that the Born and Rytov approximations have the same domain of validity by comparing the second order terms [10]. However, one could obtain an iterative solution by substituting Ψ_1 for Ψ_s in equation (8) as suggested by Ishimaru [6]. If we again write δn as $\varepsilon \delta n$ and make this substitution in equation (8), we obtain a sequence of Ψ_j's ordered by powers of ε which bear no simple relation to the Ψ_j's in equations (2) and (6), the two expressions for Ψ_2 being of similar magnitude only if

$$\int_V |\nabla \Psi_1|^2 G \, U_o \, dr \approx \int_V 2 \delta n \, U_o \, \Psi_1 \, G \, dr - \frac{\Psi_1^2}{2 U_o} \tag{9}$$

One could expect that Rytov will be preferable when fluctuations in δn are large compared to the wavelength and for larger distances of propagation [9].

In either the first Born or Rytov approximations, it follows [11, 2] that the Fourier transform of the scattered field U_1 (Born) or $U_o \Psi_1$ (Rytov) over any plane normal to the incident field direction allows one to determine δn, the Fourier transform of δn, on the surface of an Ewald sphere of radius k and whose centre is determined by the incident field direction. By varying the angle of incidence one can calculate δn on sets of semi-circular arcs and, when data are sufficiently dense, inverse transform. Alternatively, rather than interpolating in Fourier space, for each angle of incidence one can propagate either U_1 or $U_0 \Psi_1$ back into object space using the "inverse" Green function (assuming no evanescent waves) and superpose [1].

Three problems can arise with these procedures. The first concerns the consequences of having an object δn which does not conform to the requirements for the validity of the Born or Rytov approximations. There is much to be done here by comparison with exact solutions or data from known objects. The second concerns the consequences of having only limited incident angles and limited samples of the scattered field. Thirdly, the required phase unwrapping of U in order to make use of the Rytov approximation is non-trivial.

3. LIMITED COMPLEX DATA AVAILABLE

The restoration or reconstruction of an object when only limited (incomplete) discrete data are available is a common problem with any radiation. Such problems are known to be ill-posed but regularisation theory defines a set of admissible or approximate estimates for the object by imposing constraints on these estimates [5]. Even in the absence of noise, incomplete sampled data means that there are infinitely many object estimates consistent with the Fourier data and the object support, if known. This ambiguity is overcome by using prior knowledge to reduce the class of allowed solutions and to design an optimality criterion. For finite data it is always necessary to use models to find an optimum estimate $\tilde{\delta n}$ for δn, and a simple Fourier series expression may not be the best. A commonly used and quite successful technique for interpolating or extrapolating Fourier data is the iterative procedure of Gerchberg-Papoulis [12]. By repeatedly Fourier transforming between object and data spaces, imposing the anticipated object support and measured data values at each iteration, convergence is to a minimum norm solution estimate. This solution is the orthogonal projection onto the subspace of appropriately space limited and data consistent object functions. The numerical performance of such an algorithm will depend upon the noise but can be regularised and the corresponding error assessed [13]. We let the function p encode the support of the object and perhaps some profile information also. One can choose an optimum estimator for δn which incorporates this. Suppose

$$\widehat{\delta n}(K_j) = \int_V \delta n(r) e^{-iK_j \cdot r} d^3 r$$

$$= (\delta n, g_j) \tag{10}$$

If we consider H, the weighted L_2 space with inner product

$$(f,g)_H = \int_V p^{-1} f g^* d^3 r \qquad (11)$$

where p is real and positive and * denotes complex conjugate then

$$\hat{\delta n}(K_j) = (\delta n, pg_j)_H \qquad (12)$$

The object estimate incorporating p and given by

$$\tilde{\delta n} = p \sum_j a_j g_j \qquad (13)$$

minimises $\| \delta n - \tilde{\delta n} \|_H$ and is a unique estimate of minimum norm when the a_j found by solving the linear equations [14]

$$\hat{\delta n}(K_j) = \sum_m a_m (pg_m, pg_j)_H = \sum_m a_m \hat{p}(K_j - K_m) \qquad (14)$$

The $\tilde{\delta n}$ thus obtained will have features driven by p and will be data consistent wherever the data $\hat{\delta n}(K_j)$ are located. If p is a binary function encoding only the object support then equation (13) gives the closed form solution of the Gerchberg-Papoulis iterative method. Clearly, if p is incorrect or inappropriate a data consistent $\tilde{\delta n}$ will still be found. Also, any broadband noise on the data can cause spurious oscillations in $\tilde{\delta n}$ because of the requirement that it be zero outside the object support. However, a more realistic function p would take on a small value outside the expected object support and this ensures stability. This modification to p has been shown to be equivalent to a form of Miller regularisation [15].

Examples of the application of this technique to limited angle limited sampled Fourier data are shown in [15]. Extensions of this approach to provide a non-linear estimator for which Burg's maximum entropy method is a special case can be found in [16].

4.   LIMITED INTENSITY DATA AVAILABLE

If only the intensity of the field U or the scattering amplitude F, can be measured then its phase must be found; we assume in this section that some kind of holographic phase recording is not possible. The phase retrieval problem has been the subject of several recent reviews [e.g. 17,18] and can be conveniently divided into one dimensional and more than one dimensional problems. It is well known that in one dimension many equally acceptable solutions for the phase could occur even with complete and noise free data. The reason for this is that if U or F is a bandlimited function (entire function of exponential type) it can be specified by its zero locations like a polynomial (the Hadamard product). The intensity data has the zeros of U and, in conjugate positions those of

U^* leading to a zero symmetry about the real axis. If U had M complex zeros then 2^M possible phase functions could be constructed consistent with the intensity data. In higher dimensions a product formation still exists for U but now it is a product of irreducible factors (the Osgood product) having from one to an infinity of factors [18].

Clearly, with finite data and arbitrary phases it is always possible to find an infinity of δn having any support and consistent with the modulus samples of the scattering amplitude. A technique such as that described in §3 would do this. By enforcing, with the weighting function p (equation (13)), that δn should be identically zero outside some support, an extrapolation of the available data is realised regardless of the phases associated with that data. If the support information encoded by p is correct, and in general it cannot be deduced from the intensity data [19], still no constraints can be brought to bear on the choice of phases to associate with the data.

The problem becomes somewhat more tangible if we do not impose the constraint that δn is zero or very small outside its support. If a discrete model is adopted for δn of the form

$$F(x,y,z) = \sum_{j,k,l}^{M-1} \delta n_{jkl} e^{ijx} e^{iky} e^{ilz} \tag{15}$$

we can write

$$F(u,v,w) = \sum_{j,k,l}^{M-1} \delta n_{nkl} u^j v^k w^l \tag{16}$$

which is a polynomial of finite degree. We wish to find M^3 object samples δn_{jkl} from at least this number of samples of F or $|F|^2$, with some $\delta n_{jkl} = 0$ defining the object support.

In this form, it has been shown [20] that for the set of multi-variable polynomials of finite degree the set of reducible polynominals is of measure zero. This implies that F is very likely to be a single irreducible polynomial and hence that a unique phase can be associated with $|F|$. To be certain of phase uniqueness for this model there is the following necessary condition. If the non-zero coefficients δn_{jkl} of equation (16) are replaced by unity and the corresponding polynomial [F] known as the form of F [21] is irreducible then F is irreducible. In terms of objects and their scattered fields this means that if the scattering amplitude from an aperture or uniform volume is irreducible then it remains so whatever δn is introduced into that support. Actually demonstrating irreducibility is non-trivial but a suffient condition was recently suggested based on Eisenstein's criterion [22]. The constraints on the object for this criterion to hold are that two points A and B in a plane are non zero, the object support lying anywhere within the box shown in the figure. Point A may be anywhere to the right of the location shown and when sufficiently far would result in $|F|^2$ being an off-axis hologram from which δn would be found.

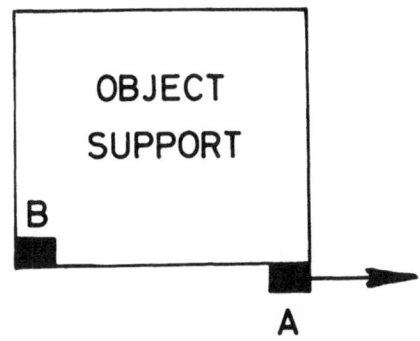

Ensuring that there is a unique phase in this way does not help us find it. An algorithm such as Fienup's [23] which repeatedly Fourier transforms between spaces, imposing the object support and intensity data each iteration, is moderately successful. This is particularly the case for object supports lacking symmetry which may relate to the comments made above. However the method only succeeds when Eisenstein's criterion is satisfied when the reference point A is large; this corresponds to an almost holographic situation, Fienup's algorithm converging because a simple initial phase guess is close to the actual phase.

Thus apart from making additional and different measurements of some kind [e.g. 18], knowing that the phase is unique does not offer any obvious way for determining it.

5. CONCLUSIONS

This paper was concerned with the problem of object reconstruction from partial information and three main aspects of this problem were discussed.

(i)   Inverse scattering modelled by the first Born or Rytov approximations is computationally feasible provided the phase unwrapping can be reliably carried out for Rytov. However, conditions for the validity of these approximations and their accuracy remain vague.

(ii)   The first Born or Rytov approximations can be regarded as requiring a Fourier inversion for object reconstruction. The difficulties associated with having sampled limited (angle) data can be reduced by constructing an estimator for the object, which can incorporate whatever prior knowledge about it is available.

(iii)   If only single intensity data sets are available, prior knowledge about the object such as its support can be incorporated but appears unhelpful unless a discrete model for the object is adopted. A unique phase can then be assured but finding that phase appears to be an open problem.

6. ACKNOWLEDGEMENTS

The author would like to acknowledge the support of NATO (in particular NATO grant 0556/82) and the U.K. S.E.R.C. He would also like to acknowledge his collaboration on the work presented here with C.L. Byrne, A.M. Darling, J.C. Dainty, S. Leeman and L. Zapalowski.

7. REFERENCES

1. Devaney, A.J., 'A filtered backpropagation algorithm for diffraction tomography', Ultrasonic Imaging, 4, 336-350 (1982).
2. Devaney, A.J., 'Inverse scattering theory within the Rytov approximation', Opt. Letters, 6, 374-376 (1981).
3. Devaney, A.J., 'Inversion formula for inverse scattering within the Born approximation', Opt. Letters, 7, 111-112 (1982).
4. Kaveh, M., Soumekh, M. Lu, Z.Q., Mueller, R.K. and Greenleaf, J.F., 'Further results on diffraction tomography using Rytov's approximation', Acoustical Imaging, 12, Eds. E.A. Ash and C.R. Hill, 599-608 (1982).
5. Bertero, M., De Mol, C. and Viano, G.A., 'The stability of inverse problems', ch.5 of Inverse Scattering Problems in Optics, Ed. H.P. Baltes, 20, Topics in Current Physics, Springer-Verlag, Berlin, 161-214 (1980).
6. Ishimaru, A., 'Wave propagation and scattering in random media', Vol. II, ch. 17, Academic Press, New York (1978).
7. Morse, P. and Feshbach, H., 'Methods of Theoretical Physics', Vol. II, McGraw-Hill, New York (1953).
8. Sancer, M.I. and Varvatsis, A.D. 'A comparison of the Born and Rytov methods', Proc. IEEE, 58, 140-141 (1970).
9. Fried, D.L., 'Diffusion analysis for the propagation of mutual coherence', J. Opt. Soc. Amer., 58, 961-969 (1968).
10. Yura, H.T., Sung, C.C., Clifford, S.F. and Hill R.J., 'Second-order Rytov approximation', J. Opt. Soc. Amer., 73, 500-502 (1983).
11. Wolf, E., 'Three-dimensional structure of semi-transparent objects from holographic data', Opt. Communications, 1, 153-156, (1969).
12. Papoulis, A., 'A new algorithm in spectral analysis and band-limited extrapolation', IEEE CAS-22, 735-742 (1975).
13. Abbiss, J.B., Defrise, M., De Mol, C. and Dhadwal, H.S., 'Regularised iterative and non-iterative procedures for object restoration in the presence of noise: an error analysis', J. Opt. Soc. Amer., 73, November issue (1983).
14. Hall, T.J., Darling, A.M. and Fiddy, M.A., 'Image compression and restoration incorporating prior knowledge', Opt. Letters, 7, 467-468 (1982).
15. Darling, A.M., Hall, T.J. and Fiddy, M.A., 'Stable, non-iterative, object reconstruction from incomplete data using

prior knowledge', J. Opt. Soc. Amer., 73, November issue (1983).

16. Byrne, C.L., Fitzgerald, R.M. Fiddy, M.A., Hall, T.J. and Darling, A.M., 'Image restoration and resolution enhancement', J. Opt. Soc. Amer., 73, November issue (1983).

17. Taylor, L.S., 'The phase retrieval problem', IEEE AP-29, 406-416 (1981).

18. Fiddy, M.A., 'The phase retrieval problem', SPIE 413, Inverse Optics, Ed. A.J. Devaney, paper 26 (1983).

19. Fienup, J.R., Crimmins, T.R. and Holsztynski, W., 'Reconstruction of the support of an object from the support of its autocorrelation', J. Opt. Soc. Amer., 72, 610-624 (1982).

20. Hayes, M.H. and McCllean, J.H., 'Reducible polynomials in more than one variable', Proc. IEEE, 70, 197-198 (1982).

21. Dainty, J.C. and Fiddy, M.A., 'The essential role of prior knowledge in phase retrieval', submitted for publication.

22. Fiddy, M.A., Brames, B.J. and Dainty, J.C., 'Enforcing irreducibility for phase retrieval in two dimensions', Opt. Letters 8, 96-98 (1983).

23. Fienup, J.R., 'Phase retrieval algorithms: a comparison', App. Optics, 21, 2758-2769 (1982).

II.7 (NM.5)

APPROXIMATION OF IMPULSE RESPONSE FROM TIME
LIMITED INPUT AND OUTPUT: THEORY AND EXPERIMENT

Tapan K. Sarkar[1]
Soheil A. Dianat[1]
Bruce Z. Hollmann[2]

Abstract.   Since it is practically difficult to generate and propagate an impulse, often a system is excited by a narrow time domain pulse. The output is recorded and then a numerical deconvolution is often done to extract the impulse response of the object. Classically, the fast Fourier transform technique has been applied with much success to the above deconvolution problem. However, when the signal to noise ratio becomes small, sometimes one encounters instability with the FFT approach. In this paper, the method of conjugate gradient is applied to the deconvolution problem. Unlike the FFT, this approach solves the deconvolution problem entirely in the time domain with any desired degree of accuracy. The conjugate gradient method has the advantage of both a direct method (i.e., the solution is obtained in a finite number of steps) and that of an iterative method (the round-off and truncation are generally limited to the last stage of iteration). Moreover, the method converges for any initial guess. Also for the application of this method the time samples need not be uniform as in the FFT. In addition, the dc component can be recovered with this method, but not with FFT. Computed impulse responses utilizing this technique are presented for measured incident and scattered fields from a sphere and a cylinder.

[1] Tapan K. Sarkar
[1] Soheil A. Dianat
  Department of Electrical Engineering
  Rochester Institute of Technology
  Rochester, New York 14623 U.S.A

[2] Bruce Z. Hollmann
  Naval Surface Weapons Center
  Dahlgren, Virginia 22448 U.S.A.

This work has been supported in part by the U.S. Office of Naval Research at Rochester Institute of Technology, and by the U.S. Naval Air Systems Command at the Naval Surface Weapons Center.

351

W.-M. Boerner et al. (eds.), Inverse Methods in Electromagnetic Imaging – Part 1, 351–364.
© *1985 by D. Reidel Publishing Company.*

1. Introduction

Consider an electromagnetic system which is excited by a causal time domain wave-form x(t), and the recorded output is y(t). Then we know they are related by the convolution integral

$$y(t) = \int_0^\infty x(t - \tau)h(\tau)d\tau = \int_0^t x(t - \tau)h(\tau)d\tau, \qquad (1)$$

where $h(\tau)$ is the impulse response of the system.

We could write (1) as an operator equation

$$\hat{A} h(t) = y(t) \qquad (2)$$

where $\hat{A}$ is the convolution operator $\int_0^\infty x(t - \tau)(\cdot)d\tau$. Even though (1) is a very simple integral equation, the numerical solution of (1) is an extremely difficult problem. The difficulty is that the solution h(t) does not depend continuously on the data.

To illustrate this point, consider two solutions to (1)

$$h_1(t) = h(t) \Rightarrow \text{exact solution} \qquad (3)$$

and

$$h_2(t) = h(t) + B \cos \omega t \qquad (4)$$

where B and ω are two arbitrary constants. Now observe that

$$|y_2(t) - y_1(t)| = |B \int_0^t x(t - \tau)\cos \omega\tau d\tau| \qquad (5)$$

As an example, consider the causal input to be

$$x(t) = e^{-t} \quad \text{for } t > 0 \qquad (6)$$

By utilizing (6) in (5) we obtain

$$|y_2(t) - y_1(t)| = \left| B \cdot \frac{\omega \sin \omega t + \cos \omega t - e^{-t}}{\omega^2 + 1} \right| \leq |B| \cdot \left[\frac{2 + \omega}{\omega^2 + 1} \right] \tag{7}$$

Equation (7) tells us that for a fixed B, we can choose ω large enough so that the absolute difference between $y_2(t)$ and $y_1(t)$ would be very small. Since we do computation with a finite number of digits, the difference in $y_2(t) - y_1(t)$ can be interpreted as the digital representation of the exact output. Therefore, we could have two inpulse responses $h_1(t)$ and $h_2(t)$ which are radically different and yet produce the same output $y(t)$ when convolved with $x(t)$. It is important to point out that the problem is not in the digital representation of the analog signal nor in the numerical technique used to obtain the solution. The problem is inherent in equation (2). Operator equation (2) belongs to the class of problems defined as ill-posed problems [1-3]. What the problem actually is, is that in (2) the operator does not have an inverse. We now give a mathematical proof for the cause of instability of equation (1) for any input $x(t)$.

Since $\hat{A}h = y$, symbollically we can write the solution

$$h(t) = \hat{A}^{-1} y(t) \tag{8}$$

However, if $\hat{A}^{-1}$ does not exist, then of course (8) has no meaning. We now explain mathematically that $\hat{A}^{-1}$ does not exist. Since $x(t)$ is the input to the system and also as the input is of finite energy, $x(t)$ is square integrable, i.e.

$$\int_0^T x^2(t)dt < \infty \tag{9}$$

In Stakgold's notation [4], $x(t)$ is the kernel of the integral operator $\hat{A}$. Since the kernel is square integrable, the operator is Hilbert Schmidt [p. 300, 4]. It is shown [p. 336, 4] that for a Hilbert Schmidt operator, the smallest eigenvalue approaches zero. In other words, zero is the accumulation point of the smaller eigenvalues of $\hat{A}$ and hence $\hat{A}^{-1}$ does not exist. This is not of any solace to a person who is interested in obtaining a solution to (1).

In summary, even though theoretically $\hat{A}^{-1}$ does not exist, numerically we can still find a solution and now we have to be very careful in what we mean by a solution. We explain in the forthcoming section how we obtain a solution and what is our definition of the solution. It is important to stress that for the deconvolution problem, the concept of a solution is subjective.

2. Historical Perspective

Interestingly enough, the convolution type integral equation arises in almost all areas of engineering and researchers in electrical engineering have been solving this problem for quite a few years. Yet there is no clearcut solution to this problem for reasons mentioned above.

A conventional way of solving (1) is to transform it to the frequency domain by the application of the Fast Fourier Transform. So we obtain:

$$H(j\omega) = \frac{Y(j\omega)}{X(j\omega)} \tag{10}$$

and then obtain h(t) by the inverse Fourier Transform

$$h(t) = \frac{1}{2\pi} \int_{-\infty}^{\infty} H(j\omega)e^{j\omega t}d\omega \tag{11}$$

Even when there is no noise in the data, the above approach for deconvolution of waveforms arising in electromagnetic systems runs into problems for three specific reasons. The reasons, as observed by other researchers [5-7], are as follows.

(a)   It is quite possible that at a particular frequency ω, $X(j\omega)$ becomes zero in (10) and $H(j\omega)$ blows up.

(b)   Often in electromagnetic systems, h(t) has an impulse at t = 0. The conventional z-transform techniques cannot handle this jump discontinuity at t = 0. Special care has to be taken to extract this impulse at the origin. Even though methods exist to do this type of extraction, they are problem dependent and sensitive to signal to noise ratio.

(c)   In most cases, the transient responses are time limited and, hence, by definition they are not band limited. Therefore, if one obtains the transform $X(j\omega)$ and $Y(j\omega)$ without paying proper attention to the sampling frequency, aliasing could be a severe problem.

It is for these reasons that researchers, for deconvolving transient waveforms, have applied techniques which are entirely in the time domain, i.e. the deconvolution is done without going to the frequeency domain [7-10]. In this paper we apply the conjugate gradient method to solve the deconvolution problem entirely in the time domain. Since we are solving the problem entirely in the time domain, we do not have to worry about the problem of aliasing. However, the problem about which we have to be careful is that the samples are close enough so we can approximate the time domain waveform by linear segments. The conjugate gradient method has been applied previously for the ill-posed image enhancement problems [11]. In [12] it has been shown that if the number of iterations in the conjugate gradient method is determined properly, then the undue oscillations in the solution do not appear.

Another advantage of the conjugate gradient method is that we always recover the impulse of the impulse response at the origin, since we do not deal with discrete transforms.

3. The Method of Conjugate Gradient

In this method, the basic philosphy is that instead of solving for $h = \hat{A}^{-1}y$ directly, we try to minimize the functional $F(h)$ given by

$$F(h) = <R, R> = <\hat{A}h - y, \hat{A}h - y> \tag{12}$$

where the inner product is defined as

$$<C,D> = \int_Z C(z)D(z)dz \tag{13}$$

and the norm of C is given by

$$\|C\|^2 = <C, C> = \int_Z |C^2(z)|dz \tag{14}$$

The conjugate gradient method starts with an initial guess h_0 and generates

$$P_0 = - b_0 \hat{A}^* R_0 = - b_0 \hat{A}(^*\hat{A}h_0 - y) \tag{15}$$

where $\hat{A}^*$ represents the adjoint operator. With reference to equation (1), the adjoint operator $\hat{A}^*$ is the advance convolution operator. We now define the adjoint operator as [4]

$$\hat{A}^*z = \int_0^\infty x(t - \tau)z(t)dt = \int_0^\infty x(t)z(t + \tau)dt \tag{16}$$

In defining (16) it has been assumed that $x(t)$ is causal. The conjugate gradient method then develops

$$h_{k+1} = h_k + \alpha_k P_k \tag{17}$$

$$R_{k+1} = R_k + \alpha_k \hat{A}P_k \tag{18}$$

$$\alpha_k = \frac{1}{\|\hat{A}P_k\|^2} \tag{19}$$

$$P_{k+1} = P_k - b_{k+1} \, \hat{A}^* R_{k+1} \tag{20}$$

and

$$b_k = \frac{1}{\| \hat{A}^* R_k \|^2} \tag{21}$$

Equations (15)-(21) are somewhat different from the conventional Hestenes and Steifel algorithms [13-14]. The above method brings about a certain saving in memory. This is because in [14] one needs to store the four vectors h_k, P_k, R_k and $\hat{A}P_k$ whereas in the modified algorithm it is not necessary to store $\hat{A}P_k$.

The method of conjugate gradient always converges for any initial guess and for any functional equations with a bounded operator as long as (15)-(21) are implemented. The actual rate of convergence is given in [15]. It is important to point out that even when y is not in the range of the operator $\hat{A}$ (this may happen when y is contaminated with noise), the method of conjugate gradient yields the minimum norm solution. In addition, it is shown in [15] that the method of conjugate gradient solves the normal equation

$$\hat{A}^* \hat{A} h = \hat{A}^* y \tag{22}$$

exactly.

The last question we address in this section is how to terminate the iteration. This is because, for ill-posed problems, too many iterations will minimize the error $\hat{A}h - y$ no doubt, but would introduce spurious oscillations in the solution. The reason for this is explained by the example of section 1. On the contrary, too few iterations may give a solution that is not good enough. The criterion that we utilize to stop the iterative process in the conjugate gradient method is that the magnitude of the largest elemtent in the residual must be below a certain value, i.e.

$$|R_n(t)| \leqslant M \text{ (a prefixed constant)} \tag{23}$$

Now we discuss the problem of choosing M.

For most problems of interest, the elements of the input and output are not sharply defined. This is because there is always some discretization or quantization error associated with each sample of x(t) and y(t). It is now assumed that all that is known about the input x(t) and output y(t) is that they are within the following intervals. Here the subscript e denotes exact quantities.

$$x_e(t) - \Delta x(t) \leqslant x(t) \leqslant x_e(t) + \Delta x(t) \tag{24}$$

$$y_e(t) - \Delta y(t) \leqslant y(t) \leqslant y_e(t) + \Delta y(t) \tag{25}$$

It is further assumed that $\Delta x(t_n)$ and $\Delta y(t_n)$ at the sampled values of the input and output are independent distinct quantities due to round-off error and, moreover, one does not depend on the other. Thus,

$$\int_0^T x_e(t - \tau)h_e(\tau)d\tau = y_e(t) \tag{26}$$

and

$$R(t) = \int_0^T x(t - \tau)h(\tau)d\tau - y(t) \tag{27}$$

Observe in (26) and (27) we have replaced the upper limit of ∞ in (1) by a finite value T. This is because in actual practice we can record the data only up to a finite time T. We further assume that

$$|\Delta x(t)| \leqslant C \text{ (constant)} \tag{28a}$$

and

$$|\Delta y(t)| \leqslant C \text{ (same constant as in (28a))} \tag{28b}$$

We then check at each interation whether

$$C\left[\int_0^T |h(\tau)|d\tau + 1\right] \geqslant |R(t_n)| \tag{29}$$

where t_n represents the sampling instances. As was shown by Oettli and Prager [16] the above inequality in (29) is both a necessary and sufficient condition for $h(t)$ to be a solution of (26) under (24) and (25). So for the conjugate gradient method, where the residuals are computed routinely, if for a certain residual $R(t)$ the inequality in (29) is satisfied, we have obtained an excellent solution under the conditions of (24) and (25). Also, note that the introduction of the conditions (24) and (25) does not make the solution of (26) unique. In fact, there are many solutions of (26) under the constraints of (24) and (25). However, only those solutions are acceptable which satisfy inequality (29).

The next question is how to determine the constant C in (28). The value of C is obtained from the number of bits used in the quantization process. In our experiments, eight bit quantization has been utilized in recording the digital form of the signal. However, the number of effective bits in the signal would be much less than eight, because of additional noise in the data. This could be stray pick-up, base line shift, time sampling jitter and so on. In all our theoretical computations we have utilized 4.5 effective bits, rather than eight. The figure 4.5 has been chosen after careful analysis of the equipment and experimental set up. For our case, the constant C is given by

$$C = \frac{1}{2^{4.5}} = .04419417 \tag{30}$$

This value of C has been utilized in all the computations reported in section 6.

A brute force universal way of making this algorithm perform well over a wide range of problems is to stop the iteration interactively when the solution starts diverging. This is because iterative methods for ill-posed problems provide asymptotic convergence, i.e. the solution improves with the first few iterations and then becomes worse as more iterations are done.

4. Description of the Experimental Setup

The experiments were conducted on an indoor aluminum time domain range, 9.14 meters (30 ft) in diameter. The conical transmitter antenna, in the center of the ground plane, has a half angle of 5.27 degrees, and is 4.19 meters (13 ft. 9 in.) tall. A sub-nanosecond Gassian pulse is transmitted and strikes an object on the ground plane. The return from the object is received by a curved-mouth TEM horn, whence it goes to a sampling oscilloscope and a microprocessor for processing and storage. A detailed description of the development of this antenna has been given elsewhere [17]. The system is bistatic with the transmitter (T) and the receiver (R) separated by 1.52 meters (60 in.). The distance from the object position (0) to the transmitter (T) and receiver (R) is 2.03 meters (80 in.). That is, TOR is an isosceles triangle.

A diagram of the apparatus beneath the ground plane is shown in Figure 1. The impulse generator is a spark tube device which generates a sub-nanosecond Gaussian pulse about 1 kilovolt in amplitude. A capacitive pickoff probe feeds a trigger pulse to the sampling oscilloscope. The signal pulse goes through a 15.24 meter (50 ft) delay line of RG-331 semirigid cable to the transmitter. The return from the object is fed to the sampling oscilloscope. The microprocessor controls the horizontal position of the sampling scope trace. The position of the trace is held in one position on the screen while the microprocessor receives as many samples as desired, computes the average, stores the result in memory, then moves the trace to the next point on the screen where the averaging procedure is repeated. The microprocessor does this at 256 equally spaced points across the sampling scope screen. As many samples as desired (from 1 to 255) may be averaged at each position. After a set of 256 averaged values is collected, the results may be plotted and/or recorded on digital tape by command from the terminal. In all recorded data runs, 50 samples were averaged at each of 256 positions on the waveforms.

FIGURE 1. EXPERIMENTAL APPARATUS

FIGURE 2. INPUT AND OUTPUT WAVEFORMS FOR A 19″ DIAMETER
CONDUCTING SPHERE

5. Numerical and Experimental Results

As a first example, we consider computing the impulse response of a 19 in. (48.26 cm) diameter sphere from measured incident and scattered waveforms. In the actual experiment, a hollow 19 in. hemisphere was used. First, the radiated pulse was received by a TEM horn facing the transmitting antenna. This waveform is marked as "input" in Figure 2. Next the hemisphere was taped to the ground plane with conducting tape. The field scattered by the hemisphere was received by the TEM horn facing the hemisphere. The recorded waveform is marked as "output" in Figure 2. Next the conjugate gradient method is applied to compute the impulse response, given the "input" x(t) and the "output" y(t). The equations (15) - (21) are successively applied until the error criterion described by (29) with C given by (30) is satisfied. The estimate of the impulse response obtained after five iterations is presented in Figure 3.

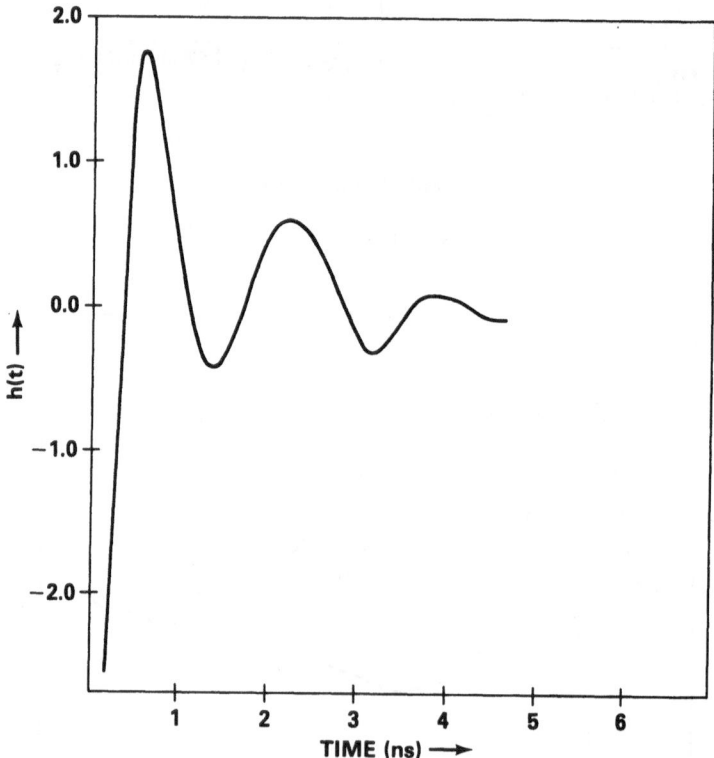

FIGURE 3. IMPULSE RESPONSE OF A 19″ DIAMETER SPHERE

The reasonances of a solid sphere are given by the following modes and their cut-off frequencies. The modes are [p.271, 18], $TM_{m,1,1}$ (543 MHz); $TM_{m,2,1}$ (766 MHz), $TE_{m,1,1}$ (889 MHz) and so on. The frequencies in brackets represent the cut-off frequencies of the modes. Observe the lowest order modes are, therefore, the three $TM_{m,k,1}$

modes. Except for a rotation in space, these three modes have the same mode pattern. It is expected that the scattered fields may have these resonant frequencies. The computed impulse response is shown in Figure 3. From the impulse response, it is seen that indeed some of these frequencies are present. It is important to point out that the impulse response indeed contains an impulse at the origin. We have been able to recover this information without incorporating any additional constraints on the solution procedure. It is important to point out that the recovery of the impulse is often not possible by the classical z-transform and FFT procedures.

As a second example, we consider computing the impulse response of a finite length conducting cylinder of diameter 15 in. (38.10 cm) and length 60 in. (152.4 cm). The half cylinder was taped to the ground plane and was irradiated by an incident field from the broadside direction with the electric field perpendicular to the axis of the cylinder. The incident and the scattered waveforms for this case are shown in Figure 4. If the cylinder were of infinite length, this type of excitation then would typically represent a TE excitation. However, since the cylinder is of finite length, it is expected that in addition to TE modes, there will be TM modes. The dominant resonant frequencies for an infinite length cylinder are TE_{11} (461 MHz), TM_{01} (603 MHz), TE_{21} (765 MHz), TE_{01} (960 MHz) and TM_{11} (960 MHz) - a degeneracy. The computed impulse response is presented in Figure 5. Again, the impulse at the origin has been reproduced. Also, the resonant frequencies mentioned earlier are present in the impulse response.

FIGURE 4. INPUT AND OUTPUT WAVEFORMS FOR A 15" DIAMETER,
60" LONG CONDUCTING CYLINDER

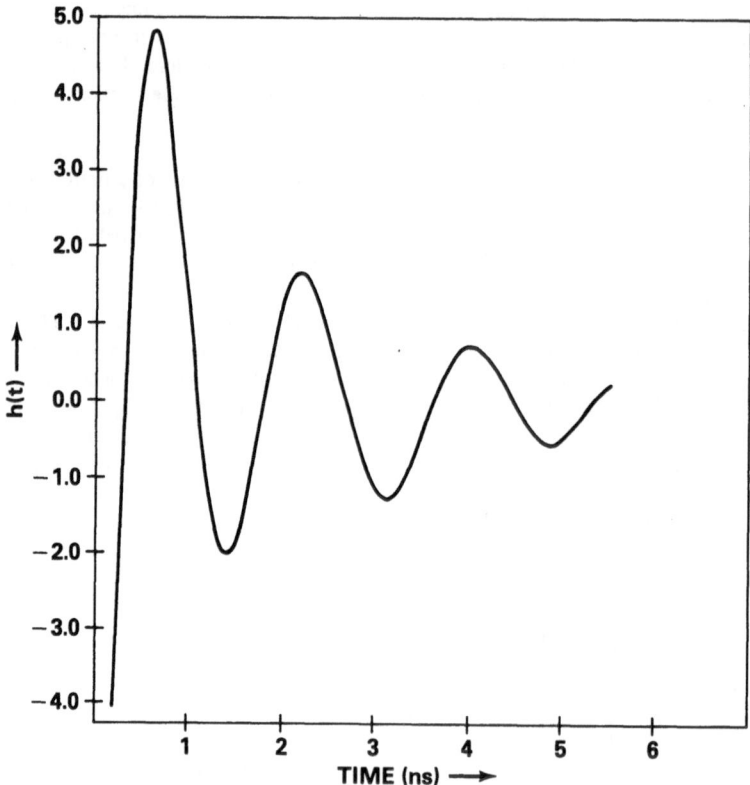

FIGURE 5. IMPULSE RESPONSE OF THE CYLINDER

6. Conclusion

The conjugate gradient method is presented to compute the impulse response from measured input and output data. Even though the deconvolution problem is ill-posed, the conjugate gradient method would yield stable results as long as the number of effective bits in the signal is correctly chosen.

Computed impulse responses from measured data have been computed to illustrate the ability of the conjugate gradient method to perform automatic stable deconvolution of noisy data.

7. References

[1] A. N. Tikhonov and V. Y. Arsenin, "Solutions of Ill-Posed Problems," New York, Wiley, 1977.

[2] M. Z. Nashed, "Some Aspects of Regularization and Approximation Solutions of Ill-Posed Operator Equations," in Proc. 1972 Army Analysis Conference, pp. 163-181.

[3] T. K. Sarkar, D. D. Weiner and V. K. Jain, "Some Mathematical Considerations in Dealing with Inverse Problems," IEEE Trans. of Antennas and Propagation, March 1981, pp. 373-379.

[4] I. Stakgold, "Green's Functions and Boundary Value Problems," John Wiley, New York, 1979.

[5] E. M. Kennaugh and D. L. Moffatt, "Transient and Impulse Response Approximations," Proc. IEEE, Vol. 53, Aug. 1965, pp. 893-901.

[6] C. L. Bennett, R. Hieronymus and H. Micras, "Impulse Response Target Study," Sperry Research Center, Subdury, MA, Final Report, F30602-76-C-0209.

[7] N. Nahman and M. E. Guillaume, "Deconvolution of Time Domain Waveforms in the Presence of Noise," N.B.S. Technical Note 1047, National Bureau of Standards, Boulder, Colorado, October 1981.

[8] F. Ba Hli, "A Time Domain Approach: Aspects of Network and System Theory," R. E. Kalman and N. Declaris, Eds., New York: Holt, Rinehart and Wisston, 1971.

[9] M. P. Ekstrom, "A Spectral Characterization for the Ill-Conditioning in Numerical Convolution," IEEE Trans. Acoustics, Speech, Signal Processing, Aug. 1973, pp. 344-347.

[10] T. K. Sarkar, D. D. Weiner, S. A. Dianat and V. K. Jain, "Impulse Response Determination in the Time-Domain Theory," IEEE Trans. on Antennas and Propagation," July 1982, pp. 657-663.

[11] C. Johnson, "Numerical Solution of Ill-Posed Problems Using the Conjugate Gradient Method," Int. Symposium on Ill-Posed Problems: Theory and Practiace, Oct. 1979.

[12] O. M. Alifanov and S. V. Rumyanstev, "Regularizing Gradient Algorithms for Inverse Thermal Conduction Problems," Inzhenerno-Fisicheskii Zhurnal," Vol. 39, No. 2, Aug. 1980, pp. 253-258.

[13] M. Hestenes and E. Stiefel, "Method of Conjugate Gradient for Solving Linear Systems," J. Res. Nat. Bur. Standards, Vol. 49, pp. 409-436. 1952.

[14] T. K. Sarkar, K. Siarkiewicz, and R. Stratton, "Survey of Numerical Methods for Solution of Large Systems of Linear Equations for Electromagnetic Field Problems," IEEE Trans. on Antennas and Propagation, Vol. 29, Nov. 1981, pp. 847-856.

[15] U. J. Kammerer and M. Z. Nashed, "On the Convergence of the Conjugate Gradient Method for Singular Linear Operator Equations," SIAM Journal Numer. Anal., Vol. 4, 1972, pp. 165-181.

[16]· W. Oettli, "On the Solution Set of a Linear System with Inaccurate Coefficients," J. SIAM Numerical Analysis, Ser. B., Vol. 2, No. 1, 1965, pp. 115-119.

[17] B. Z. Hollmann, "A New Curved-Mouth TEM Horn for Time Domain Data Acquistiion," National URSI Meeting, 5-7 Jan 1983, p. 154.

[18] R. F. Harrington, "Time Harmonic Electromagnetic Fields," McGraw Hill, New York, 1961.

II.8 (NM.4)

APPROXIMATION SCHEMA FOR GENERALIZED INVERSES:
SPECTRAL RESULTS AND REGULARIZATION OF ILL-POSED PROBLEMS

Virginia V. Jory

Microwave Systems Division
Engineering Experiment Station
Georgia Institute of Technology
Atlanta, Georgia 30332, USA

ABSTRACT

Let H_1 and H_2 denote Hilbert spaces and let B denote a closed linear operator mapping a dense subset of H_1 into H_2. Spectral methods are used to construct approximation schema for the Moore-Penrose generalized inverse B^+. The approximators are shown to provide regularization operators for the equation $Bx = y$. Some of the results are extended to dissipative operators on reflexive and general Banach spaces.

1. Introduction

We let H_1 and H_2 be Hilbert spaces and let B denote a closed linear operator mapping a dense subset of H_1 into H_2. The operator equation

$$Bx = y \qquad (1.1)$$

is said to be well-posed in the sense of Hadamard provided it has a unique solution which depends continuously on the data "y"; otherwise it is ill-posed or improperly-posed. Examples of ill-posed problems in electromagnetic theory are: (1) the antenna synthesis problem in which a desired far-field radiation pattern is specified and the aperture distribution required to yield the far-field pattern is to be determined, and (2) the meteorological reconstruction problem in which spatial distributions of temperature and water vapor are to be determined from radiometric measurements, (3) medical tomography, (4) nuclear magnetic resonance imaging, and (5) more generally, almost any problem represented by the integral equation

$$B\, x(t) = \int_a^b K(t, \xi)\, x(\xi)\, d\xi,\; t \in [a, b],\, K \in L^2\, ([a,b] \times [a,b]).$$

365

W.-M. Boerner et al. (eds.), Inverse Methods in Electromagnetic Imaging - Part 1, 365–374.
© 1985 by D. Reidel Publishing Company.

Methods of regularizing such ill-posed problems (that is, redefining an associated well-posed problem) include changing the definition of a solution, changing the space in which the problem is formulated, introducing regularization operators, and utilizing probabilistic techniques. Here emphasis is placed on changing the definition of a solution to include the notion of a generalized inverse and on introducing regularization operators.

Suppose, for example, B is an nxn matrix and y is an nx1 column vector over the real or complex numbers; then exactly three possibilities exist:

(i)   there is a unique solution to Equation (1.1),
(ii)   there is an infinite set of solutions, or
(iii)   there is no solution.

In case the first possibility holds, (that is, B is one-to-one), then it remains to calculate the inverse B⁻¹. When the second possibility occurs, (that is y is in the range of B but the null space of B is nontrivial), the problem is one of choosing the "best" solution, say the solution u of minimal norm. In case the third possibility holds, (that is, y is not in the range of B), it still might be possible and even desirable to assign a "best solution." One might, for example, assign a vector u which minimizes IBx-yI, a so-called *least squares solution*. (Again, such a vector exists since the range of B is a closed subspace.) Notice that since the null space of B also is nontrivial, the problem again reduces to choice. This final issue may be settled as before by selecting the vector u_0 of minimal norm which minimizes IBx - yI. that is, the *least squares solution of minimal norm*. The operator B† which makes the assignment is referred to as the Moore-Penrose generalized inverse of the matrix B. Of course, the definition of the Moore-Penrose generalized inverse has been extended to much more general operators B.

2.   Preliminary Results, Definitions

Again, let H_1 and H_2 denote Hilbert spaces and let B denote a closed linear operator mapping a dense subset of H_1 into H_2. Let D(B), N(B) and R(B)ᶜᴸ denote the domain, null space, and the closure of the range of B, respectively. Then, in the Hilbert space setting, the Moore-Penrose or orthogonal generalized inverse B† exists and is the unique solution to the following four equations

$$BB†B = B, \quad B†B = 1 - P, \text{ on } D(B),$$
$$B†BB† = B†, \quad BB† = Q, \text{ on } D(B†),$$

where P and Q are the orthogonal projections onto N(B) and R(B)ᶜᴸ, respectively [6]. Moreover, if y is in the domain of B†, then B†y is the least squares solution of minimal norm of Equation (1.1).

The notion of a least squares solution leads to a less restrictive definition of well-posedness than well-posedness in the sense of Hadamard.

Definition 2.1   The Equation (1.1) is said to be <u>well-posed</u> in (H_1, H_2) if for each y in H_2, (1.1) has a unique least squares solution of minimal norm; otherwise the equation is <u>ill-posed</u>.

It can be shown that the following statements are equivalent (see [5], [7]):

(i)   the operator equation (1.1) is well-posed in (H_1, H_2);
(ii)   B has closed range in H_2; and
(iii)   the generalized inverse $B^\dagger$ is a bounded operator on H_2 into H_1.

Thus, in general, ill-posed problems are characterized by noise in the data, y, causing large errors in the solution, x, of Equation (1.1).

Recall from Hilbert space theory that if B is closed and densely defined from H_1 into H_2, then $N(B) = N(B^*B)$, $N(B^*) = N(BB^*)$, $R(B)^{CL} = R(BB^*)^{CL}$, $R(B^*)^{CL} = R(B^*B)^{CL}$, $H_1 = R(B^*)^{CL} \oplus N(B)$, $H_2 = R(B)^{CL} \oplus N(B^*)$, and $D(B^\dagger) = R(B) + N(B^*)$ (see, for example, [5], [6], [9]). In the Hilbert space setting, spectral methods from the operational calculus [9] provide an excellent tool for construction of approximators to the generalized inverse $B^\dagger$. Groetsch [1] utilized the operational calculus to obtain the following result for bounded operators.

Theorem 2.2   (Groetsch [1]). Suppose T is bounded linear, $T : H_1 \rightarrow H_2$. Let S be an unbounded subset of $(0, \infty)$. Let $\{U_\beta(t) : \beta \in S\}$ be a family of real valued functions such that each U_β is continuous on $[0, \| T \|^2]$ and such that there exists M > 0 with

$$|tU_\beta(t)| \le M \quad \text{for all t and } \beta, \text{ and}$$

$$U_\beta(t) \rightarrow t^{-1} \text{ as } \beta \rightarrow \infty \text{ for each } t \neq 0. \text{ Then,}$$

$$\lim_{\beta \rightarrow \infty} U_\beta(T^*T)T^*b = T^\dagger b \text{ for all } b \in D(T^\dagger)$$

Examples of families $\{U_\beta : \beta \in S\}$ satisfying the assumptions of Theorem 1 are shown by Groetsch [1] to include

$$(i) \quad U_\beta(t) = \int_0^\beta e^{-\beta t} dt, \text{ for } t \in S = (0, \infty),$$

$$(ii) \quad U_\beta(t) = (1 + \beta t)^{-1}, \text{ for } t \in S = (0, \infty),$$

$$(iii) \quad U_\beta(t) = \sum_{k=0}^\beta (1 + at)^k, \text{ for } a > 0 \text{ and } t \in Z^+.$$

$$(iv) \quad U_\beta(t) = \sum_{k=0}^{\beta} \frac{1}{k+1} \prod_{j=1}^{k} (1 - \frac{1}{1+j} \, t), \, for \, t \in Z^+$$

Lardy [4] established the following result by use of spectral methods for unbounded self-adjoint operators.

Theorem 2.3   (Lardy [4]).   Let A be a densely defined, closed linear operator mapping H_1 into H_2. If y is in $D(A^\dagger)$, then

$$A^\dagger y = \sum_{k=1}^{\infty} A*(1 + AA^*)^{-k} y.$$

Moreover, if R(A) is closed then

$$A^\dagger = \sum_{k=1}^{\infty} A^*(1 + AA^*)^{-k}.$$

Definition 2.4   Suppose E is any Banach space and A: $D(A) \subseteq E \rightarrow E$ is a linear densely defined operator. Then A is said to be _dissipative_ only in case for each $c > 0$, $|(1-cA)| \geq |x|$ for each x in D(A) and R(1-cA) = E.

3.   Approximation Schema for $B^\dagger$ as Regularization Operators for $Bx = y$

Let E_{B*B} denote the spectral family for $B*B$ (see, for example, [9]), where B^* is the Hilbert space adjoint of B, and let S be an unbounded subset of $(0, \infty)$. We establish a representation theorem for the generalized inverse $B^\dagger$ and show that the resulting approximation schema provide regularization operators for Equation (1.1).

Theorem 3.1   Suppose for each β in S, f_β is a continuous function on $[0, \infty]$ such that for $\lambda > 0$, $\lim_{\beta \to \infty} f_\beta(\lambda) = 1/\lambda$. Suppose also that there exists $M > 0$ such that $|\lambda f_\beta(\lambda)| \leq M$ almost everywhere with respect to E_{BB^*}. Let h_β be defined by:

$$h_\beta(\lambda) = \phi(\lambda) - \lambda f_\beta(\lambda), \, where$$

$$\phi(\lambda) = 0, \quad if \, \lambda = 0$$

$$= 1, \, if \, \lambda > 0.$$

Suppose there exist a real-valued function h and a number γ such that

$$a) \quad \int_0^\infty |h(\lambda)|^2 d\|E_{B^*B} \, (\lambda) x\| \quad exists \, for \, all \, x \, in \, D(B), \, and$$

b) *if* $\beta \geq \gamma$, *then* $\|h_\beta\| \leq \|h\|$.

Suppose, finally, that

c) $B^* f_\beta(BB^*) = f_\beta(B^*B)B^*$ *on* $D(B^*)$ *and*,

d) $Bf_\beta(B^*B) = f_\beta(BB^*)B$ *on* $D(B)$

Then $\quad$ s - $\lim_{\beta \to \infty} B^* f_\beta(BB^*) = B^\dagger$.

(This result was presented at a conference on approximation theory honoring Professor George G. Lorentz on the occasion of his seventieth birthday [3]. The proof is given here for completeness and to represent the flavor of the mathematics involved.)

Proof. Let y be in $D(B^\dagger)$. Since $H_2 = R(B)^{CL} \oplus N(B^*)$, there exists n in $N(B^*)$ such that $y = BB^\dagger y + n$. Thus for each β:

$$|B^\dagger y - B^* f_\beta(BB^*)y|^2 = |B^\dagger y - B^* f_\beta(BB^*)y|^2$$

$$= |B^\dagger y - B^* B f_\beta(B^*B)B^\dagger y|^2$$

$$= |[\phi(B^*B) - B^* B f_\beta(B^*B)]B^\dagger y|^2$$

$$= \int_0^\infty |\phi(\lambda) - \lambda f_\beta(\lambda)|^2 d|E_{B^*B}(\lambda)B^\dagger y|^2 \qquad (3.1)$$

The last equality in (3.1) obtains as a result of the operational calculus for (unbounded) self-adjoint operators (see, for example, [9, Chapter XII]) Furthermore, by the Lebesgue dominated convergence theorem, the integral in (3.1) has limit zero as β approaches infinity.

Definition 3.2 Let Λ be an indexing set. A regularization family (or regularization algorithm as in [7]) for (1.1) is a family of bounded linear operators $\{F_\alpha : \alpha \in \Lambda\}$ from H_2 into H_1 with the property that for every y_0 in $D(B^\dagger)$, there is an element x in the set of all least squares solutions of (1.1) such that for every $\varepsilon > 0$, there exists $\delta > 0$ and α in Λ satisfying $|F_\alpha y - x| \leq \varepsilon$, whenever $|y - y_0| \leq \delta$.

The effect of a regularization family is illustrated in Figure 3.1.

Theorem 3.3 $\quad$ Under the hypotheses of Theorem 3.1, the family of operators

$$\{B^* f_\beta (BB^*): \beta \in S\}$$

is a regularization algorithm for Equation (1.1).

Proof. Let $\varepsilon > 0$ and y in $D(B^\dagger) = R(B) + N(B^*)$ be given. Let β be such that

$$|B(B^\dagger y - *f_\beta \,(BB^*)\, y_0| \leq \varepsilon/2.$$

Since $\lambda f_\beta(\lambda)$ is bounded almost everywhere with respect to E_{BB^*}, $B^* f_\beta (BB^*)$ is bounded. Thus there exists δ such that if $|y - y_0| \leq \delta$, then

$$|B^* f_\beta(BB^*)\, y_0 - B^* f_\beta (BB^*)\, y | \leq \varepsilon/2.$$

Realization that $B^\dagger y$ is a least squares solution of (1.1) yields the result.

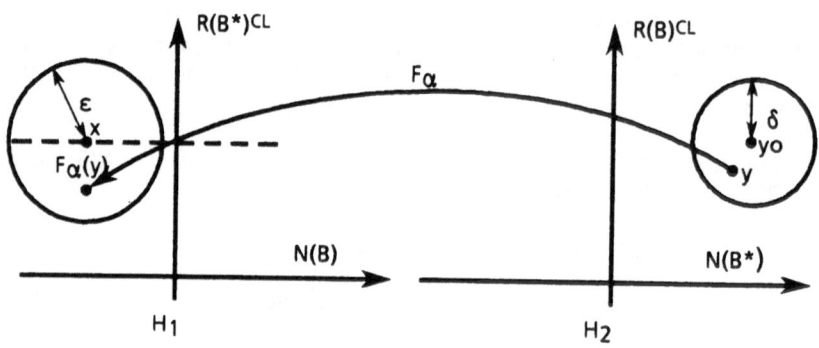

FIGURE 3.1   Mapping By a Regularization Operator.

4.   Applications

Examples of families of real-valued functions which satisfy the hypotheses of Theorem 3.1 are given next.

Theorem 4.1   The sets of functions $\{f_\beta : \beta \in S\}$ defined by the following equations provide regularization families for Equation (1.1)

$$(i) \quad f_\beta(\lambda) = \frac{1}{\beta} \sum_{j=1}^{\beta} \sum_{i=1}^{j} (1+\lambda)^{-i}, \, \beta \in S = Z^+,$$

$$(ii) \quad f_\beta(\lambda) = \int_0^\beta e^{-\xi\lambda} d\xi, \, \beta \in S = R^+$$

(iii) $f_\beta(\lambda) = \beta(1 + \beta\lambda)^{-1}, \beta \in S + R^+$

(iv) $f_\beta(\lambda) = \sum_{p=1}^{\beta} \lambda_p \sum_{j=1}^{p} (1 + \lambda_j \lambda)^{-1}, \beta \in S = Z^+,$ where

$\{\lambda_p\}$ is a positive number sequence such that $\Sigma\lambda_p = +\infty$.

For bounded operators B, the family in (ii) results in the Showalter integral formula for $B^\dagger$ and the family in (iii) yields the Tikhonov regularization algorithm of order zero. The family in (iv) includes the result of Lardy with $\lambda_p = 1$ for all p.

Remark 4.2 The convergence in Theorem 3.1 is uniform when B has closed range; moreover, the convergence is of order $1/\beta$ for the approximators to $B^\dagger$ constructed from the families in Theorem 4.1 (i) - (iii). In (iv), if $\Sigma\lambda_p^2 = +\infty$, the convergence is of order $(\Sigma_{p=1}^{\beta} \lambda_p^2)^{-1}$.

Remark 4.3 The following lemma is utilized in the proof of Theorem 4.1, (see [8] for (i), (ii), [2] for (iii), (iv)).

Lemma 4.4 Let $\beta, \lambda > 0$, z be in $D(B^*)$, x be in $D(B)$, and T_{-B^*B}, T_{-BB^*} be the C_0 semigroups generated by $-B^* B$, $-BB^*$ respectively. Then

(i) $(1 + \lambda B^*B)^{-1}B^*z = B^* (1 + \lambda BB^*)^{-1}z$

(ii) $(1 + \lambda BB^*)^{-1}Bx = B (1 + \lambda B^*B)^{-1}x$

(iii) $T_{-B^*B}(t)B^*z = B^*T_{-BB^*}(t)z$, for all $t > 0$, and

(iv) $\dfrac{1}{\beta}\displaystyle\int_0^\beta \xi T_{-B^*B}(\beta - \xi)B^*z d\xi = B^* \dfrac{1}{\beta}\displaystyle\int_0^\beta \xi T_{-BB^*}(\beta - \xi)z d\xi.$

Example 4.5 For an example of an operator whose range is not necessarily closed and to which the computational procedures of this section apply, we shall consider an integral operator of the first kind B: $H_1 \rightarrow H_1$ defined for x in $L^2[a,b]$ by

$$Bx(.) = \int_a^b K(.,\xi)x(\xi)\,d\xi$$

where the kernel K is such that

$$\int_a^b \int_a^b |K(\xi,\sigma)|^2 d\xi d\sigma < 0.$$

Since B is compact, R(B) is closed if and only if R(B) is finite dimensional. Of course $B^\dagger$ is bounded only in case R(B) is closed. To compute $B^\dagger$, let $U_\beta = -\beta(1 + \beta B^*B)^{-1}$, according to the family f_β in Theorem 4.1 (iii). Then B* is given by

$$B^*x(.) = \int_a^b K(\xi, .) x(\xi) \, d\xi$$

for all x in H_1. Thus, if f_β is in H_1 then for $\beta > 0$, $-U_\beta B^* f_\beta$ is the solution ω_β of the integral equation of the second kind

$$\omega_\beta(.) + \beta \int_a^b [\int_a^b K(\xi,.) K(\xi,o) \, d\xi] \omega_\beta(o) do = \beta \int_a^b K(\xi,.) f_\beta(\xi) d\xi. \qquad (4.1)$$

Since $-B^*B$ is dissipative in H_1, in the sense of Definition 2.4, then a unique solution to Equation (4.1) exists for each $\beta > 0$. Furthermore by Theorem 3.1 if f_β is in $D(B^\dagger)$ then $B^\dagger f_\beta = \lim_{\beta \to \infty} \omega_\beta$. Thus we see that the generalized inverse of an integral operator of the first kind may be obtained as the pointwise or strong limit of solutions of integral equations of the second kind.

5. Regularization Operators in Banach Space

With less structure in a general Banach space than in a Hilbert space, we lose the powerful operational calculus which allows us to construct regularization *operators* from *functions* of real numbers in the Hilbert space setting. Moreover, since the notion of a least squares solution may not persist in the general Banach space, it is necessary to modify Definition 3.2 slightly.

Throughout this section we assume that E_1 and E_2 are Banach spaces with A a closed linear operator mapping a dense subset of E_1 into E_2. We assume also that there are subspaces S and M so that $E_1 = N(A) \oplus S$ and $E_2 = R(A)^{CL} \oplus M$. Let P be the projector onto N(A) along S, Q be the projector onto R(A)^{CL} along M, and let $A^\dagger = A^\dagger_{P,Q}$.

Definition 5.1. Let J be an indexing set. A <u>regularization family relative to (P, Q)</u> for

$$Ax = y \qquad (5.1)$$

is a collection of bounded linear operators $\{F_a: a \in J\}$ from E_2 into E_1 with the property that for any y_0 in R(A) + M, there is an element n in N(A) such that, for every $\varepsilon > 0$, there exists $\delta > 0$ and a in J, so that $|F_a y - (A^\dagger + n)| \le \varepsilon$, if $|y-y_0| \le \delta$.

Definition 5.1 agrees with Definition 3.2 in case E_1 and E_2 are Hilbert spaces and P and Q are the orthogonal projectors onto N(A) and R(A)^{CL}, respectively; the element $A^\dagger y + n$ is then a least squares solution of (5.1).

The main result of this section follows.

Theorem 5.2 Let S be a subset of the real numbers such that $\sup_{\beta \in S} \beta = \infty$
Suppose $E = E_1 = E_2 = R(A)^{CL} \oplus N(A)$ and $\{U_\beta: \beta \in S\}$ is a family of bounded linear
operators defined on E satisfying Properties 1-4.

Property 1. The members of the family $\{1-A U_\beta: \beta \in S\}$ map E into D(A).

Property 2. For each β in S the operators A and U_β are commutative on D(A).

Property 3. The family of operators $\{1-AU_\beta: \beta \in S\}$ is uniformly bounded on
E.

Property 4. On $R(A)^{CL}$, s-lim$_{\beta \to \infty}(1-AU_\beta)$ exists and is the zero element of E.

Then $\{U_\beta Q: \beta \in S\}$ is a regularization family relative to (P, Q) for Ax = y,
where P is the projector onto N(A) and Q = 1-P.

Proof. Let y be in R(A) + N(A). Then $|A^\dagger y - U_\beta Q y| = |A^\dagger y - U_\beta A A^\dagger y| = |(1-AU_\beta)$
$A^\dagger y|$. Thus s-lim$_{\beta \to \infty} U_\beta Q = A^\dagger$, and, since for each β, $U_\beta Q$ is bounded, the result
follows.

If E is a reflexive Banach space and A is dissipative in E, then regularization families
satisfying the hypotheses of Theorem 5.2 can be constructed for Equation (5.1).

Theorem 5.3. Suppose E is a reflexive Banach space and A is dissipative on E. Then
each of the following families provides regularization operators relative to (P, Q)
for Ax = y, where P and Q are as in Theorem 5.2.

$$(i) \quad \{-\beta(1-\beta A)^{-1}Q: \beta \in R^+\},$$

$$(ii) \quad \{-\frac{1}{\beta} \sum_{j=1}^{b} \sum_{i=1}^{j} (I-A)^{-i}Q: \beta \in Z^+\}, \; and$$

$$(iii) \quad \{-\frac{1}{\beta} \int_0^\beta \xi T(\beta\xi)Q d\xi: \beta \in R^+\}, \; where$$

T is the C_0 semigroup generated by A.

6   Additional Work

Work is underway to identify linear penalty functions q_α associated with the
regularization operators F_α described herein and with others under investigation

so that in the presence of noisy data in the non-closed range case the problem becomes one of minimizing

$$|F_\alpha y - A^\dagger y_0| \le |F_\alpha y_0 - A^\dagger y_0| + |F_\alpha(y_0 - y)| \le |F_\alpha y_0 - A^\dagger y_0| + |q_\alpha(y_0 - y)|.$$

where y and y_0 are as in Figure 3.1.

ACKNOWLEDGMENTS

The author wishes to express appreciation to those who have influenced and encouraged her work in inverse problems, especially Dr. James V. Herod, Dr. M. Zuhair Nashed, and Dr. Wolfgang M. Boerner.

REFERENCES

1. Groetsch, C. W. , Generalized Inverses of Linear Operators, Representation and Approximation, Marcel Dekker, New York, 1977.

2. Jory, V. V., Approximators to generalized inverses as regularizers of ill-posed problems, Numer. Funct. Anal. and Optimiz., 3(4), 1981, pp. 477-492.

3. Jory, V. V., Spectral representation and approximation of the generalized inverse of a closed linear operator, Approximation Theory III, Editor: E. W. Cheney, Academic Press, New York, 1981, pp. 543-548.

4. Lardy, L. J., A series representation for the generalized inverse of a closed linear operator, Atti Acad. Naz. Lincei Rend. Cl. Sci. Fis. Mat. Natur., Ser. VIII, 58, 1975, pp. 152-156.

5. Nashed, M. Z., Singular operator equations, Nonlinear Functional Analysis and Applications, Editor: L.B. Rall, Academic Press, New York, 1971, pp. 311-359.

6. Nashed, M. Z. and Votruba, G. F., A unified operator theory of generalized inverses, Generalized Inverses and Applications, Editor: M. Z. Nashed, Academic Press, New York, 1976, pp. 1-109.

7. Nashed, M. Z. and Wahba, G., Generalized inverses in reproducing kernel spaces: An Approach to regularization of linear operator equations, MRC Technical Summary Report #1200, University of Wisconsin Mathematics Research Center, Madison, Wisconsin, 1972. See also SIAM J. Math. Anal. 5, 1974, pp. 974-987.

8. Riesz, F. and Sz.-Nagy, B., Functional Analysis, Frederick Ungar, New York, 1955.

9. Yosida, K., Functional Analysis, Springer-Verlag, New York, 1978.

II.9 (SP.1)

MAXIMUM ENTROPY METHODS IN ELECTROMAGNETIC/GEOPHYSICAL/
ULTRASONIC IMAGING

R. M. Bevensee

Lawrence Livermore National Laboratory
Livermore, California, USA

Maximum Entropy (ME) methods of resolving underdetermined
electromagnetic images are reviewed. A new non-Burg method of
resolving complex amplitudes of coherent radiators in wavenumber space (direction) is described. Several ME methods in
geophysics are surveyed, based on Burg's method for spatial
array data processing and a Boltzmann method for parameter
distributions in the earth. A Boltzmann-ME method is described
for resolving anomalies in materials ultrasonically, where the
data is governed by a Fredholm integral equation of the second
kind. Some issues of uniqueness, confidence, and noise are
briefly assessed.

1. ME METHODS IN ELECTROMAGNETICS

Maximum Entropy methods of resolving underdetermined
objects from image data with or without noise have proven
effective in both real space and wavenumber space. We first
review several Boltzmann-type ME methods for resolving
incoherent sources in real space and wavenumber space,
respectively. Then we outline a special Boltzmann-ME method,
unrelated to the Burg or Capon methods of spatial array data
processing (see Sec. 2.2), for resolving an angular distribution of coherent sinusoidal radiators in amplitude and phase
from receiver data or, alternatively, a spatial distribution
of coherent antennas from far-field data. We describe
numerical results from a special computer code for two-
dimensional distributions of infinitely long, parallel
radiators. We deduce an approximate "super-resolution" limit
of this method for resolving distant coherent radiating

375

W.-M. Boerner et al. (eds.), Inverse Methods in Electromagnetic Imaging – Part 1, 375–395.
© *1985 by D. Reidel Publishing Company.*

sources. Fundamental limitations on the resolution of various radiating sources are discussed.

1.1   Reconstruction From Photon Images

A Boltzmann-ME method has been applied successfully to enhance a photon image in a square aperture system [1]. If $\hat{p}_i(\bar{r}_i)$ is the estimated probability of photon emission from the i'th object pixel at position $\bar{r}_i$ and $t_{ik}(\bar{r}_i, r'_k)$ $= \sin^2(x-x') \sin^2(y-y')/[(x-x')(y-y')]^2$ (normalized coordinates) is the transmission factor to $\bar{r}' = (x',y')$ of the image, and $B(\bar{r}')$ is the measured brightness, the quantity

$$\sum_{i,\text{object}} [\hat{p}_i(\bar{r}_i)t_{ik}(r_i,\bar{r}'_k) - B_k(\bar{r}'_k)]^2 \tag{1}$$

is the squared deviation between predicted and measured brightness due to noise in the k' pixel. The quantity

$$E\{\hat{p}_i\} = \sum_{k,\text{image}} \sum_{i,\text{object}} [\hat{p}_i t_{ik} - B_k]^2 \tag{2}$$

is the noise "energy" in the image. The ME method of [1] seeks the most probable distribution of the $\hat{p}_i$ subject to the constraint of E being less than the known noise level.

Entropy H measures the logarithm of the total number of distinct combinations of photon intensities on the object,

$$H = - \sum_i \hat{p}_i(\ln \hat{p}_i - 1) \; . \tag{3}$$

In maximizing H relative to the $\hat{p}_i$ subject to the constraints it is usually convenient to employ Lagrange's method of undetermined multipliers and maximize an "objective function." In this case it is

$$F = H + \beta E\{\hat{p}_i\} + \gamma(\sum_i \hat{p}_i - 1) \; . \tag{4}$$

Here β is an a priori weight multiplier for the noise and γ is the multiplier for the normalization $\sum \hat{p}_i = 1$. The set of equations obtained by setting $\partial F/\partial \hat{p}_i = 0$ for all i and the constraint equation must be solved to obtain all the $\hat{p}_i$.

For large β this technique enables resolution of incoherent sources separated by $\simeq \lambda/2L$ along x or y, which is

half the Zernike-Van Cittert limit. The angular resolution
limit for broadside antennas is also $\leq \lambda/L$.

Figure 1 shows the rather impressive resolution possible
for a flat object on a 20 x 20 grid of binary pixels from its
diffraction-limited image. The ME resolution is extremely
sensitive to ß for large ß; the resolution shown is for ß = 10^7.

Fig. 1. ME resolution of a flat object from its
diffraction limited image. (a), object defined on a
20x20 grid of binary pixels, (b), image, (c)
resolution with ß = 10^7 [1, p. 51].

1.2   Reconstruction in Wavenumber Space

Two-dimensional digital image reconstruction is important
in many physical sciences, notably radiography, radio
astronomy, optics, and microscopy. An example is the problem
of resolving radio stars in close angular proximity against
the Fourier-transform background noise.

Let the relative intensity or brightness of a distant
point source be $f(\bar{r})$; the wavenumber spectrum S is the
transform

$$S(\bar{k}) = \int_{-\infty}^{\infty} dx \int_{-\infty}^{\infty} dy\; f(\bar{r})e^{-j\bar{k}\cdot\bar{r}} \;.\qquad(5)$$

The measured brightness $B(\bar{k})$ differs from S because of noise.

Discretization of $\bar{r}$- and $\bar{k}$-space and the assumption that
the $S(\bar{k}_m)$ are independent, zero-mean random variables with
known variances σ_m enables us to write an "energy" constraint
as [3]

$$E\{\hat{f}_i\} = \sum_{m=1}^{M} \frac{1}{\sigma_m^2} \left| B(k_m) - \Delta A \sum_{i=1}^{I} \hat{f}_i(x_i, y_i) e^{-j2\pi(u_m x_i + v_m y_i)} \right|^2 \le M, \tag{6}$$

ΔA being the object pixel area, $\hat{f}_i$ an ME estimate of the brightness, and $2\pi(u_m, v_m) = (k_{mx}, k_{my})$, the transverse wavenumber vector.

The entropy H is written

$$H = +\Delta A \sum_i \ln \hat{f}_i \tag{7}$$

because for photons $1 <$ (degrees of freedom) $<< \bar{n}_i$, $\bar{n}_i = $ average number of photons emitted per second. Equation (7) is the correct entropy for the classical wave (field) limit for Bose-Einstein statistics [2].

The objective function analogous to (4) is

$$F = H + \beta E\{\hat{f}_i\} \tag{8}$$

which has a unique local maximum. Wernecke and D'Addario [3] describe an algorithm for maximizing F relative to variations of the $\hat{f}_i$ subject to a stated initial estimate of β. An impressive example of this Fourier synthesis is the reconstruction of the radio double star Cygnus A (3C405). The direct Fourier transform resolves the star whose parts are separated about 120 arc secs somewhat ambiguously, with considerable background noise. The ME method, using only 50 data points on the nine concentric measurement ellipses in the u-v wavenumber plane, resolves them more sharply and with essentially no background noise!

1.3 Resolution of Two-Dimensional Coherent Radiators in Amplitude and Phase

The preceding sections treated the resolution of distant radiating incoherent sources by direct spatial transformation from object to image and by Fourier transformation. We now describe an ME method we have developed for resolving coherent radiators, all at sinusoidal frequency ω. Unlike the Burg method, ours does not rely on measured and extrapolated correlation coefficients. We have applied the method in

two dimensions to resolve angular distributions of distant
radiators from receiver data and--with a minor change in the
transfer functions--to resolve spatial distributions of
antennas from far-field angular measurements. All radiators
are infinitely long and parallel. We shall infer some general
properties of our ME method from synthetic data examples
studied.

We further assume that the coherent nature of the sources
is known and that receiver noise has been correlated out by
mixing with a low-noise local oscillator, for example.

Figure 2 illustrates the "distant source" problem of
resolving K possible radiators in amplitude and phase, the kth
being $A_k e^{i\psi_k}$, from measurements at I receivers, $I < K$, the ith
being $B_i e^{i\phi_i}$. In amplitude all $R_{ik} \simeq R_o$, which is absorbed
into the B_k-amplitude. The equation for the ith receiver is,
with neglect of the common phase factor $\exp(-i\omega R_o/c)$, c = light
velocity, is

$$\sum_{k=1}^{K} A_k e^{i\psi_k} e^{i\omega\rho_i \cos(\Theta_i - \Theta_k)/c} = B_i e^{i\phi_i} \quad \text{measured} \qquad (9)$$

$$1 \leq i \leq I < K,$$

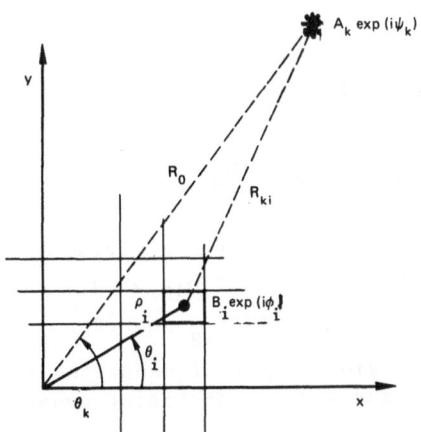

Fig. 2. Description of the "distant source" resolution
problem in two dimensions. Coherent source $A_k e^{i\psi_k}$ at
frequency ω radiates to the receivers in its far field;
receiver i records a total field $B_i e^{i\phi_i}$.

and this is separated into real and imaginary parts, $A_k e^{i\psi_k} = A_{iR} + iA_{iI}$, etc., as

$$\sum_k (A_{kR}C_{ki} - A_{kI}S_{ki}) = B_{iR} \; ; \; \sum_k (A_{kR}S_{ki} + A_{kI}C_{ki}) = B_{iI} \tag{10}$$

with

$$C_{ki} = \cos[\omega\rho_i\cos(\Theta_i-\Theta_k)/c], \; S_{ki} = \sin[\omega\rho_i\cos(\Theta_i-\Theta_k)/c] \tag{11}$$

This formulation of the "distant source" problem is identical to the reciprocal problem of resolving K coherent antennas in the receiver region of Fig. 2 from I far-field measurements, with the proviso that receiver position vector ρ_i in Fig. 2 must be changed to source vector ρ_k. The solution to this reciprocal problem by our method has been published [4] and we summarize it for both problems below.

We regard each A_{kR}, A_{kI} as constructed of an enormous number of building blocks of size Δ (real), so $A_{kR} = n_{kR}\Delta$, $A_{kI} = n_{kI}\Delta$. We choose to define the entropy of the sources without regard for phase and with respect to amplitude squared. The Boltzmann H, with no prior probabilities on the $n_k = n_{kR} + in_{kI}$ is therefore

$$H = \ln \frac{N^2!}{\prod_k |n_k|^2!} \; , \; k \leq K, \; \sum |n_k|^2 = N^2 \tag{12}$$

Stirling's approximation yields

$$H = -N^2 \sum_k p_k \ln p_k \; , \; p_k = |n_k|^2/N^2 \tag{13}$$

The objective function is F = H + all the constraint equations (10), each with its Lagrange multiplier β_{iR} or β_{iI} and including the constraint $\Sigma p_k = 1$. Then $\partial F/\partial n_{kR}$ and $\partial F/\partial n_{kI}$ set to zero for each $k \leq K$ yield a set of equations for all the n_{kR}, n_{kI}, and the I+1 additional constraint equations furnish enough information for determining all the multipliers as well.

One can show that $\partial^2 H/\partial n_{kR}^2 < 0$, $\partial^2 H/\partial n_{kI}^2 < 0$ for all k and fixed N^2 at the solution point; i.e., $H = H_{max}$ and there is only one H_{max}.

The essential equations for the A_k and β_i are

$$A_{kR} = (1/2L_k) \sum_{j=1}^{I} (\beta_{jR}C_{kj} + \beta_{jI}S_{kj}), \quad k \le K \ , \qquad (14a)$$

$$A_{kI} = (1/2L_k) \sum_{j} (-\beta_{jR}S_{kj} + \beta_{jI}C_{kj}) \qquad (14b)$$

along with

$$L_k = \ell n [A_{kR}^2 + A_{kI}^2)/A_T] \quad , \quad A_T = \sum_k (A_{kR}^2 + A_{kI}^2) \quad , \qquad (15)$$

the constraint relations (10), and an implied constraint from (10) and (14) which is useful in scaling the A_k^2:

$$-H = \sum_k (A_{kR}^2 + A_{kI}^2)L_k = \frac{1}{2} \sum_j (\beta_{jR}B_{jR} + \beta_{jI}B_{jI}) \qquad (16)$$

Matrix equations for the β_{jR}, β_{jI} are obtained by multiplying (14a) by $C_{kj'}$, (14b) by $S_{kj'}$, subtracting and taking $\sum_k$. Use of (10a) yields

$$\sum_{j=1}^{I} [F_{j'j}\beta_{jR} + G_{j'j}\beta_{jI}] = b_{j'R} \quad , \quad j' \le I \qquad (17a)$$

Similarly we obtain

$$\sum_i [-G_{j'j}\beta_{jR} + F_{j'j}\beta_{jI}] = b_{j'I} \quad , \qquad (17b)$$

in which

$$F_{j'j} = \frac{1}{2} \sum_k (C_{kj}C_{kj'} + S_{kj}S_{kj'})/L_k \quad ,$$

$$G_{j'j} = \frac{1}{2} \sum_k (S_{kj}C_{kj'} - C_{kj}S_{kj'})/L_k.$$

The algorithm iterates through a loop in which the β_j determine the A_k in (14), which are then solved to satisfy (16), after which the β_j are solved by (17), the B_k are solved by (10), and if they are not close enough to the measured B_k the loop is repeated. The initial L_k of (15) are all set equal and the initial β_j solved by (17) accordingly. Typical problems of interest require ten or so iterations to achieve a few percent accuracy in the ME distribution.

As the examples will show, this ME algorithm tends to yield the relative coherent source amplitudes. Spatial correlation coefficient data have not been incorporated as constraints; they would tend to fix the <u>absolute</u> source amplitudes.

The ME equations (14), (15) are actually solved relative to a total parameter of A_T' instead of A_T, with A_T' greater. This procedure corresponds to a convenient entropy shift which allows us to avoid the double-valued nature of the equation solved for each A_{kR}, after which A_{kI} is found. That equation is obtained by dividing (14b) by (14a) to give $\alpha_k \triangleq A_{kI}/A_{kR}$, whereupon (14a) is solved as

$$A_{kR}\ln[A_{kR}^2(1+\alpha_k^2)/A_T'] = T_{kR}(\beta) \quad . \tag{18}$$

The left side is a double-valued function of $|A_{kR}|$, but if A_T' is large enough A_{kR} may be found in the range $0 \leq |A_{kR}| < \sqrt{A_T'/(1+\alpha_k^2)}$ and the relative A_{kR} determined correctly. After converging on the ME solution A_T is obtained by (15).

The solution for the β_k in (17) is by Gauss elimination and back substitution, in which the pivot elements often are $< 10^{-5}$. However, this does not, by itself, introduce non-negligible error in the iterative solution or slow convergence.

We comment at this point about an apparent complication in the analogous situation with entropy formulated in terms of amplitude and not amplitude squared. Then the entropy is $-N \sum_k p_k \ln p_k$ of (13) but with $p_k = |n_k|/N = \sqrt{n_{kR}^2 + n_{kI}^2}/N$. The equations replacing (14) are

$$A_{kR} = A_k\cos\psi_k = (A_k/L_k') \sum_j (\beta_{jR}C_{kj} + \beta_{jI}S_{kj}) \tag{19a}$$

$$A_{kI} = A_k\sin\psi_k = (A_k/L_k') \sum_j (-\beta_{jR}S_{kj} + \beta_{jI}C_{kj}) \tag{19b}$$

with

$$L_k' = \ln(A_k/A_T'), \quad A_T' = \sum_k A_k \quad , \tag{20}$$

and these two equations for angle ψ_k (A_k cancels) may be incon-
sistent, there being more A_k than β_j, unless the algorithm is
modified nontrivially.

According to the analogy between this spatial resolution
of sources in wavenumber space and temporal resolution of
sources in frequency space by Fourier transform, our ME method
serves as an alternative to the Burg method of waveform
spectral analysis.

We studied the angular resolution capability of a linear
array of 10 receivers spaced $\lambda/2$ apart in Fig. 2 for
resolving two distant coherent sources $< \lambda/L$ apart in angle.
Figure 3 shows the resolution for four cases of two sources
separated by 6.3° = .52 λ/L : 1/$\underline{0}$° and 2/$\underline{90}$° nearly broadside and 45° to broadside and two 1/$\underline{0}$° sources nearly broadside and 45° to broadside. In each case, the scan region was
+ 12° on either side of one of the sources, in steps of 1.2°;
i.e., we assumed 21 possible sources.

Fig. 3 "Superresolution" of two coherent sources
0.52 λ/L apart in angle, by a 10-receiver array of
length L=9.5$\lambda/2$. Left: both sources 1/$\underline{0}$° in a
sector broadside and at 45°. Right: sources 1/$\underline{0}$°,
2/$\underline{90}$° in a sector broadside and at 45°.

In all four cases depicted in Fig. 3 the relative
amplitudes were well resolved. The phases of the $1/\underline{0}°$ and
$2/\underline{90}°$ sources were resolved within 12°; those of the two $1/\underline{0}°$
sources, within about 3°. In our opinion we have obtained
superresolution of coherent sources, and the nearly 3°
half—maximum of each curve indicates the resolution could be
better with this linear array. The resolution of two $1/\underline{0}°$
sources separated by .25 λ/L at broadside, scanned within a
$\pm$ 6° sector at 21 angles, showed two peaks at the actual
source positions. Although the phase resolution was excellent
(-0.7°, -0.1°) the relative peaks differed by a ratio of 1.5.

1.4 Fundamental Limitations on the Resolution of Radiating
Sources

There are four situations to consider: incoherent
sources, with and without additive noise and coherent sources,
with and without noise. Gabriel [5] has explained that
incoherent sources without noise can be resolved arbitrarily
closely in angle by conventional techniques (viz, adaptive
arrays) and gives an empirical universal curve for resolution
versus SNR with noise present. However, Steinberg and Luthra
[6,6A] show that this "universal" curve can be surpassed by
time averaging, provided that the noise power N_0 is removed
from the autocorrelation $R_n(0)$, after which the conventional
ME method (Burg method [7]) will yield angular superresolution
to $\lesssim 0.2 \lambda/L$. Such superresolution can be attained by the
Burg method for incoherent sources with no noise.

Resolution of coherent sources is more difficult. With
no noise the Burg method is comparable to conventional
resolving methods; with noise the time—averaging method which
improved the resolution of incoherent sources with noise
present won't make the Burg method preferable to conventional
ones [6]. However, our non—Burg ME method indicates that
coherent sources with or without noise can be superresolved to
perhaps 0.25 λ/L in angle without resorting to artificial
aperture modulation in order to create a Doppler shift between
sources [5]. With this method one could reduce the receiver
noise to an acceptable level by modulating the received
signals by a local oscillator of noise sufficiently low as to
make the data processing noise level satisfactorily lower than
the numerical noise from the ME method itself, as evidenced by
the background levels in Fig. 3.

Luthra and Steinberg believe that there is no fundamental
resolution limit for a receiver of finite size aperture,
either for coherent and/or incoherent sources, despite the S/N
ratio [8], provided that the noise is thermal or its
statistical equivalent.

2. ME METHODS IN GEOPHYSICS

We review the work of several researchers in resolving
the density distribution of the earth from its mass, moment of
inertia and gravity coefficients. Spatial array data
processing by the Burg and Capon methods is briefly surveyed.
We conclude by summarizing our work with a particular
Boltzmann-ME technique for resolving two-dimensional
underground parameter distributions from cross-borehole data.
An eigenvalue method for improving the numerical stability of
our iterative technique is explained.

2.1 Density Distribution of the Earth

Rietsch [9] has obtained an ME estimate of the spherically
uniform component of earth density, $\rho(r)$ from radius, mass,
and moment of inertia data. He postulates a "Gibbsian"
probability distribution $p(\rho)$ and defines the entropy as
$H = - \int p(\rho)\ln[p(\rho)/w(\rho)]dv$, $w(\rho)$ being an "invariant measure
function," which in this problem is irrelevant because no "a
priori" probability distribution is assigned to the discretized
ρ_n assigned to the equivolume shells into which the earth
is subdivided. He solves for the expectation values $\bar{\rho}_n$ over
the $p(\rho)$ and is able to incorporate assumptions about lower
and upper limits ρ_ℓ, ρ_u in the ME expression for $\bar{\rho}_n$.

We have also obtained a Boltzmann-ME solution to the
problem with no assumptions about ρ_ℓ, ρ_u. Comparison of
the two plots of Rietsch's $\bar{\rho}_n(r_n)$ for $\rho_\ell = 0$ or 1 g/cm^3
and $\rho_u = N\bar{\rho}$ (N = no. shells, $\bar{\rho}$ = average earth density)
with each other and with our curve--all for N = 100--show
surprisingly little difference down to a depth of about 4000
km below ground. The ME formulation and assumptions have the
most effect in the inner core.

A lateral, non-spherical component of earth density has
been inferred from spherical harmonic gravity field coefficients by Rubincam [10], assumed to be a small correction to
the spherically symmetric $\rho(r)$ assumed known. He too used a
Gibbsian-ME method which effectively obtains the population
numbers of indistinguishable particles following Bose-Einstein
statistics. An advantage to this method is the fact that the
(grand) partition function can be factored, yielding a concise
Bose-Einstein function for the ME lateral density distribution.
The solution implies density anomalies concentrated near the
earth's surface and decreasing with depth, in contrast to
other non-entropy studies. We believe it would be more
difficult to extend such a Gibbsian-ME treatment of density to

include seismic constraints than to extend the corresponding
Boltzmann-ME method.

2.2 Spatial Array Data Processing

ME spatial array analysis conventionally means estimation
of the wavenumber spectrum (i.e., the directions of distant
sources) from a sampled, finite spatial autocorrelation
function by the Burg method [7]. This method is analogous to
the Burg method of spectral analysis of a finite temporal
sample, with frequency analogous to wavenumber and time to the
space dimension. In spatial array processing one usually can
average over time, with the consequence that the wavenumber
spectrum is not the Fourier transform of the spatial autocorrelation function but rather its power spectrum (its Fourier
transform over time). Aside from this distinction, both array
processing and the corresponding power spectral analysis by
the Burg method are equivalent to all-pole autoregressive
modeling. We regard the Burg method as a solution to a
slightly underdetermined ME problem, in the sense that the
space or time record is extended only one more point in order
to derive the ME wavenumber or frequency spectrum,
respectively.

Lacoss [11] has investigated the MLM (Maximum Likelihood
Method) of Capon [12] and the MEM for power spectra, or
equivalently, wavenumber spectra. He reports that both
methods are particularly valuable for resolving narrow
frequency peaks, and neither is unduly sensitive to small
noise in the underlying correlation functions. This is
because both adapt to the actual noise by predicting power at
a given frequency with least disturbance from other
frequencies.

2.3 Resolution of Two-Dimensional Underground Parameter Distributions

The problem is to invert a Fredholm integral equation of
the first kind, subject to constraints, in order to infer an
underground distribution of electrical conductivity, dielectric constant or reciprocal seismic wave velocity from
cross-borehole measurements of sinusoidal-frequency
attenuation, phase shift, or transit time, respectively, of
the rays traversing the region. The description of our
Boltzmann-ME method of inferring an ME parameter distribution
in discrete cells (rather than a Gibbsian probability
distribution over possible earth models) has already been
published [13], along with synthetic data examples. We merely
summarize results here.

The discretized form of the Fredholm equation is written

$$T_i = \sum_{k=1}^{k} D_{ik}\sigma_k \quad 1 \le i \le I < K \quad , \tag{21}$$

where T_i is the ith ray observable (eg., ray attenuation), D_{ik} is the ith path distance in the kth cell, and σ_k is the unknown parameter (eg., conductivity) in the kth cell. There are n_k blocks of $\Delta\sigma$ in the kth cell, $\sigma_k = n_k\Delta\sigma$, the total number of blocks is $N = \sum n_k$ and the entropy is

$$H = \ell n \frac{N!}{\prod n_k!} \simeq -N \sum_{L} (n_k/N) \, \ell n(n_k/N) \tag{22}$$

and the objective function to be made an extremum is

$$F = H + \gamma \sum n_k + \sum_{i=1}^{I} \beta_i [\sum_{k=1}^{K} D_{ik}n_k\Delta\sigma - T_i] \quad . \tag{23}$$

The ME solution is

$$\hat{\sigma}_k = \hat{\sigma}_T \exp[\sum_i \beta_i D_{ik}]/Z \quad , \qquad \hat{\sigma}_T = \sum \hat{\sigma}_k \tag{24}$$

partition function Z being

$$Z = \sum_k \exp[\sum_i \beta_i D_{ik}] \quad , \tag{25}$$

and the β_i must be found to satisfy the I constraints of (21).

We were able to resolve anomalies rather well in various synthetic data examples with an algorithm which neglected noise explicitly in the T_i, assuming $\hat{\sigma}_T$ was known to within 10% or so, and assuming straight rays (known D_{ik}). The algorithm computed consecutively (A) estimates $\hat{\sigma}_k$ from a $\bar{\beta}$ vector by (24)--starting from a simple β_o, (B) estimates $\hat{T}_j$ from the $\bar{\beta}$ by (21), with (24) in place of σ_k, and (25), (C) a correction to $\hat{\sigma}_T$ by a factor obtained from the implied constraint

$$\sum_i \beta_i \hat{T}_i = \sum_i \beta_i T_i \qquad (\sum \hat{T}_i = \sum T_i \text{ if } \bar{\beta}_o = 0) \quad , \tag{26}$$

in which $\hat{T}_i$ is proportional to the $\hat{\sigma}_k$ by (21), (D) $\hat{T}_i$ and $\hat{\sigma}_k$ scaled by the factor just obtained, (E) values of the deriva-

tives $\partial \tilde{T}_i / \partial \beta_j = \partial \tilde{T}_j / \partial \bar{\beta}_i$ obtained analytically from (21) to
(25), (F) the changes $\Delta \beta_i$ necessary to improve the computed ray
data T_i according to

$$T_i - \hat{T}_i = \sum_{j=1}^{I} \frac{\partial \hat{T}_i}{\partial \beta_j} \Delta \beta_j \quad , \qquad 1 < i \leq I \qquad (27)$$

which in matrix form is

$$\Delta T_{Ix1} = M_{IxI} (\Delta \beta)_{Ix1} \qquad (27')$$

by LU decomposition and back substitution. The algorithm then
returned to step (A) and iterated until convergence was
obtained to three significant digits or better. $\hat{\sigma}_T$ also
converged to σ_T with this accuracy if it was specified
initially with 10% or so error. In fact, once the solution
determines the $\hat{\sigma}_k / \hat{\sigma}_T$ (21) will yield the correct $\hat{\sigma}_T$.

If we wanted to include the effect of ray bending
according to ray optics we could do so between steps (D) and
(E) above and correct the D_{ik}. Regardless of the constraints, there is only one extremum of H and it is a
maximum, since $\partial^2 H / \partial n_k^2 < 0$ for all k subject to whatever
constraints are imposed.

One can expect to resolve regions of K cells effectively
with $I \simeq K/2$ rays, provided each cell is probed by at least
one ray. In one example, a 48-cell region with one cell of
$\sigma' = 50$ and the rest of $\sigma = 5$ was resolved by 25 rays to
have $\hat{\sigma}' = 39.5$, one neighbor of 15.8 and all other cells with
$\hat{\sigma} < 8.15$.

In such problems certain rays can be removed if they
"overprobe" the region without changing the ME solution for the
$\hat{\sigma}_k$. However, the solution to (27) becomes more ill-conditioned because of the nature of the Fredholm integral equation
itself.

A typical solution to (27) tends to show very small
($< 10^{-5}$, for example) pivotal elements in the triangulariza-
tion of the M-matrix. This can be alleviated by employing
eigenvalue techniques. Generally, the best procedure is to
resolve M into eigenvector-eigenvalue form and remove the
vectors of very small eigenvalues:

$$M \rightarrow M^c = W_{IxL} \Lambda_{LxL} W_{LxI}^t \quad , \qquad L < I \quad , \qquad (28)$$

and solve for $(\Delta\beta)$ using M^c in (27'). The contraction may vary from iteration to iteration.

However, it may be necessary to use the full M-matrix in (27') for the final iterations when the norm of $(\Delta\beta)$ is very small in order to converge to the correct β!

3. ME ULTRASONIC IMAGING

Ultrasonic imaging is a nondestructive evaluation technique of inferring the presence of an anomaly such as a void or impurity within an object by irradiating it with ultrasonic waves and measuring the scattered wave field. Usually the object can be irradiated from only one side, and shear/compressional wave mixing occurs on discontinuity surfaces. We report below the results of inverting monochromatic scatter data in two-dimensional, simplified, synthetic-data problems by our Boltzmann-ME method [14].

3.1 The Method of Inverting the Fredholm Integral Equation

Gubernatis [15] has presented the exact equations for three-dimensional scattering of an incident shear or compressional wave by a volume anomaly. We assume a simplified, two-dimensional situation in which a compressional (or shear) wave characterized by one cartesian component is scattered from a homogeneous anomaly in an infinite medium, with neither coupling between components nor wave mixing. The Fredholm-II equation for the monochromatic total scalar displacement $u(\bar{r})$ is, in three dimensions,

$$u(\bar{r}) = u^\circ(\bar{r}) + \int dV' \; g(\bar{r},\bar{r}') \; v(\bar{r}')u(\bar{r}') \quad , \qquad (29)$$

with u° the incident (plane) wave and green's function g being

$$g(\bar{r},\bar{r}') = \frac{k_o^2}{4\Pi\rho\omega^2} \frac{e^{-ik_o R}}{R(\bar{r},\bar{r}')} \quad , \qquad k_o = \frac{\omega}{v_p} = \frac{2\Pi}{\lambda} \qquad (30)$$

ρ = density, ω = circular frequency ($\exp(i\omega t)$ time dependence), v_p = phase velocity, λ = wavelength. $v(\bar{r}')$ describes the anomaly as

$$v(\bar{r}') = \delta\rho\omega^2 \; \Theta(\bar{r}') \quad , \qquad (31)$$

where $\delta\rho$ = change in density from the background value and $\Theta(\bar{r}') = 0$ outside the anomaly and 1 inside.

The problem is to infer the area and shape of the anomaly via $\Theta(\bar{r}')$ from the scattered far-field displacement in direction $\bar{k}$, given by u^s.

$$u^s(\bar{k}) = S e^{-ik_o r/r} \quad , \tag{32}$$

$$S = \frac{k_o^2}{4\Pi\rho\omega^2} \int dV \, \exp(i\bar{k}\cdot\bar{r}) \, v(\bar{r})u(\bar{r}) \quad . \tag{33}$$

This synthetic problem includes the essential ingredients of the full coupled-wave problem formulated in tensor notation.

For the ME analysis we discretized the 2D region into square cells Δ on a side and interpreted (29) per unit length in the third direction so as to read

$$u(\bar{r}_i) = u^o(\bar{r}_i) + \sum_k \frac{\Delta e^{-ik_o R}}{R(\bar{r}_i,\bar{r}_k)} \, u(\bar{r}_k) \, \tilde{\Theta}(\bar{r}_k) \quad , \tag{34}$$

with $\tilde{\Theta} = k_o^2 \Delta\delta\rho\Theta/(4\Pi\rho)$ and special evaluation of the summand for the $k = i$ cell. We wrote (33), $\bar{r}$ of cell $k = \bar{r}_o - r_k$ and $\bar{k}_i = \bar{k}$, as

$$S = \Delta \sum_k e^{i\bar{k}_i \cdot \bar{r}_k} \, u(\bar{r}_k) \, \tilde{\Theta}(\bar{r}_k) \quad , \text{ the observable.} \tag{35}$$

We interpreted S as the scattered field from a "radiating source" distribution, with source $A_k \exp(i\psi_k) = u(\bar{r}_k) \tilde{\Theta}(\bar{r}_k)$ at the center of the kth cell in the anomaly, where $u(\bar{r}_k)$ is determined by the Fredholm equation (34). We then solved for the $A_k \exp(i\psi_k)$ by the same Boltzmann-ME method employed in Sec. 1.3 to solve the problem reciprocal to the "distant source" resolution problem [4]. The A_k-amplitude distribution indicates the subjective probability of $\Theta(\bar{r}_k)$ being zero (no anomaly) or one (anomaly) within the region examined. Note that A_k measures $|u(\bar{r}_k) \Theta(\bar{r}_k)|$, so several scattering experiments would resolve $\Theta(\bar{r}_k)$ more reliably.

In the synthetic data problems explored with this ME method the Fredholm equation was only used to obtain the exact S corresponding to a measurement, not in the ME solution. We

observed in Sec. 1.3 there is only one maximum entropy
solution to this problem.

3.2 Synthetic Data Examples

The very first example selected arbitrarily to test this
method is the problem of resolving the 2D anomaly shown in
Fig. 4. In the 100 cells subdividing the 2λ x 2λ region the
anomaly of $\tilde{\Theta} = -1$ occupied 10 cells. For an incident plane
wave u° propagating in the $-x$ direction ($\phi_s = 0$) we fed
into the ME algorithm scatter data u^s in 20 directions, at
equal angles over $-81° < \phi_s \leq 90°$. In solving the equivalent
of (17) for the β_i each iteration by LU decomposition and back
substituting some of the pivot elements were miniscule

($\approx 10^{-5}$) but eigenvector decomposition was unnecessary on the
CDC 7600 and the algorithm converged in about 10 iterations. For
this u° the 21 cells of largest magnitude $.0315 \leq |u_k \Theta_k| \leq .074$
are indicated by the lines slanting upward and to the right in
Fig. 4. The magnitudes in the other cells are all $< .0295$.

We repeated the analysis but for $\tilde{\Theta} = -10$ and got the
same resolution with the 21 cells of largest $|u_k \tilde{\Theta}_k|$.

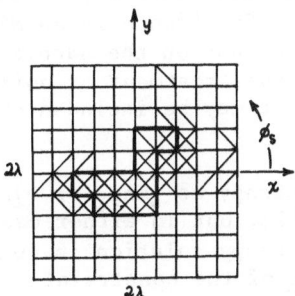

Fig. 4. Two-dimensional anomaly and its resolution
by the "radiating source" ME method [14].

We then processed the data for $\tilde{\Theta} = -1$ and u° incident
from the $\phi_s = 30°$ direction, and the 21 cells of largest mag-
nitude $.025 \leq |\hat{u}_k \tilde{\Theta}_k| \leq .102$ are indicated by the lines slanting
upward and to the left.

Note that all but one of the anomalous-region cells are
included in both ME groups of 21 cells each. Six "normal"
cells are indicated as possibly anomalous.

Such synthetic data examples as these suggest guidelines for applying this ME method to more complicated problems. If the anomaly within N_A cells is known to lie within a region of K cells, the number of scatter data (complex) J_s should be increased until reasonable convergence to a (fuzzy) shape is attained. This will occur when

$$N_A < J_s \lesssim K/2 \quad .$$

The overall size of the region picked is not crucial but the cell size should be $\Delta/\lambda \lesssim .25$. Scatter data should be taken over as wide an angle as possible, which may be $< 180°$; data from several incident waves should be processed. The value of Θ, which measures both the density anomaly $\delta\rho/\rho$ and the cell size Δ/λ, seems not to be crucial for resolution. Finally, the method does not inherently favor single-volume anomalies; disjoint ones of different Θ might very well be resolved.

4.   THEORETICAL AND PRACTICAL ISSUES IN ME INVERSION

In this section we address the issues of uniqueness, confidence, and noise in ME inversion. We have already mentioned the uniqueness of our Boltzmann-ME solution to problems in various areas. Confidence in an ME solution being close to the correct one depends on the nature of the system. There seems to be no rigorous treatment of noise within any ME framework at present, but it may be included empirically quite satisfactorily.

The measurement of entropy as $\sum \ln p_i$ or $-\sum p_i \ln p_i$ depends on the nature of the system [2], but in either case, the uniqueness of a maximum entropy solution is guaranteed by the convexity of H <u>independent</u> of the constraints. However, there is usually more than one type of ME analysis for a problem; we have contrasted to some extent the Gibbsian versus our Boltzmann-ME treatment of some problems. The Burg method is of the Gibbsian type. We have also observed that, in principle, there are at least two Boltzmann-ME philosophies to follow in solving coherent radiation problems of interest to us--one with entropy formulated in terms of amplitude squared (power) and the other based on amplitude. Objective criteria for preferring one ME analysis are simply not available.

The confidence one can place on an ME solution yielding predictions close, in some objective sense, to the correct ones depends critically on whether the system analyzed is "fluid" or not. The second law of thermodynamics applies to systems not only with an astronomical number of configurations (microstates) but also with the freedom to assume any of these states, or at least a subensemble of states with H close to H_{max}, during an observational period. And so one places high confidence in H being within ΔH of H_{max}, as given by the Concentration Theorem [15], loosely written as

$$N\Delta H = \chi^2(s) \quad , \qquad\qquad (36)$$

where the right side designates the Chi-squared distribution at significance level s. N is the number of trials of an experiment or elements in a configuration (number of molecules, symbols in a possible message, elements of luminescence, building blocks of a parameter distribution, etc.)

However, suppose the system is not "fluid," such as a parameter distribution in a solid volume. There is no guarantee that the system will be near its maximum entropy state, (36) is merely a description of concentration about H_{max}, and an ME solution merely furnishes a most probable, very rational guess about the system configuration. Since we cannot relate such a physical system to the average of an ensemble of mathematical systems we must say there is no defined confidence that the system will be within ΔH of the ME system.

Despite this unsettling situation it is of some comfort to know that, as N increases, the ME system becomes more probable according to (36). We can say more about the distribution of individual elements about their H_{max} values as $N \rightarrow \infty$ from a Darwin-Fowler (D-F) analysis, appropriate to a Boltzmann-entropy description. Suppose the system has K cells, with n_k elements in the kth cell, $\Sigma_k n_k = N$. The D-F analysis for the first two moments of n_k yields

$$\bar{n}_k = n_k^{ME} \qquad\qquad (37a)$$

$$\overline{(n_k - \bar{n}_k)^2}/\bar{n}_k^2 \cong 1/\bar{n}_k \quad , \qquad N \gg \bar{n}_k \; . \qquad (37b)$$

In the problems we have discussed all n_k were enormous, which made our Boltzmann-ME solutions overwhelmingly most probable--but not more possible.

Jaynes [16] has discussed the modification of an ME solution so as to include noise in a full Bayesian solution. In essence, noise constraints are added in the manner of Eq.

(6), and any ME solution which satisfies them as well as the other constraints would be a member of an ensenble of most probable solutions with noise included. The noise blurs the distribution of each parameter n_k about its noiseless n_k^{ME} value.

5. REFERENCES

1. Soffer, B. H. and Kikuchi, R., "Maximum Entropy Image Estimation Analysis of Image Formation with Thinned Random Arrays," Hughes Research Laboratories, Malibu, CA, 90265, 1981. Sponsored by the Air Force Office of Scientific Research.

2. Kikuchi, R. and Soffer, B. H., Maximum Entropy Image Restoration I The Entropy Expression, Jour. Opt. Soc. Am., 67, pp. 1656-1665, 1977.

3. Wernecke, S. J. and D'Addario, L. R., Maximum Entropy Image Reconstruction, IEEE Trans. Comput., C-26, pp. 351-364, 1977.

4. Bevensee, R. M., "Maximum Entropy Resolution of Two-Dimensional Antenna Distributions," IEE Conf. Pub. 195, Antennas and Propagation, Part 1: Antennas, pp. 383-387, 1981.

5. Gabriel, W. F., Spectral Analysis and Adaptive Array Superresolution Techniques, Proc. IEEE, 68, pp. 654-666, 1980.

6. Steinberg, B. D. and Luthra, A. J., "Single and Multiple Snapshot Maximum Entropy Imaging of Two Targets," Valley Forge Research Center, Quarterly Progress Report, No. 36, University of Pennsylvania, Valley Forge, PA, pp. 22-29, 1981.

6A. Luthra, A. J. and Steinberg, B. D., "The Maximum Entropy Image of Two Targets," Proc. IEEE, 70, pp. 98-99, 1982.

7. Burg, J. P., Maximum Entropy Spectral Analysis in "Modern Spectral Analysis," ed. by D. G. Childers, IEEE Press, pp. 34-41, 1978.

8. Steinberg, B. D., Private Communication.

9. Rietsch, E., The Maximum Entropy Approach to Inverse Problems, J. Geophys., 42, pp. 489-506, 1977.

10. Rubincam, D. P., "Information Theory Lateral Density Distribution for Earth Inferred From Global Gravity Field," Journal Geophys. Res., 87, pp. 5541-5552, 1982.

11. Lacoss, R. T., "Data Adaptive Spectral Analysis Methods," Geophys., 36, pp. 661-675, 1971.

12. Capon, J., "High-resolution Frequency-wavenumber Spectrum Analysis," Proc. IEEE, 57, pp. 1408-1418, 1969.

13. Bevensee, R. M., "Solution of Underdetermined Electromagnetic and Seismic Problems by the Maximum Entropy Method," IEEE Trans. Ant. Prop., AP-29, pp. 271-274, 1981.

14. Bevensee, R. M., "The Maximum Entropy Formulation of Inverse Problems of NDE," Review of Progress in Quantitative Non Destructive Evaluation, 2, D. O. Thompson and D. E. Chimenti, eds., pp. 937-949, Plenum Press, NY, 1983.

15. Gubernatis, J. E., "Elastic Wave Scattering Calculations and the Matrix Variational Pade Approximation Method," Proceedings of the DARPA/AFWAL Review of Progress in Quantitative NDE, AFWL-TR-81-4080, Rockwell International Science Center, pp. 300-310, 1981.

16. Jaynes, E. T., "On the Rational of Maximum-Entropy Methods," Proc. IEEE, 70, pp. 939-952, 1982.

Work performed under the auspices of the U.S. Department of Energy by the Lawrence Livermore National Laboratory under contract number W-7405-ENG-48.

II.10 (NM.1)

DETAILED NEAR/FAR FIELD MODELING OF COMPLEX
ELECTRICALLY LARGE THREE DIMENSIONAL STRUCTURES

Allen Taflove
Korada Umashankar

IIT Research Institute
10 West 35th Street
Chicago, Illinois 60616

ABSTRACT

The finite-difference time-domain (FD-TD) method is shown as a
means of accurately computing electromagnetic scattering by
arbitrary-shaped, extremely complex metal or dielectric objects
excited by an external plane wave. In the present method, one
first uses the FD-TD method to compute the near total fields
within a rectangular volume which fully encloses the object.
Then, an electromagnetic field equivalence principle is invoked
at a virtual surface of this rectangular volume to transform
the tangential near scattered fields to the far field. To
verify the applicability of this method, the surface currents,
near scattered fields, far scattered fields, and radar cross
section of canonical two-dimensional and three dimensional
objects are presented. For these cases, it is shown that the
FD-TD method provides magnitude of current and field
predictions which are within ±2.5% and further phase values
within ±3° of values predicted by the method of moments (MOM)
at virtually every point including in shadow regions. The FD-
TD method is presently being extended to three-dimensional
structures as large as twenty wavelengths.

This work was supported partly by Air Force Contract No.
F30602-80-C-0302, Rome Air Development Center, GAFB, NY and
partly by Internal Research of the Electronics Division, IIT
Research Institute, Chicago, IL

W.-M. Boerner et al. (eds.), Inverse Methods in Electromagnetic Imaging - Part 1, 397–430.
© *1985 by D. Reidel Publishing Company.*

1. INTRODUCTION

General electromagnetic scattering problems have been difficult
to treat with either analytical [1] or numerical methods because of the complicating effects of curvatures, corners, apertures, and dielectric loading of structures. In an attempt to
gain insight into scattering mechanisms using analytical and
numerical approaches it has been necessary to use canonical
structures, rather than realistic models. A potential alternate approach is the finite-difference time-domain (FD-TD)
method [2-17] which allows the computation of internal and external near fields by direct modeling of realistic structures.

In order to treat realistic scattering problems effectively, a
method has been developed [14,15,17] which involves combining
the FD-TD method with a near-field to far-field transformation
using field equivalences [18]. In this method, the scattering
problem is analyzed in two steps by treating the relatively
complex near-field region and the relatively simple far-field
region separately. The method involves first the determination
of equivalent electric and/or magnetic currents tangential to a
virtual surface surrounding the scatterer of interest by using
the FD-TD method for a given external illumination. The computed near-field equivalent currents are transformed then to
derive the far-field scattering pattern and the radar cross
section. Since the FD-TD method can deal with dielectric, permeable, and inhomogeneous materials in a natural manner, it is
possible to incorporate most of the physics of wave interaction
with any complex scatterer of interest.

To validate applicability of this method, the FD-TD computed
surface electric current distribution and near scattered electric and magnetic fields are presented for the case of a two
dimensional rectangular metal cylinder and a three dimensional
metal cube subject to plane-wave illumination at normal incidence. These currents and scattered fields are compared to
method of moments (MOM) computed results. The far scattered
field pattern and radar cross section are derived from the FD-
TD data using a near to far-field transformation and are also
compared to the results obtained using MOM [19]. It is shown
that a very high degree of correspondence is obtained using
this present proposed method.

The FD-TD method is presently being extended to obtain detailed
modeling of near/far fields of complex three-dimensional
scatterers as large as twenty wavelengths. Experimental veri-
fications are also being obtained for radar cross section of
complex material scatterers including composites.

2. FORMULATION

A brief enumeration of the steps to analyze scattering by complex objects is presented in this section, based on the application of the FD-TD method to obtain the near fields, and then the transformation of the near fields to far fields to obtain the scattering cross section. In Figure 1 is shown a geometry of an arbitrary-shaped scattering object. This object can be either conducting, dielectric, or permeable as well as inhomogeneous. It can have apertures/cavities containing internal dielectric loading [9-10]. The incident field($\vec{E}^i$, $\vec{H}^i$) excites the scatterer to produce simultaneously some interior field penetration and also an exterior scattered field ($\vec{E}^s$, $\vec{H}^s$).

2.1 Near-field analysis

As shown in Figure 1, the scatterer is enclosed in a rectangular volume with a boundary surface, S_a, for analyzing the near total fields based on FD-TD [9,10,14,15,17]. The finite difference formulation of the FD-TD method allows a straightforward modeling of the surfaces and interiors of arbitrary complex structures. The structure of interest is mapped into the space lattice, Figure 1, by first choosing the space increment [10] and assigning values of permittivity and conductivity to each component of total electric field $\vec{E}$. No special handling of electromagnetic boundary conditions at media interfaces is required because the curl equations generate these conditions in a natural way [10]. The various details of the structure are modeled with a maximum resolution of one unit cell, with thin surfaces being modeled as stepped-edge sheets. The explicit numerical formulation of the FD-TD method is particularly suited for programming with minimum storage and execution time using recently developed array processing computers. The required computer storage and running time increases only linearly with N, where N is the total number of unknown field components [9]. Since all FD-TD operations are explicit and can be performed in parallel, rapid array-processing techniques can be readily applied. As demonstrated [9,10], these can be used to solve for more than 10^6 field components in a single FD-TD problem (current array processing capability).

As discussed in reference [9,10] the FD-TD method is a direct numerical solution to the Maxwell's time-dependent curl equations useful for studying propagation of an electromagnetic wave into a volume of space containing an arbitrary shaped dielectric or conducting body. By time-stepping or repeatedly implementing a finite-difference analog of the curl equations at each cell (Figure 1) of the corresponding space lattice, the incident wave is tracked as it first propagates to the

FIGURE 1.   GEOMETRY OF A GENERAL SCATTERER IN FREE SPACE
MEDIUM AND FD-TD LATTICE ARRANGEMENT.

(a)

(b)

FIGURE 2. DIVISION OF FD-TD LATTICE INTO TOTAL-FIELD
AND SCATTERED-FIELD REGIONS: (a) LATTICE
DIVISION; (b) TYPICAL CONNECTING CONDITIONS
BETWEEN REGIONS.

structure and then interacts with it via penetration and
diffraction. Wave tracking is completed when the desired late-
time or sinusoidal steady-state behavior is observed at each
lattice cell. The time stepping for the FD-TD method is accomplished by positioning the components of $\vec{E}$ and $\vec{H}$ about a unit
cell of the lattice, as shown in Figure 1, and evaluating
$\vec{E}$ and $\vec{H}$ at alternate half-time steps. Centered difference
expressions can be used for both the space and time derivatives
to attain second order accuracy in the space and time incre-
ments.

The following discussion briefly summarizes important new
features of the FD-TD method. A more complete discussion is
contained in References 15 and 17.

2.2 FD-TD lattice regions and plane wave source condition [12,14-17]

As shown in Figure 2a, the latest formulation of the FD-TD
lattice involves the division of the computation space into two
distinct regions, separated by a rectangular surface which
serves to connect fields in each region. In two dimensions,
the surface has four faces; in three dimensions the surface has
six faces.

Region 1 of the FD-TD lattice is denoted as the total-field
region. Here, it is assumed that all computed field quantities
are comprised of the sum of the incident wave and the scattered
field. The interacting structure of interest is embedded within this region.

External to Region 1 is Region 2 of the FD-TD lattice which is
denoted as the scattered-field region. Here, it is assumed
that all computed field quantities are comprised only of the
scattered field. The outer lattice planes bounding Region 2,
called the lattice truncation planes, serve to implement the
free-space radiation condition.

The rectangle faces comprising the boundary between Regions 1
and 2 contain $\vec{E}$ and $\vec{H}$ field components which require the
formulation of various field-component differences across the
boundary planes for proceeding one time step. These computations serve to connect the total-field and scattered-field
regions and simultaneously generate the desired plane wave of
arbitrary polarization and angle of incidence [14-17]. An
alternative approach for incident-wave generation based on the
Huygen's source formulation is discussed in Reference 12.

There are a number of key advantages to this methodology: (a)
The total-field formalism is retained for the entirety of the

interacting structure, permitting accurate computations of low-level fields penetrating into cavities through apertures, and in the shadow regions of scatterers. (b) The scattered-field formalism is retained for the lattice truncation region, permitting a very accurate simulation of the radiation condition. (c) The incident wave contribution need be computed or stored only for the field components at the rectangular surface connecting Regions 1 and 2. This results in much less computation or storage than if the incident field were to be computed at all points within the interacting structure to implement a pure scattered-field formalism. (d) The scattered near field in Region 2 can be easily integrated to derive the far-field scattering and radar cross section, as discussed later in this paper.

2.3 Lattice truncation conditions [3-5,8,15-17,20,21].

A basic consideration with the FD-TD approach to solve electromagnetic field problems is that most such problems are usually considered to be "open" problems where the domain of the computed field is ideally unbounded. Clearly, no computer can store an unlimited amount of data, and the field computation zone must be limited. The computation zone must be large enough to enclose the structure of interest, and a suitable boundary condition on the outer perimeter of the computation zone must be used to simulate the extension of the computation zone to infinity. Outer boundary conditions of this type have been called either radiation conditions, absorbing boundary conditions, or lattice truncation conditions.

In three dimensions, an outgoing scattered-wave field component, F^S (either an electric or magnetic field) has a (r, θ, ϕ) variation of the type [22],

$$F^S = F_0 e^{j(\omega t - k_0 r)} \left[\frac{A(\theta,\phi)}{r} + \frac{B(\theta,\phi)}{r^2} + \cdots \right] \qquad (1)$$

Here, the bracketed infinite series represents in effect a multipole expansion of the scattered field, where A, B, . . . are initially unknown functions of angular position.

First-order FD-TD simulations of the outer lattice boundary condition approximate the $A(\theta,\phi)/r$ dependence only. Second and higher-order approximations simulate the $B(\theta,\phi)/r^2$ and higher-order r-dependent behavior of F^S in addition to the $A(\theta,\phi)/r$ term [15]. A typical FD-TD computation realizing a second order correct radiation condition and its accuracy is given in references 16 and 20.

2.4 Sinusoidal steady-state information

Such data can be obtained either by (a) directly programming a single-frequency incident plane wave or (b) performing a separate Fourier transformation step on the pulse waveform response. Both methods require time-stepping to a maximum time equal to several wave periods at the desired frequency. The second method has two additional requirements. First, a short-rise-time pulse suffers from accumulating waveform error due to overshoot and ringing as it propagates through the space lattice. This leads to a numerical noise component which must be filtered before Fourier transformation. Second, Fourier transformation of many lattice-cell field-versus-time waveforms would significantly add to the total requirements for computer storage and execution time [10].

Recent work has shown that very accurate magnitude and phase information for sinusoidal steady state FD-TD problems can be obtained by method (a) above and observing the peak positive and negative-going excursions of the fields over a complete cycle of the incident wave (after having time-stepped through 2-5 cycles of the transient period following the beginning of time stepping). For certain two and three-dimensional scattering problems, a dc offset of particular computed field components can be possible. This leads to the following requirements to obtain correct magnitude and phase data:

1. The peak-to-peak value of the sinusoidal response at any point must be observed to eliminate the effects of any dc offset upon the computation of the phasor magnitude.

2. The zero-crossing of the field waveform may not be useful in determining relative phase. Rather, it may be necessary to locate the zero-derivative points of the waveform for this purpose, possibly incorporating an interpolation algorithm to enhance resolution of the relative phase computation without requiring a smaller time step.

As will be shown, this methodology has been found to be successful in achieving ±2.5% (or better) and ±3° correspondence of FD-TD results with moment-method computations for two-dimensional canonical problems; and similar correspondence is obtained for three-dimensional canonical problems as well.

2.5 Derivation of far-field scattering data:

Far-Field scattering data can be obtained with the FD-TD method by applying a powerful and flexible near-field to far-field

transformation [14,17,18].   Referring to Figure 3, a rectangular virtual surface, S_a, which fully enclose the
scatterer, is located in the scattered field region (Region 2
of Figure 2).   The tangential components of the scattered
fields $\vec{E}^s$ and $\vec{H}^s$, are first obtained at S_a using FD-TD. Then,
as indicated in Figure 3b, an equivalent problem is set up
which is completely valid for Region B, external to S_a.   The
new excitation data are $\vec{J}_{S_{eq}}$ and $\vec{M}_{S_{eq}}$, the equivalent surface
electric and magnetic currents respectively, on S_a which are
obtained according to [22,23]:

$$\vec{J}_{S_{eq}} (\vec{r}) = \hat{n} \times \vec{H}^s(\vec{r}) \tag{2a}$$

$$\vec{M}_{S_{eq}} (\vec{r}) = -\hat{n} \times \vec{E}^s(\vec{r}) \tag{2b}$$

where $\hat{n}$ is the outward unit normal vector at the surface S_a.

The scattered far fields are given by the transform of the
equivalent currents of equations (2a) and (2b) over the free
space Green's function [23].   If $(\mu_0,\ \varepsilon_0)$ are the region B
medium characteristics with $k_0 = 2\pi/\lambda$ and $\eta_0 = 120\pi$, for θ and
ϕ polarizations the following scattered far-field
expressions are obtained,

$$E_\theta = (-jk_0\mu_0) \left[A_\theta + \frac{F_\phi}{\mu_0} \right] \tag{3a}$$

$$E_\theta = (jk_0\mu_0) \left[A_\phi - \frac{F_\theta}{\mu_0} \right] \tag{3b}$$

where

$$A_\theta = A_x \cos\theta \cos\phi + A_y \cos\theta \sin\phi - A_z \sin\theta \tag{4a}$$

$$F_\theta = F_x \cos\theta \cos\phi + F_y \cos\theta \sin\phi - F_z \sin\theta \tag{4b}$$

$$A_\phi = - A_x \sin\phi + A_y \cos\phi \tag{4c}$$

$$F_\phi = - F_x \sin\phi + F_y \cos\phi \tag{4d}$$

FIGURE 3. ELECTROMAGNETIC EQUIVALENCE TO TRANSFORM NEAR-FIELDS TO FAR-FIELDS.

and the potentials in the far-field region are given by

$$
\begin{bmatrix} \vec{A}, \\ \vec{F} \end{bmatrix} = (\frac{e^{-jk_o r}}{4\pi r}) \iint\limits_{S_a} \begin{bmatrix} \vec{J}_{s_{eq},} \\ \vec{M}_{s_{eq}} \end{bmatrix} e^{jk_o r' \cos \xi} ds'_a \qquad (5a)
$$

$$
r' \cos \xi = (x' \cos \phi + y' \sin\phi) \sin \theta + z' \cos \theta \qquad (5b)
$$

The radar cross section is calculated as the ratio of

$$
RCS = 4\pi r^2 \begin{bmatrix} \dfrac{E_\theta^2 + E_\phi^2}{E_\theta^{i\,2} + E_\phi^{i\,2}} \end{bmatrix}, \quad r \to \infty \qquad (6)
$$

E_θ^i and E_ϕ^i are the corresponding components of the incident plane wave excitation, for θ and ϕ polarizations.

This approach to computing the far scattered fields is very promising since 1) the near-field data for arbitrary scatterers can be obtained in a straight-forward manner using the FD-TD method; and 2) the transformation of the near-field data to the far field is computationally simple and independent of the nature of the scatterer which resides within the integration surface, S_a. The complete bi-static radar cross section is a natural result of this procedures, for a given incident angle and polarization of the illuminating wave. For problems involving variable incidence and/or polarization of the illuminating wave, new FD-TD data is required for each selected incident-wave condition.

3. BASIS OF THE MOM MODEL USED FOR FD-TD VERIFICATION

General arbitrary-shaped scattering bodies can be analyzed based on a frequency domain integral-equation/boundary-value problem approach. A popular technique for the numerical solution which implements this approach is the method of moments [19]. This method is particularly suited for low-frequency scattering problems, but can be applied to bodies spanning approximately 1 wavelength in three dimensions [26-33].

The following formulations of the method of moments have been found to be generally suited for certain scattering bodies based upon their geometry and material characteristics:

1. <u>Conducting Scatterers</u> (homogeneous, isotropic)

    a. Electric field integral equation formulation (EFIE) for closed and open bodies

    b. Magnetic field integral equation formulation (MFIE) for closed bodies

2. <u>Dielectric Scatterers</u> (homogeneous, isotropic)

    Combined field integral equation formulation (CFIE)

Recent work has concentrated on Case 1 [31, 32], especially in the development of the triangular patch model for arbitrary scatterers. Further extension to dielectric scatterers is presently being completed.

To treat arbitrary-shaped conducting bodies [31], the EFIE formulation is generally used. Referring to Figure 1, if S denotes the surface of an open or closed body, the scattered electric field is given by

$$\vec{E}^S = j\omega\vec{A} - \nabla\phi \tag{7}$$

where the magnetic vector potential and the scalar potential are defined as

$$\vec{A}(\vec{r}) = \frac{\mu_0}{4\pi} \iint_S \vec{J}(\vec{r}') \, G(\vec{r},\vec{r}') \, ds' \tag{8}$$

$$\phi(\vec{r}) = \frac{1}{4\pi\varepsilon_0} \iint_S \rho^e(\vec{r}') \, G(\vec{r},\vec{r}') \, ds' \tag{9}$$

$$G(\vec{r},\vec{r}') = \frac{e^{-jk_0\left|\vec{r} - \vec{r}'\right|}}{\left|\vec{r} - \vec{r}'\right|} \tag{10}$$

where (μ_0,ε_0) are scalar constants representing the permeability and permittivity of the surrounding medium; $\vec{J}$ and ρ^e are the unknown electric current and charge distributions induced on the conducting scatterer; and R $= \left|\vec{r} - \vec{r}'\right|$ is the distance between an arbitrarily-located observation point $\vec{r}$ and a source point $\vec{r}'$ on S. Both $\vec{r}$ and $\vec{r}'$

are defined with respect to a global coordinate origin. An integro-differential equation for $\vec{J}$ is derived [19,32] by enforcing the boundary condition $n \times (\vec{E}^i + \vec{E}^s) = \vec{0}$ on S

$$-\vec{E}^i_{tan} = (-j\omega\vec{A} - \nabla\Phi)_{tan}, \quad \vec{r} \text{ on S} \tag{11}$$

In arbitrary surface modeling the above EFIE has the advantage that it applies to both open and closed bodies, whereas the MFIE applies only to closed surfaces [31]. On the other hand, for arbitrarily-shaped objects the EFIE is considerably more difficult to apply than the MFIE.

For modeling arbitrarily-shaped surfaces, planar triangular patch models have been found to be particularly appropriate. Since triangular patches are capable of accurately conforming to any geometrical surface and boundary, the patch scheme is easily described to the computer, and a varying patch density can be used according to the resolution required in the surface geometry or current. This patching method permits the straight-forward construction of MOM basis functions defined on the triangular patches which are free of line charges.

In References [31] and [32], planar triangular patch modeling and the MOM [19] were applied to develop numerical procedures using the EFIE formulation to treat scattering by arbitrarily-shaped conducting bodies. An efficient computer code based on this formulation was described which is capable of handling either open or closed and arbitrarily-curved structures of finite extent. Discounting computer limitations, the code can, in principle, treat any conducting object whose surface is orientable, connected (or multiply-connected), and free of intersecting surfaces. An updated version of this computer program [34] was used to obtain benchmark data for the surface currents and scattered fields for the conducting cube scatterer discussed next.

4. NUMERICAL RESULTS - TWO DIMENSIONAL CASE

In order to validate the feasibility of this hybrid method to analyze electromagnetic scattering, a canoniocal two-dimensional conducting structure is studied [25]. The numerical results of the FD-TD-computed surface electric current distribution, and near electric and magnetic fields are presented for the case of a two-dimensional rectangular metal cylinder subject to a TM-polarized illumination at normal (broadside) incidence. These electric currents and near fields are compared to the MOM-computed results [19, 24]. The scattered-field pattern and the corresponding radar cross section (RCS)

are derived from the near-to-far field transformation of the
FD-TD data. These are then compared to the results obtained by
using the MOM. It is shown that a very high degree of correspondence is obtained using this method.

4.1 Square cylinder

The scattering by a two-dimensional conducting cylinder of
arbitrary cross section is considered first to validate feasibility of the present method. This cannonical problem, shown
in Figure 4, is well documented [19,22,24]. We first consider
the example of the scattering of a plane wave by a square
conducting cylinder. The cylinder has the electrical size
$k_0 A_s = 1$, where A_s is the half-width of the side of the
cylinder. The plane-wave excitation is TM polarized, with
components E_z^i and H_x^i, and propagates in the +y direction so
that it is at normal incidence to one side of the cylinder
($\phi^i = 90^0$). An 84-point MOM solution of Equation (11) is used
as the benchmark for comparison with all FD-TD results, with
pulse current expansion and point matching [19,24].

For the FD-TD analysis, the square cylinder is embedded in a
two-dimensional lattice as shown in Figure 2. Each side of the
cylinder spans 20 lattice-cell divisions. The connecting
virtual surface between the FD-TD total-field and scattered-
field regions is located at a uniform distance of 5 cells from
the cylinder surface. Figures 4a and 4b graph the comparative
results for the FD-TD and MOM analyses of the magnitude and
phase of the cylinder surface electric current distribution for
this case. Here, the FD-TD-computed surface current is taken
as $n \times \vec{H}_{tan}$, where n is the unit normal vector at the cylinder
surface, and $\vec{H}_{tan}$ is the FD-TD value of the magnetic field
parallel to the cylinder surface, but at a distance of 0.5
space cell from the surface. (The displacement of the
$\vec{H}$ component from the cylinder surface is a consequence of the
spatially-interleaved nature of the $\vec{E}$ and $\vec{H}$ components of the
FD-TD lattice, indicated in Figure 1. In Figure 4a, the
magnitude of the FD-TD-computed surface current agrees with the
84-point MOM solution to better than ±1% (±0.09 dB) at all
comparison points more than 2 cells from the cylinder edges
(current singularities). In Figure 4b, the phase of the FD-TD
solution agrees with the MOM solution to within ±3° at all
comparison points, including the shadow region. The uncertainty bars shown in this figure indicate the present level of
imprecision in using the FD-TD method to locate the constant-
phase points of the computed time-domain $\vec{H}_{tan}$ waveform
(equivalent to ±1 time step). This imprecision can be reduced
in future FD-TD programs by incorporating a simple
interpolation algorithm to achieve fractional time-step
resolution of points of constant phase.

FIGURE 4a.   COMPARISON OF MOM AND FD-TD RESULTS FOR
MAGNITUDE OF ELECTRIC CURRENTS ON SURFACE
OF SQUARE CYLINDER.

FIGURE 4b.   COMPARISON OF MOM AND FD-TD RESULTS FOR
PHASE OF ELECTRIC CURRENTS ON SURFACE
OF SQUARE CYLINDER.

Figure 5a and 5d show the comparison of the magnitude and phase of the near-scattered electric and magnetic fields computed by the FD-TD method and MOM. The electric field is tangential to a virtual surface, S_a, located at a uniform distance of h_a = 7 space cells from the cylinder surface; and the magnetic field is tangential to a virtual surface, S_a, located at 6.5 cells from the cylinder surface. Both virtual surfaces are embedded in the scattered-field region of the FD-TD lattice. The level of correspondence between the FD-TD and MOM results is ±2.5% (±0.2 dB) and ±3°.

In order to obtain the far field pattern and the scattering cross section, the near-field to far-field transformation discussed in 2.4 is followed for two-dimensional structures. The near scattered fields shown in Figures 5a to 5d are converted into the corresponding equivalent electric and magnetic current distributions along S_a according to Equations 2a and 2b.

Figure 6 shows the scattering cross section of the square conducting cylinder obtained by the near field (FD-TD data) to far field transformation. The results agree very well with the scattering cross section derived directly from the electric current on the square cylinder.

5. THREE DIMENSIONAL CASE NUMERICAL RESULTS

This section presents comparative numerical results obtained using the most recent formulations of the FD-TD [14, 15] and MOM triangular surface patch techniques [31, 32] for the surface currents and far scattered fields of a canonical three-dimensional scatterer, a conducting cube, illuminated by a plane wave. It is shown that a very high degree of correspondence exists between the results obtained by these two disparate approaches.

5.1 Scatterer geometry

Figure 7 illustrates the basic geometry of the cube scattering problem. The cube was assumed to have the electrical size $k_0 s$ = 2, where s = side width of cube. The plane-wave excitation was assumed to have the field components E_z^i and H_x^i and propagate in the +y direction (normal incidence to one side of the cube). For the FD-TD analysis, each side of the cube was spanned by 400 square cells (20 x 20 lattice-cell divisions). For the MOM analysis, each face of the cube was spanned by either 18 triangular patches or 32 triangular patches (to test the convergence of the patch model). The 18-patch/face MOM model required the filling and inversion of a 162 x 162 system matrix. The 32-patch/face MOM model required the filling and inversion of a 288 x 288 system matrix. Comparative results

FIGURE 5a.   COMPARISON OF MOM AND FD-TD RESULTS FOR
MAGNITUDE OF NEAR ELECTRIC FIELD TANGENTIAL
TO CONTOUR S_a.

FIGURE 5b.   COMPARISON OF MOM AND FD-TD RESULTS FOR
PHASE OF NEAR ELECTRIC FIELD TANGENTIAL
TO CONTOUR S_a.

FIGURE 5c. COMPARISON OF MOM AND FD-TD RESULTS FOR
MAGNITUDE OF NEAR MAGNETIC FIELD TANGENTIAL
TO CONTOUR S_a.

FIGURE 5d. COMPARISON OF MOM AND FD-TD RESULTS FOR
PHASE OF NEAR MAGNETIC FIELD TANGENTIAL
TO CONTOUR S_a.

FIGURE 6.   SCATTERING CROSS SECTION OF CONDUCTING SQUARE CYLINDER.

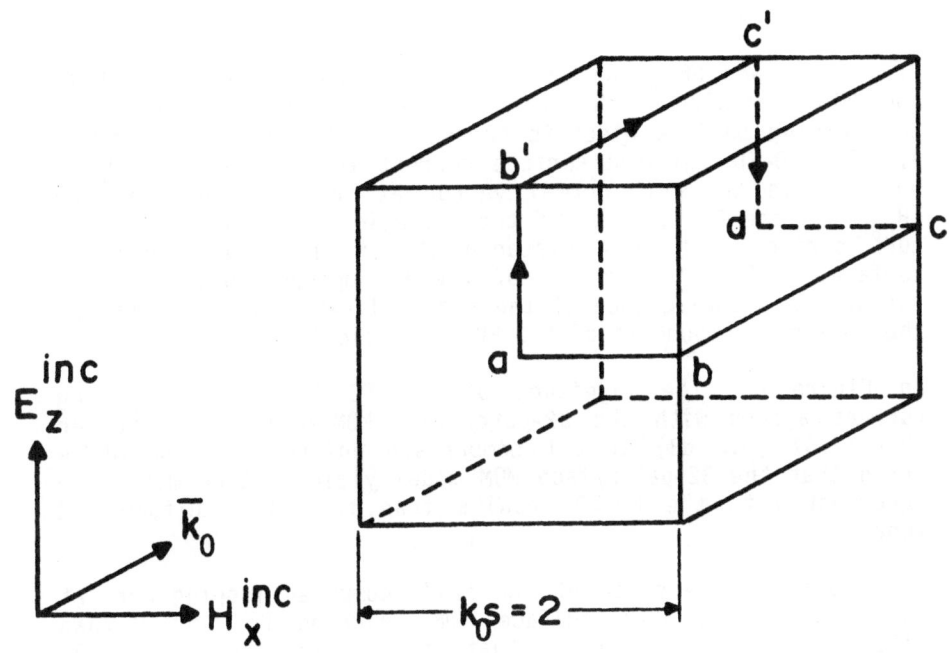

FIGURE 7. GEOMETRY OF METAL CUBE SCATTERER.

are graphed along two straight-line loci along the cube
surface, as shown in Figure 7. Locus abcd is parallel to the
incident magnetic field. Locus ab'c'd is parallel to the
incident electric field.

5.2 Comparative surface currents

Figures 8a and 8b graph the comparative results for the FD-TD
and MOM/triangular patch analyses of the magnitude and phase of
the looping surface electric current (in the direction ab'c'd
). The FD-TD computed surface current is taken as $n \times \vec{H}_{tan}$,
where n is the unit normal vector at the cube surface, and
$\vec{H}_{tan}$ is the FD-TD value of the magnetic field parallel to the
cube surface, but at a distance of 0.5 space cell from the
surface. (The displacement of the H-component from the cube
surface is a consequence of the spatially-interleaved nature of
the E- and H-components of the FD-TD lattice).

In Figure 8a, the magnitude of the FD-TD computed looping
current agrees with the 32-patch/face MOM solution to better
than ±2.5% (0.2 dB) at all comparison points. It should be
noted that the 32-patch/face MOM model yields better magnitude
agreement with the FD-TD results than the 18-patch/face MOM
model.

In Figure 8b, the phase of the FD-TD computed looping current
agrees with the 32-patch/face MOM solution to better than
±1°. It should be noted that the lower-resolution, 18-
patch/face MOM model yields a phase anomaly of almost -100° in
the shadow region. The MOM phase is corrected to coincide with
the FD-TD results upon going to the higher-resolution, 32-
patch/face MOM model. Evidently, the MOM triangular patching
approach has a slow convergence in the shadow region, even for
quite an electrically small three-dimensional object.

Figures 9a and 9b graph the comparative results for the FD-TD
and MOM/patch analyses of the magnitude and phase of the z-
directed current along the cut abcd. In Figure 9a, the FD-TD
computed magnitudes agree with the edge-corrected [34] 32-
patch/face MOM model to better than 2.5% (0.2 dB) at all
comparison points. Anomalous behavior is noted for the 32-
patch/face MOM model without edge correction, especially near
the cube edges b and c. Evidently, the MOM patching approach
requires edge correction to permit proper resolution to the
current singularity behavior for physically realizable numbers
of patches.

In Figure 9b, the phase of the FD-TD computed z-directed current agrees with the 32-patch/face MOM model to better than
±10°. The 18-patch/face MOM model is again seen to have severe

FIGURE 8a.   MAGNITUDE OF LOOPING CURRENT, CENTER VERTICAL CUT.

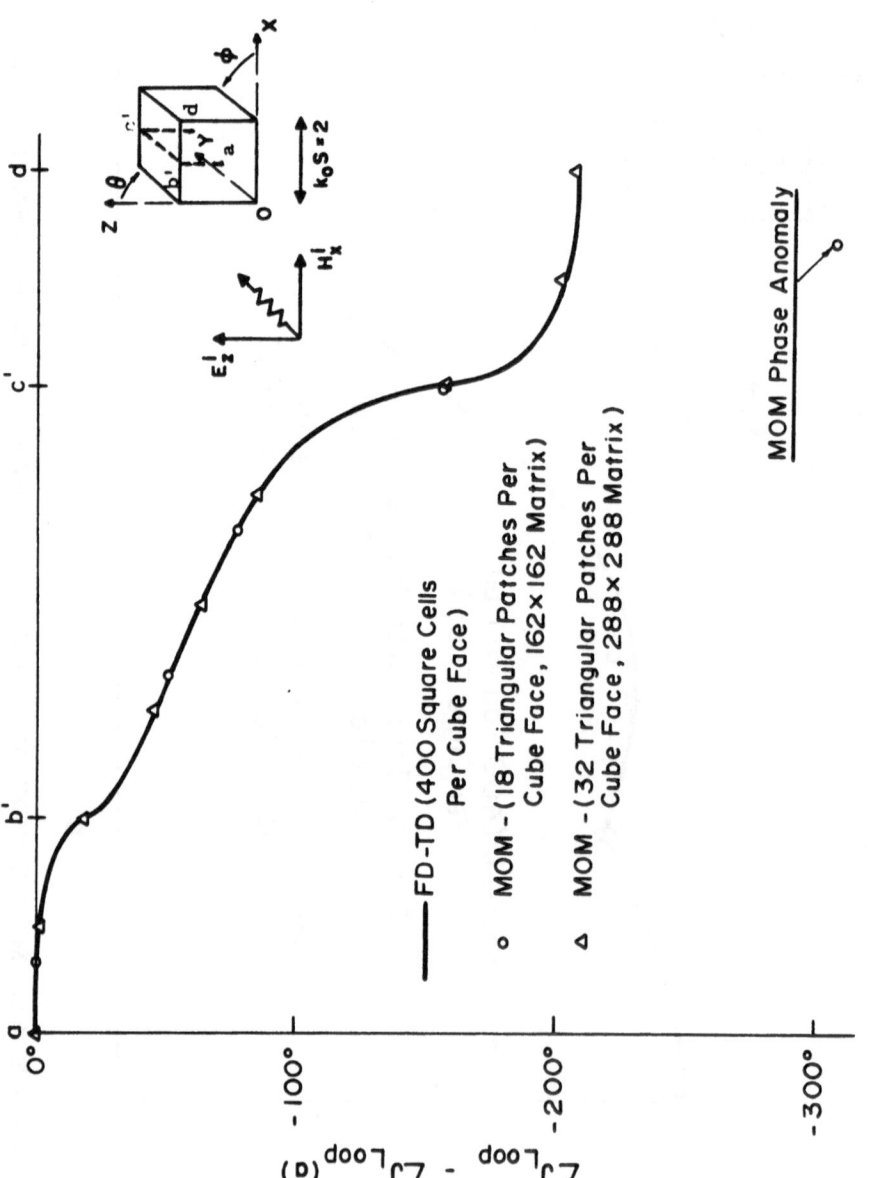

FIGURE 8b.   PHASE OF LOOPING CURRENT, CENTER VERTICAL CUT.

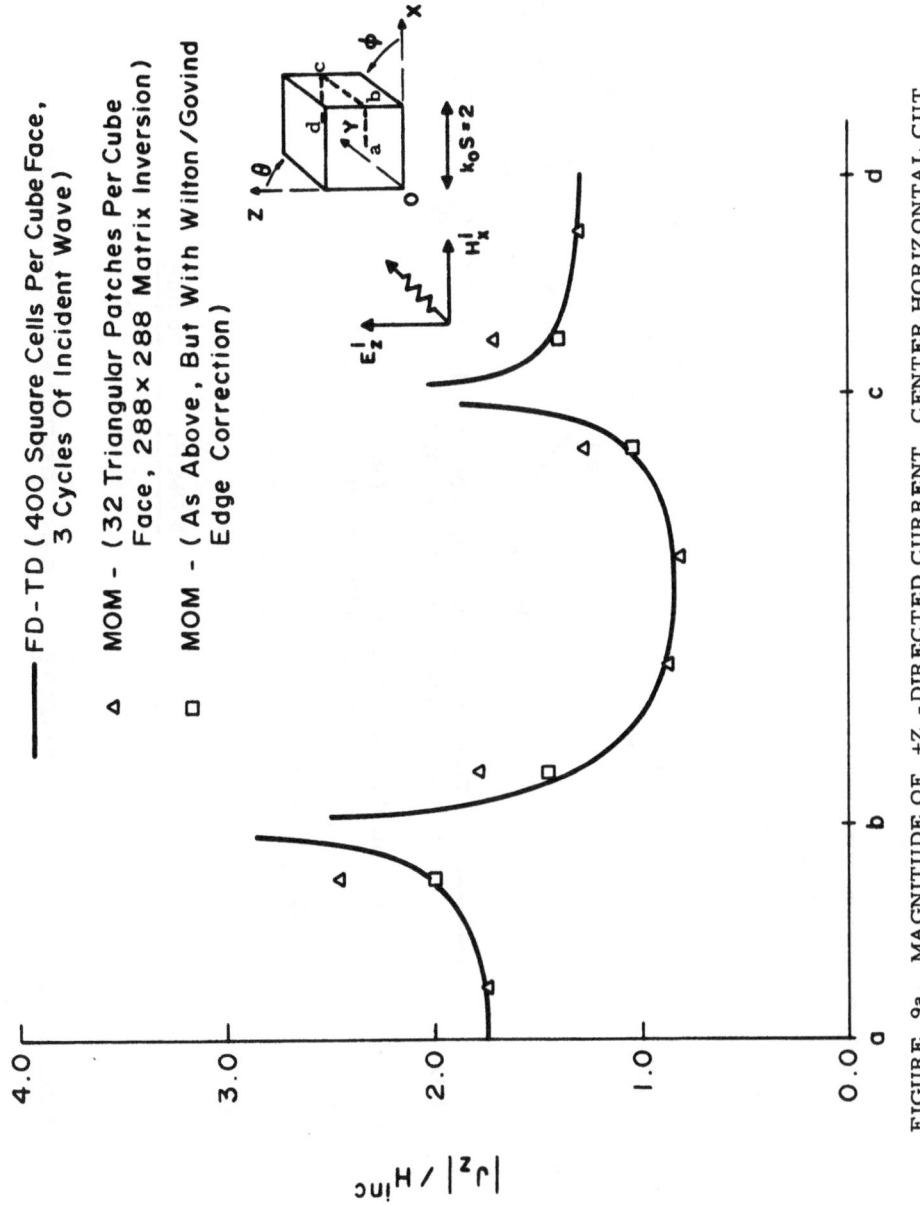

FIGURE 9a. MAGNITUDE OF +Z - DIRECTED CURRENT, CENTER HORIZONTAL CUT.

FIGURE 9b.   PHASE OF +Z -DIRECTED CURRENT, CENTER HORIZONTAL CUT.

phase problems in the shadow region. In fact, the switch from 18 patches/face to 32 patches/face is seen to result in better phase agreement with the FD-TD model at non-shadow points as well.

5.3 Comparative scattered far-fields

Figure 10 shows radar cross section (RCS) for the conducting cube scatterer with normal plane wave excitation on side of the cube. The RCS result is based on the FD-TD near field data and transformed to far field based on equations (3) to (6). The results are in well agreement with the RCS computation based on the induced surface electric currents on the cube.

5.4 Comparative computer resources

In analyzing the surface current results, it is clear that the MOM triangular patching model approaches the results of the FD-TD model as the number of triangular patches per cube side is increased. It should be noted that the 32-patch/face MOM model required a large expenditure of computer resources to fill and invert the 288 x 288 matrix (over \$200 on the Control Data Cybernet).   Application of the MOM patch model to three-dimensional objects larger than the $k_o s$ = 2 of the cube would result in increasingly severe computer resource require-ments. On the other hand, the FD-TD model of the cube required very comparable computer resources (\$200), and yet can be directly applied to three-dimensional scatterers as large as $k_o s \approx$ 25 with little loss of accuracy for the radar cross section calculations.

The MOM approach thus seems to be best suited for three-dimensional structures no larger than about 1 wavelength, so that matrix filling and solving is tractable. FD-TD, however, gives promise of modeling structures as large as 20 or more wavelengths as the capabilities of vector-array processing computers advance. This promise is currently being explored.

6. CONCLUSIONS

General electromagnetic scattering problems are difficult to treat. This paper presents a method to analyze scattering by complex objects by combining the FD-TD method to obtain near scattered fields with a near-field to far-field transformation based on electromagnetic equivalences. To validate the applic-ability of this method, canonical two-dimensional and three-dimensional structures are analyzed and the results are verified with respect to the method of moments. This poten-tially alternative approach has wide applications to analyze

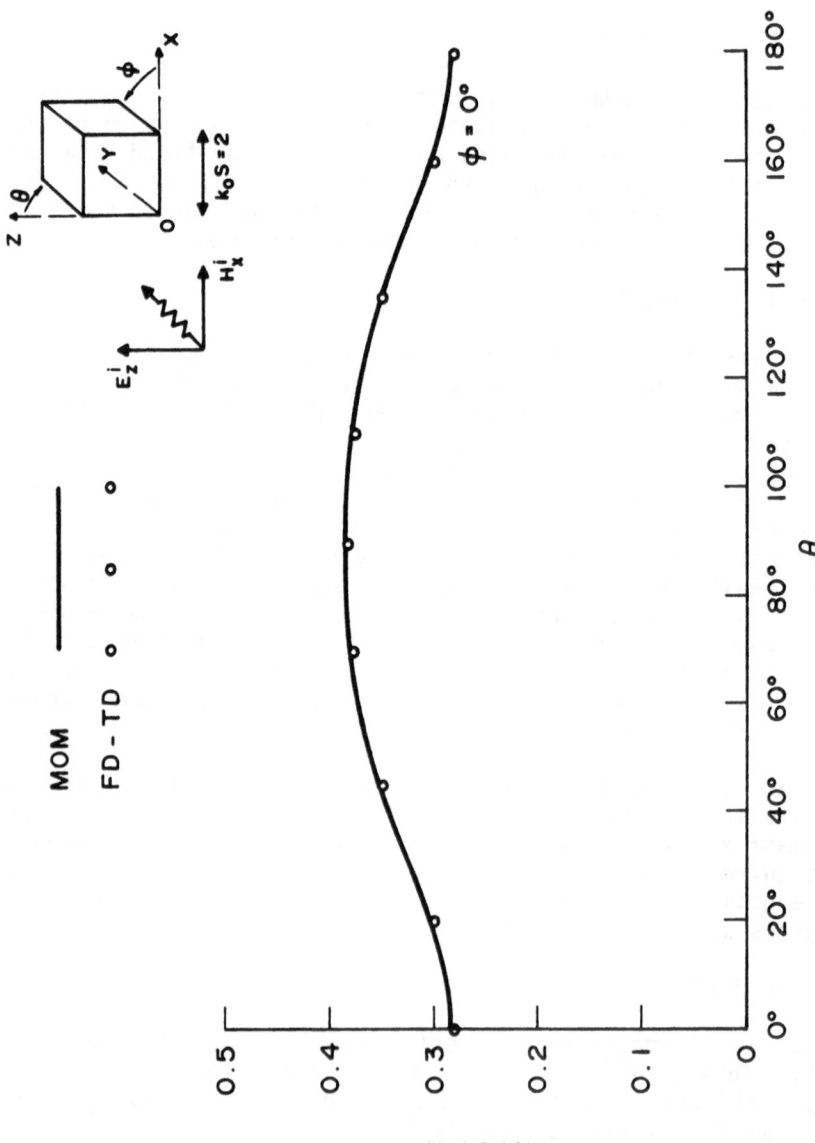

FIGURE 10. SCATTERING CROSS SECTION OF CONDUCTING CUBE SCATTERER.

scattering by structures with complex apertures and dielectric or permeable loadings.

Recently, the FD-TD method has been extended to three-dimensional structures as large as 20 wavelengths. Measurement verification of the far scattered fields are currently being obtained for such scatterers. Initial results indicate accuracy of the FD-TD computed radar cross section to within 0.5 dB.

REFERENCES

1. J.J. Bowman, T. B. A. Senior and P. L. E. UsLenghi, eds., Electromagnetic and Acoustic Scattering by Simple Shapes. John Wiley and Sons, New York, 1969.

2. K.S. Yee, "Numerical Solution of Initial Boundary Value Problems Involving Maxwell's Equation in Isotropic Media", IEEE Trans. Antennas Prop., Vol. AP-14, May 1966, pp. 302-307.

3. C. D. Taylor, D. H. Lam, and T. H. Shumpert, "Electromagnetic Pulse Scattering in Time-Varying Inhomogeneous Media", IEEE Trans. Antennas Prop., Vol. AP-17, September 1969, pp. 585-589.

4. D.E. Merewether, "Transient Currents Induced on a Metallic Body of Revolution by an Electromagnetic Pulse", IEEE Trans. Electromagnetic Compatibility, Vol. EMC-13, May 1971, pp. 41-44.

5. A. Taflove and M. E. Brodwin, "Numerical Solution of Steady-State Electromagnetic Scattering Problems Using the Time-Dependent Maxwell's Equations", IEEE Trans. Microwave Theory Tech., Vol. MTT-23, August 1975, pp. 623-630.

6. A. Taflove and M.E.Brodwin, "Computation of the Electromagentic Fields and Induced Temperatures Within a Model of the Microwave-Irradiated Human Eye," IEEE Trans. Microwave Theory Tech., Vol. MTT-23, November 1975, pp. 888-896.

7. R. Holland, "Threde : A Free-Field EMP Coupling and Scattering Code" IEEE Trans. Nuclear Science, Vol. NS-24, December 1977, pp. 2416-2421.

8. K. S. Kunz and K. M. Lee, "A Three-Dimensional Finite-Difference Solution of the External Response of an Aircraft to a Complex Transient EM Environment : Part I - The

Method of its Implementation", IEEE Trans. Electromagnetic Compatibility, Vol. EMC-20, May 1978, pp. 328-333.

9. A. Taflove and K.R. Umashankar, "A Hybrid Moment Method/Finite Difference Time Domain Approach to Electromagnetic Coupling and Aperture Penetration into Complex Geometries," Chap. 14 in Applications of the method of Moments to Electromagnetic Fields, B.J. Strait, ed., SCEEE Press, Orlando, Florida, February 1980.

10. A. Taflove, "Application of the Finite-Difference Time-Domain Method to Sinusoidal Steady-State Electromagnetic-Penetration Problems," IEEE Trans. Electromag. Compat., Vol. EMC-22, August 1980, pp. 191-202.

11. R. Holland, L. Simpson, and K.S. Kunz, "Finite-Difference Analysis of EMP Coupling to Lossy Dielectric Structures," IEEE Trans. Electromag. Compat., Vol. EMC-22, August 1980, pp. 203-209.

12. D. E. Merewether, R. Fisher and F. W. Smith, "On Implementing a Numeric Huygen's Source Scheme in a Finite-Difference Program to Illuminate Scattering Bodies", IEEE Trans. Nucl. Sci., Vol. NS-27, December 1980, pp. 1819-1833.

13. R. Holland and L. Simpson, "Finite-Difference Analysis of EMP Coupling to Thin Struts and Wires," IEEE Trans. Electromag. Compat., Vol. EMC-23, May 1981, pp. 88-97.

14. A. Taflove and K.R. Umashankar, "A Hybrid FD-TD Approach to Electromagnetic Wave Backscattering", 1981 URSI/APS - International Symposium Proceedings, Los Angeles, California, June 1981, p. 82.

15. A. Taflove and K.R. Umashankar, "Solution of Complex Electromagnetic Penetration and Scattering Problems in Unbounded Regions," in Computational Methods for Infinite Domain Media - Structure Interaction, AMD-Vol. 46, American Society of Mechanical Engineers, November 1981, pp. 83-113.

16. G. Mur, "Absorbing Boundary Conditions for the Finite-Difference Approximation of the Time-Domain Electromagnetic-Field Equations", IEEE Trans. Electromag. Compat., Vol. EMC-23, November 1981, pp. 377-382.

17. A. Taflove and K. R. Umashankar, "User's Code for FD-TD", Final Report RADC-TR-82-16 by IIT Research Institute,

Chicago, IL to Rome Air Development Center, Griffiss AFB, NY, on Contract F30602-80-C-0302, February, 1982.

18. S.A. Schelkunoff, "Field Equivalence Theorems," Comm. Pure Appl. Math., Vol. 4, June 1951, pp. 43-59.

19. R. F. Harrington, Field Computation by Moment Methods., MacMillan, New York 1968.

20. B. Engquist and A. Majda, "Absorbing Boundary Conditions for the Numerical Simulation of Waves", Math Comp., Vol. 31, July 1977, pp. 629-651.

21. G. A. Kriegsmann and C. S. Morawetz, "Solving the Helmholtz Equation for Exterior Problems with Variable Index of Refraction: I", SIAM J. Sci. Stat. Comput., Vol. 1, September 1980, pp. 371-385.

22. J. Van Bladel, Electromagnetic Fields., McGraw Hill, New York 1964.

23. R. F. Harrington, Time Harmonic Electromagnetic Fields. McGraw Hill, New York 1961, Chapter 3.

24. K.K. Mei and J. Van Bladel, "Scattering by Perfectly Conducting Rectangular Cylinders," IEEE Trans. Antennas Prop., Vol. AP-11, March 1963, pp. 185-192.

25. K. R. Umashankar and A. Taflove, "A Novel Method to Analyze Electromagnetic Scattering of Complex Objects", IEEE Trans. Electromagnetic Compatibility, Vol. EMC-24, November 1982.

26. N. C. Albertsen, J. E. Hansen, and N. E. Jenson, "Computation of Radiation from Wire Antennas on Conducting Bodies", IEEE Trans. Antennas Prop., Vol. AP-22, March 1974, pp. 200-206.

27. L. L. Tsai, D. G. Dudley, and D. R. Wilton, "Electromagnetic Scattering by a Three-Dimensional Conducting Rectangular Box", J. Applied Physics, Vol. 45, October 1974, pp. 4393-4400.

28. A. Sankar and T. C. Tong, "Current Computation on Complex Structures by Finite Element Method", Electronics Letters, Vol. 11, October 1975, pp. 481-482.

29. E. H. Newman and D. M. Pozar, "Electromagnetic Modeling of Composite Wire and Surface Geometries", IEEE Trans. Antennas Prop., Vol. AP-26, November 1978, pp. 784-789.

30. J. Singh and A. T. Adams, "A Non-Rectangular Patch Model
     for Scattering from Surfaces", IEEE Trans. Antennas
     Prop., Vol. AP-27, July 1979, pp. 531-535.

31. D. R. Wilton, S. M. Rao, and A. W. Glisson, "Electromag-
     netic Scattering by Surfaces of Arbitrary Shape", in
     Applications of the Method of Moments to Electromag-
     netic Fields, B. J. Strait, ed., SCEEE Press, Orlando,
     Florida, 1980, pp. 139-170.

32. S.   M.   Rao,   D.   R.   Wilton,   and   A.W.Glisson,
     "Electromagnetic Scattering by Surfaces of Arbitrary
     Shape", IEEE Trans. Antennas Prop., Vol. AP-30, May 1982,
     pp. 409-418.

33. E. K. Miller and A. J. Poggio, "Moment-Method Techniques
     in Electromagnetic from an Application Viewpoint",
     Chapter 9 in Electromagnetic Scattering, P.L.E. Uslenghi,
     ed., Academic Press, New York, 1978.

34. D.R. Wilton and S. Govind, "Incorporation of Edge Condi-
     tions in Moment Method Solutions", IEEE Trans. Antennas
     Prop., AP-25, November 1977, pp. 845-850.

II.11 (TS.5)

APPLICATIONS OF TIME-DOMAIN INVERSE SCATTERING TO ELECTROMAGNETIC
WAVES

W. Tabbara

Laboratoire des Signaux et Systèmes
Groupe d'Electromagnétisme
CNRS - ESE, Plateau du Moulon
91190 GIF-sur-YVETTE, FRANCE

ABSTRACT : Over the past few years, interest has developed in the
use of time-domain methods or data in order to solve inverse scattering problems. This approach has many attractive features among
them the large amount of information contained in a single pulse
response and simple physical interpretations. The improvement in
time-domain technology increases the interest in such an approach.
The purpose of this paper is to present a short review of some of
the techniques that have been used in the analysis of time-domain
inverse scattering.

I. INTRODUCTION

One may say that the time-domain approach is the most natural one to solve inverse scattering problems. The measurement of
the travel time of a pulse will help locate a target or a fault in
a material; the impulse response of a target is a characteristic
feature of its shape and constitution, and provides us with wide-
band information on the target. Furthermore, time-domain analysis
allows simple physical interpretation of scattering phenomena. In
fact, in many inverse scattering problems a rapid inspection of
the scattered pulses will provide usefull information on the number of discontinuities (or scattering centers), their position, or
in the case of dielectric bodies, their reflecting power. The
advances in digital computer capability made computations directly in the time-domain possible, and triggered the development of
powerful inverse methods (iterative ones), where a fast and accurate solution of the direct problem (computation of the scattered
field) is needed at each step.

In this paper we shall consider three large groups of inverse

431

W.-M. Boerner et al. (eds.), Inverse Methods in Electromagnetic Imaging – Part 1, 431–439.
© 1985 by D. Reidel Publishing Company.

problems : location, identification and reconstruction of a target. The target will be a perfect conductor or a dielectric (with
losses or not). It is not our purpose to provide an exhaustive
inspection of time-domain inverse scattering methods, but to present few ones illustrating the main approaches used in the past
few years.

II. TARGET LOCATION

Detection and location of a target are basic problems in inverse scattering. Pulses have been used for many years to detect
and locate targets in extended media. Once a pulse is emitted,
the reception of a pulse indicates the existence of a target (the
term is considered in a very broad sense, as a local modification
of the properties of the propagating medium); the inspection of
the shape of the returned pulse, will provide some information on
the nature of the target. Measuring the time delay between emission and reception, and considering the medium outside the target
to be known, one may obtain the distance from the emitter to the
target. One common application is the detection by means of radars
and sonars. Other applications concern the detection of voids or
buried objects [1], the location of discontinuities in dielectric media [2] or the location of faults along a propagating
device. In order to detect two or more closely located targets,
short pulses are required. The propagation in a dissipative or
dispersive medium, introduces overlaps between returns from different scatterers and deconvolution methods are needed in order to
separate the various contributions to the total pulse.

III. TARGET IDENTIFICATION

It is not necessary to have an image of a target to be able
to recognize it. One may think that the determination of some characteristic parameters of the target will allow a fairly accurate
identification.

1. Waveform identification (Pattern recognition)

A natural manner to identify a target belonging to a given
class of objects, is to compare the waveform from the unknown
target to the one of each object in the given class and identify
the target with the object which provides the waveform that looks
like the one that has been measured.

First, one builds a data set which contains as much information as possible on specific objects : waveforms as function of
aspect angle, polarization...

Next, one defines the logic used in comparing the measured
waveform with the ones taken from the data set. The simplest one

is the nearest neighbor logic where a euclidian metric is used to compute the distance between the two waveforms. More efficient methods can be developped based on various stastical approaches. These methods, while requiring more elaborate algorithms, are more accurate and fast. With the identification logic in place, one may then start the classification process. An interesting application of this approach is given in [3], where the power of this method clearly appears and suggests further developments and investigations.

2. Poles identification

Instead of keeping the whole waveform related to a given object, one may look at extracting from it some parameters which are characteristics of the object. Convenient sets of parameters are built with poles, or with the natural resonant frequencies of the object. They depend on the geometrical and electromagnetical parameters of the object.

The basic idea is that the time-domain response $f(t)$ of a target can be written in the form :

$$f(t) = \sum_{\ell=1}^{N} A_{\ell} \, e^{\gamma_{\ell} t}$$

Where N can be finite or not, γ_{ℓ} is the pole of the target and A_{ℓ} the residue at the pole γ_{ℓ}. The resonant frequencies are obtained from a frequency-domain analysis using moment method and the dominant ones coïncide with the dominant poles [4], [25].

Pole extraction is a problem in itself since, as it has been shown in [5], different approaches based on Prony's method [6], Kalman filter and maximum likelihood gives different results. Furthermore, higher order poles (large values of ℓ) are very sensitive to noise. But it seems that Prony's method is a convenient one. A catalog of poles can be built for objects of interest and stored in a computer.

Two different identification processes are possible. In the first one, when a waveform is received from an unknown target (illuminated by the same pulse that has been used to extract the poles in the catalog), the corresponding poles are extracted and compared by means of an appropriate parameter (correlation coefficient, mean square error...) to those in the catalog. The nearest ones found identifie the target. In some applications [7] it is possible to compute the poles analytically and then compute from their expression the geometrical and electromagnetical parameters of the target.This is a powerful approach but of limited scope.

In the second approach, the poles of a given object in the catalog are used to synthesize its response to the pulse used to illuminate the unknown target. The synthesized response f_S is compared to the received waveform and the target is identified with the object giving the nearest synthesized response. Comparison can be made by means of correlation techniques[8]. This approach has been extensively studied and provides very good results . The effects of polarization, the choice of the discretization step Δt and the number of poles used to compute f_s have been examinated [8, 4 , 9].

3. Natural mode identification

In the preceding section we mentionned the fact that the natural frequencies of a scatterer can be used to characterize it. To each natural frequency we associate a waveform called the natural mode. The purpose of the method developed by Chen et al. and described in [10, 11], consists in synthesizing a waveform for the incident signal which excites the target in such a way that the return signal consists of only a single-natural mode of that target in the late-time period. The incident signal is synthesized from a sum of damped sinusoïds using the values of the natural frequencies of the target. The duration T_e of the signal is taken to be equal to the period of the first natural mode. The late time-period begins for $t > t_e + T_t$ where T_t is the transit time of the scatterer. The method requires the knowledge of a large number of natural frequencies. Its usefulness seems limited to targets of simple shape.

IV. TARGET RECONSTRUCTION (Imaging)

An unambiguous method in target recognition is the formation of an image of that target form the measurement of its scattered field. Various approaches have been developped, but they usually apply to simple geometries.

1. Perfectly conducting targets

1.1. Exact method

From the investigation of time-domain scattering by means of integral equations, Bennett [12] developed, for the case of rotationally symmetric target illuminated by a wave from the axial direction, an iterative method that allows the reconstruction of the generating curve defined by the function $\rho(z,t)$. The basic equation is :

$$\rho(z,t) = [\ 2 \ r_o \ \vec{a}_H \cdot H_R^s(\vec{r},t) - \frac{1}{2\pi} \vec{a}_H \cdot \int_S (\vec{J}_R - 2\vec{a}_{n\Lambda}\vec{H}_R^i)(\vec{r}',\tau)\Lambda\vec{a}_r \ dS \ ]^{1/2}$$

$$\tau = t - R/C, \ \vec{a}_H = \vec{H}^i/|\vec{H}_i| \ , \ R = |\vec{r}_o - \vec{r}'| \ , \ \vec{a}_r = \frac{\vec{r}_o - \vec{r}'}{R} \qquad (1)$$

where $\vec{H}_r^i$ is the incident ramp field, $r_o\vec{H}_R^s$ the ramp response, $\vec{J}_R$ the induced current, r_o the distance of the far-field observer, $\vec{a}_n$: unit normal. An initial value of ρ, ρ_o, is given, then the solution of the direct problem gives J_R, which by means of (1) provide a new value ρ_1 of ρ. The process in then continued until it converges. Example in [12] for a sphere capped cylinder shows rapid convergence and good reconstruction in the shadow region. A basic requirements for a fast reconstruction is the existence of a fast algorithm for the solution of the direct problem at each step of the iterative process.

1.2. Approximate methods

An early method based on the physical optics approximation linked the backscattered ramp response $f_R(t)$ of the target to its cross-sectional area by a plane perpendicular to the direction of the incident wave [13] :

$$f_R(t) = \frac{A(r)}{\pi C^2} \qquad o < r = \frac{Ct}{2} < r_o \qquad (2)$$

$r = o$ identifies the tip of the target where $A(o) = o$ and r_o the shadow boundary. This identity requires low-frequency data, apparently violating the physical optics approximation. In order to obtain an image of the target, an intuitive approach using three look angles has been developped [14] and gave rather accurate results for complex bodies.

Later, generalizing a physical-optic frequency-domain method established by Bojarski, Bleinstein[15] developed a time-domain approach allowing the reconstruction of the characteristic function $\gamma(\vec{x})$ ($\gamma(\vec{x}) = 1$ inside the target, zero outside) from the measurement of the backscattered waveform $f(\vec{x}_o,t)$ for all directions of incidence :

$$\gamma(\vec{x}) = \frac{1}{4\pi^2} \int_\Omega f(\vec{x}_o ,- \frac{2 \vec{x}_o \cdot \vec{x}}{C}) \, d\vec{x}_o \qquad (3)$$

Where Ω denotes the unit sphere with variable $\vec{x}_o$. This method has been mostly investigated in the frequency domain, but the main results remain valid in the time-domain. It must be noted that (2) can be derived from (3). The main drawback of this method is the necessity of measuring the backscattered field for all direction in space. Both approaches described above, deserve further investigations and specially the shadow region, or those of non symmetric target are strongly dependent on polarization. Corrective terms must be introduced as mentionned in [16].

The third method is not really a time-domain one. It is a

high frequency approach based on the fact that the specular radar
cross-section is simply related to the Gaussian curvature K
through $\sigma_{sp.} = \pi/K$. In order to isolate the specular return,
short pulses are used [17] and in the case of body of revolution,
$\sigma_{sp.}$ measured at various angles of incidence leads to a simple
reconstruction of the target shape. Here too, polarization information will increase the accuracy of the method.

2. Dielectric targets (lossy or not)

2.1. Exact methods

A widely used method is the Time-domain Reflectometry (TDR).
A pulse illuminates the unknown dielectric sample placed in a
transmission line, the reflected pulse is used to determine the
permittivity and tangent loss of the sample as a function of
frequency. The reflection coefficient of the sample is first determined, then simple formulas allow the computation of the unknown parameters [18]. This broad-band tehnique can also be
usedto analyse networks or characterize discontinuities in transmission lines.

The determination of the permittivity ε and conductivity σ
of stratified medium has triggered the development of various
methods. Among them, one based on an integral representation of
the field directly in the time domain [19, 20, 21]. In [21]
a step by step approach of the integral equation gives ε or σ,
it is a very fast approach but very sensitive to noise and thus
does not allow very deep investigation. On the other hand in [19,
20] optimization thechniques are used. Initial values are given
for ε and σ , the reflected field is computed and compared by
means of a convenient distance to the measured field. The iterative process is stopped whenever computed and measured values are
sufficiently close. These techniques are powerful, accurate and
not very sensitive to noisy data. They allow fairly deep investigation of the medium and the reconstruction of ε or σ . The
determination of both parameters is not yet possible. Another advantage of the methods described in [19, 20] is the existence of
a very fast algorithm for the solution of the direct problem, this
is a fundamental point for real-time analysis.

2.2. Approximate methods

A large class of methods is based on the Born approximation,
where the field in the scatterer is assumed to be equal to the
incident field. This method has been developed in many domains :
diagnostic of strafied media, analysis of tapered transmission
lines... . The purpose is the reconstruction of a permittivity
profile (lossless medium) or an impedance profile. In the case of
stratified media, it has been shown [22] that the step response

of the medium, conveniently corrected, allows the determination
of the interfaces and the values of ε in each layer. The method
behaved accurately for values of ε up to 10 ε_0. Above this limit the shape of the profile was correctly determined but the
values of ε where false. The approach was not very sensitive to
noise. The analysis of thin layers requires short pulses.

Another class of methods uses the time of flight measurement
to retrieve the refractive index profile of the target. These
methods are based on a ray propagation hypothesis which leads to
the solution of an Abel integral equation when spherical symmetry
is assumed [23]. It has been used in the inversion of radio oc-
culation data, and in plasma diagnosis, among other applications.
More recent applications of time of flight measurement are related to ultrasonic and NMR tomography [24] where very successful
results have been obtained.

V. CONCLUSION

From this brief review one may note the wide spectrum of application of time-domain measurements in inverse problems. They
provide flexible approaches with simple and fast physical interpretation. There is a need to improve actual methods by taking into
accounts polarization effects, extension to three dimensional
structures lacking rotational symmetries, and to develop new approaches in the case of imperfectly conducting targets.

REFERENCES

[1] Moffatt D.L., Puskar R.J., Peters L. Jr. : 1973, Electro-
       magnetic pulse sounding for geological surveying with ap-
       plication in rock mechanics and the rapid excavation pro-
       gram. Final Report 3408-2, Electroscience Laboratory, The
       Ohio State University, Columbus, Ohio 43212, U.S.A..

[2] Robinson L.A., Weir W.B., Young L. : 1974, Location and
       recognition of discontinuities in dielectric media using
       synthetic RF pulses, Proc. IEEE, Vol. 62, n° 1, January,
       pp. 36-44.

[3] Bennett C.L., Auchenthaler A.M., Smith R.S., De Lorenzo
       J.D. : 1973, Space-time integral equation approach to the
       large body scattering problem. Final Report RADC-TR-73-70,
       Sperry Research Center,Sudbury, Massachussets, U.S.A..

[4] Chuang C.W., Moffatt D.L. : 1976, Natural resonances of
       radar targets via Prony's method and target discrimination,
       IEEE Trans. Aerosp. Electron. Syst., Vol AES-12, n° 5,
       Sept., pp. 583-589.

[5]   Gavel D.T., Candy J.V., Lager D.L. : Parameter estimation
        from noisy transient electromagnetic measurement, Lawrence
        Livermore Laboratory, Livermore, CA 94550, U.S.A..

[6]   Hildebrand F. : 1956, Introduction to numerical analysis,
        Mc Graw-Hill, New York.

[7]   Miller E.K., Lager D.L. : 1979, Inversion of one-dimensio-
        nal scattering data using Prony's method, Report n° UCRL-
        52667, Lawrence Livermore Laboratory, Livermore, CA 94550
        U.S.A..

[8]   Moffatt D.L., Mains R.K. : 1975, Detection and discrimina-
        tion of radar targets, IEEE Trans. Ant. Prop., Vol. AP-23,
        N° 3, May, pp. 356-367.

[9]   Moffatt D.L., Rhoads C.M. : 1982, Radar identification of
        naval vessels, IEEE Trans. Aerosp. Electron. Syst., Vol.
        AES - 18, N° 2, March, pp. 182-187.

[10] Chen K.M., Westmoreland D. : 1982, Radar waveform synthe-
        sis for exciting single-mode backscatters from a sphere
        and application for target discrimination, Radio-Science,
        Vol. 17, N° 3, May-June, pp. 574-588.

[11] Chen K.M., Nyquist D.P., Westmoreland D., Chuang C.I.,
        Drachman B. : 1982, Radar waveform synthesis for single-
        mode scattering by a thin cylinder and application for
        target discrimination, IEEE Trans. Ant. Prop., Vol AP-30,
        N° 5, Sept., pp. 867-880.

[12] Bennett C.L. : 1981, Inverse scattering in Theoretical
        methods for determining the interaction of electromagnetic
        waves with structures, J.K. Skwirzynski editor, Sijthoff
        and Noordhoff, The Netherlands.

[13] Kennaugh E.M., Moffatt D.L. : 1965, Transient and impulse
        response approximations, Proc. IEEE, vol. 53, n° 8,
        August, pp. 893-901

[14] Shubert K.A., Young J.D., Moffatt D.L. : 1977, Synthetic
        radar imagery, IEEE Trans. Ant. Prop., Vol. AP-25, n° 4,
        July pp. 477-483.

[15] Bleistein N. : 1976, Physical optics farfield inverse scat-
        tering in the time domain, Jour. Acoust. Soc. Amer., Vol.
        60, n° 6, Dec., pp 1249-1255.

[16] Boerner W.M. : 1979, Development of physical optics inverse scattering using Radon projection theory, Conference on Mathematical methods and applications of scattering theory, Washington, D.C., May.

[17] Hong S., Borison S.L. : 1968, Short-pulse scattering by a cone-direct and inverse, IEEE Trans. Ant. Prop., vol. AP-16, N° 1, Jan., pp. 98-102.

[18] Weir W.B. : 1974, Automatic measurement of complex dielectric constant and permeability at microwave frequencies, Proc. IEEE, Vol. 62, N° 1, Jan., pp. 33-36.

[19] Lesselier D. : 1981, Optimization theory and time-domain inverse scattering, Radio Science, Vol. 16, Nov. Dec 1981, pp 1059-1063.

[20] Tijhuis A.G. : 1981, Iterative determination of permittivity and conductivity profiles of a dielectric slab in the time domain, IEEE Trans. Ant. Prop., Vol. AP-29, March, pp. 239-245.

[21] Lesselier D. : 1978, Determination of index profiles by time-domain reflectometry, Jour. Optics, Vol. 9, n° 6, pp. 349-358.

[22] Tabbara W. : 1979, Reconstruction of permittivity profiles from a spectral analysis of the reflection coefficient, IEEE Trans. Ant. Prop., Vol. AP-27, n° 2, March, pp. 241-244.

[23] Weston V. : 1978, Electromagnetic inverse problem, in Electromagnetic Scattering, PLE Uslenghi editor, Academic Press, New York.

[24] Cho Z.K., Kim H.S., Song H.B., Cumming J. : 1982, Fourier transform nuclear magnetic resonance tomographic imaging, Proc. IEEE, Vol. 70, n° 10, Oct., pp. 1152-1173.

[25] Sarkar T.K., Nebat J., Weinder D.D., Jain V.K. : 1980, Suboptimal approximation/identification of transient waveforms from electromagnetic systems by pencil-of-function method, IEEE Trans. Ant. Prop., Vol. AP-28, n° 6, November, pp. 928-933.

II.12 (TS.4)

TIME-DEPENDENT RADAR TARGET SIGNATURES, SYNTHESIS
AND DETECTION OF ELECTROMAGNETIC AUTHENTICITY FEATURES

David L. Moffatt and Ted C. Lee

The Ohio State University
Department of Electrical Engineering
2105 Neil Avenue
Columbus, Ohio 43210

A rational function approximant for extracting the complex
natural resonances of a scatterer from complex measured or
calculated multiple frequency scattering data is described. The
extraction procedure incorporates preprocessing filtering and
windowing techniques as well as a sum operator. Utility of the
extraction process is illustrated for simulated data on a
spherical scatterer, to demonstrate noise sensitivity, and for
other objects using measured or calculated data. A method for
using the complex natural resonances to improve target imaging
techniques is suggested.

W.-M. Boerner et al. (eds.), Inverse Methods in Electromagnetic Imaging - Part 1, 441–460.

INTRODUCTION

It is well known that the time-dependent electromagnetic backscatter responses elicited by ramp-type interrogating signals can be used to produce three-dimensional isometric images of a target. The approach is an approximate one and handicapped because the physical optics approximation, with attendent lack of shadow region information, is exploited. This paper basically explores how complex natural resonances, related to circumferential paths on the target, can be extracted and utilized.

Currently available experimental co-polarized and cross-polarized scattering data on both classical objects such as spheres, spheroids and discs and more complicated scatterers are in the form of stepped-swept frequency complex data spanning at least 10:1 bandwidths. A rational function model for fitting these scattering data and extracting complex natural resonances is described. Filtering and windowing are utilized to minimize the problem of noisy data. A method for incorporating certain of the complex natural resonances into improved image models of the target is suggested.

Canonical Responses and an Area Function

There is no loss of generality if a perfectly conducting scatterer is assumed to be located in the forward hemisphere ($z>0$) of a rectangular coordinate frame and an incident$\wedge$plane wave field is assumed to be linearly polarized in the x direction* and propagating in the +z direction. It has been repeatedly** demonstrated that the physical optics approximation predicts a relationship [1,2]

$$A_z = -\pi c^2 F_R(2z/c)(U(z) - u(z-z_1)) , \tag{1}$$

between the target's cross sectional area function, A_z, and the backscattered ramp response waveform F_R. The cross sectional area function A_z is the projection of the scatterer surface between z=0 and a cutting plane at z onto the xy plane. The distance z_1 is the distance from the origin to the shadow boundary, assumed for simplicity to be orthogonal to the propagation direction. With normalizations as in [2], the ramp response is given using Fourier synthesis as [3]

---

* the carat denotes a unit vector.
** see [1] for perhaps the earliest derivation and [2] for a latter more readily accessible publication.

$$F_R(t'/T_0) = -T_0^2/(2\pi^2) \sum_{n=1}^{N} A(2\pi n)/(n^2)\cos(2\pi nt/T_0+\phi(2\pi n)),$$

$$(2)$$

where $A<\phi$ is the scattered field in phasor notation and T_0 is the fundamental period.

The previous paper [3] demonstrated that extending the physical optics approximation to encompass the entire scatterer generally yielded improved ramp response estimates and correspondingly improved imaging results were inferred. It is desirable however to develop more formal methods for extending the utility of ramp response waveforms for imaging as well as other applications. When an aperiodic signal interrogates a scatterer the time-dependent response waveform consists of first a forced response as the signal front moves across the object and then a free or natural response as the signal front moves beyond the scatterer. Thus the ramp response consists of an early time forced response as a ramp-type excitation moves across the scatterer and then a later time natural response. It follows that an expansion of the total ramp response waveform in terms of complex natural resonances would not be correct.* For early times the complete geometry of the scatterer has not yet been defined and the global properties are therefore inappropriate. It has been demonstrated [4] that the first electric and magnetic mode complex natural resonances of a conducting sphere can be extracted from the ramp response waveform using samples for times corresponding to returns coming from the shadow boundary and beyond. It was found however that attempts to extract resonances corresponding to higher order modes were unsuccessful, i.e., the poles were highly inaccurate. There is a good reason for this failure. As is clearly demonstrated by the remarks of Kennaugh and Moffatt in [5], the first 19 electric mode pole-pairs of the sphere serve to completely define the creeping wave of the sphere as seen in the backscattered impulse response waveform. Equating the known ratio of required bandwidths to define the ramp and impulse response waveforms to the ratio of poles which could reasonably be expected to be extracted from the creeping wave portion of the response yields approximately three for the ramp response. The inverse frequency squared weighting appropriate for the ramp response essentially negates the effects of the higher poles. This has been seen previously in another context [2], where it is stated that there is little discernable difference between the ramp response synthesized from data spanning a 10:1 bandwidth and data spanning a 60:1 bandwidth.

---

*It is assumed that finite summations without an entire function, as are inevitably used in practice, are utilized.

Rational Function Model

Numerous methods and procedures for extracting complex
natural resonances of an object from measured or calculcated
radiation or scattering data have been suggested. A rational
function model was first suggested in [2]; some initial work
using such a model was reported in [6] and later models were also
proposed [7]. The rediscovery of Prony's method and attendent
eigenanalysis [8,9] directed most effort toward time domain
approaches. A rational function model however has been found to
be sometimes advantageous because nonharmonically related data
samples can be used and this in turn permits extensions of the
model to much higher frequencies than previously tested. Also,
the restriction to a free response region is avoided. The
rational function model is written as

$$G(jx_i) = \frac{\displaystyle\sum_{n=0}^{N} a_n(jx_i)^{n+2}}{1 + \displaystyle\sum_{n=1}^{M} b_n(jx_i)^n} \qquad i = 1,2, \ldots, I \ . \qquad (3)$$

where

$$x_i = k_i L \ . \qquad (4)$$

is an electrical length, k the wavenumber and L some convenient
linear reference length for the object. It is assumed that the

$$G(jx_i) = \alpha_i + j\beta_i \ , \qquad (5)$$

are known, generally from measurements, for I values of x_i. The
order of the numerator and denominator polynomials can be
selected to make the model Rayleigh-like or not as appropriate.
In this paper, a numerator polynomial of order N+2 is called
model I and of order N model II. From Equation (3), I complex
equations in the unknown polynomial coefficients can be
generated. The equations can be solved in a least squared error
sense or in an exact sense. Some of our initial comparisons of
extracted poles using both approaches indicate that on some
occasions exact solutions can be advantageous. Similarily, the
normalization shown in Equation (3) should be arbitrary, i.e.,
mathematically any coefficient could be normalized. It is found,
however, that numerically the normalization shown achieves
superior pole results. Also, double precision computer programs
are inevitably needed.

We have found that rational function models are most
effective when sample values $(G(jx_i))$ are taken from near

relative maxima and minima of the amplitude scattering data. This means that in general nonharmonically related samples are used with other samples somewhat arbitrarily taken between the maxima and minima. When extending the rational function models to relatively high frequencies, a preponderance of the nonextremum samples should be at lower frequencies. It has also been noticed that the fits to data obtainable with rational function models fail abruptly beyond the highest frequency sample. The reason is that a rational function model is not an appropriate approximation at higher frequencies.

Consider the first few poles and residues (backscatter) of a conducting spherical scatterer. The sphere is admittedly an overworked example. We choose, however, to work with real poles and residues even when essentially using simulated data (finite number of poles) to avoid bias. Much too often, simulated data are generated to test a proposed pole extraction scheme and then the assumption is made that all of the residues are of equal magnitude. This type of assumption is unrealistic as will be seen in Table I. An alternative approach is to let the residues be quasi-random [10]. Our simulated data (finite number of poles) are generated as, $(k_i a = 0.2(0.02)4.0)$

$$E^s(jk_i a) = \sum_{i=1}^{2M'} \frac{R_i}{jk_i a - p_i} \quad , \tag{6}$$

where R_i are the residues, M' the number of pole-pairs, p_i the poles and a the radius of the sphere. A comparison of the exact and extracted poles and residues when 4 pole-pairs are used and noiseless data are assumed is shown in Table I. To 5 s there is zero error in the extracted poles and residues. Figure 1 shows the amplitude and phase fits that were achieved using the rational function model. It is not apparent but there are actually two curves in Figure 1. The fit however is so good that only one curve is visable. The crosses shown on the amplitude curve show those frequencies used for input data (samples) for the fit. The results in Table I are by no means surprising and are indeed equivalent to a number of Prony-type extraction methods [10] when noiseless data are used. As has been shown however [10], even eigenanalysis becomes ineffective as a pole extraction method when the data are contaminated by more than a moderate amount of noise. While the results discussed in this paper are much improved ones over previous efforts [10] we still maintain that the pole extraction from real world full scale measurements will usually not be feasible. However, as is illustrated in this paper, pole extraction is feasible from controlled laboratory measurements. Since exploitation of the complex natural resonances in identification and imaging need not from the various windows are listed. Table IV demonstrates two points. First, if the system poles lie close to the ends of the window then large errors due to truncation can be anticipated. Second, these errors tend to increase at higher frequencies. The reason for the truncation errors are obvious. The gradual failure at higher frequencies is due to the inadequacy of the model. Rational approximants are at best reasonable in the Rayleigh and low resonance ranges. At higher frequencies, while they can be forced to model, they fail abruptly beyond the highest frequency sample. It is also apparent from these and other examples that when windowing is used, the assumed system order should be roughly twice the suspected system order. As a general rule we discard as curve fitting poles which are not obtained by two or more windows, right half-plane poles, poles with very small residues, real poles and poles with very large negative real parts. Only when such poles consistently reappear (several windows) are they reconsidered. We also discard poles which are approximately cancelled by zeros of the rational approximant. Much of the need for double precision programs involves pole-zero cancellations.

To test the effects of noise on the pole extraction process, we add two uncorrelated pseudo-random Gaussian processes with zero mean and standard deviation σ to the real and imaginary parts of the simulated sphere scattering data. The signal to noise ratio in dB is defined as

$$S/N = 20 \log \left| \frac{1}{N\sigma} \sum_{k=1}^{N} A(k) \right| \quad , \qquad (7)$$

where $A(k)$ is the sphere scattering plus noise at the frequency jw_k and N is the total number of samples used. An example of the noisy scattering data is shown in Figure 5 for a noise standard deviation of 0.1. Figure 5 shows the signal plus noise (solid curve) and the rational approximant fit (dashed curve). The poles extracted by an exact fit and windowing for noise standard deviations of 0.01, 0.02 and 0.1 are shown in Table V. The results are not good and obviously pole extraction fails for signal-to-noise ratios much less than roughly 20 dB.

To preprocess the data we implement a tenth order zero phase shift Butterworth lowpass filter by using the product of a fifth order filter and its complex conjugate. Details of the filter and its implementation are given in [13]. The utility of the filter is shown in Figures 6, 7 and 8. Figure 6 shows the noisy amplitude scattering data of the sphere with a signal-to-noise ratio of 14.81 dB, Figures 7 and 8 compare the amplitude and phase data respectively after filtering with the original

noiseless amplitude and phase curves. Clearly much of the
deleterious effects of the noise have been removed without
seriously distorting the true amplitude and phase data.

Preprocessing with a Sum Operator

Least squared error solutions (eigenanalysis) have the
advantage that all of the available data samples can be used but
are sensitive to noise and may have a convergence problem if
iteration is used. Another way to use more of the available data
is with row echelon algorithms [13,14] as are used in Gauss
elimination to reduce N rows of a matrix to one row. In
practice, the data are first low pass filtered and then the
rational function approximant is applied using a sum operator.
an example of the result is shown in Figure 9 where the processed
result of the noisy data in Figure 6 (S/$_N$ = 14.81 dB) is compared
to the noiseless amplitude data. Improvements over the result on
Figure 7 (filtering alone) may appear slight but more of the data
with some target information have been used. The pole-pairs
extracted using filtering alone and filtering plus a sum operator
are given in Tables VI and VII respectively. Difficulty with the
highest frequency pole, as shown, could be anticipated because of
truncation which forces both the filter and sum operator to be
one-sided at the high frequency end.

The Thin Circular Disc

For scatterers of sufficient geometrical simplicity the
rational function approximation (low frequencies) can be combined
with an analytical method developed by Kennaugh [15] (high
frequencies) to obtain estimates of the complex natural
resonances of the object. Table VIII illustrates the overlap,
for broadside backscatter which can be obtained to correct the
K-pulse approach (which uses Geometrical Theory of Diffraction
estimates) at the low frequency end. There are two dominant
strings of complex natural resonances for the disc. One set of
resonances is from diffraction at the edges of the disc, the
second comes from a creeping wave around the circumference of the
disc. For backscatter the diffraction mode dominants at
broadside incidence and the creeping wave mode at off-broadside
incidence. Table IX gives poles obtained using rational function
approximants of Hodge's calculated data [16]. Table X is an
extended list of the K-pulse estimated [13] creeping wave poles
obtained by using Knott et al. creeping wave analysis [17]. In
Table IX, the angle given is the angle off broadside, PHI
polarization is in the plane of the disc and THETA polarization
is perpendicular to the plane of the disc.

necessarily involve direct pole extraction [10,11,12], the poles remain a valuable tool.

Processing

When the rational approximant in Equation (3) is applied to real scattering data, the proper order of the numerator and denominator polynomials (N,M) as well as the optimum sample points (jx_i) are unknowns. It follows that even with noiseless data, some error in the pole extraction procedure will occur. It was noted above that best results are achieved when the frequency samples are at a near relative maxima and minima of the amplitude scattering data and all of the results shown in this paper utilize this fact. In Tables II and III, the effect of numerator and denominator orders, Rayleigh (model I) or non-Rayleigh (model II) rational approximants and minor adjustment of sample positions on the error of the extracted poles and residues are shown. The simulated input data are the same as for Table I. There is one far-reaching conclusion which can be drawn from the results in Tables I, II and III. When the assumed system order (M) is not correct, curve fitting poles are obtained. In these simulated examples the proper system order and the correct poles are known a priori and can be identified. Thus while the poles are extracted with satisfactory accuracy even with improper assumed system order, a method is needed for identifying the curve fitting poles for the general problem. This procedure, windowing, and a filtering process to reduce the effects of noise and clutter are described next. We note that if six sphere pole-pairs and residues are used to calculate the simulated data over the same spectral range ($k_i a = 0.2(0.02)4.0$) then two of the poles have oscillatory parts greater than the highest input data. These last two poles are not extracted satisfactorily by any of the models. One can conclude therefore that the assumed system order should always be equal to or greater than the actual suspected (from measured amplitude spectrum) system order.

Windowing, i.e., applying the model repeatedly over overlapping spectral spans to cover the spectral range covered by the available data is a well-known technique [7]. In principle, only those poles which repeatedly appear from various windows are considered true poles of the system. The degree of effort devoted to windowing is essentially dictated by the available processing time. Since we have a priori concluded that noise dictates pole extraction in nonreal time, windowing can be a very effective procedure. We show one example using the simulated data generated by six sphere poles and residues over the $k_i a = 0.2(0.02)6.0$ spectral range. The windows are shown in Figures 2, 3 and 4, where the rational approximant fits to the amplitude data are shown. Within each window the fits, amplitude and phase (not shown) are excellent. In Table IV, the poles extracted

TABLE I
COMPARISON OF TRUE POLES AND EXTRACTED POLES, TRUE RESIDUES
AND EXTRACTED RESIDUES (M=8,N=7); FOUR PAIRS OF POLES AND
RESIDUES ARE USED IN THE SIMULATED DATA

Poles (M=8,N=7)

true	extracted	% of error real	imag.
-0.50000+/-j0.866025	-0.50000+/-j0.866025	0.E0	0.E0
-0.701964+/-j1.80740	-0.701962+/-j1.80740	1.04E-4	2.07E-4
-0.842862+/-j2.75786	-0.842859+/-j2.757866	1.03E-4	4.97E-5
-0.954230+/-j3.71478	-0.954238+/-j3.714782	2.08e-4	4.97E-5

Residues

True	extracted	real	imag.
-0.0946447+/-j0.516674	-0.094644-/+j0.516674	1.33E-4	0.E0
-0.633323-/+j0.0853256	0.633320-/+j0.08533	4.76E-4	3.75E-4
0.0802221+/-j0.733736	0.08024+/-j0.73373	2.42E-3	8.07E-4
-0.822075+/-j0.07674781	-0.822088+/-j0.07676	1.57E-3	1.48E-3

TABLE II
COMPARISON OF TRUE POLES AND EXTRACTED POLES, TRUE RESIDUES
AND EXTRACTED RESIDUES (M=20,N=19); FOUR PAIRS OF POLES AND
RESIDUES ARE USED IN THE CALCULATION OF THE SIMULATED DATA

True Poles	Extracted poles	% of error real	imag.
-0.50000+/-j0.866025	-0.50000+/-j0.86602	0.E0	4.95E-4
-0.701964+/-j1.80740	-0.70196+/-j1.80741	2.06E-4	5.16E-4
-0.842862+/-j2.75786	-0.84283+/-j2.75791	1.11E-3	1.74E-3
-0.954230+/-j3.71478	-0.95421+/-j3.71492	5.22E-4	3.65E-3

Curve fitting poles
0.19396+/-j0.78060
0.23181+/-j1.51326
-0.39652+/-j1.86546
0.12284+/-j2.97965
0.18009+/-j3.84572
0.43016+/-j2.16078

Curve fitting zeros
-0.19396+/-j0.78060
-0.23181+/-j1.51327
-0.39649+/-j1.86548
0.12284+/-j2.97967
0.18008+/-j3.84573
0.43016+/-j2.16079

Extracted zeros

-0.38086+/-j3.01887
-0.16461+/-j1.68440
-1.16102+/j0.
1.29822+/-j0.
-1.42429+/-j0.

True residues	Extracted residues	% of error	
-0.0946447-/+j0.516674	-0.09463-/+j0.51668	2.79e-3	1.15E3
0.633323-/+j0.0853256	0.63332-/+0.0854	4.76E-4	1.16E-2
0.0802221+/-j0.733736	0.08046+/-j0.7335	3.22E-2	3.197E-2
-0.822075+/-j0.07674781	0.8219+/-j0.07706	2.119E-2	3.781E-2

*all residues corresponding to the curve fitting
poles are in the order of 1.E-5or less for both
real and imaginary parts.

TABLE III
THE EXTRACTED POLES AND THE ERRORS BETWEEN THE RATIONAL
FUNCTION FIT AND THE ORIGINAL DATA (SIMULATED DATA I)

TRUE POLES	MODEL I M=14,N=13	Percent error real part	imag. part
-0.500000+/-j0.866025	-0.50000+/-j0.866022	0.E0	2.98e-04
-0.70196+/-j1.80739	-0.7019523+/-j1.80743	6.06E-4	1.55E-3
-0.842849+/-j2.75786	-0.84282+/-j2.75741	1.457E-3	1.56E-2
-0.954299+/-j3.714787	-0.9648+/-j3.7253	0.27	0.27

MAX. AMP. ERROR 0.19E-5
AVE. AMP. ERROR 0.2E-6
MAX. PHA. ERROR 0.1068E-3
AVE. PHA. ERROR 0.155E-4

MODEL II M=14,N=13	Percent error real part	imag. part
-0.499998+/-j0.86606	2.E-4	3.505E-3
-0.70196+/-j1.80738	2.06E-4	1.03E-3
-0.84295+/-j2.75778	3.05E-3	2.77E-3
-0.95515+/-j3.71550	2.4E-02	1.877E-2

MAX. AMP. ERROR 0.5E-6
AVE. AMP. ERROR 0.1E-6
MAX. PHA. ERROR 0.916E-4
AVE. PHA. ERROR 0.148E-4

Model I M=16,N=15	percent error real	imag.
-0.499997+/-j0.866023	3.01E-4	1.97E-4
-0.701981+/-j1.80739	8.76E-4	5.16E-4
-0.84286+/-j2.75810	7.03E-5	8.33E-3
-0.95048+/-j3.71478	9.78E-2	0.E0

MAX. AMP. ERROR 0.19E-5
AVE. AMP. ERROR 0.2E-6
MAX. PHA. ERROR 0.176E-3
AVE. PHA. ERROR 0.187E-4

TABLE IV
A COMPARISON OF TRUE POLES AND POLES EXTRACTED VIA
WINDOW TECHNIQUE

Exact poles of sphere	Poles extracted via RFA & windows(three) technique	PERCENT ERROR REAL PART	IMA. PART
-0.500000+/-j0.866025	-0.50038+/-j0.86610 (W I)	3.80E-2	7.50E-3
-0.701964+/-j1.80740	-0.7086+/-j1.8133 (w I)	0.342	0.304
-0.842862+/-j2.75786	-0.8766+/-j2.7483 (w I)	1.17	0.332
-0.954230+/-j3.71478	-0.9277+/-j3.7388 (w I)	0.692	0.626
-1.04764+/-j4.67641	-0.9996+/-j4.6191 (W II)	1.002	1.196
-1.12891+/-j5.64163	-1.1338+/-j5.5328 (w III)	8.50E-2	1.891

TABLE V
A COMPARISON OF EXACT AND EXTRACTED POLES OF SPHERE VIA RATIONAL FUNCTION APPROXIMATION; A GAUSSIAN UNCORRELATED WHITE NOISE WITH STANDARD DEVIATION IS ADDED TO THE BACKSCATTERING DATA OF SPHERE

| | poles extracted via RFA | |
exact poles	$\sigma = 1.E-2$	$\sigma = 2.E-2$
-0.500000+/-j0.866025	-0.5362+/-j0.9616	-0.4386+/-j0.80129
-0.701964+/-j1.80740	-0.4745+/-j1.9577	-0.6482+/-j1.8509
-0.842862+/-j2.75786	-0.38518+/-j2.643	-0.9795+/-j2.7384
-0.954230+/-j3.71478	-0.4745+/-j3.538	-0.6285+/-j3.5081

$\sigma = 1.E-1$
-0.439+/-j0.5497
-0.295+/-j1.6033
-0.2574+/-j2.784
not located

TABLE VI
POLES EXTRACTED VIA RATIONAL FUNCTION APPROXIMATION AND AVERAGING TECHNIQUES (S/N=14.81 dB)

TRUE POLES
-0.50000+/-J0.866025
-0.701964+/-J1.80740
-0.842862+/-J2.75786
-0.954230+/-J3.71378

EXTRACTED POLES ($\sigma = 2E-1$)	Percent Error TRUE	IMAG.
RUN 1		
-0.2989+/-J0.8643	20.11	0.17
-0.2300+/-J1.6802	24.34	6.56
-0.1827+/-J2.2712	22.89	16.88
-0.2455+/-J3.6937	18.48	0.55
RUN 2		
-0.4571+/-J0.9676	4.29	10.16
-0.4727+/-J2.088	11.82	14.47
-0.2651+/-J3.296	20.03	18.66
-0.1739+/-J3.674	20.34	1.063
RUN 3		
-0.4695+/-J0.9854	3.05	11.94
-0.2530+/-J2.160	23.15	18.19
-0.5323+/-J2.5157	10.77	8.397
-0.4474+/-J3.6547	13.21	1.57
AVERAGE POLES		
-0.405+/-J0.9391	9.5	7.307
-0.318+/-J1.976	19.80	8.70
-0.3267+/-J2.6943	17.90	2.20
-0.2889+/-J3.674	17.35	1.06

TABLE VII
POLES EXTRACTED VIA RATIONAL FUNCTION APPROXIMATION
USING DIGITAL FILTER AND SUM OPERATOR (S/N=14.81 dB)

poles extracted via digital filter and sum operator M=16,N=15	percent error real	imag.
run 1		
-0.568+/-j0.915	6.8	4.89
-0.664+/-j1.790	1.96	0.89
-0.565+/-j3.001	9.63	8.43
run 2		
-0.392+/-j0.927	10.8	6.10
-0.455+/-j1.927	12.73	6.17
-0.453+/-j3.08	13.52	11.17
run 3		
-0.568+/-j0.8235	6.80	4.25
-0.789+/-j1.538	4.49	13.89
-0.455+/-j2.902	13.45	4.99
run 4 M=14,N=13		
-0.531+/-j0.926	3.10	6.037
-0.446+/-j1.730	13.21	3.992
-0.321+/-j2.784	18.08	0.893
-0.246+/-j3.587	18.46	3.33
Average		
-0.515+/-j0.897	1.499	3.097
-0.589+/-j1.759	5.826	2.455
-0.429+/-j2.923	14.35	5.72

TABLE VIII
COMPARISONS OF POLES OF DISK AT BROADSIDE BACKSCATTERING
USING RATIONAL FUNCTION APPROXIMANT AND GTD METHOD

	RFA		GTD method
	String I	String II	
1.	-2.22+/-j1.656	-0.49+/-j1.23	
2.	-1.04+/-j4.87		-1.04+/-j5.0542
3.	-1.16+/-j8.10		-1.1616+/-j8.2116
4.	-1.24+/-j11.28		-1.2418+/-j11.3611
5.	-1.27+/-j14.4		-1.3024+/-j14.5075

TABLE IX
POLES EXTRACTED BY RATIONAL FUNCTION APPROXIMANTS
THE RATIONAL FUNCTION APPROXIMANTS ARE FITTED TO ka=6.5

poles

PHI POL	P1	P2	P3	P4
0.		-1.04+/-J4.87	-1.16+/-J8.10	-1.24+/-J11.28
	-0.49+/-J1.23			
15.	-0.49+/-J1.23	-0.64+/-J2.33	-1.38+/-J3.3	-0.98+/-J4.82
30.	-0.49+/-J1.23	-0.61+/-J2.32	-1.39+/-J3.71	-0.95+/-J4.52
45.	-0.49+/-J1.23	-0.62+/-J2.32	-0.67+/-J3.40	-0.75+/-J4.3
60.	-0.49+/-J1.23	-0.63+/-J2.3	-0.97+/-J3.8	
75.	-0.49+/-J1.25	-0.61+/-2.33	-0.725+/-J3.35	-0.67+/-J4.5
90.	-0.49+/-J1.23	-0.60+/-J2.31	-0.74+/-J3.42	-0.65+/-J4.28

THETA POL.

	P1	P2	P3	P4
15.	-0.49+/-J1.22	-0.64+/-J2.27	-1.08+/-J3.30	-0.82+/-J5.02
30.	-0.49+/-J1.23	-0.62+/-J2.33	-1.23+/-J3.52	-0.94+/-J4.57
45.	-0.49+/-J1.23	-0.64+/-J2.34	-0.87+/-J2.9	-0.86+/-J4.8
60.	-0.49+/-J1.23	-0.62+/-J2.32	-0.65+/-J3.21	-1.52+/-J5.4
75.	-0.49+/-J1.23	-0.63+/-J2.32	-0.64+/-J3.24	-1.26+/-J5.8

TABLE X
POLES OF DISK USING SENIOR'S CREEPING WAVE MODE

1.	(-0.5057350,-1.208781)
2.	(-0.6030402,-2.282039)
3.	(-0.6746498,-3.331739)
4.	(-0.7328901,-4.370643)
5.	(-0.7826892,-5.403155)
6.	(-0.8265909,-6.431364)
7.	(-0.8660754,-7.456467)
8.	(-0.9021299,-8.479179)
9.	(-0.9354153,-9.499999)
10.	(-0.9664097,-10.51928)
11.	(-0.9954736,-11.53727)
12.	(-1.022884,-12.55416)
13.	(-1.048859,-13.57012)
14.	(-1.073574,-14.58525)
15.	(-1.097172,-15.59966)
16.	(-1.119781,-16.61343)
17.	(-1.141471,-17.62662)
18.	(-1.162356,-18.63929)
19.	(-1.182501,-19.65149)
20.	(-1.201968,-20.66326)
21.	(-1.220811,-21.67463)
22.	(-1.239077,-22.68564)
23.	(-1.256811,-23.69632)
24.	(-1.274049,-24.70668)
25.	(-1.290820,-25.71675)
26.	(-1.307164,-26.72656)
27.	(-1.323088,-27.73611)
28.	(-1.338675,-28.74543)
29.	(-1.353844,-29.75452)
30.	(-1.368698,-30.76340)

Figure 1a. Comparison of amplitude of the simulated data (___)
           and the rational function approximant (...) obtained
           with ka=0.2 to ka=0.4, M=8,N=7. The x's are the __
           selected data points.

Figure 1b. Comparison of phase of the simulated data (___)
           and the rational function approximant (...) obtained
           with ka=0.2 to ka=0.4, M=8,N=7. The x's are the __
           selected data points.

Figure 2. Amplitude plot of a rational function approximant and
          the window technique at M=14,N=13. X's are the data
          points used in the exact solution.

Figure 3. Amplitude plot of a rational function approximant and the window technique at M=14,N=13. X's are the data points used in the exact solution.

Figure 4. Amplitude plot of a rational function approximant and the window technique at M=14,N=13. X's are the data points used in the exact solution.

Figure 5. A plot of the amplitude of both original data (solid line, an uncorrelated Gaussian white noise is added to the backscattering data of sphere) and the rational function fit (dotted line). σ=0.1.

Figure 6. A plot of the amplitude of both original data (solid line, an uncorrelated Gaussian white noise is added to the backscattering data of sphere) and the rational function fit (dotted line). σ=0.1.

Figure 7. A comparison of amplitude of the exact data and the noisy data filtered by a 10th order digital filter. The original data have a S/N=14.81 dB.

Figure 8. A comparison of phase of the exact data and the noisy
data filtered by a 10th order digital filter. The
original data have a S/N=14.81 dB.

Figure 9. The application of rational function approximant
to the prefiltering noisy data of sphere where 10th
digital filter and sum operator are used. (S/N=14.81
dB) x's' are the selected data points for the exact
fit.

Poles and Ramp Response

An approximate model

$$G(s) = \frac{a + bs}{1 + ds + es^2} e^{-\gamma s} , \tag{8}$$

for incorporating complex natural resonances into imaging methods utilizing the ramp response waveform is suggested. This model has a time domain response waveform

$$g(t) = \left[\frac{b}{e} e^{-\alpha(t-\tau)}\cos(\beta(t-\tau)+\left(\frac{a-b\alpha}{e\beta} e^{-\alpha(t-\tau)}\sin(\beta(t-\tau)\right)\right]u(t-\tau) .$$

$$\tag{9}$$

where $\alpha = \dfrac{d}{2\alpha}$

and $\beta = \dfrac{1}{2}\sqrt{\dfrac{4}{e} - \left(\dfrac{d}{e}\right)^2}$.

If the delay, τ, is taken as roughly the time to the maximum value of the ramp response (always negative) then the unknown constants a,b,d and e can be determined by fitting g(t) to four values of the ramp response between τ and the first negative to positive zero crossing. The dominant pole $-\alpha+j\beta$ so determined, when compared to a target pole library, will reveal at least to first order the dominating circumferential shape of the target shadow geometry.

Conclusions

A rational function approximant for extracting the complex natural resonances of an object from multiple frequency complex scattering data for the object has been developed. The approximant utilizes preprocessing using low pass filtering and a sum operator. Windowing is used to eliminate pattern fitting poles. Applied to simulated noisy data for a spherical scatterer, dominant complex natural resonances are satisfactorily extracted with signal to pseudo random noise levels as low as 14.82 dB. Based on these results we conclude that pole extraction from measured scattering data is only feasible under controlled conditions, i.e., using laboratory measurements. This in no way detracts from use of the poles in targe identification and possibly in target imaging. It has also been demonstrated that poles obtained using rational function approximants can be used to correct, at low frequencies, poles obtained from a K-pulse approach. This was demonstrated for a thin circular disc.

An emperical technique for using the dominant complex natural resonances of a scatterer to deduce the basic geometrical shape of the scatterer in the shadow region has been suggested.

REFERENCES

[1] J.W. Eberle, "Extension of the Physical Optics Approximation to Small Bodies", Report 827-6, The Ohio State University ElectroScience Laboratory, Department of Electrical Engineering; prepared under Contract No. AF 19(604)-3501 for Bistatic Reflection Characteristics, Air Force Cambridge Research Center, November 1, 1959.

[2] E.M. Kennaugh and D.L. Moffatt, "Transient and Impulse Response Approximations", Proc. IEEE, Vol. 53, No. 8, pp. 893-901, 1965.

[3] D.L. Moffatt, "Ramp Response Radar Imagery Spectral Content", IEEE Trans. Antenna Propagat., Vol. AP-29, No. 2, pp. 399-401, 1981.

[4] C.W. Chuang and D.L. Moffatt, "Natural Resonances of Radar Targets via Prony's Method and Target Discrimination", IEEE Trans. Aerospace Electron Sys., Vol. AES-12, No. 5, pp. 583-589, 1976.

[5] E.M. Kennaugh and D.L. Moffatt, "Comments on Impulse Response of Conducting Sphere Based on Singularity Expansion Method", Proc. IEEE, Vol. 70, No. 3, pp. 294-295, 1982.

[6] D.L. Moffatt, "Time Domain Electromagnetic Scattering from Highly Conducting Objects", Final Report 2971-2, The Ohio State University ElectroScience Laboratory, Department of Electrical Engineering; prepared under Contract No. F19628-70-C-0125 for the Air Force Systems Command, Laurence G. Hanscom Field, Massachusetts, May 1971.

[7] J.N. Brittingham, E.K. Miller and J.L. Willows, "Pole Extraction from Real Frequency Information", Proc. IEEE, Vol. 68, No. 2, pp. 263-273, 1980.

[8] M.L. Van Blaricum, "Techniques for Extracting the Complex Natural Resonances of a System Directly from its Transient Response", Ph.D. dissertation, Department of Electrical Engineering, University of Illinois at Urbana-Champaign, 1976.

[9]   ElectroScience Laboratory, "Time Domain Interpretations of
       Scattering Phenomena", Final Technical Report 2144-2, The
       Ohio State University ElectroScience Laboratory, Department
       of Electrical Engineering; prepared under Grant No. GK-708
       for the National Science Foundation, Washington, D.C., July
       11, 1968.

[10]   D.L. Moffatt, J.D. Young, A.A. Ksienski, H.C. Lin and C.M.
       Rhoads, "Transient Response Characteristics in
       Identification and Imaging", IEEE Trans. Antennas
       Propagation, Vol. AP-29, No. 2, pp. 192-205, 1981.

[11]   L.C. Chan, D.L. Moffatt, and L. Peters, Jr., "A
       Characterization of Subsurface Radar Targets", Proc. IEEE,
       Vol. 67, No. 7, pp. 991-1000, 1979.

[12]   L.C. Chan, L. Peters, Jr. and D.L. Moffatt, "Improved
       Performance of a Subsurface Radar Target Identification
       System through Antenna Design", IEEE Trans. Antenna
       Propagation, Vol. AP-29, No. 2, pp. 307-311, 1981.

[13]   Ted C. Lee, "Approximate Methods for Obtaining the Complex
       Natural Electromagnetic Oscillations of an Object", Ph.D.
       dissertation, The Ohio State University, Department of
       Electrical Engineering, Columbus, Ohio, 1983.

[14]   J.R. Auton and M.L. Van Blaricum, "Investigation of
       Procedures for Automatic Resonance Extraction from Noisy
       Transient Electromagnetics Data", Vols. I, II and III,
       Final Report on Contract N00014-80-C-0299, Effects
       Technology Inc., 1981.

[15]   E.M. Kennaugh, "The K-Pulse Concept", IEEE Trans. Antenna
       Propagation, Vol. AP-29, No. 2, pp. 327-331, 1981.

[16]   D.B. Hodge, "Spectral and Transient Response of a Circular
       Disc to Plane Electromagnetic Waves", IEEE Trans. Antennas
       Propagation, Vol. AP-19, pp. 558-562, 1971.

[17]   E.F. Knott, T.B.A. Senior, and P.L.E. Uslenghi, "High-
       Frequency Backscattering from a Metallic Disc", Proc. IEEE,
       Vol. 118, No. 12, 1971.

II.13 (SR.3)

SINGULARITIES IN QUASI-GEOMETRICAL IMAGING[*]

G. Dangelmayr & F.J. Wright

Institute for Information Sciences
University of Tübingen
Koestlinstr. 6, 7400 Tübingen, FRG

Various geometrical and quasigeometrical techniques for determining the shape of an unknown scatterer by analyzing backscattered acoustic or electromagnetic signals are reviewed. A geometrical imaging method is proposed for localizing edges by means of mid field data taken along a curve in the observation surface. The focusing phenomena associated with edge diffracted rays are discussed in terms of catastrophe theory. Closed surfaces are reconstructed from far field travel times by utilizing Minkowski's support fuction and the far field focusing directions are identified with singularities of the spherical mapping. A quasigeometrical imaging method is presented which is based on electromagnetic cross-polarized amplitude measurements.

1. INTRODUCTION

This paper reviews various geometrical and quasigeometrical techniques developed in [7] and [8] for the reconstruction of scattering surfaces from scalar (acoustic) or electromagnetic backscattered (zero-offset) signals. For high frequencies of the incident field the principal radii of the surface are large compared with the wavelength. Then the backscattered field at a given observation point depends not on the field over the entire scattering surface but only over restricted portions of that surface centered around the specular points. The contribution to the field coming from a neighborhood of a specular point is embodied in an asymptotic expansion in inverse powers of the frequency which consists of a phase function (the eikonal) and a hierarchy of amplitudes corresponding to the sequence of increasing inverse powers. In homoge-

[*] Work supported in part by the Stiftung Volkswagenwerk, FRG

461

W.-M. Boerner et al. (eds.), Inverse Methods in Electromagnetic Imaging - Part 1, 461–472.
© 1985 by D. Reidel Publishing Company.

neous media the phase function is essentially the specular distance, i.e., the spatial distance from the observation point to the specular point on the surface which determines the geometric travel time. Any inversion technique which uses only this phase information is referred to as "geometrical imaging". The term "quasigeometrical imaging" stands for techniques which incorporate information about geometric optics or higher order amplitudes into the inversion process.

The paper is organized as follows. In Sec. 2 the basic scattering formulae are established. In the scalar case the physical optics or Kirchhoff-approximation is presented for the backscattered wave in the mid field. In the electromagnetic case we confine ourselves to far field backscattering and perform a two term approximation of the backscattered field where the physical optics approximation is supplemented by a non-Kirchhoff correction. In Section 3 we consider a surface with an edge and present a geometrical imaging method which permits one to reconstruct the edge alone from mid field data taken along a curve in the observation surface. A geometrical imaging method for the smooth part of the surface is reviewed in [11]. Special consideration is given to the caustics associated with specular points on the edge which are identified with topological singularities of the distance function and classified by means of catastrophe theory. In Section 4 a geometrical imaging method based on far field travel times is presented which relies on Minkowski's support function. The far field focusing directions are interpreted as singularities of the spherical mapping and their local geometry is elucidated. In Section 5 we present a quasigeometrical imaging method based on electromagnetic first order far field amplitudes. The complete quadratic approximation of the surface at a specular point is derived from cross-polarized amplitude measurements which allows for a pointwise reconstruction of the scattering surface.

2. BASIC DIFFRACTION FORMULAE

2.1 Scalar Diffraction

A time periodic point source with frequency ω is localized at $\underline{x}_0$ and emits a spherical wave

$$u_I(\underline{x},\underline{x}_0,\omega)=(1/4\pi R)\exp(i\omega R/c), \quad R=\left|\underline{x}-\underline{x}_0\right| \tag{1}$$

which propagates through a homogeneous medium with velocity c and is incident upon an acoustically hard scatterer bounded by a surface S, where a scattered field $u_S(\underline{x},\underline{x}_0,\omega)$ is induced. We confine ourselves to zero offset and denote the backscattered field by $U(\underline{x}_0,\omega)=u_S(\underline{x}_0,\underline{x}_0,\omega)$. In the high frequency limit the field U can be approximated by the physical optics or Kirchhoff approximation [2]

$$U(\underline{x}_o,\omega) \sim -(i\omega/8\pi^2 c)\int\limits_{L(\underline{x}_o)} dS (\partial R^{-1}/\partial n)\exp(2i\omega R/c) \tag{2}$$

where $L(\underline{x}_o)$ is the geometrical lit side of S relative to the source point $\underline{x}_o$ and $\partial/\partial n$ denotes the outward normal derivative. The relation (2) must be understood in a true asymptotic sense, i.e., the leading terms of the asymptotic series of U and the r.h.s. of (2) in inverse powers of the frequency coincide, irrespectively of the types of the underlying specular points. If one attempts to improve (2) for higher order asymptotics, non-Kirchhoff correction terms must be taken into account [7] resulting from an iterative solution of the integral equations of diffraction theory [14]. For our purposes the leading order asymptotic relation (2) is sufficient.

2.2 Electromagnetic Far Field Diffraction

Consider an electromagnetic plane wave with magnetic field vector

$$\underline{h}_I = h_o\underline{a} \exp(i\omega\underline{x}\cdot\hat{\underline{x}}/c), \quad |\hat{\underline{x}}|=|\underline{a}|=1, \quad \hat{\underline{x}}\cdot\underline{a}=0 \tag{3}$$

which propagates in the direction of $-\hat{\underline{x}}$ (the time harmonic factor $\exp(i\omega t)$ has been suppressed in (3)) and is incident upon a perfect conducting target bounded by a closed surface S. The origin of the coordinate system is assumed to be inside the target. For large r the magnetic vector of the backscattered field $\underline{h}_s(\underline{x}_o,\omega)$ received at $\underline{x}_o = r\hat{\underline{x}}$ can be written as [5]

$$\underline{h}_S(\underline{x}_o,\omega) = (1/4\pi r)\exp(i\omega r/c)\underline{m}(\hat{\underline{x}},\omega)+0(r^{-2}) \tag{4}$$

where $\underline{m}$ is given by [13]

$$\underline{m}(\hat{\underline{x}},\omega)=(i\omega/c)\hat{\underline{x}}\times\int\limits_S dS\underline{J}(\underline{x},\hat{\underline{x}},\omega)\exp(-i\omega\underline{x}\cdot\hat{\underline{x}}/c) \tag{5}$$

with J being the surface current induced on S by the incident field. Defining

$$V_K(\hat{\underline{x}},\omega)=(-2i\omega/c)\int\limits_{L(\hat{\underline{x}})} dS (\hat{\underline{x}}\cdot\underline{n})\exp(-2i\omega\underline{x}\cdot\hat{\underline{x}}/c), \tag{6}$$

$$\underline{m}_1(\hat{\underline{x}},\omega)=h_o\hat{\underline{x}}\times\int\limits_{L(\hat{\underline{x}})} dS (\kappa_2-\kappa_1)\{(\underline{a}\cdot\underline{t}_2)\underline{t}_1+(\underline{a}\cdot\underline{t}_1)\underline{t}_2\}\exp(-2i\omega\underline{x}\cdot\hat{\underline{x}}/c) \tag{7}$$

where $L(\hat{\underline{x}})$ is the lit side relative to the direction $\hat{\underline{x}}$, the reduced far field magnetic vector $\underline{m}$ is approximated by

$$\underline{m}(\hat{\underline{x}},\omega) \sim h_o\underline{a}V_K(\hat{\underline{x}},\omega)+\underline{m}_1(\hat{\underline{x}},\omega). \tag{8}$$

In Eqs. (6) and (7) $\underline{n}$ is the outward unit normal, $\underline{t}_1$ and $\underline{t}_2$ are unit vectors pointing in the principal directions of S at $\underline{x}$ such that $\underline{t}_1 \times \underline{t}_2 = \underline{n}$ and κ_1, κ_2 are the principal curvatures corresponding to the directions $\underline{t}_1, \underline{t}_2$, respectively. The relation (8) is a high frequency approximation in the sense that the first two terms of the asymptotic expansions of $\underline{m}$ and the r.h.s. of (8) in inverse powers of ω coincide, i.e., (8) is a first order asymptotic identity. The first term $h_0 a V_\kappa$ in (8) is the Kirchhoff approximation which results from the standard physical optics approximation of the current $\underline{J}$ and determines the leading order asymptotics. The second term $\underline{m}_1$ is obtained from a first order correction [4,6,7] to the physical optics current derived from the integral equations of diffraction theory [14]. For further details we refer to [7]. Contrary to the scalar leading order asymptotic relation (2) we have utilized a first order asymptotic approximation in the electromagnetic case because $\underline{m}_1$ contains useful information about the surface's curvatures at the specular points (Section 5).

3. MAPPING OF EDGES

In this Section it is assumed that the scatterer has the form of a wedge as shown in Fig.1 where the observation point $\underline{x}_0$ varies on the surface Σ above the wedge. The scattering surface S is identified with the upper boundary of the scatterer because the lower one does not contribute to the asymptotic backscattered field. We assume further that S is only slightly curved so that for any $\underline{x}_0$ on Σ the lit side $L(\underline{x}_0)$ is the whole surface S. The edge E of the wedge is parametrized via $\underline{x} = \underline{x}_e(1)$ where 1 is the arclength along E.

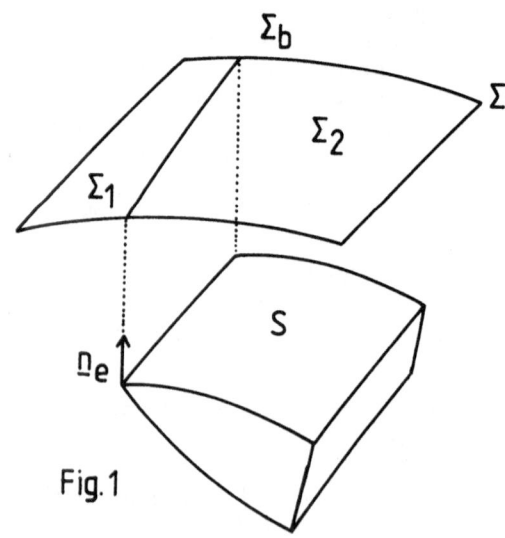

Fig. 1

The unit tangent is denoted by $\dot{\underline{x}}_e(1) = d\underline{x}_e(1)/dl$ and the surface's upward unit normal at $\underline{x}_e(1)$ by $\underline{n}_e(1)$ which, in general, differs from the edge's principal normal.

The task is to reconstruct the edge from measurements of backscattered scalar fields on Σ where the frequency response $U(\underline{x}_0, \omega)$ is given by (2) with $L(\underline{x}_0)$ replaced by S. The leading order asymptotics of U are governed by the specular points on S for which the distance $R = |\underline{x}_0 - \underline{x}|, \underline{x} \epsilon S$, has a stationary value. Due to the surface's edge the integration domain in the integral (2) has a boundary. Consequently, two types of specular points contribute to U, viz., interior ones and specular points on the boundary. Properties of the former and their use for imaging the smooth part of S have been discussed in [8,11]. Here we confine ourselves to specular points on the boundary and present a method which permits one to reconstruct the edge.

3.1 Geometric Imaging of the Edge

The stationarity condition for the distance $R_e = |x_0 - x_e|$ from x_0 to the edge $x_e = x_e(1)$ requires that the distance vector $R_e = x_0 - x_e$ lies in the normal plane of the edge point x_e. For any point x_0 on Σ there may be several specular points $x_{es} = x_{es}(x_0)$ on E and we label them and further geometrical quantities evaluated at x_{es} by the subscript s, e.g., $R_{es} = x_0 - x_{es}$, $R_{es} = |R_{es}|$. We assume now that the specular edge-distances $R_{es} = R_{es}(x_0)$ have been inferred from back-scattered signals. Then the geometric reconstruction of the edge is based on the identity

$$R_{es}(x_0) \nabla_{x_0} R_{es}(x_0) = x_0 - x_{es}(x_0). \tag{9}$$

For a generalization of (9) to rays in an arbitrary inhomogeneous medium see [15]. Since $|\nabla R_{es}| = 1$, only two components of ∇R_{es} are needed. Moreover, all points x_0 on the curve C_e defined by the intersection of the normal plane of x_e and Σ yield the same edge point. It suffices, therefore, to vary x_0 on a curve C_0 on Σ which does not coincide with one of the curves C_e. Let C_0 be parametrized by the arclength σ, $x_0 = x_0(\sigma)$, and let $\dot{x}_0(\sigma)$ be its unit tangent. Let $t(\sigma)$ be a unit vector in the tangent plane of $x_0(\sigma)$ perpendicular to $\dot{x}_0(\sigma)$. Then, defining

$$\alpha = dR_{es}/d\sigma, \quad \beta = \frac{\partial}{\partial \rho} R_{es}(x_0 + \rho t)\Big|_{\rho=0} \tag{10}$$

where all quantities in (10) are evaluated along C_0, the edge is reconstructed via

$$x_e = x_0 - R_{es}\{\alpha \dot{x}_0 + \beta t + (1 - \alpha^2 - \beta^2)^{1/2} n_0\} \tag{11}$$

n_0 being the downward unit normal of Σ at $x_0(\sigma)$. Hence, measurements of R_{es} in a small neighborhood of a curve C_0 on Σ allows a reconstruction of the edge E.

3.2 Edge-Focusing

In general, focusing occurs if two or more specular points (resp. their rays) coalesce in a degenerate specular point x_{sc} which may be in the interior of S or on the edge. The codimension of a focusing point $x_0 = x_{0c}$ is defined as the number of coalescing specular points minus one and is identical with the catastrophe theoretic codimension of the topological singularity the distance $R=|\underline{x}_{oc}-\underline{x}|$, $\underline{x}\epsilon S$, exhibits at $\underline{x}=\underline{x}_{sc}$ [12,16,18]. The highest possible codimension encountered generically in multi-parameter families of functions - here the distance function $R=|\underline{x}_o-\underline{x}|$, $\underline{x}\epsilon S$ - is the number of parameters involved which in our case is just the number of degrees of freedom the observation point $\underline{x}_o$ possesses. Because $\underline{x}_o$ varies on a two-dimensional observation surface Σ, only singularities with codimension less than or equal to two can generically occur [12,16,18]. We consider here only singularities on the edge. For degenerate specular points in the interior of S see [8,11]. Rather than listing the catastrophe normal forms we confine ourselves to a geometric description of the various singularities. For a more detailed discussion including codimension-three singularities see [19].

(i) Shadow boundary. The straight lines in the normal directions $\underline{n}_e$ sweep out the geometrical shadow boundary which generically intersects the observation plane Σ in a curve Σ_b. We denote by Σ_1 and Σ_2 those parts of Σ lying to the right and to the left of Σ_b, respectively, (cf. Fig. 1). Hence, in Σ_1 there are interior and edge speculars while in Σ_2 interior speculars fail to exist. On Σ_b an edge- and an interior specular point coalesce, i.e., Σ_b is a line of codimension one singularities. The specular distances R_s and R_{es} in the vicinity of Σ_e are sketched in Fig. 2 where $R_s=|\underline{x}_o-\underline{x}_s|$ is the distance associated with an interior specular $\underline{x}_s$. The R_s-branch terminates tangentially in the R_{es}-branch at the shadow boundary Σ_b.

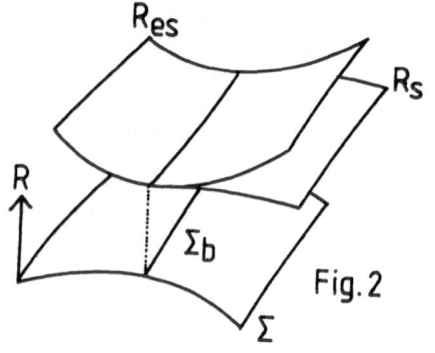

Fig. 2

(ii) Edge-caustic. We define $\overline{EV}$ to be the envelope of the normal planes of the edge, i.e., the surface in three-dimensional space where neighboring normal planes intersect. EV is a ruled surface with rulings made up by straight lines in the normal planes which are parallel to the binormal directions and pass through the curvature centres of the underlying edge points. Generically EV intersects the observation surface Σ in a curve Σ_e. On Σ_e two specular edge points coalesce, i.e., two branches of specular edge distances join along a cusped line above Σ_e, similar to what happens when two interior speculars coalesce [8,11]. Hence, EV plays the role of the edge's caustic and Σ_e is a line of codimension-one singularities.

(iii) Caustics on the shadow boundary. A codimension two singularity takes place if EV or the interior caustic (the surface's evolute) meets the shadow boundary giving rise to the coalescence of three rays, two from the edge and one from the interior in the former and conversely in the latter case. One possible arrangement of the various branches of R_s and R_{es} is sketched in Fig. 3 for an edge caustic and in Fig. 4 for an interior caustic meeting the shadow boundary. In Fig. 4, Σ_i denotes the line where the surface's evolute intersects Σ.

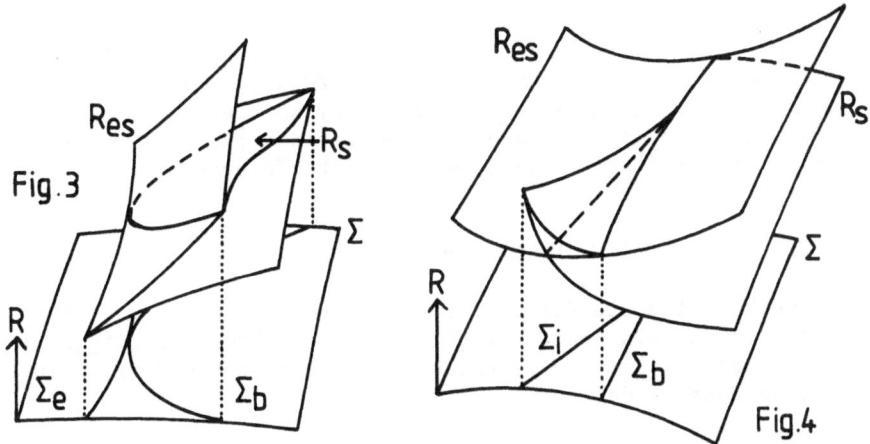

(iv) Cusps of the edge caustic.   The edge of regression ER of the envelope of the normal planes of E is defined as the curve on EV which is everywhere tangent to the rulings. Alternatively, ER may be defined as the curve formed by the centres of the osculating spheres of E [10]. Along ER two smooth sheets of EV are glued together, i.e., ER plays the role of a rib line [8,15] of the edge caustic EV. Generically ER intersects the observation plane in isolated points where Σ_e exhibits a cusp. Near this cusp point the specular edge distances behave qualitatively like the swallowtail bifurcation set, similar to cusp points associated with interior caustics [8, 11].

The phenomena described so far exhaust all generic singularities up to codimension two which can occur on the edge. Special examples of surfaces where these and higher singularities take place are presented in [19].

3.3 Contributions to the Backscattered Field from Specular Edge Points

Let $x_{es} = x_{es}(x_0)$ be a non-degenerate specular point on the edge. The contribution U_{es} to the backscattered field U coming from a neighborhood of x_{es} is given by [7]

$$U_{es} \sim \gamma A_{es} \left| 2c/\omega \right|^{1/2} \exp\{2i\omega R_{es}/c + \tfrac{1}{4}\pi\mu\,\mathrm{sgn}\omega\}(1 + 0(\omega^{-1})) \tag{12}$$

with amplitude

$$A_{es} = (8\pi R_{es})^{-3/2} (R_{\shortparallel}/R_{\perp}) |1 - \underline{R}_{es} \cdot \underline{\ddot{x}}_{es}|^{-1}. \tag{13}$$

Here, $\gamma = 1(-1)$ if $\underline{x}_0$ is on Σ_2 (Σ_1) and $\mu = \text{sgn}(1 - \underline{R}_{es} \cdot \underline{\ddot{x}}_{es})$. We have used the notation $R_{\shortparallel}$ and $R_{\perp}$ for the components of the distance vector R_{es} parallel and perpendicular to $\underline{n}_{es}$, respectively, and $\underline{\ddot{x}}_e = d^2 \underline{x}_e / dl^2$ denotes the curvature vector pointing in the direction of the edge's principal normal. The delta-response $G'(\underline{x}_0, t)$, obtained by Fourier inversion of $U(\underline{x}_0, \omega)$, diverges like $(t - t_{es})^{-1/2}$ at the geometric two way travel-time $t_{es} = 2R_{es}/c$. This permits one to detect the travel-times t_{es}, e.g., from an aperture limited Fourier inversion of the backscattered field U [7] or from the spacing of peaks in the high frequency interference spectrum [1].

4. GEOMETRIC INVERSION IN THE FAR FIELD

We assume here the standard far field configuration, i.e., the scatterer occupies a bounded region which contains the origin of the coordinate system and is enclosed by a smooth surface S. In the scalar case we define

$$V(\underline{\hat{x}}, \omega) = \lim_{r \to \infty} \{(4\pi r)^2 \exp(-2i\omega r/c) U(r\underline{\hat{x}}, \omega)\}, \tag{14}$$

so that the far field approximation of U for large r is $U \sim \exp(2i\omega r/c) V/(4\pi r)^2$. Since asymptotically, up to leading order in $1/\omega$ [7]

$$V(\underline{\hat{x}}, \omega) \sim V_K(\underline{\hat{x}}, \omega) \tag{15}$$

where the Kirchhoff approximation V_K is defined by (6), the leading order asymptotics of both, the electromagnetic and the scalar backscattered far fields are governed by the same function V_K. In order to deduce the actual leading order asymptotic term of V_K we have to perform a stationary phase integration of the integral in (6). The stationary condition selects those points $\underline{x}_s = \underline{x}_s(\underline{\hat{x}})$ on the lit side $L(\underline{\hat{x}})$ for which the outward normal points in the direction of $\underline{\hat{x}}$. We use the same notation as in Sec. 3, i.e., the subscript s refers to a geometric quantity evaluated at $\underline{x}_s$. Assuming then that the Gaussian curvature of S at $\underline{x}_s$ does not vanish, the contribution V_{Ks} to V_K coming from a neighborhood of the specular point $\underline{x}_s$ takes the form

$$V_{Ks} \sim \nu A_s \exp(-2i\omega \rho_s/c)(1 + 0(\omega^{-1})) \tag{16}$$

where

$$\rho_s = \underline{x}_s \cdot \underline{\hat{x}}, \quad A_s = 2\pi |\kappa_{1s} \kappa_{2s}|^{-1/2} \tag{17}$$

and $\nu =1(-1)$ or $-i\text{sgn}\omega$ if the principal curvatures κ_{1s} and κ_{2s} are both negative (positive) or of opposite signs corresponding to a convex (concave) or saddle-type region of S around $\underline{x}_s$, respectively. From (14) and (16) we discover that the geometric two-way travel-time for a wave travelling from the source $r\hat{\underline{x}}$ to S and then back to the source again is given by $t_s=2(r-\rho_s)/c$. Using methods described in [1,8,11] (e.g., an aperture limited Fourier inversion of the frequency response) permits one to infer t_s and consequently ρ_s from backscattered signals. The task is now to reconstruct S from this phase information.

4.1 Utilizing Minkowski's Support Function

A genuine surface S can be divided into several regions S_j where S_j is either convex, concave or of saddle-type. In general, the regions S_j are separated by certain smooth lines where the Gaussian curvature vanishes. Let Ω_j be the spherical image of S_j on the unit sphere (the outward normal directions). Because the spherical image of S_j is single valued (the Gaussian curvature does not vanish), S_j can uniquely be parametrized in the form $\underline{x}=\underline{X}_j(\underline{n})$, $\underline{n}\epsilon\Omega_j$, where $\underline{n}$ is the outward unit normal of S_j at $\underline{x}$, i.e., the polar angles of $\underline{n}$ are treated as independent variables. Let C_j be the cone in three dimensional space subtended by the directions in Ω_j. Then Minkowski's support function [17]

$$M_j(\underline{\xi})=\underline{\xi}\cdot\underline{X}_j(\underline{\xi}/|\underline{\xi}|), \quad \underline{\xi}\epsilon C_j \tag{18}$$

associated with S_j is a single valued homogeneous function of degree one defined in C_j. From M_j the region S_j can be reconstructed via [17]

$$\underline{x}=\nabla_\xi M_j(\underline{\xi})=\underline{X}_j(\underline{\xi}/|\underline{\xi}|), \quad \underline{\xi}\epsilon C_j \tag{19}$$

We note that if the gradient in (19) is written out in polar coordinates, then the angular and radial derivatives give the components of $\underline{X}_j$ tangent and normal to S, respectively.

Recalling that the far field specularity condition requires that $\hat{\underline{x}}$ coincides with the outward unit normal at the specular points we may label these by $\underline{x}_{sj}(\hat{\underline{x}})=\underline{X}_j(\hat{\underline{x}})$ where $\hat{\underline{x}}\epsilon\Omega_j$ and $\underline{x}_{sj}\epsilon S_j$. From (17) and (18) we find that $M_j(\xi\hat{\underline{x}})=\xi\rho_{sj}(\hat{\underline{x}}),\xi>0$, where $\rho_{sj}=\underline{x}_{sj}\cdot\hat{\underline{x}}$ is the phase function corresponding to the specular point $\underline{x}_{sj}$. The regions S_j are then readily reconstructed by virtue of (19).

4.2 Topological Far Field Singularities

Let the surface be parametrized by two coordinates, $\underline{x}=\underline{X}(\xi,\eta)$, e.g., the polar angles of the rays coming from the origin and intersecting S at $\underline{x}$. Then the phase function $\rho=\underline{x}\cdot\hat{\underline{x}}$ in the integral (6) becomes

a function of two "state variables" (ξ,η) and two "control parameters" parametrizing the unit sphere, $\rho=\rho(\xi,\eta,\hat{x})$. The solutions
of the equations $\partial\rho/\partial\xi=\partial\rho/\partial\eta=0$ are clearly the parameter values
(ξ_s,η_s) of the specular points $\underline{x}_s=\underline{X}(\xi_s,\eta_s)$. In the terminology of
catastrophe theory, ρ exhibits a singularity at (ξ_s,η_s). Further-
more, catastrophe theory tells that the only singularities which
can generically occur are Morse singularities and degenerate singularities of the fold- and cusp-type [8,11,16,18]. The underlying
topology on the surface is the following. There are smooth closed
lines on S, called L-lines, where one of the two principal curvatures vanishes, i.e., the spherical map is locally not invertible.
These L-lines are the boundaries between adjacent regions S_j and
their normals constitute the focusing directions. For $\underline{x}$ on L let
$P(\underline{x})$ be the principal curvature line corresponding to that principal curvature which vanishes at $\underline{x}$. For "almost all" points $\underline{x}$ on
L the curvature line $P(\underline{x})$ intersects L transversely and ρ exhibits
a fold-singularity. For isolated points $\underline{x}_c$ on L, however, $P(\underline{x}_c)$
has a tangential contact with L (the zero principal curvature di-
rection is tangent to L) and ρ encounters a cusp-singularity. In
Fig. 5a we have sketched an L-line (solid) with a cusp-point $\underline{x}_c$,
two principal curvature lines $P_j=P(\underline{x}_j)$ and $P_c=P(\underline{x}_c)$ (broken).
The images L' and P_j', P_c' of resp. L and P_j, P_c on the unit sphere
under the spherical mapping are shown in Fig. 5b. The various
branches of the "specular distances" ρ_s near $\underline{n}(\underline{x}_c)$ behave qualitatively like the bifurcation set of the swallowtail catastrophe.
For further details about cusps of the spherical mapping we refer
to [3]. The geometry of the additional singularities which generically occur in a one parameter family of surfaces are discussed in
[9].

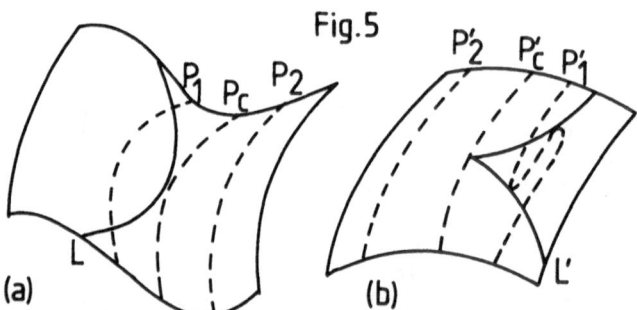

Fig.5

5. INVERSION BY MEANS OF FIRST ORDER ASYMPTOTIC ELECTROMAGNETIC
   FAR FIELD AMPLITUDES

Recalling that (8) is a first order asymptotic relation for the
reduced far field magnetic vector $\underline{m}(\hat{x},\omega)$, Eq. (5), the first two
terms of the asymptotic contribution $\underline{m}_s$ to $\underline{m}$ coming from a neigh-

borhood of the specular point $\underline{x}_s$ are found by a stationary phase integration of (6) and (7). The result is [7]

$$\underline{m}_s \sim h_o \nu A_s \exp(-2i\omega\rho_s/c)\{\underline{a}+(c/2i\omega)(A_1\underline{a}+B\underline{b})+O(\omega^{-2})\} \qquad (20)$$

where A_s, ρ_s and the sign factor ν are defined in Sec. 4, Eq. (17), and $\underline{b}=\hat{x}\times\underline{a}$, i.e., $\underline{a}$ and $\underline{b}$ are orthogonal unit vectors in the tangent plane. The amplitude A_1 results from the first order asymptotic expansion of V_{Ks} [7] and needs not concern us here further. The amplitude B is given by

$$B=(\kappa_{1s}-\kappa_{2s})\sin(2\alpha) \qquad (21)$$

where α is the angle enclosed by $\underline{a}$ and the principal direction $\underline{t}_1$. By measuring the polarizations in the directions of $\underline{a}$ and $\underline{b}$ one infers ν, the absolute value of the Gaussian curvature and the first order amplitude B. If the same experiment is performed with an incident polarization direction $\tilde{\underline{a}}$ obtained by a 45°-rotation of $\underline{a}$, the corresponding first order amplitude $\tilde{B}$ is given by

$$\tilde{B}=(\kappa_{1s}-\kappa_{2s})\cos(2\alpha) . \qquad (22)$$

Eqs. (21) and (22) permit one to determine the curvature difference $\kappa_{1s}-\kappa_{2s}$ and the angle α, i.e., the principal directions, from measurable quantities. Using this and the geometric optics amplitude A_s one infers the principal curvatures κ_{1s}, κ_{2s} [7].

The method described above allows for a deduction of the complete quadratic approximation of the surface at the specular points from cross-polarized amplitude measurements. The task is now to reconstruct the shape of the scattering surface which is achieved as follows. Assume first that the curvature information is given for a specular point in a convex region where the principal curvatures are negative. Fix a (x_1,x_2,x_3)-coordinate system and consider an ellipsoid $E:(x_1/c_1)^2+(x_2/c_2)^2+(x_3/c_3)^2=1$ with yet unknown semi-axes c_j. There is exactly one point $\check{x}$ on E which has the same outward normal direction as the prescribed one $\hat{x}$. Equating the quadratic approximation of E at $\check{x}$ with the measured one yields three equations which determine the three semi-axes c_j. From this one finds the coordinates of the point $\check{x}$ on E which is identified with the underlying specular point $\underline{x}_s$. The technique resembles the methods of equivalent ellipsoids introduced by Chaudhuri and Boerner [6] for relating the time domain response of convex scatterers to the projected area function. If the principal curvatures are positive (the specular point is in a concave region) one perfoms the same computation with $(\hat{x},\kappa_1,\kappa_2)$ replaced by $(-\hat{x},-\kappa_1,-\kappa_2)$ and then reflects the resulting point on E at the origin. In saddle-type regions the same technique may be utilized as in convex regions, but with a hyperboloid rather than an ellipsoid.

REFERENCES

[1] Achenbach, J.D., and Norris, A., "Crack characterization by the combined use of time-domain and frequency-domain scattering data", 1981 AF/DARPA-Review of Quantitative NDE, in press.

[2] Baker, B.B. and Copson, E.T., "The mathematical theory of Huggens' principle", 2nd ed., Oxford University Press 1950.

[3] Banchoff, T., Guffney, T. and McCrory, C., "Cusps of Gauss mappings", Pitman, London 1982.

[4] Bennet, C.L., Auckenthaler, A.M., Smith, R.S. and De Lorenzo, J.D., "Space time integral equation approach to the large body scattering problem", Sperry Rand Research Centre, Sudbury, Mass., Final Report on Contract F 30602-71-C-0162, RADC-CR-73-70, AD76394 (May 1973).

[5] Boerner, W.-M., "Polarization utilization in electromagnetic inverse scattering", in H.P. Baltes (ed.), "Inverse scattering problems in optics", Springer 1980.

[6] Chaudhury, S.K. and Boerner, W.-M., Appl. Phys. 11, pp. 337-350 (1976).

[7] Dangelmayr, G., "Asymptotic inverse scattering", submitted for publication (1983).

[8] Dangelmayr, G. and Güttinger, W., Geophys.J.R.Astr.Soc. 71, pp. 79-126 (1982).

[9] Dangelmayr, G. and Armbruster, D., "Singularities in phonon focusing", submitted for publication (1983).

[10] Goetz, A., "Introduction to differential geometry", Addison Wesley 1970.

[11] Güttinger, W., "Topological approach to inverse scattering in remote sensing", this Volume.

[12] Güttinger, W. and Eikemeier, H. (eds.), "Structural stability in physics", Springer 1979.

[13] Jackson, J.D., "Classical Electrodynamics", 2nd ed., Wiley 1975.

[14] Maue, A.W., Zeitschr. Phys. 126, pp. 601-618 (1949).

[15] Norris, A.N. and Achenbach, J.D., "Mapping of a crack by ultrasonic methods", to be published in J.Acoust.Soc.Am.

[16] Poston, T. and Stewart, I., "Catastrophe theory and its applications", Pitman, London 1978.

[17] Stoker, J.J., Comm. Pure Appl. Math. 3, pp. 231-257 (1950).

[18] Thom, R., "Structural stability and morphogenesis", Benjamin 1975.

[19] Wright, F.J., and Dangelmayr, G., in preparation.

II.14 (RS.5)

DECONVOLUTION OF MICROWAVE RADIOMETRY DATA

R. H. Dittel

Deutsche Forschungs- und Versuchsanstalt
für Luft- und Raumfahrt
Institut für Hochfrequenztechnik
8031 Oberpfaffenhofen, FRG

ABSTRACT

The methods of deconvolution by multiple convolution, maximum-likelihood and matrix inversion have been investigated for the purpose of data reconstruction for multispectral microwave radiometry data. The principle objective was the use of a-priori-information in order to reduce the variety of the solution manifold, to force the process to converge and to decrease computer time. By application of some algorithms on real data, a contrast enhancement by the factor 5 could be reached for objects 1/10 of the 3-dB-halfwidth of the antenna wide.

1. INTRODUCTION

The microwave signal of a body, measured for the purpose of earth-scientic investigations is altered in a characteristic manner on the way from the radiation source to the output of the detection equipment. The different steps are as follows:

a)   addition of reflected radiation energy from the environment to the emitted radiation energy of the object

b)   alteration of the thus integral generated radiation energy by obsorption and attenuation on passing through the atmosphere.

c)   convolution of the radiation patterns of the resolution elements of the scene detected by the imaging characteristics of the antenna system.

W.-M. Boerner et al. (eds.), Inverse Methods in Electromagnetic Imaging - Part 1, 473–492.
© 1985 by D. Reidel Publishing Company.

d)   addition of systematical and statistical intensity fluctu-
        ations to the detected radiation energy by the electronic
        components of the detection equipment.

In order to determine the nature of a detected object from its
radiation signature, it is necessary to eliminate or at least
to reduce the influence of the distortions pointed out under
items a) to d) and thus obtain the genuine radiation signature
of the object. The processing in general is performed in a way
reverse to the signal sequence, starting with the elimination of
the noise and systematic errors and ending with the atmospheric
correction of the signal.

The purpose of this paper is to demonstrate the results of an
investigation concerning the practical possibilities of signal
reconstruction by deconvolution and thus correcting the influence
indicated under item c).

The necessity for such a reconstruction is mandatory for the following reasons:

a)   the convolution effect leads to unsharp imaging of the scene
        and thus to difficulties in object boundary identification.

b)   the convolution effect leads to contrast reduction and thus
        to the impossibility of discriminating geometrically small
        objects and object induced intensity fluctuations. The term
        "small" is assumed for object sizes representing a filling
        factor < 0.1. The filling factor is defined as the ratio of
        the object projection area relative to the projection area
        of the 3-dB-halfwidth of the antenna pattern.

c)   the convolution effect leads to signatures at the boundaries
        of areas with different radiation characteristics. The gene-
        ration of radiation intensities not really existing at the
        different locations leads to wrong identification of object
        signatures and properties and in consequence to misclassifi-
        cations of objects and their location in the scene.

2. BASIC METHODS OF DECONVOLUTION SOLUTIONS

There are basically two different approaches for the reconstruction of the scene T from the image B, generated by convolution

        $A * T = B$

with the antenna characteristics A applying deconvolution operations.

a)   in the time domain by multiplying the above term with the inverse of the antenna characteristics:

$$T = A^{-1} B$$

b)   in the frequency domain, by multiplying the transformed terms

$$A(\omega) \cdot T(\omega) = B(\omega)$$

with the term $1/A(\omega)$

$$T(\omega) = \frac{B(\omega)}{A(\omega)} .$$

The transformation of $T(\omega)$ into the time domain leads to the temperature map T of the scene.

The application of the deconvolution algorithms for either one method involves the following problems:

a)   the reconstruction in the time domain requires even for modest scene sizes and geometrical resolutions data matrices of several hundred elements in line and column direction.

The matrix operations are characterized by a large number of multiplications. These lead to rounding errors and in consequence to numerical inaccuracies of the inverse matrix. The effects of the rounding error in a digital computer are enhanced by the special property of the antenna characteristics. Its function is approximately normal distributed and contains therefore small weighting values in the region off the symmetry axis. It is this set of numerically small values that lead to relatively large rounding errors during the process of matrix inversion. The result is an enhancement of just those elements, whose value in the inverse matrix should have the value 0.

The consequence for the reconstruction is an exceeding enhancement of the small deviations from the ideal undisturbed result of a convolution operation. An increase in the numerical accuracy (by increasing the number of bits) only results in higher memory requirements and computer time, but does not necessarily lead to higher stability of the deconvolution process, as statistical numerical fluctuations are not eliminated by an increase of the number of digits. A comparable problem also arises at the reconstruction of images in the frequency domain.

b)   A general point of controversy, concerning the deconvolution process, is the assumption that an exact reconstruction of a scene cannot be obtained due to the effects of missing knowledge about the data elements outside the image boundaries.

An analysis of the antenna – /scene-relation, however, shows, that
the real deconvolution of a scene does not require the full information content of the infinite large projection area of the
antenna characteristics, but only needs a minor part thereof
around the symmetry axis. Outside this central region, the weighting
factors rapidly decrease to the value 0 and can therefore be fully
ignored for the deconvolution process. Of importance is the requirement, that a large number of independent data elements of
the scene are available.

3. DATA ERRORS AND SELECTION OF ALGORITHM

For the derivation of the convolution and deconvolution terms,
the presence of undistorted data has been assumed. The real data,
however, are disturbed by a variety of systematical and statistical errors, whose origin is not necessarily induced by and related to the radiation properties of the object detected. The influence of the errors can be defined in the form

$$B = A(T + n_1) + n_2 + n_3$$

which leads to the deconvolution term

$$A^{-1} B = A^{-1} A(T + n_1) + A^{-1} n_2 + A^{-1} n_3.$$

The different contribution habe the following properties:

a)   the contribution n_1 which represents the incertainty ΔT of
     the radiation temperature prior to the deconvolution process
     by the receiving antenna. It is influence by the convolution
     process. Typical examples for n_1 are system external distur-
     bances of technical radiation sources and atmospheric phe-
     nomena

b)   the contribution of the system induced systematical error n_2
     and the statistical error n_3. Examples for n_2 is the swit-
     ching cycle to the reference radiation source in the radio-
     meter system, n_3 is the thermal noise of the receiver section.
     They are not convolved by the system.

In order to facilitate the deconvolution process, as many as
possible error contributions should be eliminated prior to the
data reconstruction process. The problem is to properly identify
and eliminate the different contributions in such a way, that
the object information is distorted only to a minor degree.
In most cases, n_1 is characterized by weak shorttime effects, which
are eliminated through the digitation procedure. Only influences
of exterme radiation intensity and long duration, such as signals
from radar stations, cannot be eliminated by a simple filtering
process and therefore requires special treatement. In general,
$n_1 \ll T$ and therefore $T + n_1 \sim T$ can be assumed.

No difficulties occur for the elimination of the contribution n_2, as the properties of systematic errors in general are very well known and can be replaced from the data by application of filter procedures. Care has to be taken on the necessary of the filter characteristics.

Under this assumption the deconvolution term is reduced to

$$T = A^{-1} B + A^{-1} n_3.$$

For practical cases, the reproducibility of the deconvolution process depends on the statistical properties of the term $A^{-1} n_3$, which generally leads to diverging behaviour of the deconvolution process. If n_3 is of minor influence to the signal value T, then the application of recursive methods many lead to exceeding radiation temperature values.

For most of the algorithms described in textbooks and applied for practical cases absolute ignorance of the object and data properties and the image distortions is assumed. In general there is, however, no complete ignorance of the scene properties, but there exists in most cases basic knowledge on the approximate average radiation temperature, kind and extension of statistical data errors and the approximate temperature profiles along the flight path. This knowledge is being used as a first approximation to the most probable solution of the algorithm applied. The a-priori-knowledge not only reduces the amount of computer time, but also the high variety of possible solutions, which otherwise would be generated due to the statistical errors of the individual data elements. Thus the application of a-priori-knowledge also sets the preliminaries for divergence of the reconstruction process. The question for the optimum method of deconvolution for a given data set can be answered in such a way, as the best algorithm is the one, which best uses the a-priori-knowledge offered. An for all possible cases best algorithm for the solution of all kinds of problems does not exist.
Depending on the properties of the data, a-priori-knowledge can be used to quantify deterministical and statistical processes. The parameter may be single values, or value sets for the characterization of single or global data properties. Typical examples of a-priori-parameters may be:

- the arithmetic mean or the median as the most probable value of a physical quantity

- the standard deviation as a value for the most probable degree of data, fluctuation (e.g. thermal noise)

-   upper and lower limit of a data range to define the dynamics
    of deterministical and statistical properties (e.g. inten-
    sity dynamics of a scene)

-   the degree of derivation between single data elements of a
    time or spatial data sequence (e.g. for the characterization
    of common temperature fluctuations caused by statistical
    processes).

4. INVESTIGATED DECONVOLUTION ALGORITHMS

The following algorithms have been investigated for the purpose
of deconvolving radiometry data:

a)   deconvolution by multiple convolution,

b)   deconvolution by maximum likelihood,

c)   deconvolution by matrix inversion.

4.1 Deconvolution by multiple convolution

The deconvolution process is maintained on the basis of the convolution term

$$B = A * T$$

This form does not indicate the presence of noise, which statistically influence the original image

$$B = A * T + N$$

and thus the result of the deconvolution process. For practical
applications a distinction between noise and undisturbed signal
and their seperation is not possible. Therefore the algorithm has
to be designed in such a way, as to converge when applied on noise
disturbed data.

The best approximation of the first iteration step n = 0 is the
original noise distorted data set T itself.

$$T^{(n=0)}(x) = B^{(o)}(x).$$

The following iterations steps are maintained as demonstrated in
the following loop [1]:

$$B^{(n)}(x) = T^{(n)}(x) * A(x)$$

$$T^{(n+1)}(x) = T^{(n)}(x) + K(x)[B^{(0)}(x) - B^{(n)}(x)]$$

$$|T^{(n+1)}(x) - T^{(n)}(x)| < \varepsilon \rightarrow \text{End}$$

$$n = n + 1$$

The contribution of the different iteration steps leads to a solution in the form

$$T^{(n)} = (-1)^{n-1} \binom{n}{1} B^{(n-1)}.$$

This term indicates, that the reconstruction is a non-linear operation. Convergence is only obtained if the elements of the deconvolution term,

are $\quad \sum_i |a_i| \leq 1$

for all weighting elements a_i of the antenna characteristics, which exactly is the case. This performance does, however, give no support for the depression of noise in the data. Of importance for this intention is the term

$$K(x) = C\left[1 - 2\left(\frac{T^{(n)}(x) - \dfrac{H(x) + L(x)}{2}}{H(x) - L(x)}\right)\right],$$

where $H(x)$ and $L(x)$ are the upper and lower boundary value of the intensity dynamics [1]. The purpose of the term is also to decrease ascillatory fluctuations at the location of abrupt intensity changes and the enhancement of statistical data fluctuation such as those generated by noise. C is considered as on amplification factor. The disadvantage of the corrective term $K(x)$ is its deterministic restriction to the boundary values. This leads to a depression of naturally excessive intensities and problably to diverging resolution if values $T > H(x)$ or $T < L(x)$ are present. In order to prevent this phenomenon, an alternative corrective coefficient

$$K(x) = C\left[1 - e^{-1/2}\left(\frac{T^{(n)}(x) - \dfrac{H(x) + L(x)}{2}}{V(x)\,\sigma(x)}\right)^2\right]$$

was defined. This term has a probabilistic characteristic as it continuously decreases the step width with increasing deviation from the most probable intensity value without deterministic restriction by the now most probable treshold values $H(x)$, $L(x)$ of the intensity dynamics. $V(x)$ is a parameter to control the influence of stochastic events.

The   investigation of this and the other algorithms were performed
on two different types of intensity profiles:

a)   a temperature profile, consisting of only one abrupt inten-
       sity step varying between 100 K and 400 K

b)   a highly periodical fluctuating to test the feasibility of
       the algorithms to reconstruct objects which require high
       geometrical resolution capabilities for recognition.
       (The figures present the convolved and the deconvolved data
       for comparison. The synthetic and the real data are used to
       demonstrate the capabilities of the algorithms on one-di-
       mensional intensity profiles.)

Legend:
DV: Sum of absolute values
    for solution enhance-
    ments for succeeding
    iteration steps
DU: Sum of absolute values
    of deviation of the so-
    lution from unconvolved
    data set
DG: Sum of absolute values
    of deviation from con-
    volved data set
(All values refer to 80
data elements)

Fig. 1

Fig. 1 shows the convergence properties. They demonstrate, that
the first 25 reconstruction steps bring significant improvement,
so that further trials to improve the result can be abandoned.
The good convergence behaviour for noise free data (Fig. 2) is

Fig. 2

Fig. 3

due to the fact, that even for missing a-priori-knowledge the
image values are the best estimation for the reconstruction process. Of significance is the oscillatory effect at the location
of abrupt intensity changes. This effect can be suppressed by
using a-priori-knowledge for the upper and lower boundary of the
most probable dynamic range of the radiation intensity, symmetric
or asymmetric to the real intensity value (Fig. 3, Fig. 4).

Fig. 5 shows deconvolved noise distorted data. The error reducing
effect is demonstrated by variation of V in the Fig. 6a - 6c. The
influence of varying parameter C to the transition behaviour is
presented in Fig. 7a - 7d.

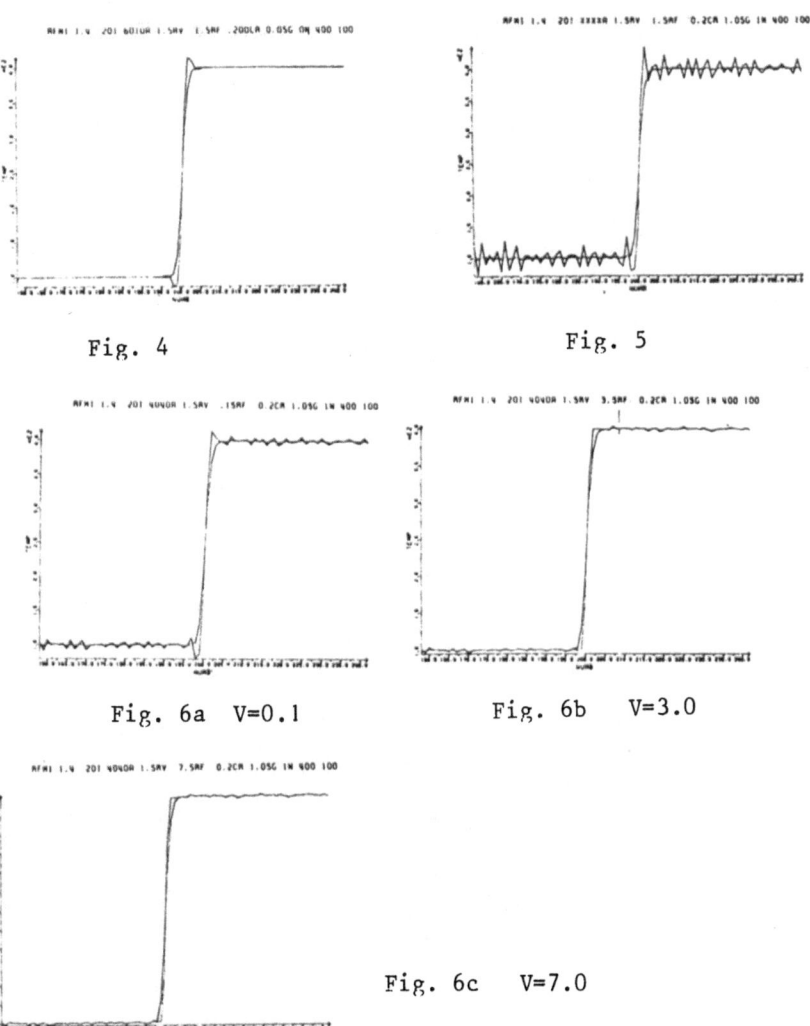

Fig. 4   Fig. 5

Fig. 6a   V=0.1   Fig. 6b   V=3.0

Fig. 6c   V=7.0

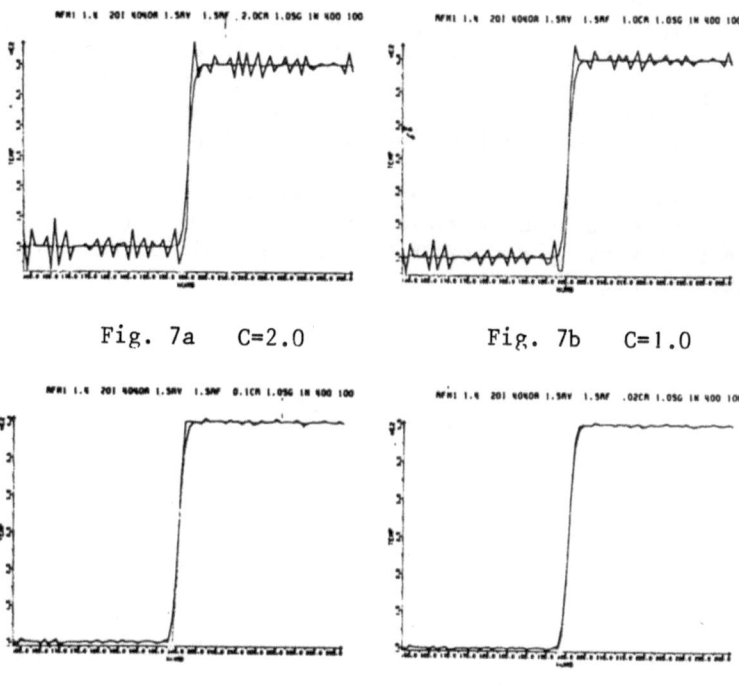

Fig. 7a   C=2.0   Fig. 7b   C=1.0

Fig. 7c   C=0.1   Fig. 7d   C=0.02

Fig. 8

Good reconstruction could be obtained with a set of data, consisting of intensity fluctuations only two elements wide, although the width of the antenna characteristics over the region from −10 to +10 degrees consisted of 81 resolution elements, each representing an angular resolution of 0.25 degrees (Fig. 8).

Fig. 9a   C=0.2   Fig. 9b   C=2.0

Difficulties only occured for noise disturbed data. In these cases, it was not always possible to successfully suppress the noise influence to the data elements. A proper selection of the control parameters leads to an increase of the quality of the result (Fig. 9a, 9b).

Fig. 10   C=0.2

The fact, that high periodicity fluctuations do not permit expressed transient behaviour of the reconstruction curve between the individual elements, causes sometimes a signigicant deviation from the ideal result (Fig. 10).

4.2 DECONVOLUTION BY MAXIMUM LIKELIHOOD

Checking the convolved data it is very difficult to distinguish
between object- and noise-generated intensity peaks. Only the form
and the texture of the neighbour data element permit to identify
the reason for an expressed peak.

The method of principle likelihood [2] in contrary to the multiple
convolution method makes a probability investigation for every
data element and its vicinity. The antenna characteristics can be
compared to the probability function with an overall probability
of 1.

The probability $P(T_i)$ is determined by the Bayes-relation

$$P(T_i) = \sum_K \frac{P(B_K/T_i)\ B(T_i)\ P(B_K)}{\sum_j P(B_K/T_j)\ P(T_j)} \ ,$$

where i, j, k are the indices of the vectors B, T. This term indicates, that the profile T can interatively be determined to a
probability $P(T_j)$ by application of a Bayes-Operator.

The iteration process derives the temperatur T_i for the iteration step (r + 1) from the value T_i at the preceding step r:

$$T_{i,\ r+1} = T_{i,\ r} \sum \frac{A_{i,k}\ B_k}{\sum_j A_{j,\ k} \cdot T_{j,\ r}} \ .$$

The convergence behaviour and the quality of the results are
identical to those obtained with the multiple convolution method.
For this reason, reference can be made to the different figures
published there. It should, however, be pointed out, that the
algorithm is not strong enough, to sufficiently suppress intensity excesses at locations of abrupt intensity changes or due to
noise distortions. For this reason a correcting factor K(x), as
described for the multiple convolution method has also been used
here.

For some examples, the algorithm showed for the first few iteration cycles a slight divergend trend, but after not more than
the third step switched to converging behaviour. The main advantage of this algorithm compared to the multiple convolution method
is evidently its greater stability and trend in convergence, even
for the case of very uncertain a-priori-knowledge.

4.3 Deconvolution by Matrixinversion

The term

$$T = A^{-1} AT = A^{-1} B + A^{-1} N$$

is definitely only converging in the cases $N = 0$.

The reason for solution instability are stochastic data behaviour and numerical problems due to quasi-singular properties of the convolution matrix. The effect are highly oscillating solutions even for data with on N/S-ratio of 10^{-4} to 10^{-5}.

The solution by matrix inversion was obtained with the term [3]

$$T = (A^t A + \gamma G)^{-1} (A^t B + \gamma W).$$

The individual parameters are:

G   matrix to reduce the solution variety and to prevent oscillatory solution,

B   matrix of convolved data,

W   matrix of a-priori-knowledge,

γ stabilization term of the solution,

A, A^t antenna matrix and its transposed, (smoothed by second and higher difference terms of neighbour matrix elements).

For the general case, the convolution matrix A is point symmetric normal distributed and all elements are $|a_i| < 1$. The best solutions were obtained for $G = I$ (unitary matrix).

The algorithm is no iterative method, so that the deconvolution result is obtained in one step. (This could be verified for different examples.) Fig. 11 shows the results of the deconvolution, using undisturbed data. They are characterized by a very good reconstruction of the unconvolved data set. The investigation showed, however, that the result is very much depending on the quality of the a-priori-information. Unfortunately the solution tends in any case to the assumed, estimated temperature profile, and if the assumption is not correct, to a wrong deconvolution result.

Of general interest is the effect of the value of the parameter γ for the solution stability. A decrease of the value γ shows an increase in the solution stability even for undistorted data. The effect of instability is expressed in form of a high-amplitude oscillation (Fig. 12a - 12n).

Fig. 11

1 Iterationstep

Fig. 12a $\gamma = 10^{-20}$

Fig. 12b $\gamma = 10^{-15}$

Fig. 12c $\gamma = 10^{-10}$

Fig. 12d $\gamma = 10^{-9}$

Fig. 12e $\gamma=10^{-8}$

Fig. 12f $\gamma=10^{-7}$

Fig. 12g $\gamma=10^{-6}$

Fig. 12h $\gamma=10^{-5}$

Fig. 12i $\gamma=10^{-4}$

Fig. 12j $\gamma=10^{-3}$

Fig. 12k $\gamma=10^{-2}$ Fig. 12l $\gamma=10^{-1}$

Fig. 12m $\gamma=5$ Fig. 12n $\gamma=10$

5. PRACTICAL DECONVOLUTION RESULTS

The capibilities of the algorithms, investigated on synthetic data, has been demonstrated with real data [4]. For this purpose, intensity profiles, taken with fixed antenna position, were used. Although the methods were applied to scenes, taken at 11 GHz, 32 GHz and 90 GHz, here only the deconvolution results of the 11 GHz band are presented. For the reasons indicated, only the multiple convolution and the maximum likelihood method, not the matrix inversion method, were verified. For the purpose of data reconstruction, only the segment of the antenna characteristic just over the profile was used.

Fig. 13 shows a boundary region of a water/land area. The deconvolved profile unveils features, that in the convolved signal cannot be recognized. The objects of interest are a group of trees at the rim of the waterbody and on area of minor snow coverage. In the

bank region an area of relatively to the water surface elevated
intensity can be identified, probably consisting of a thin ice
cover over the water. As a significant drop appears the supposed
water/ice boundary. The waterbody shows not the expected constance
in radiation behaviour. This is definitely not due to statistical
errors, but to stochastic variations of the radiation temperature.

Fig. 13

Fig. 14 shows the radiation profile of an partial snow covered
area. The flight path went over a street and a parking lot. On both
sides of the street was no snow, so that this region appears at
a higher radiation intensity than the snow covered area. The low
intensity of the street is due to a ice cover. Of particular interest is a snow covered area at the left side of the profile.
Part of this region was illuminated by the sun and is evidently
covered with a thin film of melting water. It therefore appears,
due to the higher degree of reflexion, at a lower radiation temperature than the region at the extreme left which was in the shadow
region of a forest and had therefore a greater surface roughness.
On the extreme right the radiation structure of a parking lot can
be seen.

Fig. 15 shows the bank area of a lake. Of particular interest is
here a small path which could not be seen in the convolved region.
The path was only one picture elements wide and can in the convolved data not be recognized as an object different to snow.
The deconvolved region also shows a small area of ice covered
water area immediately right to the bank.

Fig. 14

Fig. 15

Fig. 16 shows a scene, where the flight path, crossing a water body, is approximating a land area. Although the central line of the flight profile is still over the water, the side lobes are already detecting the land area. Although the radiation pattern of an housing area is not visible in the convolved data and although the flight path did not go directly over this region, the patterns of the housing area appears as zig-zag line in the deconvolved profile. This example demonstrate the influence of objects outside a scene to the pattern at the rim region in a deconvolved image. Special care has to be taken at the interpretation of such phenomena.

Fig. 16

6. SUMMARY

The investigation of three algorithms (multiple convolution, maximum likelihood, matrix inversion) for the deconvolution of microwave radiometry data showed that converging solution could (expect for the third algorithm) always be obtained by use of a-priori-knowledge on data properties. Here, global information, such as the dynamic range or the approximate treshold values, were already sufficient. At the same time the computer time could be reduced up to the factor 10 for the first algorithms. For the matrix inversion method, a comparison of the computer time is irrelevant, as the method consists only of the step of matrix multiplications and matrix inversions. This algorithm reacts very sensitive on any kind of a-priori-knowledge in such a way that it has a large tendency also towards wrong information. It is therefore not in any case applicable for earth scientific purposes.

The methods of deconvolution by multiple convolution and maximum likelihood obtain good results even than, if the geometrical resolution of a single event was in the size of 0.25 degrees for an antenna halfwidth of 2.8 degrees. A contrast enhancement to obtain the correct intensity profile up to the factor 5 could be observed. The needed computer time was thus, that real time deconvolution can be considered as possible.

A significant characteristics of the first and second algorithm was their capability to reconstruct high periodicity events which require high resolution gain, even for random distorted data. In these cases a-priori-information are, however, mandatory. The matrix inversion method tends to oscillatory effects, which could be reduced ba application of a stabilization parameter. At the same time all other fluctuations are also suppressed, so that this algorithm is difficult to handle for earth scientific purposes.

LITERATURE

1   Frieden, B.R.   Image Enhancement and Restoration in:
                         Topics in Applied Physics, Vol. 6
                         (Picture Processing and Digital Filtering),
                         pp. 21-203, 227-229.

2   Richardson, W.H.   Bayesian-Based Iterative Method of
                         Image Restoration
                         I. Opt. Soc. Am., Vol. 62, Jan. 1972,
                         pp. 55-59.

3   Twomey, S.   Introduction to mathematics of Inversion
                         in Remote Sensing and Indirect Measure-
                         ments
                         (Development in Geomathematics, Vol. 3)
                         Amsterdam: Elsevier, 1977, pp. 122-144.

4   Rudolf H. Dittel   Basic Investigation on the feasibility
                         of multispectral radiation information
                         from microwave radiometry measurements
                         for earth scientific objectives of
                         remote sensing.
                         Köln: DFVLR Forschungsbericht 83-27, 1983.

Data Source:
Remote Sensing Radiometry Measurements of DFVLR, 1974-1977.

III.1 (RP.1*)

POLARIZATION DESCRIPTORS. UTILIZATION OF SYMMETRIES

Georges A. Deschamps

Department of Electrical Engineering, University of
Illinois, Urbana, Illinois 61801, USA.

ABSTRACT. This paper reviews the tools used for handling polarization problems in electromagnetics and shows how they can be
used to take into account symmetries of a scatterer. Various
representations of polarization states are described and related.
They are applied to plane waves and to radiation fields due to an
antenna or to scattering of a plane wave. Symmetries of a
scatterer help in measuring its pattern. Conversely symmetries
of the pattern can be a means of identification. A few general
remarks are made about the bearing of polarization on inverse
scattering problems.

0. INTRODUCTION

    The first part of this paper consists of a brief description
of several known representations of polarization states and their
relations. It is mostly expository but it brings out a few
subtle points that are frequently ignored. This presentation is
of necessity brief and the reader should consult the references
for more detailed discussions. The second part treats polarization transformations and their representation. The third part
discusses various "pattern" descriptions of the far field of an
antenna or of a scatterer. Part four treats the effect of
symmetries of a scatterer on its scattering properties and the
utilization of these symmetries in measurements and target identification problems. Part five offers a few general comments
about the importance of polarization for inverse problems.

    The reader should consult Appendix A for a selected
bibliography and Appendix B for a list of some notations, most of
them conventional, but a few peculiar to this paper.

W.-M. Boerner et al. (eds.), Inverse Methods in Electromagnetic Imaging - Part 1, 493–520.
© 1985 by D. Reidel Publishing Company.

1. POLARIZATION DESCRIPTORS AND THEIR RELATIONS

1.1 Concept of Polarization

Any field represented, at point r and time t, by a vector $V \in R^n$ can be specified by its magnitude $|V|$ and its direction. The latter, taken alone, is sometimes called the polarization of vector V. This is done in particular if the direction of the vector (vertical for instance) is independent of t and/or of r within some region of space: the field would be called "vertically polarized" over that region.

For a real vector field however it is hardly justified to introduce a special name synonymous of direction. In contrast, for a time harmonic field, $V(t)$ at some point r describes in general an ellipse as a function of t, and therefore has a variable direction. Expressing $V(t)$ as usual by

$$V(t) = Re \left[V \exp (-i\omega t) \right] \qquad (1.1.1)$$

we consider the direction of the complex vector

$$V = V' + iV'' \in C^n \qquad (1.1.2)$$

as defining the polarization of V.

Since V is complex an appropriate definition of direction (or parallelism) is required. We say that V_1 and V_2 have the same direction[1] if they differ only by a complex factor $z \in C$

$$V_2 = zV_1 \qquad (1.1.3)$$

which may also be expressed by

$$V_2 \in CV_1 \qquad (1.1.4)$$

where CV_1 is the set of products of V_1 by all numbers $z \in C$. The vectors $V_1(t), V_2(t)$ (and by 'abus de langage' V_1, V_2) are said to be in the same polarization state if V_1 and V_2 have the same direction.

---

[1] In other contexts parallelisms defined by $V_2 = RV_1$ or $V_2 = R_+V_1$ might be appropriate (WANG and DESCHAMPS, 1974). Those are not relevant to the present considerations.

In this article we shall consider only time-harmonic fields and very briefly quasi-harmonic and random fields. The time convention will be that of equation 1. Although most considerations apply to vectors in an n-dimensional linear space C^n, whose elements may be collection of tensors,[2] we shall mainly discuss the case of an electric field E in C^2 or C^3.

To isolate a description of the polarization state of E one can at first divide E by its magnitude $|E|$. The magnitude $|E|$ is the positive square root of $E^*\!E$ and represents an energy density. The ratio

$$\hat{E} = E \big/ |E| \qquad\qquad (1.1.5)$$

is a unitary vector ($|\hat{E}| = 1$) which has the same polarization as E.

It has been called the "unit polarization vector". One should note that any vector $E\, e^{i\alpha}$ ($\alpha \in R$) is also a unit polarization vector and one should take the set of these vectors as the proper description of polarization alone. Such sets have been called rays (not be confused with rays in geometrical optics) in the context of quantum mechanics where a ray defines the state of a quantum system [Bargman, 1964].

At this point a general remark is in order, namely that the techniques for handling polarization states are but a special case of techniques for handling quantum states [e.g. Fano, 1957]. The analogies are particularly clear with representations of the spinning electron by a pair of complex-valued wave functions.

1.2 Polarization Ellipse

Instead of considering the complex vector V = V' + iV" one may examine the locus of

$$V(t) = \mathrm{Re}\left[V \exp\left(-i\omega t\right)\right] = V' \cos\omega t + V" \sin\omega t \qquad (1.2.1)$$

It is easily seen that this point describes an ellipse $\mathcal{E}$. The characteristics of this ellipse that are common to all field vector having the same polarization are (1) the polarization plane Π defined by vectors (V',V"); (2) the direction of the major axis, which makes an angle β with some reference direction L (being an angle between two straight lines, the tilt angle β is defined

---

[2] For instance in a warm magneto plasma V may represent electric and magnetic field, the velocity field and the scalar pressure. Then $V \in C^{10}$.

modulo π or can be chosen between 0 and π); (3) the <u>axial ratio</u> minor over major axis, which can be expressed as tan α with $0 < |\alpha| < \frac{\pi}{4}$; (4) the <u>sense of travel</u>, which is called + or - by comparison with a reference sense of rotation which makes Π into an oriented plane Π. This orientation of Π can also be specified by a vector V normal to Π and is called right-hand or left-hand according to the well known rule. If the vector E is multiplied by a complex number m $e^{i\alpha}$ the ellipse $\mathcal{E}$ is expanded in the ratio m and the vector $V(0)$ is moved on the ellipse so as to sweep an area $\alpha/2\pi$ times the total area of the ellipse. These transformations leave the characteristics (1) to (4) invariant. For the limiting case of linear polarization: $\alpha = 0$, Π and sense are undetermined, and for circular polarization: $\alpha = \pm \frac{\pi}{4}$, and tilt angle β is undetermined.

1.3 Poincaré Sphere

As shown for instance in [Deschamps, 1951] the Poincaré sphere $\mathcal{P}$ provides a one-to-one representation of states of polarization. Using geographic notations, and (β, α) defined in 1.2, the longitude of P, measured from a reference point L on the equator is 2β, the latitude of P is 2 where α has been given the sign $\pm$ according to the sense of polarization. The pairs

$$(\alpha, \beta) \in [- \frac{\pi}{4}, + \frac{\pi}{4}] \times [0, \pi] \tag{1.3.1}$$

are mapped uniquely on $\mathcal{P}$. The limiting cases fall in place without disrupting the continuity.

The Poincaré sphere is at the crossroad of various polarization descriptions. A point $P \in \mathcal{P}$ is related to the slope (or polarization ratio) W = Y/X by sterographic projection (1.4). It is related to 2x2 Hermitian matrices $H_0(2)$ as shown in (1.5). Then through a map $H_0(2) \to \mathbf{R}^4$ to a vector space, in 4 dimension endowed either with the Minkowski metric or, for some computations with the Euclidean metric (see Stokes parameters in 1.5).

The main properties of Poincaré sphere are: 1) It provides a one-to-one representation of the states of polarization defined by the polarization ellipse. 2) It relates through formulas of spherical trigonometry various parameters of this ellipse: tilt, axial ratio, polarization ratio, etc. 3) It displays a "natural" measure of distance between polarization states P_1 and P_2, namely the arc $P_1 P_2$. 4) Linear transformation of a vector field (with values on $\mathbf{C}^2$) correspond to inversions of the sphere. 5) When these transformations are lossless their matrices belong to SU(2) which correspond to rotation O(3) of the sphere $\mathcal{P}$.

An inner product between polarization states (or quantum system states) can be defined in general when ψ_1 and ψ_2 belongs to C^n or to some Hilbert space where the inner product $(\psi_1, \psi_2) = \psi_1^* \cdot \psi_2$ [Bargman, 1964]. This is done by setting

$$P_1 \cdot P_2 = |(\psi_1, \psi_2)| \tag{1.3.2}$$

where ψ_1, ψ_2 are unitary vectors [their phases do not matter]. For $n = 2$ it can be shown [Deschamps, 1951] that this product, if P_1 and P_2 are the representative points on the Poincaré sphere, is

$$P_1 \cdot P_2 = \cos \delta \tag{1.3.3}$$

if 2δ is the arc length of $P_1 P_2$. This gives the magnitude of the orthogonal projection of state ψ_2 along state ψ_1. The square of this quantity

$$\cos^2 \delta = \frac{1}{2}(1 + \cos 2\delta) \tag{1.3.4}$$

can be recovered by using properties of the Hermitian matrix representation (1.5)

$$(P_1 \cdot P_2)^2 = \psi_1^+ \psi_2 \, \psi_2^+ \psi_1 = \text{tr}\,(\Psi_2 \Psi_1) = \frac{1}{2}\vec{\Psi}_1 \cdot \vec{\Psi}_2 \tag{1.3.5}$$

The last term is half the Euclidean scalar product of the Stokes' vectors $\vec{\Psi}_1 \, \vec{\Psi}_2$

$$\vec{\Psi}_1 \cdot \vec{\Psi}_2 = I_1 I_2 + M_1 M_2 + C_1 C_2 + S_1 S_2 = 1 + \vec{OP}_1 \cdot \vec{OP}_2 \tag{1.3.6}$$

which recovers (1.3.4). When $\delta = \frac{\pi}{2}$ the vectors ψ_1, ψ_2 are orthogonal or <u>cross polarized</u>. The points P_1 and P_2 are antipodal in $\mathscr{P}$.

1.4 Polarization Ratio

Let the complex vector E in a plane Π be expressed as

$$E = X\bar{x} + Y\bar{y} \tag{1.4.1}$$

in terms of its coordinates

$$\psi = \begin{bmatrix} X \\ Y \end{bmatrix} \in C^2 \tag{1.4.2}$$

and a reference frame $\bar{f}$ consisting of two (non parallel) vectors in

$$\bar{f} = (\bar{x}, \bar{y}) \tag{1.4.3}$$

Equation (1) may be written as a matrix product of (1.4.3) by (1.4.2)

$$E = \bar{f}\psi \tag{1.4.4}$$

A natural way of expressing the direction, hence the polarization, of E is to form the <u>polarization ratio</u>

$$W = Y/X \tag{1.4.5}$$

This a faithfull (one-to-one) representation of the polarization state of E (provided $W = \infty$ is added to the set C of complex numbers). Although $\bar{x}$ and $\bar{y}$ are arbitrary (non parallel) and may be complex it is convenient to take them as orthonormal and even real. A change of reference frame

$$\bar{f}' = \bar{f}M \tag{1.4.6}$$

where $M = GL(2,C)$ corresponds to the change of coordinates

$$\psi' = M^{-1}\psi \tag{1.4.7}$$

so that

$$E = \hat{f}'\psi' = \bar{f}\psi \tag{1.4.8}$$

The effect of M on the polarization ratio is of the form

$$W \rightarrow W' = \frac{aW + b}{cW + d} \tag{1.4.9}$$

if

$$M^{-1} = \begin{bmatrix} d & c \\ b & a \end{bmatrix} \tag{1.4.10}$$

Given the polarization ratio W in an orthonormal frame, a unitary vector having the same polarization as E may be taken as $\bar{f}\psi_W$ with

$$\psi_W = \frac{1}{1 + |W|^2} \begin{bmatrix} 1 \\ W \end{bmatrix} \tag{1.4.11}$$

Of course the phase of E is lost in the construction, that of ψ_W has no particular meaning.

The transition from polarization ratio W to a point P on Poincare sphere $\mathcal{P}$ is through a stereographic projection from the pole P_X of $\mathcal{P}$ that represents the linear polarization $\bar{x}$ ($W = \infty$) onto the equatorial plane. The projection is one-to-one, conformal, transforms circles into circles. The cross-ratio of any four points equals that of their images.

1.5 Representation by Hermitian Matrices

Let $\psi = \begin{bmatrix} X \\ Y \end{bmatrix} \in \mathbb{C}^2$ be the representation of the complex electric field $\bar{E}$ in an orthonormal reference frame $\bar{f}$. We associate to ψ the 2×2 matrix

$$\Psi = \psi\psi^+ = \begin{bmatrix} XX^+ & XY^+ \\ YX^+ & YY^+ \end{bmatrix} \tag{1.5.1}$$

This matrix is Hermitian

$$\Psi = \Psi^+ \in H(2) \tag{1.5.2}$$

and has zero determinant

$$\Psi \in H_0(2) \tag{1.5.3}$$

Any Hermitian 2×2 matrix $Q \in H(2)$ can be expressed in terms of 4 real numbers I, M, C, S called Stokes' parameters, by letting

$$Q = \frac{1}{2}\begin{bmatrix} I + M & C - iS \\ C + iS & I - M \end{bmatrix} \tag{1.5.4}$$

This defines a map

$$\mu: \quad H(2) \to \mathbb{R}^4: \quad Q \mapsto \vec{Q} \tag{1.5.5}$$

We have

$$\det Q = \frac{1}{4}(I^2 - M^2 - C^2 - S^2) = \frac{1}{4}\check{Q}^2 \tag{1.5.6}$$

where $\check{Q}$ is the "length" of vector $\vec{Q}$ according to Minkowski metric. A vector $\vec{Q}$ corresponding to a matrix $Q \in H_0(2)$ is on the positive light cone Γ_+.

$$I^2 = M^2 + C^2 + S^2 \tag{1.5.7}$$

The trace of Q, $I = \psi^+\psi = X^2 + Y^2$ is positive.

If ψ is a unitary vector, $I = 1$ and (M, C, S) are precisely the coordinates of the point P on Poincare sphere $\mathcal{P}$ that represent the polarization of ψ. We note that Ψ, hence $\Psi = (I, M, C, S)$, are insensitive to a change of phase.

The representation by matrices in $H(2)$ is particularly convenient for quasi-harmonic field vectors (i.e. those that occupy a narrow frequency band about ω), or statistical mixtures of purely polarized waves. The average matrix $\langle \Psi \rangle$ becomes a sum of

pure state matrices $<\psi>$ with positive weight ρ_i whose sum is 1
(ρ_i = probability of, or relative time spent in, state ψ_i). The
vectors $\overline{\langle \psi_i \rangle}$ are combined in the same manner hence $\overline{\langle \psi \rangle}$ lies
inside the cone Γ_+. The point P of coordinate $\frac{1}{T}$ (M,C,S) lies
inside the Poincaré sphere. See Deschamps and Mast [1973].
Similar considerations apply for $\psi \in C^n$ or for wave functions in
quantum mechanics [Fano, 1957].

A useful property of the Hermitian matrices Q and associated
vectors $\overline{Q} \in R^4$ is that they form a Hilbert space through the two
equivalent definition of a scalar product. In H(2)

$$Q_1 \cdot Q_2 = \text{tr} (Q_1^{\;T} Q_2) = \text{tr} (Q_1^* Q_2) = \Sigma Q_1^{ij} Q_2^{ij} \quad (1.5.8)$$

This equals half the Euclidian scalar product of the
corresponding vectors

$$Q_1 \cdot Q_2 = \frac{1}{2} \overline{Q}_1 \cdot \overline{Q}_2 \quad\quad\quad\quad (1.5.9)$$

This may be applied to Hermitian forms of a vector C^2

$$\psi^+ Q \psi = \text{tr}(\psi^+ Q \psi) = \text{tr}(Q \psi \psi^+) = \text{tr}(Q \Psi) \quad\quad (1.5.10)$$

2.   EFFECT OF A LINEAR TRANSFORMATION OF A FIELD VECTOR ON ITS POLARIZATION

We shall consider only the case of fields in a plane since
they are most important from a practical point of view. This
analysis applies to the situation where a plane wave incident on
a scatterer produces at large distances, in some direction, a
scattered field which is locally a plane wave.

Another example of linear transformation of a vector in C^2
amenable to the same analysis, is provided by the junction of two
wave guides each supporting two propagating modes. If (X,Y) are
the (complex) amplitudes of these modes incident on one side of
the junction while (X',Y') are the amplitudes of the transmitted
modes emerging from the other side we have a relation

$$\begin{bmatrix} X' \\ Y' \end{bmatrix} = \begin{bmatrix} a & b \\ c & d \end{bmatrix} \begin{bmatrix} x \\ y \end{bmatrix} , \; \psi' = T \psi \quad\quad (2.1.1)$$

where T is part of the junction scattering matrix (2.1.2a).

a. Junction   b. Termination

A similar relation describes the reflection at the end of a two modes waveguide (2.1.2b).

These networks are an exact model of a linear field transformation. They are most convenient when the modes are chosen to represent linear polarizations.

The description of the field E by its components $\begin{vmatrix} X \\ Y \end{vmatrix}$ which form a vector $\psi \subseteq \mathbf{C}^2$ depends of course on the choice of a reference frame $\bar{f} = (\bar{x}, \bar{y})$

$$E = \bar{f}\psi = X\bar{x} + Y\bar{y} \tag{2.1.3}$$

and similarly for the transformed field E' leading to the representation (2.1.1)

$$\psi \to \psi' = T\psi \tag{2.1.4}$$

A change of reference frame where $E \to E' = E$ leads to a relation of the same type between the coordinates (see 1.4.7).

We shall consider the effect of T on the field polarization in its various representations – namely the slope $W = Y/X$, the point P on the Poincaré sphere $\mathcal{P}$, the matrix $\Psi = \psi^+$, and the Stokes parameters (I,M,C,S) components of a vector $\widetilde{\Psi}$ in $\mathbb{R}^4$.

To the matrix T in (2.1.1) corresponds the fractional linear (or Möbius) transformation for the slope W.

$$\widetilde{T}: \mathbf{C} \to \mathbf{C}: W \mapsto W' = \frac{dW + c}{bw + a} \tag{2.1.5}$$

(Here **C** must be understood as the complex axis completed by a point at infinity.) The transformation T is the projective aspect of

$$T: \mathbf{C}^2 \to \mathbf{C}^2: \psi \mapsto \psi' \tag{2.1.6}$$

It is a conformal map that transforms circles into circles. Passing to the Poincaré sphere

$$C \to \mathcal{P}: \quad W \mapsto P \tag{2.1.7}$$

by stereographic projection, leads to a transformation of the Poincaré sphere onto itself

$$T: \mathcal{P} \to \mathcal{P} : \quad P \mapsto P' \tag{2.1.8}$$

which is also circular, conformal, and preserves the cross-ratio of four points. All the T form the <u>group $\mathcal{J}$ of inversions</u> of the sphere $\mathcal{P}$.

The group $\mathcal{J}$ operating on $\mathcal{P}$ can be extended to a group $\bar{\mathcal{J}}$ operating on the <u>Poincaré space</u> $\bar{\mathcal{P}}$ i.e., the space $\mathbf{R}^3$ into which $\mathcal{P}$ is imbedded. This is done by associating with each circle on $\mathcal{P}$ the plane that contains it. A point in $\mathcal{P}$ is then associated with the family of circles whose planes contains it. Transformations in $\bar{\mathcal{P}}$ are projective: they carry straight line into straight lines, planes into planes, preserve the cross ratio. Furthermore they transform the sphere $\mathcal{P}$ into itself.

If we identify the space $\bar{\mathcal{P}} = \mathbf{R}^3$ with the hyperplane $I = 1$ in the space $\mathbf{R}^4$ of Stokes vectors (I, M, C, S) every point of $\mathcal{P}$ defines a line through the origin and a unit vector (with Minkowski metric) along that line. In this manner it is seen that the Lorentz group, which leaves $I^2 - M^2 - C^2 - S^2$ invariant, is isomorphic to the group $\mathcal{J}$ and its extension $\bar{\mathcal{J}}$.

Since there is a one-to-one correspondence between Hermitian matrices and points of the space $\mathbf{R}^4$ (see 1.5) this tells us how the transformation T acts upon the Hermitian matrices $\Psi = \psi\psi^+$, that describe states of polarization.

A direct representation of transformation T acting of $\psi \in \mathbf{C}^2$ by a transformation T acting on $\underline{\psi} = \psi\psi^+ \quad H_0(2)$ is

$$T: \quad \Psi \to \Psi' = \psi'\psi'^+ = T\psi\psi^+T^+ = T\underline{\psi}T^+ \tag{2.1.9}$$

This extends naturally from $H_0(2)$ to $H(2)$ being linear with respect to Ψ. If T is unitary $T^+ = T^{-1}$

$$\Psi' = T\Psi T^{-1} \tag{2.1.10}$$

In general Ψ being expressed in term of IMCS by

$$\Psi = \frac{1}{2} (I\sigma_0 + C\sigma_1 + S\sigma_2 + M\sigma_3) \qquad (2.1.11)$$

where the σ_i are Pauli matrices, the coefficients of the matrix that acts on the Stokes vector result from the $T\Psi T^+$ expanded in the basis (σ_i) [Mueller 1943, Soleillet, 1929].

3. FAR FIELD PATTERNS OF AN ANTENNA OR A SCATTERER

3.1 Plane Electromagnetic Waves

In an homogeneous isotropic medium a plane electromagnetic wave is defined by its wave vector k and its electric field E(0) at some reference point taken as the origin. This field is a unit plane wave if E(0) is a unitary vector u. Using subscripts 1, 2, .., i, .. to distinguish various unit waves we shall say that (u_1,k_1) defines the plane wave $| 1\rangle$ whose components at point r are

$$\langle r \,|\, 1\rangle = \begin{bmatrix} E_1 (r) \\ H_1 (r) \end{bmatrix} = \begin{bmatrix} u_1 \\ Y_0 \, k_1 \times u_1 \end{bmatrix} \exp (k_1 \cdot r) \qquad (3.1.1)$$

where $Y_0 = (\varepsilon/\mu)^{1/2}$ is the medium admittance and $k_1 = k_1/k_0$ is the unit vector in direction of k_1. All fields $C \,|\, 1\rangle$ are said to be in the same state of <u>direction</u> and <u>polarization</u>, u_1 may be any point of the ray $C \, u_1$. If u_1 is chosen equal to $E(0)/|E(0)|$ it will give besides the polarization, some phase information by telling that the direction of the real field at time 0 is that of Re u_1. Note that the phase "information" carried by a choice of u_1 is not expressible by a number. In other words it is not possible to give a natural meaning to the phase difference between two field vectors having different polarizations. The phase information becomes important when interferences are considered. It cannot be separated from the polarization information. Any planewave in state $| 1\rangle$ being $a \,| 1\rangle$ with $a = Ae^{i\alpha}$ the plase α can be absorbed in whole a part by u_1

$$Ae^{i(\alpha -\phi)} \,|\, u_1 e^{i\phi}, k_1\rangle \qquad (3.1.2)$$

3.2 Far Field Patterns of an Electric Current Distribution

Let us assume that the antenna or the scatterer are of finite extent and that their far fields (for a given excitation) is that of some equivalent current distribution J(r) in free space. The radiated field at point kr (distance r is direction of some wave vector k) is given by

$$E(\hat{kr}) = ik_0 Z_0 G(r) p(k) \qquad (3.2.1)$$

where $k_o = 2\pi/\lambda$, Z_0 is the free space impedance,
$G(r) = (\exp ik_0 r)/4\pi r$ and p is the <u>current moment</u>, or strength,
of a equivalent electric dipole. This dipole depends on k as
follows: it is Fourier transform $\tilde{J}(k)$ of $J(r)$, restricted to the
sphere S: $|k| = k_0$ and projected on the plane $T_k S$ tangent to the
sphere S at point k.

The far field pattern is completely described by the complex
vector field p(k) over the sphere S. One can also write (3.2.1)
in the form

$$E(\hat{k}r) = \sqrt{Z_0}(e^{ikr}/r)A(k) \tag{3.2.2}$$

where

$$A(k) = \sqrt{Z_0} \frac{ik_o}{4\pi} p(k) \tag{3.2.3}$$

is now such that $|A|^2$ is the radiation intensity in direction of
vector k.

Expressions similar to (3.2.1) (3.2.2) also hold when the
source of the field contains magnetic currents $K(r)$. Then the
contribution of these currents to p(k) is obtained by forming the
Fourier transform K(k) evaluated on the sphere S, projecting it
on the tangent plane $T_k S$, multiplying it by the admittance Y_0 and
rotating by $-\frac{\pi}{2}$ about k. Thus

$$p(k) = -Y_0 \hat{k} \times K(k) \tag{3.2.4}$$

3.3 Radiation Intensity and Phase Patterns

Sometimes a complete description of the far field by a
complex vector such as p(k) or A(k) is not needed. This is the
case if we desire only to know the radiation intensity $|A|^2$
(power radiated per unit angle) as a function of direction. This
lead to concepts of gain and directivity.

If we attempt to describe the phase by a function of direction we encounter the difficulties already mentioned (3.1).
There is no naturally defined number that can be called the phase
of the polarized field or that can be called phase difference
between two field vectors of different polarizations.
Polarization and phase information are linked in the unit vector
$\hat{E} = E/|E|$.

One may however consider phases of a pair of components of the field having specified polarization (linear for convenience) and plot two phase patterns. The problem of specifying reference polarizations for different direction is considered in 3.4 and 3.5.

3.4 Polarization Patterns – The Poincaré Bundle

It is not easy to measure a complete scattering or radiation pattern in magnitude, polarization, and phase. The phase is particularly elusive while the intensity measured through a polarizor allows to extract easily the polarization. This is essentially a measurement of Stokes parameters or of a set of Hermitian forms which are represented in Poincaré space (The space R^3 that contains the Poincaré sphere $\mathcal{P}$) by distances to a set of planes. See Deschamps [1952, 1953], Deschamps and Mast [1973]. Of course we encounter here a difficulty already discussed: How to compare the polarization of non coplanar fields?

Using the Poincaré sphere representation we should plot the point P on $\mathcal{P}$ as a function of the vector $k \in S$. However for two different directions k_1 and k_2 the field vectors p_1 and p_2 lie in different planes $\Pi_1 = T_{k_1}S$ and $\Pi_2 = T_{k_2}S$. Thus we need a convention to transcribe polarizations in Π_1 and in Π_2 to a single sphere $\mathcal{P}$.

This convention, as described in (1.2) (1.3) consists of attaching to every point $k \in S$ an oriented plane $\overline{\Pi}$ (This will be the plane T_kS oriented by the normal vector k), and a reference direction L (a particular linear polarization lying on Π). On the sphere we choose a point L representing polarization L and we use the hemisphere $\mathcal{P}_+$ for ellipses travelled in the sense associated with k by the right hand rule. In brief we can state that the convention requires a field of straight lines tangent to the sphere S. The set of all such straight lines form a manifold PS which is known as a bundle over the base space S.

$$
\begin{array}{ccc}
L & \in & PS \\
\pi\downarrow & & \downarrow\pi \\
k & \in & S
\end{array}
\qquad\qquad (3.4.1)
$$

To each $L \in PS$ corresponds a projection πL on S which is the point k where L touches S. All lines tangent to S at point k form the fiber over k. The field of reference lines or polarizations is in fact a section of the bundle i.e., a map

$$
L_0\colon\ S \rightarrow PS\colon k \mapsto L_0(k) \qquad\qquad (3.4.2)
$$

such that π_0L_0 is the identity over S.

We have here a simple example of the bundle concept which has
come to play an important role in physics. It should be noted
that PS is not a cartesian product of the base space S by a fiber
(projective space P(1)). The field L_0 is not a map S → P(1). No
such continuous map exists because "one cannot comb a sphere!"
The bundle is not trival. To handle this situation one can cover
the sphere by two overlapping spherical caps S_a and S_b. The
bundle is trivial over each of these caps and some transition
function relates the reference directions in the overlap region.

A variant of this construction would use Poincaré spheres as
fibers leading to the Poincaré bundle $\mathcal{P}$S, a structure simply
related to PS. A polarization pattern is a section of $\mathcal{P}$S.

3.5 Concept of Copolarization

Two vectors that have the same polarization are said to be
copolarized (or copolar). This concept creates no problem when
applied to field vectors that lie in the same (oriented) plane $\bar{n}$.
However this term is more currently used with reference to a
polarization pattern (3.4) where statements are made about the
"sameness" of the field polarization in different directions,
i.e., for fields that lie in different planes Π. We have
discussed in (3.4) the necessity of specifying a field of
straight lines L (section of the bundle LS) which serves as
reference linear polarizations and determines how polarizations,
for each particular direction, are plotted on a single Poincare
sphere. Then two fields are copolar if they are plotted at the
same point. This insure that they have the same axial ratio and
the same sense of travel (relative to the oriented plane Π). But
their tilt angles have to be measured from some reference line
$L_0(k)$ which depend on the direction of propagation, described by
the wave vector k. The function

$$L_0: S \to LS: k \mapsto L_0(k) \qquad\qquad (3.5.1)$$

is a section of the bundle LS. It is largely arbitrary, except
for a smoothness condition, and there is no "natural" way to
choose it. There would be no problem if some parallelism could
be defined on the sphere S but this unfortunately is impossible.
One must resort to some arbitrary convention such as taking
direction tangent to θ (or φ) circles in spherical coordinates.
Even this cannot be done globally on the sphere as there exist no
continuous field of vectors on S. One might think of using
parallel transport of the vector u_1 at k_1 to the vector u_2 at k_2.
The result however would depend on the path from k_1 to k_2. It
seems natural to chose the shortest path i.e., an arc of great
circle joining k_1 to k_2 and to call u_2 copolar to u_1.

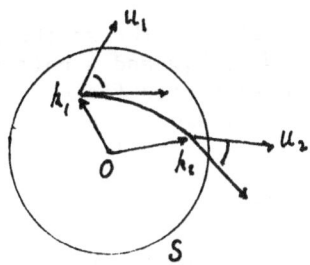

Then u_1 is also copolar to u_2. The relation is reflective. To
extend it to directions in the main beam of an antenna one can
choose k_1 along the axis of the beam, direction of maximum radiation or direction of interest for other particular reason. We
take u_1 arbitrary as defining $L_0(k_1)$. For any direction k_2
within the beam, we construct $L_0(k_2)$ as support of u_2 (k_2). This
construction can also be described by considering the
stereographic projection of the vicinity of k_1 from the point
k_2 on the sphere then the direction of u_2 projects on a parallel
to the direction of u_1. It can be shown, with a bit of trigonometry, that this is essentially the solution proposed by Ludwig
[1973]. Ludwig emphasizes the concept of cross polarization
which in fact is subordinate to that of copolarization: $| u_2,k_2 \rangle$
is cross polarized to $| u_1,k_1 \rangle$ if it is cross polarized to $| u_3,k_2 \rangle$
and u_3 is copolarized to u_1. Ludwig's solution is reasonable
that it is good in the vicinity of k_1 where the sphere does not
differ much from its tangent plane. It is made definite by the
particular choice of k_1 and has the basic flaw that if $| u_2,k_2 \rangle$
and $| u_3,k_3 \rangle$ are copolar to $| u_1,k_1 \rangle$, $| u_3,k_3 \rangle$ is <u>not</u> copolar to
$| u_2,k_2 \rangle$ (unless k_1 k_2 k_3 are coplanar).

3.6 Scattering Operator

The scattering operator, or scattering matrix, of a target
defines the scattered far field due to any incident plane wave
and by integration, the field due to any incident field expanded
into plane waves.

Consider first the incident plane waves $| E_1,k_1 \rangle$ with a fixed
wave vector k_1 and the scattered far field in direction k_2. This
field can be expressed as discussed on 3.2 by (3.2.1) or (3.2.2).
At point k_1 and in its vicinity this field is a plane wave
$| E_2,k_2 \rangle$, where E_2 depends lineary on E_1. Thus

$$E_2 = \bar{\bar{S}}(k_2,k_1) \ E_1, \tag{3.6.1}$$

Choosing reference frames in planes $\Pi_1 \perp k_1$ and $\Pi_2 \perp k_2$, $\bar{\bar{S}}$ is
represented by a 2x2 matrix. The effect of (3.6.1) on the
polarization were discussed in (2).

An alternative description of the scattering is to take a
unit incident field $|\,1\rangle = |\,u_1,k_1\rangle$ and a unit reference scattered
field $|\,2\rangle = |\,u_2,k_2\rangle$. The scattered field due to $|\,1\rangle$ being
$E_S = \bar{\bar{S}}(k_2k_1)u_1$, its component along state 2 will be $u_2^* \cdot E_S$. The
scattered field in state $|\,2\rangle$ is $S(2,1)\,|\,2\rangle$ with

$$S(2,1) = u_2 \cdot \bar{\bar{S}}(k_2,k_1)u_1 \tag{3.6.2}$$

Varying $|\,1\rangle$ and $|\,2\rangle$ over the manifold of plane wave states, S is
the scalar-valued function

$$S: \mathcal{M} \times \mathcal{M} \rightarrow \mathbf{C}: (i,j) \mapsto S(i,j) \tag{3.6.3}$$

4. UTILIZATION OF SYMMETRIES

This part of the paper analyzes the effect of symmetries of
the scatterer on its scattering properties. The general principle that symmetries of the causes imply symmetries of the
effects (P. Curie) is applied to the scattering problem direct
and inverse. Knowing some symmetries of the scatterer leads to
some simplification in the measurement of its scattering operator. Conversely observing the symmetries of the scattering
operator leads to a presumption of corresponding symmetries of
the scatterer. Unfortunately, one cannot draw this conclusion
with certainty. This is shown by simple counter-examples, e.g.
nonidentical cross dipoles may scatter as if they were identical.
Symmetries of the scattering operator however remain a means of
identifying or classifying targets.

The idea of making use of symmetries in fairly obvious, and
is used extensively in Physics and Engineering. In relation to
the scattering problem see a recent paper by Davidovitz and
Boerner [1983].

The present discussions could be considered as an extension
of this paper to arbitrary groups of isometries.

4.1 Symmetries of the Scatterer

A symmetry is defined by a transformation g of space onto
itself which carries a point r of coordinates (x,y,z) to the
point $r' = gr$ of coordinates (x',y',z'). Thus:

$$g: R^3 \rightarrow R^3: r \mapsto r' = gr: (x,y,z) \mapsto (x'y'z') = g(x,y,z) \tag{4.1.1}$$

We use the same letter g for the map $r \mapsto r'$ and for the associated transformation of coordinates. To be strictly correct one
should introduce the chart

$$\phi : r \mapsto (x,y,z) \tag{4.1.2}$$

from points to their coordinates. The coordinates would then
transform according to

$$\bar{g} = \phi \circ g \circ \phi^{-1} \tag{4.1.3}$$

which is represented by the commutative diagram

Points

$$\begin{array}{ccc} r & \xrightarrow{\;\;g\;\;} & r' \\ \downarrow & & \downarrow \\ (x,y,z) & \xrightarrow[\bar{g}]{} & (x'y'z') \end{array} \tag{4.1.4}$$

Coordinates

We shall omit the overbar, relying on the context to tell us
which is meant, g or $\bar{g}$.

The transformation g is extended naturally to a set of points
A by taking gA as the set of the transforms (or images) gr of all
points $r \in A$.

If a scatterer is a perfect conductor occupying the volume V,
its image occupies the volume gV. The image of its boundary
$\partial V = S$ is the boundary of its image $\partial(gV)$.

$$\begin{array}{ccc} V & \xrightarrow{\;\;g\;\;} & V' \\ \partial \downarrow & & \downarrow \partial \\ S & \xrightarrow{\;\;g\;\;} & S' \end{array} \tag{4.1.5}$$

The scatterer admits the symmetry g, or g in a symmetry of
the scatterer, if V is invariant under g:

$$gV = V \tag{4.1.6}$$

Then in view of (4.1.5)

$$gS = S \tag{4.1.7}$$

A scatterer may be a region of finite extent surrounded by
free space and characterized by scalar functions ε, μ, σ that
reduce to ε_o, μ_o, 0 outside of the scatterer. We like to extend
g to these functions. Let f be any scalar-valued function of
point r in space, the action of g on the function f.

$$g: f \mapsto f' = (gf) \qquad\qquad (4.1.8)$$

is defined by

$$f'(r) = f(g^{-1}r) \qquad\qquad (4.1.9)$$

This is the <u>pullback</u> of functions f under g^{-1} and could be denoted by

$$f' = (g^{-1})^* f \qquad\qquad (4.1.10)$$

or when g is a diffeomorphism by the <u>push forward</u>

$$f' = g_* f \qquad\qquad (4.1.11)$$

See Deschamps [1981].

The scatterer defined by ε, μ, σ <u>admits the symmetry</u> g when these three scalar functions are invariant under g, e.g. $(g\varepsilon) = \varepsilon$ (or with the push forward notation $g_* \varepsilon = \varepsilon$). Note that in these formulas ε, $(g\varepsilon)$, $g_* \varepsilon$ are functions, not numbers.

4.2 Isometries

The symmetries that are of interest in this paper are those that leave Maxwell's equations invariant. It can be shown that only isometries, i.e. transformations that preserve distances, have this property. For any two points r_1, r_2 and the isometry g

$$\left| gr_1 - gr_2 \right| = \left| r_1 - r_2 \right| \qquad\qquad (4.2.1)$$

Isometries also preserve angles. They map straight lines and planes respectively into straight lines and planes. Isometries comprise translations, rotations and reflections through a point or a plane. They form a group $\mathcal{J}$. Those that leave a point 0 invariant are represented by orthogonal 3x3 matrices (in Euclidean coordinates). They form a subgroup of $\mathcal{J}$ designated by O(3). By definition an orthogonal matrix equals the inverse of its transpose. Consequently its determinant equals $\pm$ 1. The matrices in O(3) having determinant +1 are called positive (or proper). They form the subgroup SO(3). The other isometries, for instance reflections through a plane or a point have determinant −1 and are called negative or improper.

It is obvious that the symmetries of a scatterer, or of any object, form a group. If these symmetries are isometries, and if the object is of finite extent, it can be shown that this group G leaves some point 0 invariant. Thus G is a subgroup of the group O(3). These groups have been extensively studied and classified, in crystallography. An easy and entertaining introduction of this topic is a set of lectures by H. Weyl [1952].

4.3 Isometric Transformations of a Vector Field

A vector field V is a function

$$V: R^3 \to TR^3: r \mapsto V(r) \qquad (4.3.1)$$

where TR^3 the tangent bundle to R^3 is essentially $R^3 \times R^3$.

An arbitrary smooth transformation g of R^3 in R^3 can be extended to vectors ($TR^3 \to TR^3$) by considering them as directional derivatives or infinitesimal vectors. If g: $r \mapsto r'$ and V is a vector at r, V' at r' = gr depends linearly on V via the Jacobian matrix of g

$$g_*: V \mapsto V' = MV \qquad (4.3.2)$$

When g is an isometry - which is the case of interest here - the matrix M is the same as that which describes g. Thus the vector field may be considered as the geometrical object, a collection of vectors each attached to a point r of space R^3. The action of g on V is to produce a field V' = (gV) whose value at point r is

$$V'(r) = (gV)(r) = g \circ V \circ g^{-1}r \qquad (4.3.3)$$

Thus

$$V' = (gV) = g \circ V \circ g^{-1} \qquad (4.3.4)$$

The parenthesis about (gV) is to avoid confusion with the composition $g \circ V$.

4.4 Action of Isometries on an Electromagnetic Field

Consider the field F defined by the pair of an electric field E and a magnetic field H. In a source free region these two vectors satisfy Maxwell's equation

$$\begin{aligned} \text{curl } H + i\omega\epsilon \ E &= 0 \\ \text{curl } E - i\omega\mu \ H &= 0 \end{aligned} \qquad (4.4.1)$$

It is readily verified that if one applies an isometry g to the curl of a vector field V one obtains $\pm$ the curl of (gV)

$$(g \text{ curl } V) = \pm \text{ curl } (gV) \qquad (4.4.2)$$

according to the sign of the isometry.

If ε and μ are invariant and if g is a positive isometry the system 4.3.1 remains unchanged when E,H are replaced by E' = (gE) and H' = (gH). When g is a negative isometry however the system becomes

$$\begin{array}{l} \text{curl } H' - i\omega\varepsilon E' = 0 \\ \text{curl } E' + i\omega\mu H' = 0 \end{array} \qquad (4.4.3)$$

The field (E',H') is not Maxwellian. To remedy this situation we define the action of g on F by

$$F \rightarrow F': \quad \begin{bmatrix} E \\ H \end{bmatrix} \mapsto \begin{bmatrix} E' = (gE) \\ H' = -(gH) \end{bmatrix} \qquad (4.4.4)$$

If the field F satisfies the boundary conditions $E_{tan} = 0$ at the surfaces of a conducting volume V, and if V is invariant under the isometry g (positive or negative), it follows from invariance of angles that the field F' = (gF) also satisfies the same boundary condition.

4.5 Alternative Approach Through Exterior Differential Forms

The quantities associated with an electromagnetic field are conveniently represented by differential forms [Deschamps, 1981]. Their relations are then expressed by means of the exterior differential operator d and through the Hodge's star operator $*$, the latter being combined with the material constants ε, μ, and σ. This makes it particularly simple to express the effect of space transformations, or changes of variables, on the electromagnetic quantities and their relations. The final results agree with those of preceding sections so that this new approach is not essential. However it still presents some interest being formally simpler and more general since it is not restricted to isometries. For the sake of brevity we shall rely heavily on references to Deschamps [1981].

A point-to-point transformation of space

$$g: \quad r \mapsto r' = gr \qquad (4.5.1)$$

is expressed in terms of coordinates by

$$g: \quad (x,y,z) \mapsto (x'y'z') \qquad (4.5.2)$$

(We omit the overbar, as explained in 4.1) If a field quantity is represented by a differential p-form β in the (x,y,z) system of coordinates it will become β' in the system of coordinates $(x',y',z') = g^{-1}(x,y,z)$. Thus β' is a pull-back of β under g^{-1}

$$\beta' = (g^{-1})^*\beta \qquad (4.5.3)$$

When g is non singular (diffeomorphism) this is also expressed by
the push forward:

$$\beta' = g_* \beta \qquad\qquad (4.5.4)$$

Such an expression has already been stated (4.1.10, 4.1.11) in
the case where β is a scalar function f. It does carry over to
p-forms with the extremely useful properly that g_* commutes with
d. That is, if $\beta = d\alpha$ and $\beta' = g_*\beta$, $\alpha' = g_*\alpha$, then $\beta' = d\alpha'$.
This is also expressed by the commutative diagram

For example the relation B = dA between the magnetic field B
(2-form) and its potential 1-form A becomes B' = dA' under an
arbitrary transformation. In contrast to this the vector relative B = curl A transforms in a simple manner B' = $\pm$ curl A' but
only when g is an isometry.

    For the differential forms approach the special role played
by isometries reappears when considering the star operator. A
relation such as D = *E between electric field one-form E and
displacement two-form D becomes D' = *'E' where D' = g_*D,
E' = g_*E and *' can be deduced from the Jacobian matrix of g.
For an isometry, in Euclidean coordinates, the Jacobian being $\pm$ 1
the star operator * is transformed into $\pm$ *. The positive sign
occurs for positive (proper) isometries such that the determinant
of the Jacobian is + 1. The positive isometries that leave a
point 0 fixed form the group SO(3), "Special orlhogonal in 3
dimensions," a subgroup of O(3). Negative (improper) isometries
such as reflection in a point or in a plane transform * unto $-*$.

    The conclusion is that a positive isometry that leaves
invariant a given medium (e.g. a scatterer) will transform a
field in that medium, due to sources or incident fields, into
another field due to the transform of the sources or of the incident fields.

    For a negative isometry the transformed field satisfies
Maxwell's equation in a medium where ε, μ are replaced by $-\varepsilon$,
$-\mu$. As one wishes to keep the medium invariant one may define
the transform of some of the electromagnetic quantities as
$+g_*$ and that of others by $-g_*$.

For instance consider the equation

$$dE + i\omega\mu H = 0 \qquad\qquad (4.5.5)$$

where $\mu = \mu_0*$ contains the star operator. Applying g_* to E, H, gives the equation

$$dE' - i\omega\mu H' = 0 \qquad\qquad (4.5.6)$$

between $E' = g_*E$, $H' = g_*E$. To recover the original equation (4.3.5) we take the transform of H as

$$H' = -g_*H \qquad\qquad (4.5.7)$$

calling H an odd quantity (while E is called an even quantity). Referring to Deschamps [1981] it can be verified that the entire set of equations of electromagnetics is preserved provided quantites in table I are even and those in table II are odd.

As an illustration consider the effect of a mirror reflection $g = g^{-1}$

$$g: \quad (x,y,z) \mapsto (x,y,-z) \qquad\qquad (4.5.8)$$

on an electric field

$$E = X(x,y,z)dx + Z(x,y,z)\ dz \ \cdot \qquad\qquad (4.5.9)$$

The pull-back

$$g^*E = E' = X(x,y,-z)dx - Z(x,y,-z)dz \qquad\qquad (4.5.10)$$

This corresponds to the vector transformation

$$(g\bar{E}) = g_0\bar{E}_0 g^{-1} = g_0\bar{E}_0 g \ . \qquad\qquad (4.5.11)$$

The effect on the two-form

$$D = U(x,y,z)dydz + V(x,y,z)dzdx + W(x,y,z)dxdy \qquad (4.5.12)$$

$$g^*D = -U(x,y,-z)dxdz - V(x,y,-z)dzdx + W(x,y,z)dxdy \qquad (4.5.13)$$

We are led to define $D' = -g^*D$ to make the star operator invariant and this corresponds to taking

$$(g\bar{H}) = -g_0\bar{H}_0 g \qquad\qquad (4.5.14)$$

as shown in (4.3) and (4.4).

4.6 Symmetries of Scattered Fields

This section discusses the conclusions that can be drawn about the scattering operator $S(i,j)$ when the scatterer has known symmetries. The general result is simply expressed: if the scatterer is invariant under an isometry g and if the incident field F_i produces a scattered held F_s, then the incident field (gF_i) will produce the scattered field (gF_s). This applies in particular when F_i is a plane wave $|1\rangle = |u_1,k_1\rangle$ and F_s is the far field $S(2,1)|2\rangle$ radiated in state $|2\rangle = |u_2,k_2\rangle$. We note that the application of g to a plane wave $|1\rangle$ gives a plane wave $|1'\rangle$ such that

$$u'_1 = g\ u_1$$
$$k'_1 = g\ k_1$$
(4.6.1)

and similarly for the plane wave $|2\rangle$. The symmetry of the scattering operator is expressed

$$S(2',1') = S(2,1)$$
(4.6.2)

To obtain a similar simple statement about the 2x2 matrix $S(k_2,k_1)$ which relates the field $\Pi_1 \perp k_1$ to the field in $\Pi_2 \perp k_2$ we can take arbitrary reference frame $\bar{f}_1$ in Π_1, $\bar{f}_2$ in Π_2. Then if we use their images under the isometry g

$$f'_1 = gf_1, \ f'_2 = gf_2$$
(4.6.3)

the matrices $S(k_2,k_1)$ and $S(k'_2, k'_1)$ gk_1 will be <u>identical</u>

$$\bar{\bar{S}}(gk_2,gk_1) = \bar{\bar{S}}(k_2,k_1)$$
(4.6.4)

A particular case of the general statement 4.6.2 and 4.6.3 results when $|1'\rangle = |1\rangle$ i.e. $gu_1 = u_1$, $gk_1 = k_1$. Then the scattered field is also unvariant under g. As an illustration of that remark consider the back scattering of a target having the following symmetries: (1) invariance under reflection P through a plane P that contains an axis A through 0, (2) invariance under a rotation $A^{\pi/2}$ through angle $\pi/2$ about A. A consequence of the group property is that

$$Q = A^{\pi/2}{}_o P{}_o A^{-\pi/2}$$
(4.6.5)

reflection though the plane $Q = A^{\pi/2}(P)$,
is also a symmetry of the target. Take
the incident field $|$ 1$>$ defined by
k_1 along A, u_1 in P. It is linearly
polarized. The back scattered field $|$ 2$>$
admits the reflection P, hence it is also
linearly polarized in plane P. The inci-
dent field $|$ 3$> = A^{\pi/2} |$ 1$>$ has similar
properties. Using (u_1,u_3) as reference
frame the matrix $S(-k_1,k_2) = 1$ as for an
infinite reflecting plane. Its backscattering cross section for circular
polarization is zero.

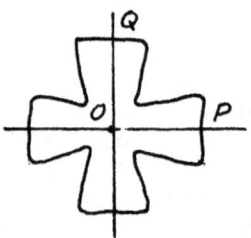

4.7 Application to Measurements of the Scattering Operator

A consequence of the results in 4.6 is that if the target
admits known isometries the measurement of its scattering properties can be simplified accordingly. The symmetries of an object
obviously form a group G that acts on the manifold $\mathcal{M}$ of plane
wave states. This group defines a fundamental domain $\mathcal{M}_o$ such
that $G\mathcal{M}_o$ (union of g $\mathcal{M}_o$ for g $\in$ G) covers exactly $\mathcal{M}$. Note that
because of linearity $\mathcal{M}$ and $\mathcal{M}_o$ may be limited to states of linear
polarization and the scattering operator defined by

$$S: \quad \mathcal{M} \times \mathcal{M} \rightarrow \mathbb{C}: (i,j) \mapsto S(i,j) \qquad\qquad (4.7.1)$$

The measurement may be limited to that of S over $\mathcal{M}_o \times \mathcal{M}$.
Combined with reciprocity this may produce substantial reduction
of labor.

5. POLARIZATION AND INVERSE PROBLEMS

One particular but still typical inverse problem is the
determination of the sources (electric and/or magnetic currents)
that produce an observed radiation pattern. The far field being
given otherwise this becomes an antenna synthesis problem
[Deschamps and Cabayan, 1972]. Referring to 3.2 the problem may
be to find J(r) knowing the function $J_S(k)$ restricted (pulled
back) to the sphere S.

A more general problem is to deduce shape, size or other
parameter of a scatterer from the observation of its scattering
properties, expressed for instances by the scattering operation
S(i,j). An essential and more difficult step is to formulate the
relation between the incident field $|$ j$>$, the scatterer's para-
meter and the induced currents: J(r,j). At this point some
approximations are usually necessary even to solve the direct
problem and to make its inversion (finding the parameters)
possible.

The difficulties of the inverse problems are many and will be discussed in depth by other members of the workshop. Prior knowledge and approximations must be used. One of these is to treat the fields as if they were scalar-valued. It is clear that something is lost in that process. One should make every effort to look at the vector problems, hence to take polarization into account. This of course does not mean that only polarization should be observed. Neglecting intensity (and phase!) is throwing away information. The only excuse for doing this may be the difficulty of the measurements, of phase in particular and the fact that sometimes polarization alone (easier to obtain) may be revealing of some characteristics of the target. In that case identification may be possible, at least if aided by prior knowledge.

There are difficulties intrinsic to holographic solutions connected with the concept of phase front reconstruction. What may be done is to "look" at the field through different polarizors and to observe the images thus obtained to gain some familiarity with their meaning. Whatever is done quantitatively it is hoped that its techniques reviewed in this paper may be of some use.

APPENDIX A. SELECTED BIBLIOGRAPHY

1. Bargman, V. (1964). Note on Wigner's Theorem on Symmetry Operations. J. Math Phys., vol. 5, pp. 862-868.

2. Boerner, W. M. (1980). Polarization Utilization in Electromagnetic Inverse Scattering, pp. 237-297 in Inverse Scattering Problems in Optics, (R. H. Baltes ed., Springer Verlag).

3. Davidovitz, M. and W.-M. Boerner (1983). Reduction of bistatic scattering matrix measurements for inversely symmetric radar targets. IEEE Transactions, vol. AP-31, pp. 237-242.

4. Deschamps, G. A. (1951). Geometrical representation of the polarization of a plane electromagnetic wave. Proc. IRE, vol. 39, pp. 540-544.

5. Deschamps, G. A. (1952). Geometric viewpoints in the representation of waveguides and waveguide functions. Proc. of Symposium on Modern Network Synthesis PIB, New York.

6. Deschamps, G. A. (1953). "A Hyperbolic Protractor for Microwave Impedance Measurements and Other Purposes." International Telephone and Telegraph Co., New York.

7. Deschamps, G. A. and H. Cabayan (1972). Antenna synthesis and solution of inverse problems by regularization methods. IEEE Trans. Antennas Propagat., vol. AP-20, pp. 268-274.

8. Deschamps, G. A. and P. E. Mast (1973). Poincare sphere representation of partially polarized fields. IEEE Proceedings, vol. AP-21, pp. 474-478.

9. Deschamps, G. A. (1981). Electromagnetics and differential forms, Proc. IEEE, vol. 69, pp. 676-698.

10. Fano, U. (1957). Description of states in quantum mechanics by density matrix and operator techniques. Rev. of Modern Physics, vol. 39-1, pp. 74-93.

11. Feynman, R. P., F. L. Vernon and R. W. Hellwarth (1957). Geometrical representation of the Schrodinger equation for solving Maser problem. J. Apply. Phys., vol. 28-1, pp. 49-52.

12. Jones, R. C. (1942). A new calculus for the treatment of optical systems. JOSA 31, pp. 443-503.

13. Ludwig, A. C. (1973). The definition of cross polarization. IEEE Trans. Antennas Propagat., vol. AP-21, pp. 116-119.

14. Mueller, H. (1948). Foundations of optics. J. Opt. Soc. Am., vol. 38, pp. 661(A).

15. Neumann, J. von. (1932). Mathematische Grunlagen der Quantenmechanik. (Berlin).

16. Perrin, F. (1926). Jour. de Physique, vol. 7, p. 390.

17. Perrin, F. (1942). Polarization of light scattered by isotropic opalescent media. J. Chem. Phys., vol. 10, pp. 415-427.

18. Soleillet, P. (1929). Ann. de Physique, vol. 12, p. 23.

19. Wang, W. D. and G. A. Deschamps (1974). Application of complex ray tracing to scattering problems. Proc. IEEE, vol. 62, pp. 1541-1551.

20. Weyl, H. (1952). Symmetry. Princeton University Press.

21. Wiener, V. (1929). Harmonic analysis and the quantum theory. J. Franklin Institute, vol. 207, pp. 525-539.

APPENDIX B. NOTATIONS

Vectors. We have abstained from using the vector sign over
quantities that have been defined as vectors such as V, E, the
wavevector k, the position vector r, etc. Length of k and r are
written k and r. The vector sign (or the unit vector sign ($\hat{}$))
are used occasionally when the same letter is used to designate
both a vector (or a unit vector) and some other associated quantity e.g. matrix Q, or Ψ and Stokes vector Q or ψ , vector E and
unit vector $\hat{E}$.

Vector Spaces. $\mathbf{R}^n$ and $\mathbf{C}^n$, sets of n-tuples of real and
complex numbers, respectively

> unit vector $u \in \mathbf{C}^n$, $u \cdot u = 1$
> unitary vector $\in \mathbf{C}^n$ $u^* \cdot u = 1$
> > (called unit vector if no confusion results)

For a matrix M the transpose is M^T, complex conjugate M^*, Hermitian transpose M^+, H(2) Hermitian matrices over C^2, $H_0(2)$ the
same with zero trace.

Groups of Transformation and nxn matrices.
$GL(n,\mathbf{R})$, $GL(n,\mathbf{C})$, non singular, nxn matrices with elements
in $\mathbf{R}$, $\mathbf{C}$.

> U(n) unitary matrices $M^+ = M^{-1}$
> O(n) orthogonal matrices $M^T = M^{-1}$

Suffix S in SL, SU, SO means subgroup of matrices with determinant 1.

A statement "T: A $\rightarrow$ B: a $\mapsto$ b" means that T maps the set A in
B, carrying a $\in$ A to b $\in$ B.

Composition. If g and g' are two transformations such that
the range of g' is in the domain of g, $g_\circ g'$ designates their
compositions

> $g_\circ g'$: $r \rightarrow g(g'(r))$

Special Symbols.

> $\mathcal{O}$ Poincaré sphere (1.3)
> $\bar{\mathcal{O}}$ Poincaré space (1.5)
> S sphere of radius k in k-space
> $T_k S$ plane tangent to S at point k
> TS tangent bundle to S
> LS bundle of lines tangent to S (3.5)
> $\mathcal{O}$ S Poincaré bundle (3.4)

III.2 (RP.5)

POLARIZATION INFORMATION UTILIZATION IN PRIMARY RADAR: An Introduction and Up-date to Activities at SHAPE Technical Centre

A.J. Poelman, M.Sc., Mem. NERG
and
J.R.F. Guy, Ph.D., Mem. IEEE

SHAPE Technical Centre, The Hague, The Netherlands

Relatively little has been published on how polarization information in the backscattered radiation from targets and clutter
could be exploited to enhance the detection capability of primary
radar systems.   The historical development of a long-term STC
research project based upon this topic is described.

The main theoretical analyses and the vector-signal processing schemes developed for practical real-time applications are
summarized and explained. The principal features of the project's
experimental polarimetric X-band radar system are described, and
typical performance figures given for the methods developed for
processing in the polarization domain; the adverse environment was
represented by experimentally-derived rain data.   Finally there
is a brief outline of future study areas.

1.   INTRODUCTION

1.1 Background

In the early seventies STC began a long-term research project
which had, as its broad terms of reference, the investigation of
the applicability of wave polarization as a source of information
for improving the performance of radar systems of the future. This
task was undertaken because of the continuing need to improve the
detection capability of radar systems which operate in adverse
environments whilst emitting a minimum of energy. Such an improvement would undoubtedly result if more of the information contained
in the backscattered and/or interference fields were utilized.

W.-M. Boerner et al. (eds.), Inverse Methods in Electromagnetic Imaging - Part 1, 521–572.
© *1985 by D. Reidel Publishing Company.*

Therefore, it was felt that in addition to studies dealing with
the use of the information associated with the frequency and
spatial domains, attention had to be devoted to the applicability
of the use of the information inherently available in the
polarization state of an electromagnetic wave.

The utilization of polarization information is a topic that
is currently receiving attention not only as a potential means of
improving target detection, but also as an aid in target imaging
applications [1].   The operational radar systems of the early
seventies made no attempt to exploit the information available in
the polarization state of electromagnetic waves, with the exception of the use of a circularly-polarized antenna to enhance
target detection in precipitation-clutter environments [2, 3].
All these systems in essence apply scalar-signal processing,
since at the antenna the incident vector-field is firstly converted to a scalar-field prior to further processing, and as a
consequence the polarization vector is lost.   To take advantage
of the vectorial nature of an electromagnetic field, there has to
be a means of enabling vector-signal processing, which implies a
system that has mutually-coherent orthogonally-polarized receive
antennas and channels.   Such a configuration is referred to here
as a dual-channel receiver system.

In general, an optimum detection receiver is one which uses
all of the information carried by an electromagnetic wave, and
this information is contained in the basic wave characteristics:
amplitude (or power), phase (or direction), frequency, and
polarization.   To date the STC project has dealt only with
polarization-vector signal processing, in order to obtain a deep
insight of its full potential.   Furthermore, the project has
tended to concentrate on those radar problem areas where this
type of processing is complementary to the well-established
processing techniques used in the other domains.   Examples of
such areas are homogeneous clutter (weather, sea, dipole-clouds)
and main-beam interference.   Because of the applications-oriented
nature of the research, it was the intention from the beginning
to support the theoretical studies with experimental work.
However, from the outset the project was faced with two main
problems:

    (a)   a lack of vector-signal processing algorithms, and

    (b)   a lack of suitable measured polarization data.

Conceivably, the lack of promising vector-signal processing
algorithms leading to practically realizable systems may well
be the reason that over the years little attention has been paid
to signal processing in the polarization domain relative to
processing in the frequency and spatial domains.

1.2 Basics

The usefulness of polarization information in processing schemes which attempt to improve target enhancement is based upon the fact that the response of an antenna system is polarization sensitive, i.e., the antenna acts as a filter.   This phenomenon has been mathematically formulated in [4].   In [5], it is shown that the response of an antenna system in terms of polarization to a monochromatic or completely-polarized incident field (i.e., a field whose polarization state does not vary with time) can simply be represented by the spherical distance between two points on the surface of a sphere which represents polarization space and is known as the Poincaré sphere.

> The polarization state of a field or a transmit antenna is defined at a point in space when looking into the propagation direction: the receive antenna polarization is the polarization which would obtain if the antenna were transmitting.

Figure 1 shows the response of an antenna in terms of its polarization at receive (M_r) to an incident field with polarization M_i.

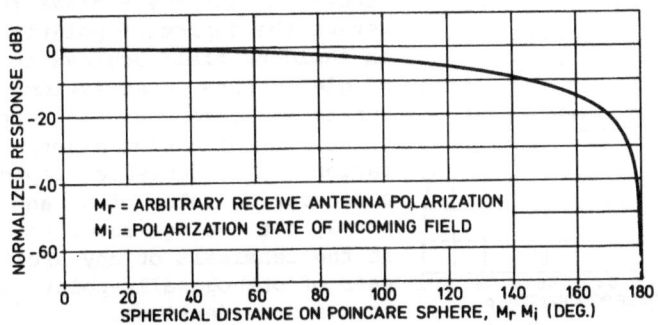

Fig.1 Incident-field polarization and receive antenna response [5]

It follows from Fig. 1 that:

(a) the maximum response is obtained when the polarization states of the incident field and the receive antenna are parallel;

(b) the maximization process is a very insensitive process;

(c) the minimum (i.e., zero) response is obtained when the polarization states of the incident field and the receive antenna are orthogonal;

(d) the minimization process is a very sensitive process; and

(e) loci of constant response are circles on the surface
     of the Poincaré sphere, concentric around the receive
     polarization.

It can be concluded from these observations that in adverse
environments polarization filtering techniques are far more
effective when they concentrate on the suppression of unwanted
signals rather than upon maximizing the response of target
returns. Further, unwanted signal suppression by means of a
single antenna polarization, or a single polarization filter (the
conventional approach used, for example, for rain-clutter suppression), is a single-notch nulling process and as such is
inherently sensitive. Hence, when the backscatter/interference
waves have fluctuating polarizations (i.e., the polarization
state varies with time) the average amount of achievable suppression is limited.   Such waves are said to be partially-polarized,
and a decomposition theorem [6] states that any partially-
polarized wave may be uniquely regarded as the sum of a completely-polarized wave and a completely-unpolarized (or randomly-
polarized) wave.   The ratio of the average power contained in
the completely-polarized portion to the average power contained
in the wave is called the "degree of polarization", p, of the
wave $(0 \leqslant p \leqslant 1)$.   Figure 2 shows the extreme average responses

versus the degree of polarization of
an incident field achievable with a
single antenna polarization. The
curves indicate that the suppression
capability of a single-notch polarization filter decreases rapidly with
a decreasing value of p, and when
p = 0, the average received powers
at the terminals of any arbitrary
pair of orthogonally-polarized antennas are identical.

Fig. 2 Extreme average responses at receive against degree of
        polarization (p) of incident field [6]

By analogy with the use of frequency (i.e. the Doppler
effect) as a discriminant for target enhancement, the use of the
backscatter-field polarization as a discriminant to improve target
detection in (homogeneous) clutter environments implies that
basically the quality of the detection of a target depends on the
extent to which the polarization characteristics of the target
and clutter returns deviate. Independent of the actual locations
of the target and clutter returns in polarization space, vector-
signal processing schemes based upon polarization filtering can,
in principle, be developed to take advantage of such deviations
provided that a dual-channel receiver is available. The polarization state of the return from an object is determined by its
backscatter operator, the scattering matrix [7,8] (a 2 x 2 matrix

defined in terms of polarization) and the polarization state
of the illuminating wave (i.e., the radar's transmit polarization). Consequently, it is obvious that the use of polarization as a discriminant for target enhancement is particularly
attractive in those environments where the transmit polarization
can either be chosen arbitrarily or selected on the basis of
"a priori" knowledge, because no on-line learning-process is
then required. The conditions under which the use of polarization information is extremely beneficial are illustrated on
the "modified polarization charts" of Fig. 3.

> The polarization chart is an orthogonal projection of
> the Poincaré sphere on a plane, having polar coordin-
> ates $\rho = \cos(2\tau)$ and $\Phi = 2\varphi$, where τ is the ellipticity
> angle and φ the orientation angle of the polarization
> ellipse. The modified polarization chart has polar co-
> ordinates $\rho = 1 - |r|$ and $\Phi = 2\varphi$, where $r (= \tan\tau)$ is the
> ellipticity ratio of the polarization ellipse $(-1 < r < 1)$;
> here $0 < r \leqslant 1$ represents right-handed elliptical polari-
> zation (RHEP), $r=0$ linear polarization (LP), and $-1 < r < 0$
> left-handed elliptical polarization (LHEP).

(a)

(b)

(c)

TARGET RETURNS ◯ WITHIN
OBSERVATION-TIME WINDOW ARE
WIDELY SPREAD IN POLARIZATION
SPACE

TARGET RETURNS WITHIN
OBSERVATION-TIME WINDOW
FORM A CLOSED SET ☐ ,
BUT WIDELY SEPARATED FROM
THAT OF THE CLUTTER IN
POLARIZATION SPACE

TARGET RETURNS WITHIN
OBSERVATION-TIME WINDOW
FORM A CLOSED SET ▨ ,
BUT THESE SETS ARE SPREAD
IN POLARIZATION SPACE FROM
WINDOW TO WINDOW

● LEFT -SENSE POLARIZED RETURNS
+ RIGHT-SENSE POLARIZED RETURNS

Fig. 3   Conditions under which the use of the polarization state of
the returns as a discriminant for target enhancement is
extremely beneficial; clutter returns within observation-
time window form a closed set and the transmit polarization
used is arbitrary, or is based on a-priori knowledge, or is
derived from an on-line learning-process

However, in the case of an 'unknown' clutter environment, a
suitable transmit polarization would be one obtained from a
"learning-process"-derived time/window-averaged characteristic
null polarization of the clutter (e.g., in the case of homogeneous clutter, averaging over a range window would be the obvious
method, since any cell in the window might contain a target).
This approach is based upon the physical phenomenon that an
object's characteristic null polarizations [7,9] (see Fig. 4) are
the two polarizations which most closely characterize the object
as far as its actual (or apparent) geometry, or structure, is
concerned. These character-
istics are reflected in its
scattering matrix. Conse-
quently, the selected average
characteristic null polarization, which can be termed
the "minimum" polarization,
would, in the absence of "a
priori" knowledge of the
target polarization signatures,
be the "best" with regard to
discriminating between target
and clutter returns.

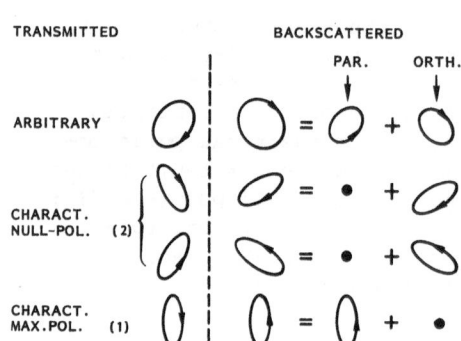

Fig. 4 Single radar object characteristic polarizations [7,9]

        As the derivation of a null polarization requires scattering
matrix data, the learning process requires the matrix data to
be in principle continuously available.   In turn this implies
that the radar system has to have mutually-coherent orthogonally-
polarized receiver channels and a specific transmit mode.   With
regard to uncoded pulsed waveforms, the scattering matrix is
measured in the most efficient way (i.e., fast and with no loss in
backscattered power) when there is sequential transmission of
two orthogonally-polarized waveforms [10] e.g., horizontal- and
vertical-polarization (HP and VP), right-hand circular- and
left-hand circular-polarization (RHCP and LHCP), etc.

        The basic differences between scalar-signal and vector-signal
processing are summarized in Table 1.

1.3 Approach

        The aim of the early studies initiated at STC was to demon-
strate the flexibility of a system with orthogonally-polarized
receiver channels, on the basis of a simple signal processing
concept not feasible with a conventional single-channel receiver
system.   A theoretical study [11,12,13], analogous to the
approach used in frequency-diversity radar systems, examined the
effectiveness of the linear addition of the video signals in the

Table 1 Comparative summary between vector- and scalar-signal
processing

	IN VECTOR APPROACH	IN SCALAR APPROACH
POLARIZATION INFORMATION	is used	is not used
OBSERVED	scattering matrix (+ Doppler shift)	cross-section (+ Doppler shift)
ANTENNA SYSTEM	dual-polarized	single/dual polarized (single in principle)
RECEIVER	dual-channel	single-channel
TRANSMIT MODE	(a) learning alternate ortho-gonally-polarized uncoded pulse waveforms (b) operating waveform polari-zation adapted to unwanted environ-ment	fixed polarization (for example: HP, VP, or CP) or switching between a limited number of polari-zations for example: HP ◄──► CP

two channels prior to the detection stage.   The antenna polarization was considered to be fixed, but adapted to a clutter
environment for minimum response. The point of interest was the
effectiveness of this processing concept in range cells outside
the clutter, that is, in clear as well as quiet environments.
However, there was no exploitation of polarization-vector information, as the only parameter processed was the total target
backscattered power.

A second area of study concentrated on a theoretical assessment of the effectiveness of "adaptive antenna polarization" for
target enhancement in a moving (i.e. time-varying), homogeneous-
type of distributed clutter. In the early seventies, because there
was no model suitable for describing the scattering properties of
homogeneous clutter in terms of polarization (with the exception
of an incomplete model for a rain-cloud), a simple model of a
random dipole-cloud was developed at STC [14]. The first study
using this model was devoted to the effectiveness of adapting
identical transmit and receive antenna polarizations for a minimum
dipole-cloud response, using as a variable the probability space

of individual dipole-axis direction [14].   A second study concerned the antenna polarization at transmit and receive which
gives rise to maximum correlated and uncorrelated orthogonally-
polarized backscatter components in the receiver channels [15].
The observations from this latter study may have significance
in processing schemes which utilize the correlation between
orthogonally-polarized backscatter components from targets and
clutter as a discriminant for obtaining target enhancement.   In
addition to the above-mentioned dipole-cloud model, a simple
model of a moving multiple-component body was investigated.

As the studies progressed into the mid-seventies it became
apparent that in practice "actual" adaptation of the antenna
polarizations for target enhancement had severe limitations in
the practical situation where the polarization properties of the
"dynamic" adverse environment are varying in both the spatial
and temporal domains [16,17,18].

These limitations prompted the next avenue of study, which
concentrated on the concept of "virtual polarization adaptation"
(VPA) for target enhancement.   This technique, which is equivalent to appropriately adapting the polarization states of the
transmit and receive antennas to the unwanted environment for
minimum response, allows optimal adaptation of the transmit and
receive polarizations to each individual radar cell, since the
adaptation takes place within the processor and not at the antenna.   Of practical importance is the fact that the VPA concept
allows the sensor head to be relatively simple in the sense that
the system need only be capable of emitting two specific
orthogonally-polarized uncoded pulsed waveforms, instead of every
type of polarization.

As investigation progressed the power and flexibility offered
by the general VPA concept became apparent, and a natural consequence was the development of algorithms for vector signal
processing within the VPA concept.   These algorithms exploit the
polarization information as a discriminant for target enhancement
to an extent not realizable by adaptation of the antenna's actual
polarization.   One of these algorithms is principally concerned
with increasing the ability to suppress unwanted signals that have
fluctuating polarization, and so enhance target detection.   As the
conventional single-notch nulling process is inherently sensitive
(see Fig. 1), the average amount of achievable suppression is
limited when the unwanted incident signals have fluctuating
polarizations (see Fig. 2).   This was considered such a serious
drawback that a means of desensitizing the nulling process was
sought, and resulted in the concept of a vector-processor called
"the multinotch logic-product polarization suppression filter",
or MLP filter.   With this filter the location, the form, the
size of the region of deep suppression in polarization space, and

the average level of deep suppression are all under design control [18,20].

The location of the MLP filter in polarization space (as determined by the average incident-field polarization of closed sets of unwanted signals) will vary in both the spatial and temporal domains in a typical adverse environment. Consequently, its adaptation to the dynamic environment would place a heavy load on the processor in terms of processing time, and therefore an algorithm which is economical in terms of processing time has been devised. This algorithm considers the MLP-filter at an arbitrary, fixed location in polarization space and involves the mathematical transformation of the polarization states of the incident-field samples to be suppressed, so that they are in effect brought to the location of the filter for processing. The process is called "polarization-vector translation" (PVT), and because it is so economical it is eminently suitable for real-time implementation [19].

Studies into the design of an effective MLP filter have revealed that to cover only a moderate area of polarization space (typically 2% to 3%) at least 61 notches are required to make up a filter giving, on average, a 40 dB suppression level in this area. The real-time realization of such a filter would, using current technology, result in quite a bulky piece of hardware. This practical limitation stimulated the need to devise a processing algorithm that would reduce the area of spread (in polarization space) of the sets of incident-field samples that were to be suppressed, such that the MLP filter could be configured with a reduced number of notches but the desired average level of suppression maintained. The resulting vector-process is also a PVT type. However, since it is essentially non-linear it is referred to as the non-linear PVT-process [21] to distinguish it from the linearly operating PVT-process.

It was evident from an early stage that it would be necessary to do experimental work in support of this project, in order to:

(a) assess the polarization signatures of clutter and targets (there was, and still is, a lack of measured data in terms of scattering matrices);

(b) study, off-line, the effectiveness of advanced vector-signal processing schemes with live data; and

(c) demonstrate the usefulness of performing signal processing in the polarization domain in real-time.

For this reason work began in 1971 on the realization of a

multi-purpose experimental X-band radar facility for polarization
studies. Since then this facility has gradually evolved and at
present has the capability to obtain and record mass live
polarization data from targets and clutter, and perform off-line
evaluations of vector-signal processing schemes with the aid of
this data.

From this brief overview the emphasis of the STC studies so
far should be clear. The ultimate objective of this project has
been, and currently still is, the investigation of new vector-
signal processing schemes suitable for real-time implementation.

1.4 Layout of Paper

The format of this paper reflects the historical development
of the principal topics, and accordingly the evolution of the
ideas involved.

Chapter 2 contains a summary of the evaluation of a simple
system which exploits polarization information by using a dual-
channel receiver system. In such a system the transmit antenna
polarization is considered to be matched to suppress the clutter,
and linear video recombination is applied in clutter-free (and
quiet) areas to maintain the system detection performance.

Prompted by the lack of a suitable model describing the
signatures of general scatterers in terms of polarization, the
study described in Chapter 3 was initiated, and derived a simple
random dipole-cloud model which subsequently enabled studies into
the applicability of discriminants useful for improving target
detection.

A major development which has had a significant influence
upon the direction of this project is the concept of virtual
polarization adaptation and the related vector-signal processing
algorithms. The freedom of not having to adapt the actual
polarization of the antenna has enormous advantages when attempting to improve target detection by means of unwanted signal
suppression. Chapter 4 highlights these advantages and an outline
of the associated adaptive vector-signal processing algorithms is
given in Chapter 5.

In order to evaluate the vector-signal processing methods
which had been devised, experimentally-derived data had to be
obtained. Therefore, an X-band experimental radar facility was
realized, and its main features are discussed in Chapter 6. The
subsequent evaluation and examples of the performances which can
be realistically expected from the processing methods (using data
derived from rain) are contained in Chapter 7.

To conclude, Chapter 8 discusses future studies on the utilization of polarization information for target enhancement in primary radar systems.

2.   USE OF LINEAR VIDEO RECOMBINATION IN DUAL-CHANNEL RECEIVERS

2.1 General

Radar systems that use linear polarization at transmit and receive generally have difficulty in detecting airborne targets in precipitation clutter.   However, a well-known technique [2] for improving detection capability in rain clutter is to use circular polarization (CP) or an almost-circular elliptical polarization (EP). The physical explanation for this improvement is that the drops in a rain cloud have a more or less spherical geometry, and thus the specular reflections by far dominate the non-specular reflections, while for aircraft targets these reflections are on average of the same order.

A property of any conventional radar system is that at receive it cancels incident waves which have a polarization orthogonal to the polarization used on transmission.   This means that although the use of CP or EP gives an improvement in rain clutter visibility of typically 15 to 30 dB [2] (because most of the scattered power is in the orthogonal polarization), there is also always a loss in target echo strength due to the cancellation by the system of part of the target reflected power (the part in the orthogonal polarization).   This means that for a radar operating in a CP mode, the target-detection performance in precipitation-free sectors is approximately 3 dB [3] less than that obtained in an HP (horizontal polarization) mode.

The conflicting requirements to obtain the best performance of target detectability in precipitation and precipitation-free sectors indicate that some form of polarization mode switching is desirable.   However, instantaneous switching between HP and CP at transmit is difficult to achieve in practice, mainly because of the high power levels involved.   Moreover, mode switching at transmit has the disadvantage that there is a loss in target-detectability in precipitation free-sectors on the same bearing as the cloud.   Therefore, faced with these problems switching at receive is a better approach.

2.2 Circular Polarization Transmission and Processing in Precipitation-free Sectors

As a means of reducing the loss in target detection in precipitation-free sectors, the study reported in [11] proposed

a system which uses fixed CP on transmit and a dual-channel receiver system.   A selector determines whether detection-perfor-
mance is better with the video signal derived from the received
parallel-polarized backscatter component, or with the recombined
video signal obtained from the linear addition of the video
signals derived from the received parallel-polarized and orthogonally-polarized backscatter components.   Since selective
switching between the two types of video is now performed at low
power levels it can be done electronically, and hence instantan-
eously.   The application of linear video addition here is optimum
(from the detectability point of view) when the received orthogonally-polarized components of the target returns (RHCP and LHCP)
are on average of equal strength.

    To evaluate the proposed system operating in the linear video
recombination mode it was compared with a conventional system that
had a single-channel receiver and operated in the CP transmit
mode.   Schematic diagrams of these two systems are shown in
Figs. 5(a) and 5(b).

a) POLARIZATION-MODE SWITCHING AT THE TRANSMITTING SIDE (HP ↔ RHCP)

b) FIXED RHCP EMISSION AND SWITCHING AT THE RECEIVING SIDE (RHCP ↔ RHCP+LHCP)

Fig. 5   Techniques to exploit circular polarization for precipi-
         tation clutter suppression [11]

Both systems were assumed to have n_h hits per channel, square
law detection in the channels, and a threshold after integration.
Additive internal, narrowband, stationary, Gaussian noise was
assumed to be present in the channels, and in the dual-channel
receiver case the noise components were independent and had equal
average power.   Although [11] treats Swerling Case-1 and Case-2
target scintillation models, the results presented here are only
for Swerling Case-1 targets (slow scintillation) for the situation
where the orthogonally-polarized components of the target returns
are statistically independent.

    The threshold levels, Y_b, are related to the probability
of false alarm, P_f, by

$$P_f = 1 - \left[I \frac{Y_b}{\sqrt{mn_h}}, mn_h - 1 \right] \quad , \; m = 1, \, 2 \qquad (1)$$

where I is the incomplete gamma function as defined by Pearson and m is an index equal to 1 for the single-channel receiver and equal to 2 for the dual-channel receiver. The probablity of detection, P_d, is given by

$$P_d = \int_{Y_b}^{\infty} f_x(x) \; dx = 1 - \int_0^{Y_b} f_x(x) \; dx \qquad (2)$$

where x is the signal after integration and, where relevant, recombinination, and $f_x(x)$ is the probability density function (pdf) of x.

An efficiency parameter, E, was used to directly compare the two systems. This is defined as the gain in the required signal-to-noise power ratio per hit over the signal-to-noise power ratio per hit in a conventional HP radar system; the transmitted power being the same for both systems.

The efficiency E_1 for the single-channel receiver is given by

$$E_1(dB) = 10 \log_{10} \left[\frac{\bar{x}_r}{\bar{x}_o} \right] = 10 \log_{10} \left[\frac{K_t^4 \; K_s^2 + 1}{(K_t^2 + 1)^2} \right] \qquad (3)$$

where $\bar{x}_r$ is the average signal-to-noise power ratio per hit when operating in the CP mode and $\bar{x}_o$ is the average signal-to-noise power ratio per hit when operating in the HP mode. Parameter K_t is introduced to take account of the practical observation that an antenna polarization different from CP would result in maximum suppression of rain clutter, and

$$K_t = \frac{\text{vertically-polarized power transmitted}}{\text{horizontally-polarized power transmitted}}$$

Also, a target parameter, K_s, relating to practical observation is introduced:

$$K_s = \frac{\text{average target cross-section to vertically-polarized waves}}{\text{average target cross-section to horizontally-polarized waves}}$$

Since the efficiency E_2 for the dual-channel system is dependent on n_h, P_d and P_f, a simple expression similar to (3) cannot be obtained. However, E_2 is related to:

$$10 \log_{10} \left[\frac{\overline{x}_a}{\overline{x}_o} \right] \text{ (dB)} = 10 \log_{10} \left[\frac{K_t^2 K_s^2 + 1}{2(K_t^2 + 1)} \right] \tag{4}$$

where $\overline{x}_a = \overline{x}/2$ and $\overline{x}$ is the average recombined signal-to-noise power ratio per hit.   A detailed derivation of the above re-lationships and of the efficiency E_2 over a wide range of n_h, P_d, and P_f is given in [11].

For an objective comparison it is sufficient to consider average values $\overline{E}_1$ and $\overline{E}_2$ of the efficiencies, with parameters K_t and K_s uniformly distributed over a practical range of 0.5 to 2.   Table 2 shows average efficiency values for both systems under the same conditions of target type, P_d, P_f, etc.   These figures do not include system losses, which typically are approxi-mately 2 dB (polarizer loss) for the single-channel system and 3 dB (2 dB polarizer loss and 1 dB video recombination loss) for the dual-channel receiver.   The performance figures obtained showed that, relative to the conventional radar system switched to the CP mode at transmit, the dual-channel receiver system should give an overall improvement of roughly 3 dB for $P_d = 0.8$ and $n_h = 10$ in precipitation-free sectors.

Table 2   Efficiency ($\overline{E}_m$) of the two systems of Fig. 5 in preci-pitation-free sectors relative to horizontally-polarized radar systems (system losses not included) [11]

$P_f = 10^{-6}$, $n_h = 10$, $0.5 \leqslant K_t^2 \leqslant 2$, $0.5 \leqslant K_s^2 \leqslant 2$		
Swerling Case-1 scintillating target	$P_d = 0.8$	$P_d = 0.5$
Single-channel receiver system in the RHCP mode (m = 1)	$\overline{E}_1 = -2.6$ dB	$\overline{E}_1 = -2.6$ dB
Dual-channel receiver system in the linear video addition mode (m = 2)	$\overline{E}_2 = 1.4$ dB	$\overline{E}_2 = 0.0$ dB
Gain of the dual-channel system over the single-channel system	4.0 dB	2.6 dB

2.3 Arbitrary Polarization Transmission and Detection in Gaussian Noise

The dual-channel receiver of Section 2.2 can be reconsidered [12,13] in a more general manner, this time with an arbitrary transmit polarization which is matched to the unwanted part of the environment for minimum average response.   Figure 6 shows the system in more detail. It is non-coherent, processes both orthogonally-polarized received backscatter components, and uses linear video addition (i.e., non-weighted addition).   The assumptions made about the receiver in the previous section are assumed to apply again, but to maintain generality the mathematical model uses the average "orthogonally-polarized" (or "cross-polarized") cross-section of the target normalized to its average "parallel-polarized" (or "co-polarized") cross-section as a parameter. This is denoted by F. Thus, for equal average noise powers in each channel,

$$F = \overline{x}_\perp \, / \, \overline{x}_{\parallel} \tag{5}$$

where $\overline{x}$ is the average target signal power to average noise power and ' $\parallel$ ' and ' $\perp$ ' denote parallel and orthogonal respectively.

Fig. 6 Block diagram of the dual-channel detection receiver; target detection in unpolarized Gaussian noise [12,13]

The range of F considered was 0 to 1, because of the wide variation of target returns in terms of polarization relative to the transmit polarization.   In [12] and [13] Swerling Case-1, -2, -3 and -4 target-scintillation models are treated, but again the results presented here are only for Swerling Case-1 for the case where the orthogonally-polarized components of the target returns are statistically independent.

The probability of false alarm and detection are again determined by equations (1) and (2) respectively. A detectability factor D_s was derived to demonstrate the performance of the system. For a single-channel receiver the detectability factor is denoted by D_{s1} and defined as the required ratio of average signal power to average noise power at the input to the threshold-device, for specified values of P_f, P_d and n_h. For the dual-channel receiver, the detectability factor is denoted by D_{s2}, but the power ratio is considered in the parallel channel at the input to the video addition unit for specific values of not only P_f, P_d, n_h, but also F.

To compare the performance of the dual-channel system to the single-channel receiver, a "figure-of-merit", ΔD_s, defined as the gain in D_{s2} relative to D_{s1}, is used.   Figure 7 shows, for Swerling Case-1 targets, ΔD_{s1} plotted against n_h for a P_f of 10^{-6}, two values of P_d (0.5 and 0.8), and values of F ranging between 0 and 1.   The results indicate that when the background

NO. OF HITS PER CHANNEL, n_h

noise is Gaussian the use of the orthogonally-polarized component in systems operating with a small number of hits per target ($n_h \leqslant 3$) and $P_d \geqslant 0.8$ is advantageous (for Swerling Case-1 targets) when $F \geqslant 0.2$.   It appears that if $F = 0$, the relative loss in target detectability is a maximum of 1 dB under the conditions investigated and for $n_h \leqslant 20$.   Thus, it may be concluded that the noncoherent processing scheme considered which applies 'unweighted' video addition is near optimal over the full range of F.

Fig. 7   Gain in D_s versus n_h for the dual-channel system relative to the single-channel system ($P_f = 10^{-6}$; non-coherent integration) [12,13]

3.   POLARIZATION SIGNATURES OF A RANDOM DIPOLE CLOUD

3.1   Dipole Cloud Model

A theoretical assessment of the effectiveness of 'adaptive antenna polarization' with respect to target enhancement in a dipole-cloud clutter environment is of importance to the design of radar systems with improved sub-clutter visibility.   The first step in such an assessment is the determination of which polarization signatures of a dipole cloud are suitable for exploitation.   A simple statistical model was developed at STC [14] because no model of the scattering properties of a random dipole cloud in terms of polarization could be found.

The main assumptions under which the model was derived are as follows:

(a)   The space angle describing the probability space of dipole axis direction is identical for all dipoles and is determined by $0 \leqslant \chi \leqslant \pi$ and $\eta_o \leqslant \eta \leqslant \pi - \eta_o$ (see Fig. 8)

FICTITIOUS
DIPOLE-AXIS
DIRECTION

χ = AZIMUTH ANGLE

η = POLAR ANGLE

Fig. 8 The dipole in its coordinate frame [14,15]

(b) Each dipole is excited only by that component of the illuminating electric field which is linearly polarized in the plane formed by the dipole axis direction and direction of propagation. In other words, its normalized scattering matrix for $\chi = \eta = \pi/2$ on a basis defined by a pair of horizontally - and vertically - polarized unit vectors is given by,

$$T(\chi = \pi/2, \eta = \pi/2) = e^{j\beta}\begin{bmatrix} 1 & 0 \\ 0 & 0 \end{bmatrix}$$

(c) The far-field backscattered pattern of each dipole is proportional to $\sin^2(\gamma)$, where γ is the angle between the dipole axis direction and the direction of propagation of the illuminating field. The normalized general scattering matrix $T(\chi,\eta)$ of the individual dipole can be derived by means of simple matrix algebra, and due to this assumption it takes the form

$$T(\chi,\eta) = \begin{bmatrix} t_{yy} & t_{yx} \\ t_{yx} & t_{xx} \end{bmatrix} = e^{j\beta}\begin{bmatrix} \sin^2(\chi)\sin^2(\eta) & \sin(\chi)\sin(\eta)\cos(\eta) \\ \sin(\chi)\sin(\eta)\cos(\eta) & \cos^2(\eta) \end{bmatrix} \quad (6)$$

(d) The axis direction of the individual dipole is uniformly distributed over the specified space angle. Under this assumption the joint probability density function of the fictitious dipole axis direction, see Fig. 8, is

$$f_{\chi,\eta}(\chi,\eta) = \sin(\eta)/\{2\pi\cos(\eta_o)\} \quad, (0 \leqslant \chi \leqslant \pi, \text{ and } \eta_o \leqslant \eta \leqslant \pi - \eta_o)$$

Further, it is assumed that the absolute phase-angle β is uniformly distributed between $-\pi$ and π, and is independent of χ and η. Under this assumption the first-order statistics (mean values) of the elements of $T(\chi,\eta)$ are zero, and the statistical properties of the scattering matrix elements can be determined in terms of second-order statistics.

(e)   The axis directions of the dipoles are independent of one another.   The number of dipoles with positions uniformly distributed through the cloud is large, and the effects of multiple scattering can be neglected. Due to these assumptions the statistical properties of the matrix elements of the dipole cloud can be readily derived in terms of second moments (first moments all equal zero), using the well-known central-limit theorem.

The dipole cloud model is characterized by

$$\overline{\left| t_{yy} \right|^2} = \sigma_{yy}^2 = \frac{1}{8} \left\{ 15 - 10\cos^2(\eta_o) + 3\cos^4(\eta_o) \right\} \tag{7a}$$

$$\overline{\left| t_{xx} \right|^2} = \sigma_{xx}^2 = \cos^4(\eta_o) \tag{7b}$$

$$\overline{\left| t_{yx} \right|^2} = \overline{t_{yy} t_{xx}^*} = \sigma_{yx}^2 = \frac{1}{6} \left\{ 5\cos^2(\eta_o) - 3\cos^4(\eta_o) \right\} \tag{7c}$$

$$\overline{t_{yy} t_{yx}^*} = \overline{t_{xx} t_{yx}^*} = 0 \tag{7d}$$

where σ_{yy}^2 and σ_{xx}^2 are the normalized average cross-section for horizontally- and vertically-polarized illumination respectively,

σ_{yx}^2 is the normalized average cross-polarized cross-section for horizontally- and vertically-polarized illuminations,

and η_o, where $0 \leqslant \eta_o \leqslant \pi/2$, is the model parameter.

The dipole cloud model has an average axis of symmetry which is oriented along the x-axis (see Fig. 8).   Although in a practical case this axis may have an arbitrary orientation, this specific orientation has been chosen to simplify the analysis without affecting the basic signatures.

3.2   Polarization Dependence of Received Backscattered Power

One potentially-exploitable polarization signature of any radar object is that antenna polarization which is identical at transmit and receive and gives a minimum response; the two possibilities are referred to as the average null-polarizations. Obviously such polarizations can be exploited to improve target detection by suppressing the unwanted part of the environment. The effectiveness of adapting identical transmit and receive antenna polarizations for minimum dipole-cloud response, with parameter η_o, has been assessed as follows [14].

The instantaneous normalized received backscattered power from the random dipole cloud is determined by [7,10]:

$$P = \left| v \right|^2 = \left| \underline{Th}_T \cdot \underline{h}_R \right|^2 = \left| \underline{Th} \cdot \underline{h} \right|^2 = \left| \underline{E}_s \cdot \underline{h} \right|^2 \tag{8}$$

where subscripts 'T' and 'R' refer to transmit and receive respectively, the generalized scattering matrix T is given by (6), and the normalized identical transmit and receive antenna polarizations $(\underline{h})$, on a basis defined by a pair of horizonally-polarized (H) and vertically polarized (V) unit vectors $(\underline{i}_y = \underline{i}_H$ and $\underline{i}_x = \underline{i}_V)$, is given by:

$$\underline{h} = \begin{bmatrix} \cos(\theta) & ; \ \underline{i}_H \\ \sin(\theta) \ \exp(j\delta); \ \underline{i}_V \end{bmatrix} \tag{9}$$

with $0 \leqslant \theta \leqslant \pi/2$ and $-\pi < \delta \leqslant \pi$ and with right-handed elliptical polarization (RHEP) defined by $0 < \delta < \pi$. The average received power $\overline{P}$ follows from substitution of (6) and (9) into (8) and the use of (7), namely:

$$\overline{P} = \sigma_{yy}^2 \cos^4(\theta) + \sigma_{xx}^2 \sin^4(\theta) + 2\sigma_{yx}^2 \left[2 + \cos(2\delta) \right] \cos^2(\theta) \sin^2(\theta) \tag{10}$$

The antenna polarizations (θ, δ) which give minimum and maximum values for $\overline{P}$ in (10) are determined by the following conditions:

$$\frac{\partial \overline{P}}{\partial \theta} = 0 \quad \text{and} \quad \frac{\partial \overline{P}}{\partial \delta} = 0 .$$

The results obtained are presented in Table 3.

Table 3 Dipole-cloud model: the parameters of the 'minimum' antenna polarization and maximum average backscattered power suppression relative to horizontal polarization; dipole orientation parameters $0 \leqslant \chi \leqslant \pi$ and $\eta_o \leqslant \eta \leqslant \pi - \eta_o$ [14]

Dipole-cloud parameter η_o	Complex polarization factor representing the minimum antenna polarization $R_{min} = [\tan(\theta)]_{min} \exp(j\delta_{min})$		Maximum average suppression for "minimum" antenna polarization
(degrees)	$[\tan(\theta)]_{min}$	δ_{min}	(dB)
0	1	$\pm\pi/2$	1.8
15	1.14	$\pm\pi/2$	2.1
30	1.92	$\pm\pi/2$	3.4
41.5	∞	–	6.2
45	∞	–	7.3
60	∞	–	14.0
75	∞	–	26.0
90	∞	–	∞

From this table it can be concluded that 'adaptive antenna polarization' is very promising for cases where the dipole-cloud parameter η_o is larger than 60°.

3.3 Cross-correlation of Orthogonally-polarized Backscatter Components

The cross-correlation of received orthogonally-polarized backscatter components depends both on the transmit and receive antenna polarizations. Extreme values for the cross-correlation obtainable with controllable antenna polarizations can be regarded as radar object polarization signatures, and therefore find application in radar systems which utilize cross-correlation to improve target detection.

The derivation of the antenna polarizations at transmit and receive that give rise to maximum correlated and uncorrelated orthogonally-polarized backscatter components in the receiving channels was the subject of the study presented in [15]. Two mathematical models of complex radar objects were considered: one representing the random dipole cloud and the other a moving, multiple-component body.

In this area of study coherency theory is applied, and so a brief review is given here together with some results obtained for the dipole cloud model.

According to (8) and (9) the electric-field vector of the backscatter wave $\underline{E}_s$ is given by,

$$\underline{E}_s = \begin{bmatrix} E_y & ; & \underline{i}_H \\ E_x & ; & \underline{i}_V \end{bmatrix}_s = T \, \underline{h}_T \tag{11}$$

and when the object is time-varying (moving), a time-varying elliptically-polarized backscatter wave is observed with components $E_y(t)$ and $E_x(t)$.   In this study the time variation of the backscatter field are attributed to random processes in which the absolute phases of the components $E_y(t)$ and $E_x(t)$ are uniformly distributed and independent of the moduli. Hence the time-averaged values $\langle E_y \rangle$ and $\langle E_x \rangle$ are zero, and in accordance with [6] it is usual to define a time-varying partially-polarized plane wave in terms of the coherency matrix of the complex field components E_y and E_x, i.e.,

$$J = \begin{bmatrix} \langle E_y E_y^* \rangle & \langle E_y E_x^* \rangle \\ \langle E_x E_y^* \rangle & \langle E_x E_x^* \rangle \end{bmatrix} = \begin{bmatrix} J_{yy} & J_{yx} \\ J_{xy} & J_{xx} \end{bmatrix} . \tag{12}$$

The average backscattered power $\langle P_s \rangle$ is then,

$$\langle P_s \rangle = \langle \underline{E}_s \cdot \underline{E}_s^* \rangle = \langle (\underline{Th}_T) \cdot (\underline{Th}_T)^* \rangle = J_{yy} + J_{xx}. \tag{13}$$

As far as the polarization state of the backscattered wave is concerned, two extreme cases [6] are of special interest:

(a) the wave is completely-unpolarized or randomly-polarized if $J_{yx} = J_{xy} = 0$ and $J_{yy} = J_{xx}$; and

(b) is completely-polarized if $|J| = J_{yy}J_{xx} - J_{yx}J_{xy} = 0$.

In general the backscattered wave will be partially-polarized, and can be uniquely expressed as the sum of two independent waves: one completely-polarized and the other completely-unpolarized. The ratio of the average power of the completely-polarized portion to the total average backscattered power is called the degree of polarization 'p' of the backscattered wave. In terms of coherency-matrix components this ratio [6] is

$$p = 1 - \left\{ [4|J|/(J_{yy} + J_{xx})^2] \right\}^{1/2} , \text{ with } 0 \leqslant p \leqslant 1. \tag{14}$$

Note that the value of 'p' given by (14) depends on $\underline{h}_T$, but is independent of the orthogonal reference frame against which the backscatter wave is measured.

The complex correlation factor (or complex cross-correlation coefficient) of the received voltages (v and $v_\perp$) at the terminals of the orthogonally-polarized receive antennas is determined by

$$\mu = \langle vv_\perp^* \rangle / \left\{ \langle P \rangle \langle P_\perp \rangle \right\}^{1/2} = |\mu| \exp(jv), \quad 0 \leqslant |\mu| \leqslant 1 \text{ and } -\pi < v \leqslant \pi, \tag{15}$$

where

$$v = \underline{Th}_T \cdot \underline{h}_R \text{ and } v_\perp = \underline{\dot{Th}}_T \cdot \underline{h}_{R_\perp}, \qquad (\underline{h}_R \cdot \underline{h}^*_{R_\perp} = 0) \tag{16a}$$

and

$$\langle P \rangle = \langle vv^* \rangle \text{ and } \langle P_\perp \rangle = \langle v_\perp v_\perp^* \rangle \tag{16b}$$

Furthermore, it is shown in [6] that,

$|\mu| \leqslant p,$

$|\mu| = p \neq 1,$ if $\langle P \rangle = \langle P_\perp \rangle,$ (17)

$|\mu| = 1,$ if $p = 1;$ generally indepedent of $\underline{h}_R$.

Expressions (11) to (17) provide the mathematical basis for the coherency analysis, which is mainly concerned with the determination of the conditions of antenna polarization that result in extreme values of $|\mu|$ for the cases $\underline{h}_R = \underline{h}_T$ and $\underline{h}_R \neq \underline{h}_T$.

By applying (6), (7), (9), (11), (12) and (14), an expression for the degree of polarization of the backscattered wave from the random dipole cloud is obtained as a function of θ_T and δ_T (the parameters which define the transmit antenna polarization) and dipole-cloud parameter η_0.

Subsequently the conditions of θ_T and δ_T for $p = p_{max}$ and $p = p_{min}$ are determined from

$$\partial[|J|/(J_{yy} + J_{xx})^2]/\partial\theta_T = 0 \text{ and } \partial[|J|/(J_{yy}+J_{xx})^2]/\partial\delta_T = 0. \quad (18)$$

The extreme degree-of-polarization values $p_{max}(\eta_0)$ and $p_{min}(\eta_0)$ are determined by substituting the solutions of (18) into the equation derived for p. Note that p determines the upperbound for $|\mu|$, see (17). In. Fig. 9 curves of p_{max}, p_{min}, and p for horizontally polarized (p_{HP}) and vertically polarized (p_{VP}) emission versus η_0 are presented. In the case where $\underline{h}_R = \underline{h}_T = \underline{h}$, conditions of $\underline{h}$ that give rise to $\mu = 0$ always exist if $p \neq 1$, and these can be determined analytically by setting $\langle vv_{\perp}\rangle$ in (15) equal to zero. Alternatively, $|\mu|_{max}$ and the corresponding antenna polarizations can be determined, by means of a numerical optimization process. For comparison the graph of $|\mu|_{max}$ versus η_0 is also shown in Fig. 9.

Fig. 9   Dipole-cloud model: particular values of the degree of polarization of the backscattered wave, and for the case $\underline{h}_R = \underline{h}_T$, extreme values of the complex cross-correlation coefficient of received orthogonally-polarized backscatter components versus dipole-cloud parameter η_0; dipole orientation parameters $0 \leqslant \chi \leqslant \pi$ and $\eta_0 \leqslant \eta \leqslant \pi-\eta_0$ [15].

In [15] the same analysis that was applied to the dipole cloud model was performed for a model of a moving, multiple-component body representing a simple target. It was concluded that these two models were totally different as far as the cross-correlation of received orthogonally-polarized backscatter components and their dependence on transmit antenna and receive antenna polarization are concerned. This observation might stimulate further studies related to the applicability of this 'cross-correlation' as a controllable parameter for the purpose of improving target detection.

4.   VIRTUAL POLARIZATION ADAPTATION FOR TARGET ENHANCEMENT

4.1 Comparison of Actual and Virtual Adaptation

In practice a radar system can be confronted with:

(a)   spatially-distributed clutter, either fixed or moving,
       for which the polarization properties differ from
       area (volume) to area (volume) or even from resolution
       cell to resolution cell, and

(b)   a number of spatially-distributed interference sources,
       which moreover have time-varying polarization charac-
       teristics.

Therefore in such an adverse environment actual adaptation
of the antenna polarizations has severe limitations because:

(a)   the severe time-constraints imposed on the adaptation
       process would be extremely difficult to meet; a
       particularly stringent condition would be separate
       adaptation of the transmit and receive-antenna polari-
       zations;

(b)   it allows only optimum adaptation to one resolution
       cell per beam direction, or at best to one group of
       resolution cells per beam direction when the polari-
       zation properties of the cells are strongly correlated
       (e.g., homogeneous clutter);

(c)   when there is no a-priori knowledge available on
       the optimum antenna polarization to be used, adaptation
       cannot take place during the periods that the system
       is in its "learning" mode; and

(d)   the antenna sub-system must be capable of generating,
       under control, every type of wave polarization.

These observations strongly indicate that in a "complex"
adverse environment actual adaptation of the antenna polarization
is impractical, and so an appropriate concept to overcome the
limitations was developed.

In the case of a "complex" clutter environment, the solution
is to give the transmit polarization a fixed pattern (in the
case of uncoded pulsed waveforms, e.g., alternate horizontal and
vertical polarization) and to use orthogonally-polarized receiving
channels.   Then measured scattering-matrix data are available,
which allows the exploitation of the property that when the
scattering matrix is known the response for any combination of

transmit and receive polarizations is determined.   Consequently
it allows the adaptation to each individual resolution cell
for minimum clutter response to take place within the processor
and not at the antenna [16,17,18], and therefore this concept
is called "virtual polarization" adaptation (VPA).   Consequently, the VPA concept is "learning-process" based, where the
learning process runs concurrently with the adaptation process.
The essential differences between actual and virtual polarization
adaptation are presented in the block diagrams of Fig. 10.
It should be remembered here that the improvement in target-to-
clutter ratio obtainable through suppression of the clutter
returns using polarization filtering depends on the degree
of polarization of the set of clutter returns and on the extent
to which target polarization signatures differ from those of
the clutter (see Section 1.2).

(a) actual adaptation   (b) virtual adaptation

Fig. 10   Adaptive polarization methods for obtaining target
          enhancement [18]

4.2   Nature of Vector-signal Processing

The simplest form of VPA in a clutter environment involves a
type of vector-signal processing which is realized as the measurement and storage of scattering-matrix data (T_i), the derivation
of a "minimum polarization (h_{min})" from this data (see Section
1.2), and the combination of the stored (or updated) scattering
matrix data with the derived minimum polarization to give minimum
unwanted signal response (v_i).   By way of illustration, Fig. 11
presents a simplified blockdiagram outlining the principles of a
VPA processor which adapts identical transmit and receive polarizations to suppress homogeneous clutter.   The stage at which
virtual adaptation takes place can be described mathematically by

$v_i = T_i h_{i-min}^T . h_{-min}^R$ (see Section 3.2) (where superscripts 'T' and
'R' signify transmit and receive respectively);   the term $T_i h_{i-min}^T$
represents the backscatter vector signals (E_i) resulting from
the transmit polarization h_{-min}^T and is the input to the polarization suppression filter represented by the receive polarization
h_{-min}^R; and the mathematical condition for minimum average response
is $<E_i> . h_{-min} = 0$.

Fig. 11 Simplified block diagram illustrating the original 'vir-
         tual polarization adaptation' (VPA) concept; virtual
         adaptation of identical transmit and receive polari-
         zations to suppress homogeneous clutter [19]

     A more detailed mathematical description of the processing
steps involved in VPA processing is presented in [16] and [17].
VPA as an operational technique is still in the development phase,
since fast algorithms for the calculation of optimum transmit and
receive polarizations suitable for real-time implementation have
yet to be developed.

     From the discussion in Section 1.2 it will be clear that
optimum suppression of unwanted returns is obtained when the
transmit and receive polarizations are adapted separately. The
transmit polarization has to be set such that the degree of polarization of the set of returns is a maximum, so that the
average power contained in the completely-unpolarized component
of the returns is a minimum (as it is physically impossible to
suppress this component); the receive polarization is then set
orthogonal to the polarization of the "optimized" completely-
polarized component of the returns. The optimum and sub-optimum
methods in adaptive clutter suppression are summarized in Table 4.

     Note: the optimum adaptive-polarization suppression tech-
            nique does not necessarily lead to a maximum value
            of the ratio of average wanted-signal power to
            average clutter-signal power.

Table 4   Adaptive-polarization clutter suppression;   the optimum
          and sub-optimum methods [16,17,18]

OPTIMUM METHOD	* adaptation of $\underline{h}^T$ for minimum average power in completely-unpolarized component of the average backscattered field-vector
	* adaptation of $\underline{h}^R$ orthogonal to completely-polarized component of the average backscattered field-vector for minimum average power reception
SUB-OPTIMUM METHOD	* adaptation of $\underline{h}^T = \underline{h}^R$ for minimum average power reception.

4.3   Effectiveness against Model of a Dipole Cloud

An example of the effectiveness of the optimum clutter-suppression method is shown in Fig. 12 for the model of a random dipole cloud reviewed in Section 3.1 [16].   For comparison, the results for the sub-optimum suppression method from Section 3.2 are also included in this figure.   From the graphs it can be concluded that considerable suppression of dipole-cloud clutter can be achieved when the dipole-cloud parameter (η_o) is sufficiently large, and that the difference in effectiveness between the optimum and sub-optimum suppression becomes less with increasing values of η_o.   As an illustration assume that the dipole cloud is characterized by $\eta_o = 60°$, and further assume that the adaptation process results in an average loss of 3 dB in target echo strength relative to the use of horizontal polarization. Then from Fig. 12 it can be seen that optimum adaptation results in a value of 14 dB for the improvement factor.

Fig. 12   Dipole-cloud model: suppression factors relative to
          horizontal polarization versus dipole-cloud parameter
          η_o [16,17,18]

5.   VECTOR SIGNAL PROCESSING ALGORITHMS WITHIN THE VPA CONCEPT

5.1   General

The vector-signal processing inherent in the VPA concept calls for algorithms which are realistically realizable as functional hardware.   As well as having to manipulate the states of the signals in polarization space, the algorithms must be fast, and able to cope with system errors.   Also, where no "a priori" information about the polarization characteristics of the un-wanted background is available, the nature of the algorithm must be based upon a learning-process.

The three processes described in this chapter were derived with the aim of improving the detectability of targets by sup-pressing backscatter returns from clutter and/or interference signals in the polarization domain.   Although the processes are quite general in nature, precipitation is the clutter medium used here to demonstrate their effectiveness.

5.2   The Multinotch Logic-product (MLP) Polarization Suppression Filter

The basis for the design of a filter to suppress unwanted signals in the polarization domain is the fact that the antenna's polarization response to a monochromatic, or completely-polarized wave, can be made equal to zero (Fig. 1).   However, when the backscatter/interference signals have fluctuating polarization and the radar system imperfections affect the polarization purity, the average amount of suppression achievable is limited, because the sharp roll-off of this single-notch response makes it a very sensitive nulling process, as is obvious from Figs. 1 and 2.

The solution to this drawback is to desensitize the nulling process by using a filter which covers a much larger area in polarization space.   This filter's response should allow a specified minimum average level of suppression over a specified coverage area to be obtained under design control.   This has been accomplished [18] by a processor which uses multiple polari-zation filters and a logic-circuit arrangement.   The composite configuration is called a multinotch logic-product polarization suppression filter (or MLP filter) and a schematic diagram of it is shown in Fig. 13.   The arrangement consists of a bank of single-notch filters, each matched to a slightly different receive polarization.   The backscatter/interference vector-signal samples fed to the bank are processed and presented to a logic-product device to select the smallest response for further consideration.

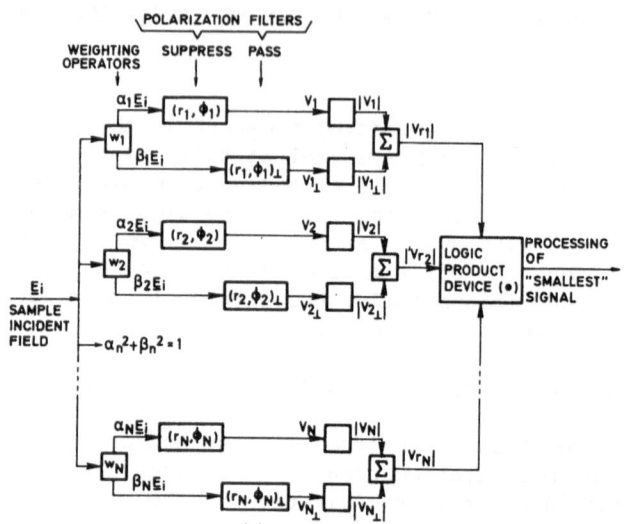

Fig. 13 Schematic representation of a general MLP filter [18]

In general, the location of the MLP filter in polarization space is flexible, in that it can be made to coincide with any 'closed' set of incident field samples to be suppressed. However, in the study reported in [20] precipitation was the principal clutter medium to be suppressed, so the most effective polarization was then CP (as discussed in Chapter 2); therefore in this case the most logical choice for the centre (i.e., location in polarization space) of the MLP filter was either LHCP or RHCP. A further advantage is that this choice simplifies the physical interpretation of the mathematics involved in the design, and so in [20] LHCP was chosen as the filter's centre; RHCP then became the centre of the area of receive polarizations.

Ultimately, the design aim is to achieve an optimum improvement factor (i.e., improvement in the ratio of average wanted-signal power to average unwanted-signal power). This is obviously determined by the spread of the unwanted returns in polarization space about the centroid of the set of samples and the variation of the location of the target returns in polarization space. However, the study described in [20] only addressed the design of an MLP filter in relation to its size and shape in polarization space in order to achieve a specific average suppression level, and the resulting effort required for practical implementation.

Although the shape of the filter in polarization space is essentially free, investigations with returns experimentally derived from rain illuminated with RHCP waves have shown the samples to be approximately circularly distributed about the centroid of the set. This shape may well be regarded as a natural phenomenon for all sorts of homogeneously distributed clutter. Hence, it seems a logical initial choice to consider a filter in which the notches are equi-spaced and circularly distributed in equi-spaced rings about LHCP.

The average level of suppression within the circular area centred on LHCP is determined by the required improvement factor. The size of the MLP filter, the number of notches per ring, and the number of rings needed to cover a specific area, are inter-related under the constraint 'desired level of suppression'. The design aim is to maintain an area in polarization space where the response is not greater than a desired level below the maximum MLP filter response, where the maximum response results from a RHCP incident field.

A cross-section of the normalized logic-product power response on the half great-circle from RHCP through HP to LHCP in polarization space for a typical multi-ring (p-ring) filter is given in Fig. 14. A suppression level of P_s can be achieved over the circular area around LHCP and bounded by the circle of ellipticity ratio r_f.

Fig. 14 Logic-product power response for receive polarizations
$$M_{r_q}(\tau_{r_q} = \pi/4 - q\alpha/2, \varphi_{r_q} = \pi/2; q = 0, 1, 2, \ldots, p)$$
and M_i located on half-great circle through HP [20]

One possible layout of the notches making up a MLP filter is shown in Fig. 15. This consists of 5 rings with 6q notches in each ring (where q = 1, ... p; p = 5). It is obvious from Fig. 15 that, since P_s is to be maintained over the whole area of deep suppression, the critical design parameter is the longest of the inter-cardinal distances between notches in adjacent rings. Coupled with this fact and because the roll-off of the MLP filter is governed by the fundamental response depicted in Fig. 1, the area covered in polarization space and the level of suppression achievable tend to be fixed for each ring. Thus, to cover a larger area, but maintain the value of P_s, one or more rings have to be added.

Fig. 15   Location of notches of MLP filter on the Poincaré sphere [20]

To demonstrate the effectiveness of the MLP filter relative to that of a single-notch filter located at LHCP, the 5-ring filter of Fig. 15 was evaluated in a deterministic manner for different conditions of target and clutter characteristics [21]. An improvement 'factor I_{MF} was defined as,

$$I_{MF} = \frac{\left[\dfrac{\text{target-plus-clutter signal power}}{\text{clutter signal power}}\right] \begin{array}{l}\text{at output of the}\\\text{MLP filter}\end{array}}{\left[\dfrac{\text{target-plus-clutter signal power}}{\text{clutter signal power}}\right] \begin{array}{l}\text{at output of the}\\\text{single-notch filter}\end{array}} \qquad (19)$$

and used to quantify the comparison. For simplicity the ellipticity ratio of the fixed clutter return (r_c) was chosen at the edge of the MLP filter (i.e., $r_c = r_f$) and that of the target return (r_t) was chosen outside of the area of deep suppression (i.e., $r_f < r_t \leqslant 1$).

Figure 16 shows I_{MF} versus r_t for a typical set of parameters. Values for the effective "insertion loss" of the MLP filter are also given. This loss is expressed by the ratio of total target-plus-clutter signal power at the output of the system to that at the input, and is an indicator of the usefulness of the processor with regard to signal detectability after processing. The results show that under the stated conditions a 5-ring MLP filter gives a performance in clutter suppression typically 20 dB better than a

single-notch filter, and is relatively insensitive to changes in
the input signal-to-clutter power ratio (S/C). However, it must

be remembered that the
5-ring MLP filter consists
of 91 notches, and there-
fore with currently-
available devices is only
realizable by a tremendous
amount of hardware. When
such a filter is to be
implemented, this hardware
factor will have to be
considered even though
difficult to quantify [20].

Fig. 16   Theoretical improvement of MLP filter over single-notch
          filter for different values of the ratio of target
          signal power and clutter signal power (S/C) at the input
          to the systems [21]

    (i)   MLP filter:- 5-ring filter centred about LHCP
                         (see Fig. 15)
                       - edge of filter r_f = -0.648
                       - average response in deep suppression
                         area is -39.13 dB

    (ii)   single-notch filter at LHCP

    (iii) ellipticity ratio of the clutter return r_c = -0.649

5.3 Linear Polarization-vector Translation (Linear PVT)

      The previous section stated that the location of a MLP filter
can be varied in an adaptive manner to coincide in polarization
space with closed sets of incident field samples to be suppressed.
Conversely, however, the filter's location can remain fixed and
the data sets translated to the area in polarization space that
it covers (to achieve the same end).   A suitable process to
perform this manipulation has been developed called linear polarization-vector translation (PVT) [19].   This is a vector-signal
processing concept which translates the polarization of a signal
from one state to another whilst conserving total power.

      A diagram illustrating the action of linear PVT in translat-
ing polarization states on the Poincaré sphere is shown in Fig. 17.
Three closed sets of incident field samples to be suppressed by
the MLP filter are shown.   The transformation operator for each
set is determined by:

    i)   either h_{min} or the average incident-field polarization
          of the set (since $\langle \underline{E}_i \rangle \cdot \underline{h}_{min}$=0, see Fig. 11); and

   ii) the polarization state corresponding to the fixed centre
       of the MLP filter, denoted by $(\underline{h}_{MLP})_{\perp}$.

This operator is a 2 x 2 unitary matrix (M) that has a de-
terminant $|M| = \pm 1$ (condition of power conservation), giving two
solutions: one for each of the two possible directions of
rotation.   A more detailed description of the mathematical
representation is given in [19], but the effectiveness of this
simplistic yet elegant process will become apparent in Chapter 7,
where it is applied to experimentally-derived rain data.

Fig. 17   Poincaré sphere representation of the linear PVT of closed
           sets of incident-field samples (to be suppressed by the
           MLP filter at a fixed location in polarization space) [19]

5.4 Non-linear Polarization-vector Translation (Non-linear PVT)

     It was pointed out in Section 5.2, that to cover only a moder-
ate circular area on the Poincaré sphere a large number of notches,
and hence a large amount of hardware, is required to make up a MLP
filter.   If this could be reduced, but the degree of suppression
maintained, target enhancement would be achieved for a more optimum, acceptable balance between the amount of hardware and the
operational benefits.   On this premise the requirement is for a
vector-process that

     i)   reduces the area of spread in polarization space of
            any data set to be suppressed;

     ii)   separates polarizations such that any difference
            between the polarization states of wanted and un-
            wanted signals can be further accentuated and
            exploited (at least when target and clutter are
            spatially separated); and

     iii)   is not restricted by the fixed roll-off of the basic
            response (Fig. 1).

A process that fulfils these requirements is discussed in [21] and is briefly described here.

The desired transfer function of the envisaged vector-process was such that the overall signal processing system would have an equivalent normalized power response of the form shown in Fig. 18. This gives control over the gradient of roll-off, and therefore the process is non-linear in nature. Because the process is still a PVT type, it has been termed the non-linear PVT process.

Fig. 18   Desired normalized power response of polarization suppression filter [21] with control of:

(a)   'size' of suppression area (transition $\hat{r}$)
(b)   average level of suppression
(c)   gradient of roll-off

It is assumed that any data sets to be processed by the non-linear PVT have already been manipulated by the linear PVT such that they are circularly distributed about LHCP (the centre of the MLP filter). This circular distribution means that the design of any process which compresses the set of unwanted samples in polarization space around LHCP need only be governed by the ellipticity ratio of the samples (independent of orientation angle). The non-linear PVT process therefore was designed to translate the ellipticity ratios of signals at the input of the system to:

values closer to LHCP when $-1 < (r_i)_{T_L} < \hat{r}$, and

values closer to RHCP when $\hat{r} < (r_i)_{T_L} < 1$.

It is obvious that $\hat{r}$ is a process control parameter which determines the transition

$$(r_i)_{T_L}, \ T_{NL} = (r_i)_{T_L},$$

where subscript T_L denotes the linear-PVT processed sample and subscript T_{NL} the non-linear-PVT processed sample. The

translation gradient of the process is controlled by a parameter m
(see [21]).   The characteristics of this process for m = 2 are
shown in Fig. 19.

Fig. 19   Theoretical $(r_i)_{T_L}$ to $(r_i)_{T_L, T_{NL}}$ translation characteristics of non-linear PVT process for translation gradient
parameter m = 2 and transition constraint parameter
$\hat{r} = 0(-0.2)-0.8$ [21]

A theoretical evaluation of the non-linear PVT combined with
a single-notch filter system, relative to the use of a single-notch
filter alone, was again performed in a deterministic manner similar
to that described in Section 5.2.   Improvement factors typical of
those that can be expected are shown for two representative sets of
parameters in Fig. 20.   The non-linear nature of the process means
its performance is heavily dependent upon (S/C).   These graphs show
that an improvement can be expected when using the process compared
to a single-notch filter alone, given that $(r_c)_{T_L} < (r_t)_{T_L} \leqslant 1$ and
(S/C) > -10 dB.   When (S/C) > -5 dB the improvement is roughly
20 dB, i.e. comparable to that obtained with the 5-ring MLP filter
of Section 5.2

Figures 20(a) and 20(b) suggest that a larger and larger improvement factor (especially for (S/C) ⩾ 0 dB) can be obtained if $\hat{r}$
is further and further increased, tending in the limit to 1.   However, this improvement is obtained at the expense of a reduction
in the level of target-plus-clutter signal power at the output of
the system.   This is shown in Fig. 20 by the "insertion loss" of
the system, from which it is clear that for fixed values of (S/C)
the area in polarization space for which the insertion loss is
less than the indicated levels (-5 dB, -10 dB and -20 dB) in each
case becomes smaller as $\hat{r}$ is increased towards 1.   This means
that although a significant improvement factor has been achieved,
the output from the system might become so low that in a practical
system target detection would become unreliable.

(a) $\hat{r} = -0.2$ (b) $\hat{r} = 0$

Fig. 20   Theoretical improvement factor of non-linear PVT pro-
          cessing combined with the single-notch filter over the
          single-notch filter alone, for different values of the
          ratio of target signal power to clutter signal power
          (S/C) at the input to the systems [21]

          (i)   non-linear PVT:-   translation gradient parameter
                                    m = 2
                                  - transition constraint parameter
                                    $\hat{r}$ = -0.2 and 0

          (ii)   single-notch filter at LHCP

          (iii) ellipticity ratio of clutter return $(r_c)_{T_L}$ = -0.649

5.5   A Typical VPA Processor

     The culmination of the three processes described in Sections
5.2 to 5.4 is the proposed VPA processor shown as a schematic
diagram in Fig. 21.   It would use the linear and non-linear PVTs
in conjunction with a MLP filter.   The vector-signals derived
from a dual-channel receiver would be fed in parallel to algorithms which derive the control parameters that best configure and
match the system to the data to be processed.

     Section 5.4 suggested that under specific conditions the use
of such a processor in conjunction with a single-notch filter
should give an improvement of about 20 dB in the wanted-to-
unwanted signal power ratio.   Thus, a further improvement should
be possible when a MLP filter is incorporated.   The application of
a 2-ring MLP filter would require typically 19 notches, which with
current technology probably represents an optimum compromise
between the detection performance in adverse environments and the
amount of hardware needed for realization.

Fig. 21   Schematic diagram of a VPA processor which uses a MLP
          filter in conjunction with the linear PVT and the non-
          linear PVT

6.   STC's EXPERIMENTAL X-BAND RADAR FACILITY

6.1 Major Features

The general requirements for a radar system with the ability
to measure the polarization characteristics of backscatter-returns/
interference have been discussed in the earlier sections of this
paper.   A block diagram of the system which has evolved at STC
to meet these needs is shown in Fig. 22.   It is a multi-purpose
X-band radar system which has two 6-ft diameter dish antennas, one
with a dual-polarized horn feed, the other a monopulse feed.   The
former transmits and receives
signals from which the experimental data is derived; the
latter operates only in a
receive mode and is used to
track targets.

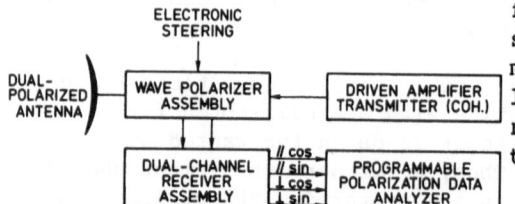

Fig. 22   Simplified block diagram of experimental X-band radar
          facility for polarization studies [18]

The antennas are mounted side-by-side on a 3-D pedestal which permits full steering in azimuth and elevation.   A photograph of the antenna system is shown in Fig. 23.

Fig. 23   The X-band antenna system; dual-polarized antenna and
          monopulse antenna mounted side-by-side on a 3-D pedestal

The heart of the system is the wave polarizer which allows the radar to operate with many different states of polarization. Figure 24 is a simplified block diagram of the "RF head", with the wave polarizer assembly and control.   This microwave portion of the system is housed in a box mounted directly behind the dual-polarized antenna.   At transmit the signal passes through a duplexer and is split equally in phase by a hybrid tee.   The two

Fig. 24   Block diagram of RF head with wave polarizer assembly
          and control

outputs are fed to the polarizer, in each arm of which there is a 4-bit ferrite phase shifter covering the range 0° to 90° in steps of $6\frac{3}{7}^{\circ}$. The phase-shifted signals are then cross-coupled by a hybrid sidewall coupler, and each of its two outputs connected to a 5-bit ferrite phase shifter. This has a range of 0° to 180°, again covered in steps of $6\frac{3}{7}^{\circ}$.

The final two outputs (horizontally-and vertically-polarized) feed an orthomode transducer which launches power into the dual-polarized horn feed. With this arrangement combinations of phase shift produced by the phase shifters result in weighted (in amplitude and relative phase) horizontally- and vertically-polarized components that recombine in the orthomode transducer and antenna. The result is a transmitted wave whose polarization state can be controlled.

The situation is similar at receive, except that the phase shifters are non-reciprocal devices. This can, however, be compensated at the control stage such that at the output of the hybrid tee two channels are available: one parallel-polarized with respect to the setting of the polarizer at transmit, the other orthogonally-polarized. These two channels, having passed through the duplexer and transmit/receive switches to achieve transmit/receive isolation, are amplified and down-converted to a 120-MHz IF. Subsequently, these signals are down converted to a second IF of 30 MHz and the video signals detected and digitized such that their phase and amplitude are obtained. This results in four 8-bit I and Q channels.

The principal characteristics of the facility are listed in Table 5. With the polarizer arrangement described, 302 different polarization states are implemented. The digital control of the polarizer is achieved via the console shown in Fig. 25. The design of this unit is based upon the modified polarization chart projection of the Poincaré sphere, with CP in the centre as described in Section 1.2. It effectively maps the 302 polarization states on to two charts, using a "shift-key" to switch between left- and right-handed states. Such a console gives a direct appreciation of the polarization state in which the system is operating. Also incorporated in the control unit are several pre-selected automatic modes of operation which switch between specific polarization states at variable rates up to 10 kHz. Several of these modes are required when deriving scattering matrix data from targets and clutter.

Table 5(a) The STC multipurpose experimental X-band radar facil-
            ity: features of the antenna system and the transmitter

---

ANTENNA SYSTEM

* Dual-polarized antenna : pencil-beam 3 dB beamwidth   ~1.2°

                         : max. sidelobe level   -25 dB

                         : gain   42 dB

                         : cross-polarization isolation > 25 dB

                         : controllable polarization

                             (a) manual   : 28 linear (LP)

                                            : 137 RH/LH elliptical
                                              (RHEP/LHEP)

                             (b) automatic : alternate HP/VP

                                            : alternate RHCP/LHCP

                                            : all LP/RHEP/LHEP in
                                              sequence

                                            : all LP/EP agile

* Tracking antenna   : monopulse feed

* 3-D pedestal   : antennas mounted side-by-side, allow-
                           ing target tracking and controllable
                           2-D/3-D scanning modes

TRANSMITTER

* Peak/average power   : 100 kW/600 W

* Frequency   : 9532 (4) 9568 MHz

* Pulsewidth   : 0.4 to 2.4 µs, variable

* Pulse repetition time : 0.1 (0.001) 10 ms

* Frequency modes   : fixed (diversity and agility to be
                           implemented in near-future)

Table 5(b) The STC multipurpose experimental X-band radar facility: features of the receiver and the polarization
data analyser/processor

RECEIVER

* Type : dual-channel (with orthogonally-
polarized channels)

* Video outputs : I and Q linear analogue and digitized 8-bit

: analogue logarithmic

POLARIZATION DATA ANALYSER/PROCESSOR

* Coherent and non-coherent processing

* Measurement window controllable in range and azimuth/elevation

Sampling rate : currently 650 kHz max. (5 MHz max.
to be implemented in near-future)

Window measurement : 0, 5, 10 or 20 s
repetition time

Number of sweeps in : 1(1)99
window

Samples in range : 8, 16 or 32

* PDP 11/34 minicomputer

Memory capacity : 256 kbytes

Disk storage : 60 Mbytes
capacity

Plotter and hardcopy outputs

Colour graphics facility for dynamic display of output

Fig. 25   STC developed 'polarization control console' (2nd
generation)

6.2   Some Early Experimental Results

By early 1977 the STC X-band radar facility had evolved to
the stage where the 30-MHz IF channel responses could be displayed
on an A-scope, and thus were available for preliminary investigations regarding the polarization dependence of received backscattered signals from radar objects.   Of principal interest for
the project at that stage was the "practical" confirmation of the
theory [7,9] stating that any single radar object (i.e., an object
of which the backscattering properties can be represented by a
scattering matrix) possesses so-called "characteristic polarizations".   These are defined as those polarizations of identical
transmit and receive antennas which produce maximum and minimum
(i.e. zero) responses at the receiver terminals;   in general for
any object there exists two characteristic minimum polarizations
and one characteristic maximum polarization.   At that time experimental data of this nature were very scarce in the open
literature (and currently still are).

Some experimental results have been published in [22].   As a
typical example, the responses for specific polarizations obtained
from an isolated boresight tower are presented in Fig. 26.   The
characteristic polarizations were obtained by trial and error using
the polarization control console of Fig. 25, and the responses
appeared to be stable over a prolonged period (many hours, depending on weather conditions).   It should be noted that, in contrast
to the maximum response, the minimum response is very sensitive to

DISTANCE
875 m

a change of antenna polarization. Furthermore, since the boresight tower represented a swaying multiple-component body, the back-scattered waves were partially-polarized and thus a zero response could not be obtained. Moreover it should be noted that since only discrete polarizations were available, deep nulling of the response in the parallel channel was limited, even when the incident field was completely polarized.

These results indicated that for man-made objects (such as bore-sight towers, buildings, etc.) the maximum to minimum response ratio in the parallel channel was between 15 and 30 dB.

Fig. 26   Boresight tower and responses at 30 MHz IF level for specific polarizations; the upper-trace presents the parallel-channel response and the lower-trace the orthogonal-channel response [22,17]

(a)   Horizontal polarization (reference)
(b)   Maximum polarization
(c)   Null-polarization 1
(d)   Null-polarization 2

It was concluded that the results were therefore encouraging enough to justify further investigations of the applicability of adaptive polarization for clutter cancellation and consequent target enhancement.

With the implementation in late 1979 of video channels (analogue and digital), a digital core store, and a limitedly-programmable desk calculator (as an intermediate solution to the inclusion of a PDP-11 minicomputer in 1982), the facility was capable of small-scale measurements and subsequent analysis of the data in order to derive object characteristic polarizations. Characteristic polarizations calculated from single-cell scattering-matrix measurements made at a fixed antenna bearing are presented on modified polarization charts in Figs. 27 and 28 for an isolated building and a foliage area respectively; they give an impression of how object polarization signatures change with increasing observation time.

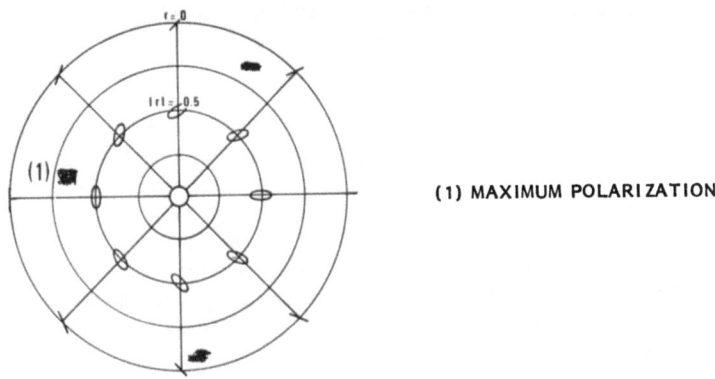

Modified Polarization Chart

(1) MAXIMUM POLARIZATION

Fig. 27 Characteristic polarizations of isolated apartment build-
         ing; 50 consecutive polarization forks, observation time
         0.99 s (transmit mode HP↔VP alternately, 100 Hz prf,
         2.0 μs pulse duration, 3.15 km sampling range)

Modified Polarization Chart

+ RIGHT-HANDED NULL POLARIZATION
- LEFT-HANDED NULL POLARIZATION
R-RIGHT-HANDED MAXIMUM POLARIZATION
L-LEFT-HANDED MAXIMUM POLARIZATION

Fig. 28 Characteristic polarizations of fluctuating clutter; slow
         and fast moving foliage (short and long tracks respec-
         tively), observation time T = 32 ms (fast) and T = 48 ms
         (slow), 2.5 KHz prf, 0.4 μs pulse duration [18]

        The results relating to the isolated building (Fig. 27) to-
gether with those for the boresight tower (Fig. 26) indicate that
unwanted returns from an urban area dominated by fixed objects
can be significantly suppressed by adaptive polarization.   It
should be noted that Fig. 27 also shows the mutual-coherent dual-
channel receiver system to be very stable in operation, at least
in the short-term sense, taking into account that the building
represents a multiple-component body.   The results for foliage
clutter in Fig. 28 [18] indicate that single-cell clutter-
suppression based on long-term averaged adaptive polarization

would result in only a moderate improvement.   However, if there
was also spatial polarization-correlation a useful amount of
suppression could probably be obtained via adaptive "window-
averaged" polarization processing.

    The general conclusion from these preliminary experiments
is that man-made objects and ground clutter exhibit distinctive
characteristic polarizations.   In the case of time-varying
clutter they are moving rapidly in polarization space, following
apparent characteristic "trajectories".   However, the exploitation of these phenomena for target discrimination, classification
and identification seems to be in its infancy and much more data
will have to be collected and analysed before an extensive range
of algorithms can be derived and evaluated.

7.   PERFORMANCE OF VECTOR-SIGNAL PROCESSING SCHEMES WITH
      EXPERIMENTALLY-DERIVED RAIN DATA

7.1   The Data Set

    To study the effectiveness and degree of suppression which
can be achieved with the signal processing schemes described in
Chapter 5, many closed sets of data have been collected using the
STC experimental X-band radar facility.   These sets were derived
from the backscatter radiation from rain illuminated with a RHCP
transmitted signal.   A plot of the incident field polarizations
of one typical set is shown in Fig. 29(a).   The transmitted signal
had a prf of 200 Hz with a pulse duration of 2 μs.   The data is
for 255 sweeps (hence the total observation time was approximately

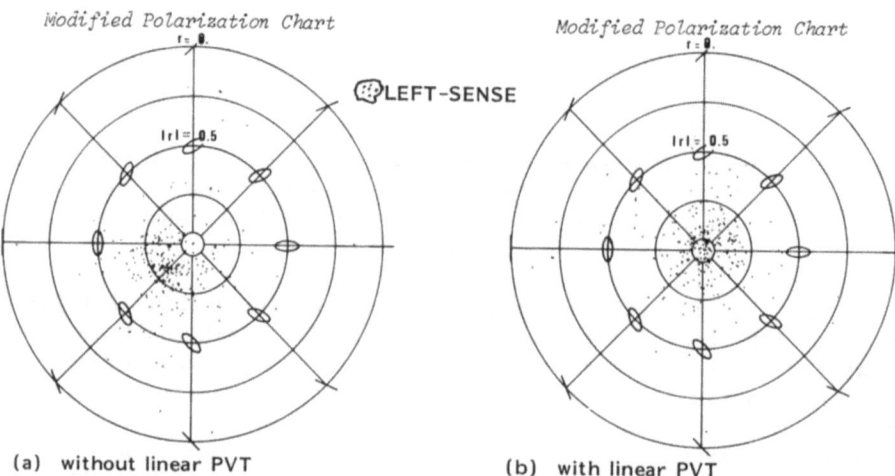

(a) without linear PVT   (b) with linear PVT

Fig. 29   Polarization state of returns from rain illuminated
          with RHCP; 255 consecutive single-sample time-windows
          (200 Hz prf, 2 μs pulse duration)

1.2 s) of a single range-cell located some 5 km from the radar, at an elevation angle of 17°.

The plots show that the polarization of the returns is predominantly left-handed (i.e., orthogonal to the polarization of the transmitted signal), but off-set from LHCP. Since this off-set is determined by environmental conditions and can be distributed anywhere around LHCP, the natural choice for the location of the centre of the suppression filter when attempting to suppress rain clutter is LHCP.

7.2 Configuration of Linear PVT plus MLP Filter

It is obvious from the plot of Fig. 29(a) that a sub-optimum value of suppression would be achieved if this data set were applied directly to only the MLP filter centred at LHCP, because of the circular layout of the notches and the off-set of the data from LHCP. The suppression can be increased by applying the linear PVT, whose controlling algorithm calculates the centroid of the set and translates this to LHCP with the data samples following accordingly. Figure 29(b) shows the centralizing effect of the linear PVT. The spread of the returns in polarization space is now more discernible and readily demonstrates the need for a filter with an area of deep suppression much wider than that of the antenna's response alone. It is also evident that for this data set the returns are approximately circularly distributed, with the majority contained in an area around LHCP and bounded by an ellipticity ratio of -0.7. Thus, "a priori", knowledge of the data set is in this case available, and so derivation of the filter parameters is easy.

Table 6 gives a summary of the performance which was achieved when processing this same data set with the linear PVT and a MLP filter of varying dimensions, based upon the typical design example of Section 5.2. The best obtainable improvement in suppression relative to the use of a single-notch filter alone at LHCP, was approximately 15 dB.

Table 6 Configuration of linear PVT and MLP filter; summary of results with the experimentally-derived data set from rain of Fig. 29(a)

Ellipticity ratio at edge of deep suppression of MLP filter (r_f)	Suppression due to use of linear PVT	Additional suppression due to the q-th ring					Increased suppression of 5-ring filter over single-notch filter at LHCP	Overall suppression relative to total received power
		q=1	q=2	q=3	q=4	q=5		
- 0.59	3.34 dB	3.20 dB	3.33 dB	2.26 dB	1.68 dB	1.34 dB	15.15 dB	32.69 dB

7.3 Configuration of the Linear PVT plus the Non-linear PVT and a Single-notch Filter

· The data set referred to in Section 7.1 was applied to a system using all three processes described in Chapter 5, but with only a single-notch filter for suppression.   The unprocessed data, having been centralized around LHCP by the linear PVT, were applied to the non-linear PVT with translation gradient parameter m of order 2 and different values of transition constraint parameter $\hat{r}$.   Figures 30(a) and (b) show two typical plots of the polarization states of the processed data for $\hat{r}$ = -0.8 and -0.2, respectively.   The nature of the non-linear PVT is well-demonstrated by Fig. 30(a). Because many of the returns lie outside the circle of ellipticity ratio -0.8, processing by the non-linear PVT when $\hat{r}$ = -0.8 translates them away from the single-notch filter; on the other hand, the returns inside this circle are compressed to LHCP.   In fact the overall effect for this data set is to reduce the degree of suppression, as is evident from Table 7.   However, as $\hat{r}$ is increased, more and more samples are translated towards LHCP, and therefore suppressed.   Figure 30(b), for $\hat{r}$ = -0.2, shows almost all the returns as being close to LHCP, and Table 7 gives the corresponding degrees of suppression.

For this particular data set, the linear PVT plus the non-linear PVT and the single-notch filter configuration gave an overall suppression, relative to that with a single-notch filter alone, of 16.8 dB.   This was approximately 1.6 dB more than was obtained with a linear PVT plus MLP filter configuration processing the same data set.

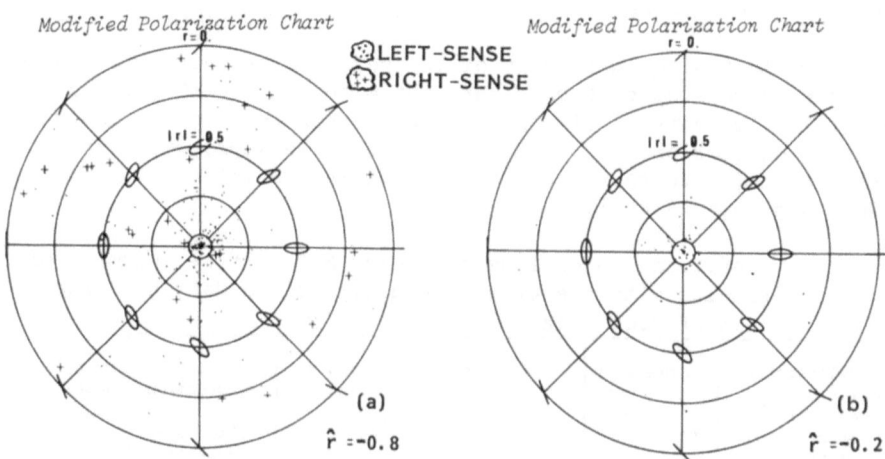

Fig. 30   Data samples of Fig. 29(b) after processing by non-linear PVT with translation gradient parameter m = 2

Table 7   Configuration of linear PVT plus non-linear PVT and a
single-notch filter; summary of results with the
experimentally derived data set from rain of Fig. 29(a)

Suppression due to use of linear PVT	Additional suppression due to use of non-linear PVT of order m = 2 for constraint values of $\hat{r}$ ranging from –0.8 to 0					Increased suppression over single-notch at LHCP; m = 2 and $\hat{r}$ = –0.2	Overall suppression relative to total received power; m = 2 and $\hat{r}$ = –0.2,
	–0.8	–0.6	–0.4	–0.2	0.0		
3.34 dB	–5.70 dB	2.09 dB	9.23 dB	13.43 dB	18.48 dB	16.77 dB	34.31 dB

7.4   Discussion of Processor Applications

It should be emphasised that in the application related to
the detection of targets in an environment containing clutter
and/or other unwanted signals, it is the ratio of average target-
signal power to average unwanted-signal power which is important.
If the polarizations of the wanted returns (i.e., those from a
target) are close to those of the unwanted, the target signals
will also be suppressed, and hence little improvement would result.

It was stated in Chapter 2 that approximately 3 dB loss in
target detectability can be expected when using fixed circular
polarization in a clear and quiet environment, relative to the
use of fixed horizontal polarization.   In conjunction with
Fig. 1 this fact suggests that the returns from targets are, on
average, linearly polarized. It follows also from Fig. 14 that
target returns then suffer an additional loss when processed by
the MLP filter centred about LHCP, and this loss increases with
increasing number of rings, to reach about 2.3 dB for the 5-ring
system of Section 5.2.   Hence, the improvement-in-suppression
figures in Table 6 should be reconsidered taking this additional
loss into account; it follows then that if the target returns
are, on the average, linearly polarized, an improvement of
approximately 13 dB can be expected in target-to-rain clutter
power ratio, relative to the use of a single-notch filter at
LHCP.

In the non-linear PVT case the situation is more complex.
The evaluation in [21] indicates that in a target-plus-rain
environment where the signal-to-clutter ratio (S/C) > –5 dB, the
effectiveness of a system using non-linear PVT characterized by

$\hat{r} = -0.2$ and m = 2 plus a single-notch filter is of the same order as that of the MLP filter, but incurring an extra "insertion loss" of approximately 5 dB.   However, this performance can be improved by selecting $\hat{r} > -0.2$.   Furthermore, in the less problematic environment where the target and the rain are spatially separated, the non-linear PVT incurs minimal "insertion loss", because the target returns are more distinguishable from the unwanted returns due to the separating action of the non-linear PVT, where the degree of separation is governed by $\hat{r}$ and m.

Since it is reasonable to expect a real-time adaptive system to react in a much shorter time than the observation periods discussed here (Fig. 29), the effective spread of the returns could well be less, and a larger overall improvement in suppression could be obtained with the processing schemes discussed in Chapter 5. Moreover, a smaller spread can mean reductions in the large amount of hardware and the lengthy processing times involved.

The realization of a 61-notch MLP filter designed to operate in real-time already involves a considerable amount of hardware, whereas the non-linear PVT approach requires less hardware and can achieve approximately the same performance as an equivalent MLP filter when (S/C) > -5 dB.   However, the use of the non-linear PVT in conjunction with, say a 2-ring MLP filter, may result in an even better performance (see Section 5.5).

Encouraged by the results of these studies [19,20,21], preliminary investigations have been made regarding the realization of real-time practical processors based upon the MLP filter, linear PVT, and non-linear PVT.

8.   FUTURE STUDIES

STC's future research work related to the utilization of polarization information in primary radar will be devoted to such of the following subjects as resources permit:

(a)   The large scale measurement of complete scattering matrix data of aircraft targets-of-opportunity, ground clutter, and weather clutter; these measurements will be used to assess the statistics of the object's polarization signatures at fixed and multiple frequencies.

(b)   The development of statistical models for target/ clutter backscattering properties in terms of polarization, from measured data.

(c)   Off-line analysis (using large-scale live data) of the effectiveness of the developed vector-signal processing algorithms when used for target enhancement in precipitation clutter.

(d)   Theoretical assessment of the effectiveness of the developed vector-signal processing algorithms when used for target enhancement in a random dipole cloud.

(e)   Feasibility studies related to the real-time implementation in primary radar systems of the vector-signal processing algorithms developed.

(f)   Development of algorithms for the adaptation of the transmit polarization in a VPA processor, and suitable for real-time implementation.

(g)   Assessment (using live data) of the compatibility of frequency agility and vector-signal processing in a weather clutter environment.

(h)   Off-line analysis (using live precipitation data) of the applicability/effectiveness of combined frequency-domain and polarization-domain signal processing.

(i)   Assessment (using live data) of the applicability of ortho-coherent MTI-like signal processing (see Fig. 31) for target enhancement [17,18].

Fig. 31   Simplified block diagram of ortho-coherent MTI processor [18]

(j)   Investigation of the applicability of the utilization of 'invariant' properties of radar object scattering matrices [23] in vector-signal processing algorithms for target enhancement.

9.   ACKNOWLEDGEMENT

     The authors would like to take this opportunity to acknow-
ledge their colleagues who have been, and still are, involved with
this project.   These are: Mr. G.J. de Zwart, Mr. K.W.P. Krom,
Mr. L. van Slingerland, Mr. A.J. Lipman, and Mr. J.P. van der Voort,
all of whom provided the technical support, and the members of the
Drawing Section and the Workshop Section.   Further, thanks are
expressed for the secretarial assistance and to colleagues in the
Editorial Services Branch, in particular Mr. R. Cameron (who has
been involved with the editing of the majority of the publications
related to this project) and Miss R. Brown who typed the final
version of this presentation.

10.   REFERENCES

[1]   BOERNER, W.M., EL-ARINI, M.B., CHAN, C.J., and MASTORIS, P.M.:
      'Polarization dependence in electromagnetic inverse problems',
      IEEE Trans., 1981, AP-29, 2, pp. 262-271

[2]   WHITE, C.D.: 'Circular radar cuts rain clutter', Electronics,
      1954, 27, pp. 158-160

[3]   SKOLNIK, M.I.: 'Radar Handbook', McGraw-Hill, 1970, Section
      28.4

[4]   SINCLAIR, G.: 'The transmission and reception of elliptically
      polarized waves', Proc. IRE, 1950, 38, pp. 148-151

[5]   DESCHAMPS, G.A.: 'Geometrical representation of the polari-
      zation of a plane electromagnetic wave', Proc. IRE, 1951, 39
      pp. 540-544

[6]   BORN, M., and WOLF E.: 'Principles of optics', Pergamon
      Press, New York, 1965 (3rd edition), Section 10.8

[7]   KENNAUGH, E.M.: 'Polarization properties of radar reflec-
      tions', Antenna Lab, The Ohio State University Research
      Foundation, Columbus, Rept. 389-12 (AD 2494), March 1952,
      RADC Contract AF28(099)-90

[8]   GRAVES, C.D.: 'Radar polarization power scattering matrix',
      Proc. IRE, 1956, 44, pp. 248-252

[9]   HUYNEN, J.R.: 'A new approach to radar cross-section
      measurements', IRE Internat. Conf. Record, Pt. 5, 10, pp. 3-
      11.

[10] HUYNEN, J.R.: 'Measurement of the target-scattering matrix',
      Proc. IEEE, 1965, 53, pp. 936-946

[11]   POELMAN, A.J.:   'Performance evaluation of two types of
       radar system having a circular-polarization facility',
       SHAPE Technical Centre, The Hague, The Netherlands,
       Technical Memorandum TM-276 (AD 883 960), April 1971
       (NATO Unclassified, limited distribution)

[12]   POELMAN, A.J. and TIMMERS, F.L.:   'Detection performance
       of an incoherent radar with two orthogonally-polarized
       receiving channels', SHAPE Technical Centre, The Hague,
       The Netherlands, Technical Memorandum TM-435 (AD 921 015),
       June 1974 (NATO Unclassified, limited distribution)

[13]   POELMAN, A.J.:   'On using orthogonally-polarized receiving
       channels to detect target echoes in gaussian noise', IEEE
       Trans., 1975, AES-11, 4, pp. 660-663

[14]   POELMAN, A.J.., and van der VOORT, J.I.:   'The polarization
       dependence of received back-scattered power from a random
       dipole cloud', SHAPE Technical Centre, The Hague, The
       Netherlands, Technical Memorandum TM-335 (AD 520 934),
       May 1972 (NATO Unclassified, limited distribution)

[15]   POELMAN, A.J.:   'Cross-correlation of orthogonally-
       polarized backscatter components', IEEE Trans., 1976,
       AES-12, 6, pp. 674-682

[16]   POELMAN, A.J.:   'Reconsideration of the target detection
       criteria based on adaptive antenna polarizations', AGARD
       Conf. Proc. No. 197 on new devices, techniques and systems
       in radar, February 1977

[17]   POELMAN, A.J.:   'The applicability of controllable antenna
       polarizations to radar systems', Tijdschrift van het
       Nederlands Elektronica- en Radiogenootschap (The Nether-
       lands), 1979, 44, pp. 93-106 (in English)

[18]   POELMAN, A.J.:   'Virtual polarization adaptation, a method
       of increasing the detection capability of a radar system
       through polarization-vector processing', Proc. IEE, Pt. F,
       1981, 128, 5, pp. 261-270.   (The substance of this paper
       was first presented at the Military Microwave '80 Confer-
       ence held in London in October 1980 under the title 'A
       study of controllable polarization applied to radar'.)

[19]   POELMAN, A.J.:   'Polarization vector translation in radar
       systems', Proc. IEE, Pt. F. 1983, 130, 2, pp. 161-165

[20] POELMAN, A.J., and GUY, J.R.F.: 'Multinotch logic-product
      polarization suppression filters; a typical theoretical
      design example and its performance using live rain data',
      submitted to Proc. IEE, Pt. F, in March 1983; accepted
      September 1983

[21] POELMAN, A.J., and GUY, J.R.F.: 'Non-linear polarization
      vector translation in radar systems; a promising concept
      for real-time polarization-vector signal processing via a
      single-notch suppression filter', submitted to Proc. IEE,
      Pt. F, in April 1983; accepted September 1983

[22] POELMAN, A.J.: 'Complex radar objects and polarization
      dependence of received backscattered echoes', Elec. Lett.
      (UK), 1977, 13, 15, pp. 433-434

[23] BICKEL, S.H.: 'Some invariant properties of the polari-
      zation scattering matrix', Proc. IEEE, 1965, 53, 8, pp.
      1070-1072

III.3 (MM.2)

POLARIZATION MEASUREMENTS IN RADIOASTRONOMY: DETERMINATION
OF THE POLARIZATION - REFLECTION - MATRIX

Manfred A.F. Thiel

Angewandte Mathematik, Maschinenbauwesen
Fachhochschule Köln, Am Reitweg 1
D-5000 KOLN 1, FR Germany

Synopsis: The following represents a summary of the paper presented by Prof. Manfred Thiel during the Workshop (MM.2) which
in great detail is contained in an excellent monograph (Thiel,
1970) containing the findings of his diploma thesis (1968).

Determination of the Polarization - Reflection - Matrix

When electromagnetic radiation is reflected by some device, its
state of polarization is, in general, changed.   In the quasi-
monochromatic approximation, this can be described (Thiel, 1968)
by a linear transformation, the operator of which is chatacter-
istic for the device and does not depend on the incident radiation.

Let us represent the state of polarization by the Stokes vector $\vec{s}$,
unprimed for the incident state and primed for the transformation
state.   Then we have:

$$\vec{s}' = [T] \cdot \vec{s}$$

In order to determine the transformation matrix $[T]$, we have to

573

W.-M. Boerner et al. (eds.), Inverse Methods in Electromagnetic Imaging - Part 1, 573–575.
© 1985 by D. Reidel Publishing Company.

measure four (4) different test states (Thiel, 1970), which we
shall choose to be linearly independent.

$$
[\overset{1}{\vec{s}'}, \overset{2}{\vec{s}'}, \overset{3}{\vec{s}'}, \overset{4}{\vec{s}'}] = [T] \cdot [\overset{1}{\vec{s}}, \overset{2}{\vec{s}}, \overset{3}{\vec{s}}, \overset{4}{\vec{s}}] \quad ,
$$

the matrix $[T]$ being the same for each transformation.

Combining formally the different Stokes-vectors to one 4x4 matrix,

$$
[s'] \equiv [\overset{1}{\vec{s}'}, \overset{2}{\vec{s}'}, \overset{3}{\vec{s}'}, \overset{4}{\vec{s}'}] \text{ and } [s] \equiv [T] \cdot [\overset{1}{\vec{s}}, \overset{2}{\vec{s}}, \overset{3}{\vec{s}}, \overset{4}{\vec{s}}],
$$

we have

$$
[\vec{s}] = [T] \cdot [s] \quad .
$$

Let us now compute the inverse matrix $[s]^{-1}$ of matrix $[s]$. This
is possible, because its column-vectors are linearly independent,
as we have chosen above.

Now, right-hand-multiplication with the inverse matrix $[s]^{-1}$
yields the transformation matrix:

$$
[T] = [s'] \cdot [s]^{-1}
$$

The transformation matrix $[T]$ is the matrix product of the matrix
$[s']$, containing the transformed states, and of matrix $[s]^{-1}$,
which is the inverse matrix $[s]$, containing the incident states as

column-vectors.

Acknowledgement: The author wishes to express his gratitude to
Prof. Wolfgang-M. Boerner for inviting him to this inspiring
Workshop, for providing the opportunity of presenting his paper,
and for accepting this summary for publication in the Proceedings.

References

M.A.F. Thiel, Ein Polarisationsmodulator und seine Anwendung in
einem Radioastronomischen Polarimeter, Diplomarbeit, 1968,
Max-Planck-Institut für Radioastronomie, Auf dem Hügel 69, D-53
Bonn 1, FR Germany.

M.A.F. Thiel, Darstellungs- und Transformationstheorie Quasi-Mono-
chromatischer Strahlungsfelder, Beiträge zur Radioastronomie, Max-
Planck-Institut für Radioastronomie, Bonn, Band I, Heft 5, März
1970.

M.A.F. Thiel, Error Calculation of Polarization Measurements,
Max-Planck-Gesellschraft zür Förderung der Wissenschaften, e.V.,
MPIFR, Bonn, Sonderdruck Nr. 87, Ser.A, June 23, 1975 (also see:
J. Opt. Soc. of Amer. Vol. 66(1), pp. 65-67, Jan. 1976.

III.4 (RS.4)

POLARIZATION-DEPENDENCE IN MICROWAVE RADIOMETRY

K. Grüner

Deutsche Forschungs- und Versuchsanstalt
für Luft- und Raumfahrt
Institut für Hochfrequenztechnik
8031 Oberpfaffenhofen, FRG

ABSTRACT

Polarization plays more and more an important role also in the
field of microwave radiometry. It helps to find new signatures
and to improve contrast or image quality. So ambiguity proplems
can be solved in a better way. Disadvantages because of worse
angle resolution in comparison to passive remote sensing methods
in the IR- and optical region are partly compensated.

At first after some fundamentals a series of examples obtained by
stationary measurements at 32 GHz and 90 GHz and airborne measurements at 32 GHz demonstrates the influence of polarization on
natural microwave radiation. Various media or objects were observed
at vertical and horizontal polarization for variing nadir angles.
Then some remarks with respect to the complete state of polarization
of the natural microwave radiation are made. It turns out, that
also the circular polarization should be taken into consideration.
Finally a proposal for a correlation measuring method is made
which in connection with polarization switching could contribute
to a better solution of the inverse radiation problem involved
in microwave radiometry.

1. INTRODUCTION

The principle objetives of microwave radiometry are the detection
of natural thermal microwave radiation of matter. It may therefore
be considered as consistent development of passive remote sensing
methods, successfully established in the optical and IR region,
towards the cm- and mm wavelength region of the electromagnetic
spectrum. By the transition to longer wavelengths disadvantages

W.-M. Boerner et al. (eds.), Inverse Methods in Electromagnetic Imaging - Part 1, 577–597.
© *1985 by D. Reidel Publishing Company.*

and advantages arise. The main disadvantages may be the essential deterioration of the angular resolution and much smaller frequency bandwidths which to a high degree determine the achievable data rate for an assumed detection sensitivity or signal to noise ratio. But because of a minor weather sensitivity of microwave radiation, a higher penetration depth and an extraordinary variety of contrast in the microwave region, microwave radiometry nevertheless has gained great interest for many applications.

The variety of contrast is the consequence of strong variations of natural microwave radiation caused by strong variations of ground reflectivity and a background (sky) radiation decreasing very hard with decreasing zenith angle especially at lower micro-wave frequencies. Thereby the reflectivity mostly not only depends on the material but also on the look angle of the radiometer and the polarization of the received radiation. These latter charac-teristics, not so significant in the IR- and the optical region, appear because many surfaces are smooth in the sense of micro-waves. By that way further independent informations with respect to an object can be gained, very important when contrast must be improved to solve ambiguity problems or new signatures are looked for.

In the following after some fundamentals a series of examples is given which shows the polarization dependence of different media for vertical and horizontal polarization.

2. FUNDAMENTALS AND EXAMPLES FOR THE POLARIZATION-DEPENDENCE

The radiation received by a radiometer from a given object is composed of a part of emitted thermal radiation and a part of reflected or scattered radiation originating from the surroundings. Usually a "brightness-temperature" (T_B) - more correct an "apparent brightness-temperature" - is attached to the received power.

For the most simple case of a homogeneous semiinfinite medium with an uniform temperature distribution bounded by a horizontal specular surface (Fig. 1) we have:

$$T_B(\Theta_n) = e(\Theta_r) \, T_{PHYS} + r(\Theta_i) \, T_{SKY}(\Theta_i) \qquad (1)$$

where:

T_B : brightness-temperature of the medium as a function of the angle of observation Θ_n (nadir angle)

T_{PHYS}: true temperature of the medium

T_{SKY} : brightness-temperature of the sky as a function of the angle of incidence $\Theta_i = \Theta_n$

$e(\Theta_r)$: emission coefficient of the medium as a function of the angle of reflection $\Theta_r = \Theta_i$

$r(\Theta_i)$: power reflection coefficient of the medium as a function of the angle of incidence.

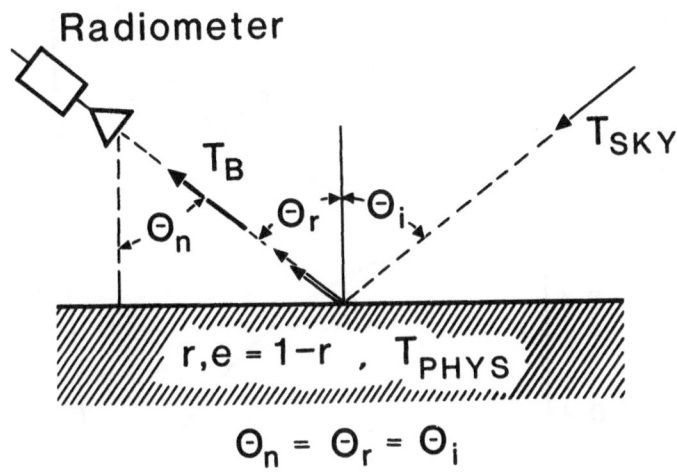

Figure 1. Sketch of the measuring configuration.

The first product at the right side of equation (1) corresponds to the equivalent blackbody temperature T_{EMIT} of the medium at a lookangle of $\Theta_n = \Theta_r$, and the second to a reflected temperature T_{REFL} which is the equivalent blackbody temperature of the sky at an angle of incidence Θ_i weighted by the power reflection factor for the same angle. By second law of thermodynamics a simple relation between the emission coefficient and the power reflection coefficient exists:

$$e(\Theta_r) = 1 - r(\Theta_i). \tag{2}$$

Thus equation (1) becomes:

$$T_B(\Theta_n) = T_{PHYS} + r(\Theta_i)\,(T_{SKY}(\Theta_i) - T_{PHYS}). \tag{3}$$

This formula is valid for either horizontal or vertical polarization whereby only $r(\Theta_i)$ and not $T_{SKY}(\Theta_i)$ shows normally a polarization-dependence. The radiation emitted by the atmosphere is said to be randomly polarized that means unpolarized.

In Fig. 2 the reflection coefficient for a water surface for
horizontal and vertical polarization – electric field perpendicular
and parallel respectively to the plane of incidence – is given as
a function of the angle of incidence.

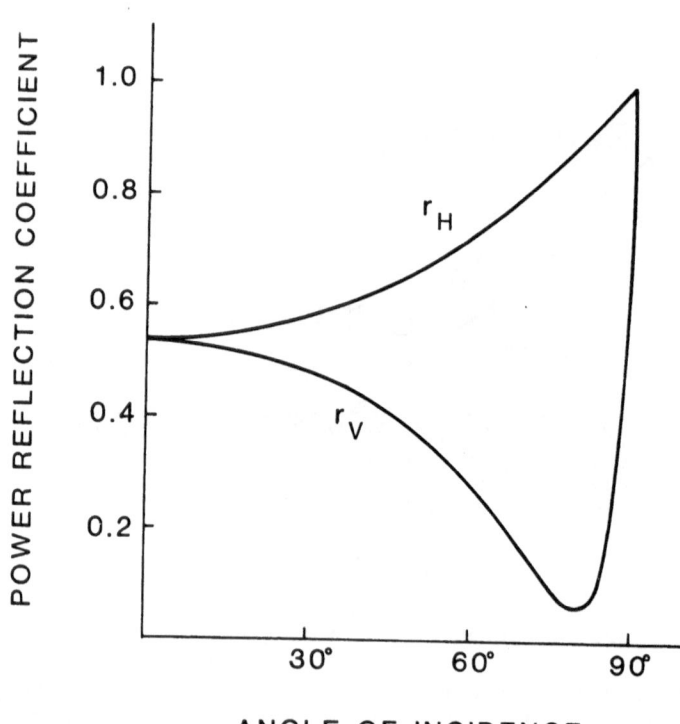

Figure 2. 32 GHz power reflection coefficient for a
water surface as a function of the angle of incidence
for horizontal and vertical polarization
theoretical values, rel. permittivity 17–j27,
water–temperature 10° C.

At 0° a value of 0.54 is obtained for both polarizations. Increasing the angle of incidence the curve for horizontal polarization
rises continuously until a value of 1 reached, while the curve for
vertical polarization decreases at first. At the Brewster angle
(≈ 80°) a minima is obtained and then a steep ascent also to a
value of 1 can be registrated.

The characteristics obtained by evaluation of the wellknown
Fresnels equation (4a) and (4b), [1] are typical for homogeneous
dielectric media (relative permittivity $\varepsilon = \varepsilon_r - j\,\varepsilon_i$)

$$r_v(\theta_i) = \left| \frac{\varepsilon \cos \theta_i - \sqrt{\varepsilon - \sin^2 \theta_i}}{\varepsilon \cos \theta_i + \sqrt{\varepsilon - \sin^2 \theta_i}} \right|^2 \tag{4a}$$

$$r_h(\theta_i) = \left| \frac{\cos \theta_i - \sqrt{\varepsilon - \sin^2 \theta_i}}{\cos \theta_i + \sqrt{\varepsilon - \sin^2 \theta_i}} \right|^2 \tag{4b}$$

The Brewster angle is influenced in the main by the real part of the relative permittivity of the medium and increases with an increasing value. No value below $45°$ is observed.

When we use the characteristics of Fig. 2 for the evaluation of eq. (1) curves are obtained for T_{EMIT} and T_{REFL} as shown in Fig. 3 (solid- and dashed lines). A clear atmosphere and T_{PHYS} equal 293 K are assumed.

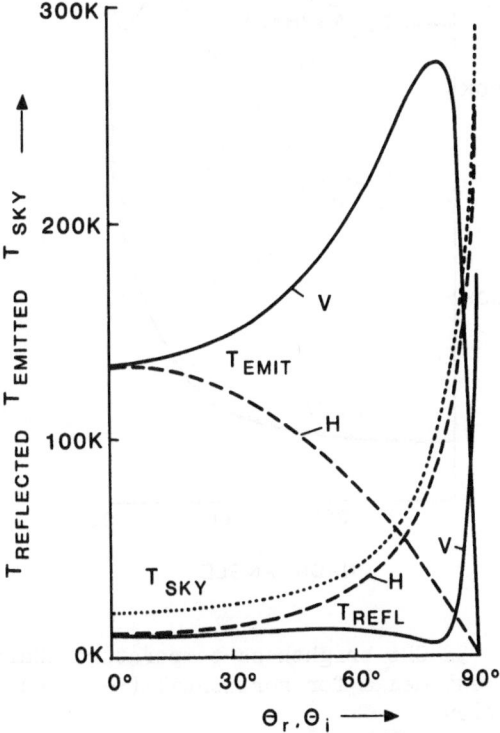

Figure 3. Brightness-temperatures at 32 GHz attached to the emitted- and the reflected radiation for a water surface and the sky radiation.

The brightness-temperature of the sky was measured at 500 m above sea level.

The polarization-dependences for both, the emitted and the reflected part of the radiation can clearly be detected, whereby T_{EMIT}, which shows reverse behaviour in comparison to Fig. 2 because of condition (2), dominates for angles up to 72° for horizontal and 88° for vertical polarization. This phenomenon depends on the background conditions however.

The radiometer measures the sum of T_{EMIT} and T_{REFL}. It is represented together with the (apparent) brightness-temperature characteristics of othder media in Figure 4.

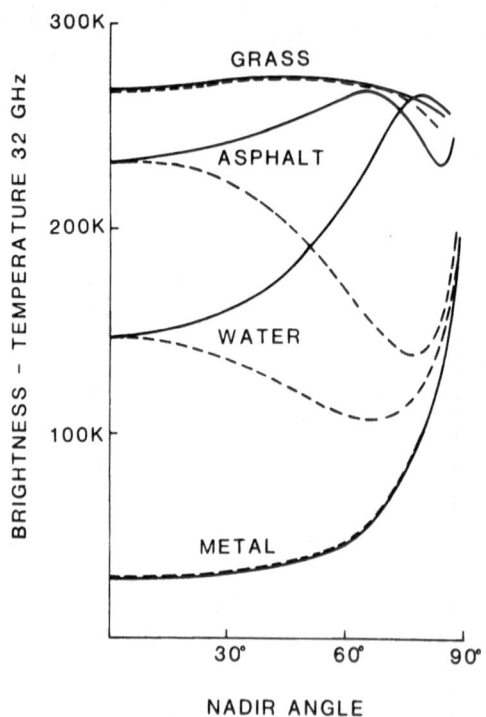

Figure 4. 32 GHz brightness-temperature characteristics of different media for horizontal (---) and vertical (——) polarization
measurements DFVLR 1980.

Discussing the characteristics for water first, we again observe the typical divergence of the characteristics for horizontal and vertical polarization, whereby the brightness-temperature for ver-

tical polarization increases with increasing nadir angle, until
near the Brewster angle a maximum is reached. On the other side
the brightness-temperature for horizontal polarization decreases.
The steep ascent above 70^0 is effected by the strong rise of the
skytemperature near the horizon. The brightness-temperature behaviour of an ideal specular reflecting metal surface also represented in Fig. 4 corresponds to that of the sky and vice versa.
No polarization effect occurs.

For an asphalt surface principally similiar curves as for water
are obtained but at a higher temperaturelevel. The maximum for
vertical polarization is already reached for 65^0.

The different behaviours for vertical and horizontal polarization
disappear more and more by increasing heterogeneity or losses of
a medium or increasing roughness of its surface, characteristics
which must be partly seen relatively to wavelength yet as already
stated above. The differences disappear in our case almost totally
for a media as grass. For inhomogenious media or rough surfaces
equation (1) must be replaced by a more complicate integral-
expression [2].

Still four general comments should be given with respect to Fig. 4
important for ground mapping in the mm wavelength region.

-    An increase of the dynamic range at the radiometer output for
     natural surfaces can be obtained by measuring the horizontally
     polarized component at higher nadir angles.

-    The media specified in Figure 4 are typical representatives
     of four classes which are observed in the main. a: metallic
     surfaces; b: water, wet smooth surfaces; c: asphalt, concrete,
     gravelly soil, wet rough surfaces; d: vegetation, rough dry
     soil.

-    At a nadir angle of about 50^0 the contrast between these
     classes is linearized for horizontal polarization by which
     the image quality can be improved as long as the ground reso-
     lution is still sufficient. For vertical polarization we
     obtain worse results.

-    Ambiguity problems because of different background conditions
     and insufficient ground resolution might be solved by polari-
     zation switching at certain nadir angles.

In Fig. 5 passive 32 GHz terrain mapping at nadir angles of 45^0,
60^0 and 78^0 for vertical and horizontal polarization is demonstrated.

Figure 5. 32 GHz images of the same scene taken at different nadir angles and for vertical and horizontal polarization, sketch of the scene at the left-hand side.

The same scene was overflown with a linescanner at short intervals. The scene contains roads, fields, meadows, small forests,
single trees, small areas with a snow cover and buildings (see
sketch at the left-hand side). Black corresponds to brightness-
temperatures near T_{PHYS}, white to lower brightness-temperatures
(220 K at 45°, 210 K at 60° and 210 K/190 K at 78° for verti-
cal/horizontal polarization). The statements given above are confirmed. The dynamic range increases with increasing nadir angle;
a higher contrast is obtained for horizontal polarization; roads
for instance can much better be detected at this polarization;
small areas with snow cover are much better detected at 60° and
78° although the ground resolution became worse; at low depression
angles for horizontal polarization distinct signatures for single
trees are obtained (black lines).

This short discussion of Fig. 5 already proves the extraordinary
importance of polarization in the field of passive microwave imaging.

We now return to further examples of stationary measurements. Fig.6
still shows an example where the typical characteristics obtained for water or asphalt respectively are no longer observed.
The diagrams were obtained for a dry hard frozen snow layer of
about 10 - 15 cm thickness on a grassy soil.

By reflection processes at the lower boundary interferences occur.
Probably also (volume) scattering at the snow crystals happens,
which would be an explanation for the low brightness temperatures
at 90 GHz but no polarization dependences. This example gives a
feeling for the complex radiation behaviour of already three layer
dielectric media. In comparison to the Fresnels formulas much more
complicate expressions are obtained. When we assume a layer of
thickness d (rel. permittivity = ε_2) bounded by two semiinfinite
media (rel. permittivity $\varepsilon_1 = 1$ (air), ε_3) then the formulas (5a)
and (5b) are obtained for the effective power reflection coeffi-
cient at the boundary between media 1 and 2:

$$r_v(\theta_i) = |-\frac{Z_{in,v} - \cos\theta_i}{Z_{in,v} + \cos\theta_i}|^2 \tag{5a}$$

and

$$r_h(\theta_i) = |\frac{Z_{in,h} - \sec\theta_i}{Z_{in,h} + \sec\theta_i}|^2 \tag{5b}$$

with

$$Z_{in,v} = \frac{\varepsilon_2'}{\varepsilon_2} \frac{1 - \frac{\varepsilon_2 \varepsilon_3' - \varepsilon_3 \varepsilon_2'}{\varepsilon_2 \varepsilon_3' + \varepsilon_3 \varepsilon_2'} e^{-j\, 4\pi/\lambda_o\, \varepsilon_2/\varepsilon_2'\, d}}{1 + \frac{\varepsilon_2 \varepsilon_3' - \varepsilon_3 \varepsilon_2'}{\varepsilon_2 \varepsilon_3' + \varepsilon_3 \varepsilon_2'} e^{-j\, 4\pi/\lambda_o\, \varepsilon_2/\varepsilon_2'\, d}}$$

$$Z_{in,h} = \frac{1}{\varepsilon_2'} \frac{1 + \dfrac{\varepsilon_2' - \varepsilon_3'}{\varepsilon_2' + \varepsilon_3'} e^{-j\, 4\pi/\lambda_o\, \varepsilon_2/\varepsilon_2'\, d}}{1 - \dfrac{\varepsilon_2' - \varepsilon_3'}{\varepsilon_2' + \varepsilon_3'} e^{-j\, 4\pi/\lambda_o\, \varepsilon_2/\varepsilon_2'\, d}}$$

where

$$\varepsilon_2' = \sqrt{\varepsilon_2 - \sin^2\theta_i} \quad , \quad \varepsilon_3' = \sqrt{\varepsilon_3 - \sin^2\theta_i} \quad .$$

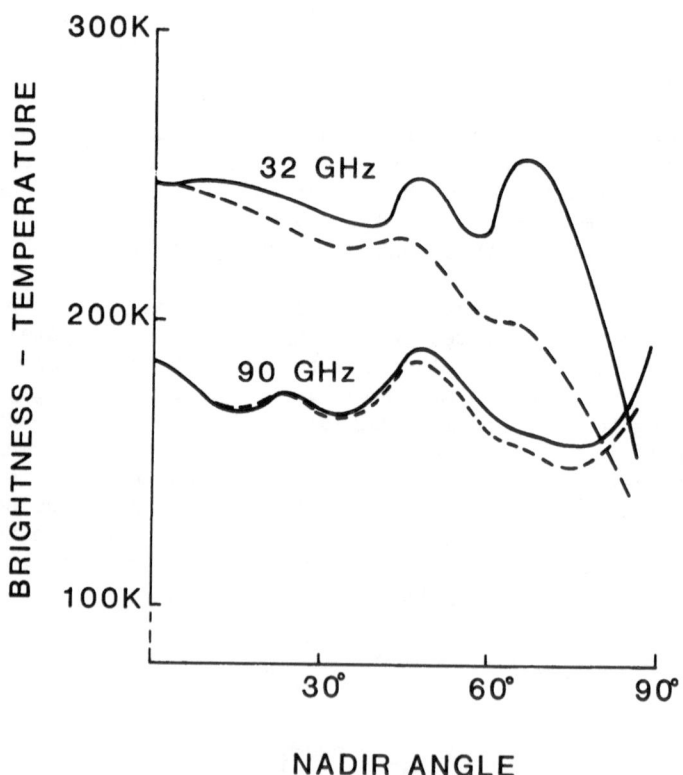

Figure 6. 32 GHz and 90 GHz brightness-temperature characteristics for a snow layer on a meadow for horizontal (---) and vertical (——) polarization
snow thickness 10-15 cm, snow hard frozen and dry,
$T_{PHYS} = -6^o$ C, measurement DFVLR 1982.

The wavelength of the incident wave is given by λ_o. For the case of an uniform temperature profile the expressions (5a) and (5b) can immediately be used together with equation (3) for the computation of the brightness-temperatures.

Another method for the representation of measurements for horizontal and vertical polarization is chosen in Fig. 7.

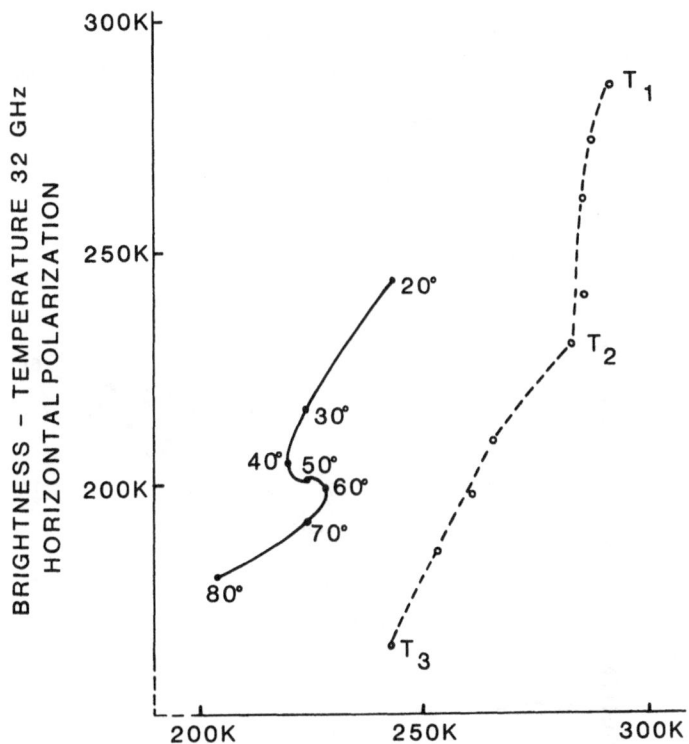

Figure 7. 32 GHz signatures of different objects by brightness-temperature measurements at horizontal and vertical polarization —— snow layer on a meadow, thickness ~10 cm, atmospheric temperature -3 $^\circ$C, variation of nadir angle (20° - 80°), --- foam on a water surface, variation of time, multilayers of bubbles at T_1, single layer at T_2, pure water surface at T_3.

The results are drawn in a plane defined by the brightness-temperatures for the two orthogonal polarizations. The diagramm at the left-hand side was obtained looking again to a layer of dry but now fresh snow (thickness ~ 10 cm) on a meadow at different nadir angles. The characteristic at the right-hand side was obtained during a period $T_1 - T_2 - T_3$ looking at a constant nadir angle of $68°$ to a water surface with foam on it. At the time T_1 we had multilayers of air-bubbles. In the following, the bubbles bursted continuously until at the time T_2 only a single layer covered the water surface. Afterwards also this single layer disappeared. The pure water surface was observed at the time T_3.

Both examples show very distinctly the different behaviour for vertical and horizontal polarization. Characteristic signatures are obtained which confirm once more the extraordinary role of polarization for microwave radiometry.

A last example is finally given in Fig. 8.

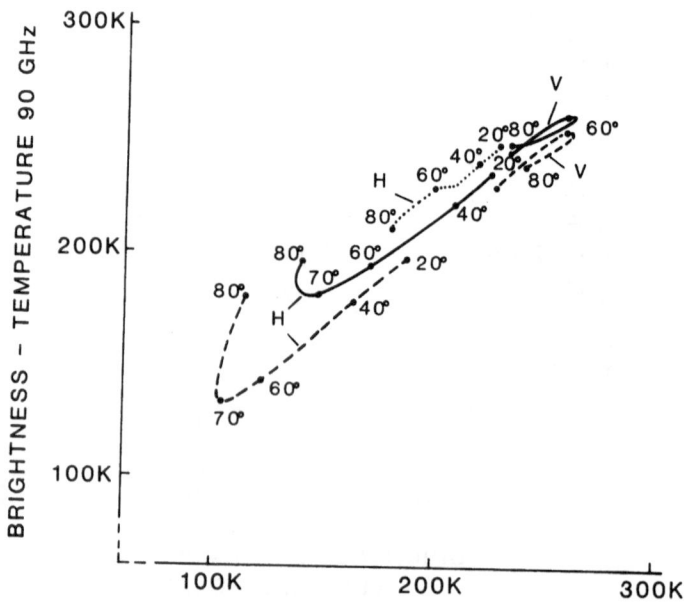

Figure 8. Signatures of asphalt for several surface conditions and horizontal and vertical polarization by simultaneous brightness-temperature measurements at 32 GHz and 90 GHz; surface dry (——), wet (---) and with a snow cover of 3.5 cm (...), result for snow and vertical polarization not drawn, measurement DFVLR, 1980.

Simultaneous measurements at 32 GHz and 90 GHz for horizontal and vertical polarization are represented once more, but now in a

32 GHz/90 GHz-plane. The curves were obtained, when looking to a asphalt surface at various nadir angles and for several surface conditions (dry ——, wet ---, covered with a 3.5 cm thick snow layer ...). The variations obtained for horizontal polarization are much more distinct than those for vertical polarization where it was renounced to draw the curve for snow cover.

The examples presented should give a first feeling what big variety of signatures can be obtained in the microwave region by using the informations contained in the natural microwave radiation when receiving the horizontally and vertically polarized components at various nadir angles and frequencies. At the end of this chapter, however, we must ask us the question whether we have extracted really all the polarization dependent informations hidden in the natural microwave radiation, when we measure the horizontally and vertically polarized components. By depolarization effects for instance there should be a (weak?) correlation between the horizontally polarized and vertically polarized electric-field components giving further independent informations. A short discussion is carried out by help of the Stokes parameter of a partially polarized wave.

2. SOME REMARKS WITH RESPECT TO THE COMPLETE STATE OF POLARIZATION

Natural microwave radiation generated by thermal processes of matter is completely randomly polarized or completely unpolarized at first. By reflection and refraction at the boundaries of different media a partially polarized wave results however within the frequency bands considered. In a certain sense, the boundaries act as a polarization-filter or -transformer.

A partially polarized wave can be regarded as the sum of a completely unpolarized wave and a completely polarized wave. To deal with partial polarization it is convenient to use the normalized Stokes parameters S_0, S_1, S_2, S_3 which completely describe the state of polarization of a wave [3] and which can be defined as a matrix of the column vector-type (Stokes vector $\underline{g}$).

The rules which are with Stokes parameters or vectors are very simple [3, 4]:

- To determine the Stokes parameters only power measurements and no phase measurements are necessary.

- If several independent waves propagating in the same direction are superimposed, the Stokes vector of the resultant wave is the sum of the Stokes vectors of the individual waves.

- The Stokes vector of a reflected or a refracted wave is related to the Stokes vector of an incident wave by a 4x4 real matrix (Müller matrix, Stokes reflection -/refraction matrix) which characterizes the object and from which its scattering matrix can be derived e.g.:

$$g^r(\Theta_r) = [M(\Theta_r, \Theta_i)] \, g^i(\Theta_i) .$$

(6)

Following [3], the normalized Stokes parameter of a plane wave (frequency f), which can be completely described in a xyz-coordinate-system (wave traveling in positive z-direction, point of measurement z = 0) by the orthogonally polarized components of the electric field

$$e_x(t) = E_x(t) \, \sin(2\pi \, ft - \delta_x(t))$$
$$e_y(t) = E_y(t) \, \sin(2\pi \, ft - \delta_y(t))$$

(7)

are:

$$S \begin{bmatrix} S_o \\ S_1 \\ S_2 \\ S_3 \end{bmatrix} = S \begin{bmatrix} 1 \\ (\langle E_x^2 \rangle - \langle E_y^2 \rangle)/ZS \\ 2 \, \langle E_x E_y \, \cos\delta \rangle/ZS \\ 2 \, \langle E_x E_y \, \sin\delta \rangle/ZS \end{bmatrix}$$

(8)

where

$\langle \ \rangle$ indicates the time average; the time variations of $E_x(t)$, $E_y(t)$, $\delta_x(t)$, $\delta_y(t)$ are assumed slow compared to frequency f which is only valid for a limited bandwidth,

δ phase difference $\delta_x - \delta_y$,

Z, S characteristic impedance of the medium and total power density($|$Poynting vector$|$)of the wave at the point of measurement.

The coordinate-system could be selected in that way that the x- and y-directions coincide with the "horizontally" and "vertically" directed electric field components as defined in chapter 1.

The polarization state of a partially polarized wave is completely described by S_1, S_2 and S_3. S_1 is proportional to the difference of the powers received at horizontal and vertical polarization. S_2 and S_3 which are <u>independent</u> quantities are like different weighted cross-correlation functions of E_x and E_y. S_1, S_2 and S_3 are equal zero in the case of a totally unpolarized wave because $\langle E_x^2 \rangle = \langle E_y^2 \rangle$ and E_x and E_y are uncorrelated. S_1, S_2 and S_3 determine a (time averaged) polarization ellipse for the completely polarized part of the wave. Also a "degree of polarization" (d) which is the ratio of the power of the completely polarized wave to the power of the total wave can be derived.

$$0 \leq d = \sqrt{S_1^2 + S_2^2 + S_3^2} \leq 1 . \qquad (9)$$

The determination of the Stokes parameter requires power measurements only, as stated above. Four independent measurements would be enough (S, S_1, S_2, S_3), but in order to simplify the following discussion six measurements are done [3]. We can carry out these by the help of a radiometer which has the same effective antenna aperture for all polarizations giving the responses: T_x, T_y, T_x', T_y', T_1, T_r (brightness-temperatures) where:

- T_x, T_y: obtained for linear x- and y directed polarizations

- T_x', T_y': obtained for the linear x- and y directed polarizations when the xyz coordinate system is turned $-45°$ around the z-axis

- T_1, T_r: obtained for left- and right-hand circular polarization.

With these values, the following expressions are obtained for S_1, S_2 and S_3:

$$S_1 = \frac{T_x - T_y}{T_x + T_y} , \qquad S_2 = \frac{T_x' - T_y'}{T_x + T_y} \qquad (10)$$

$$S_3 = \frac{T_1 - T_r}{T_x + T_y} \qquad (S \sim T_x + T_y).$$

A short analysis shows, that the condition $S_1 \neq 0$ is valid for the examples presented in the last chapter for nadir angles > 0. $S_2 \neq 0$ is only true, when the surface or the medium observed are not longer smooth or homogeneous and certain unsymmetries are given. Finally $S_3 \neq 0$ would be valid, if the surface or the medium influences a left and a right circular polarized wave in a different way. A certain inhomogeneity in depth is necessary that this might be possible. Only small quantities with respect to S_2 and S_3 can be awaited because generally polarization-dependence decreases for rough surfaces or inhomogeneous media.

To verify this, brightness-temperatures of some natural media were measured for right- and left circular polarization. The results are drawn in Figure 9. Only very small differences between the two polarizations could be registrated, which are in the order of magnitude of the temperature resolution of the radiometer used and might be errors of measurements therefore.

Figure 9. 32 GHz measurements for right- and left
circular polarization.
DFVLR measurements 1980, 1981.

Also for the sky normally no differences were observed exept for
one single case (X) several years ago when looking towards the sun
(TS) or looking to the opposite side (OS). At time a complete physical explanation of this effect cannot be given. The results also
could not be confirmed at a later time. Reproducable results were
obtained on the other hand for artificial structures as gratings
arranged crosswise at a certain distance (⊓).

In Fig. 10 a synopsis of all formulas important when dealing with
partially polarized waves is given once more.

As we have seen, there is a physical basis for further signatures
hidden in the natural microwave radiation. But the greatest variety
of signatures are an useless phenomenon, as long as sensor inability
does not allow the extraction of these informations. It is supposed,
that in the case of natural objects accuracy of measurements must
be improved very much to come to significant results (S_2, $S_3 \ll 1$).
Low noise detectors are necessary and the properties of the antennas
as the effective aperture (inclusive receiving characteristic and
input impedance behaviour) must be known very exactly.

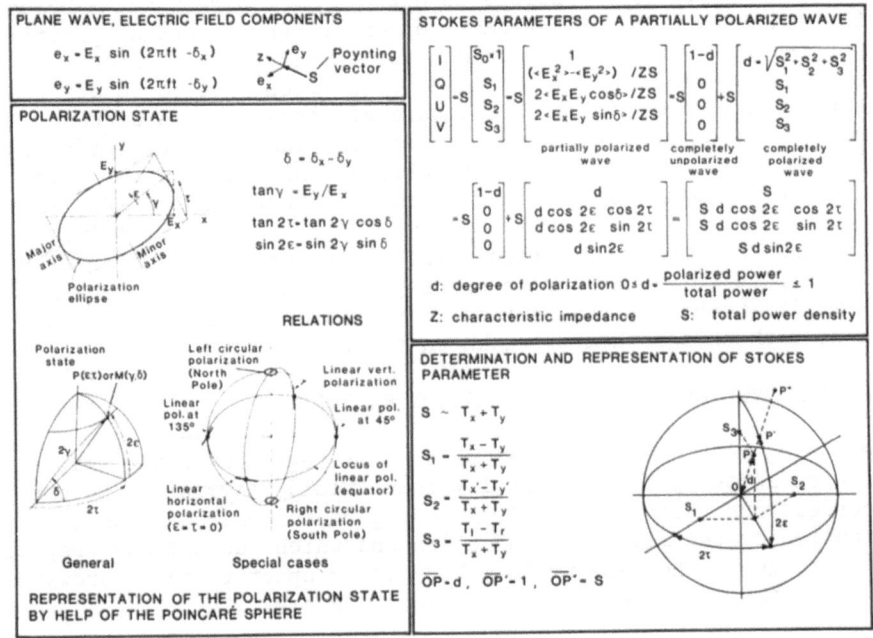

Figure 10. Polarization-relations.

If this could be realized we would have a very valuable instrument for sensing inhomogeneities transversal or in depth. Using the conventional measuring mehtod as shown in Fig. 1, the inner of the normalized Poincaré sphere (radius of 1), which is normally used to describe definitely an arbitrary polarization state of a completely polarized wave (d = 1) by a point on its surface, could immediately function as a normalized three dimensional cluster space for representation of S_1, S_2 and S_3 (see Fig. 10). When a normalization (S = 1) is not desired the true value of S could be a further criteria for discrimination (see distance OP" in Fig. 10) to reduce ambiguity problems.

It must be clear however, that this signature space must be filled at first by many values gained by aid of experimental and theoretical investigations before it can help to solve the inverse "scattering" problem of microwave radiometry.

Generally there is a great disadvantage with the measuring method outlined in Fig. 1. The radiometer normally receives at the same time emitted and reflected radiation. In the case of a complex terrain medium the power density is influenced by the physical temperature distribution of the ground as well as by the brightness-temperature distributions of the environment (weighting functions). So it will be almost impossible to extract in a direct way - by analytic methods and not on the basis of experiences - media specific properties from the measured brightness-temperatures or Stokes parameters respectively. It would be even impossible to relate the measured values to those of a standard atmosphere for a better comparison of different measurements on the same object. A series of independent (and known) weighting functions would be necessary which however cannot be realized at short intervals.

In the last chapter a measuring method is proposed therefore which could help to solve the remote sensing problem in a more direct way, if its technical realization is possible. Experimental experiences with this method are unknown to the author.

3. AN IMPROVED METHOD OF POLARIZATION RADIOMETRY

A measuring method is desirabel which only registrates the emitted or the reflected part. Principally this should be possible by use of correlation radiometers [5]. A configuration which allows the extraction of the reflected radiation and which was already proposed in [5] to solve the same problem but with respect to the microwave radiation from the moon surface is shown in Fig. 11.

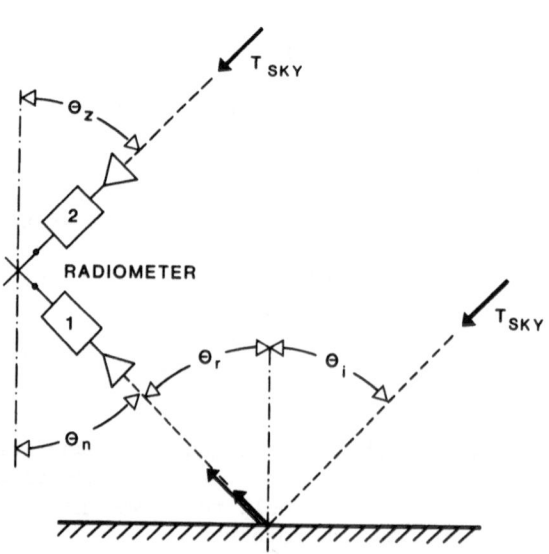

Figure 11. Correlation radiometer for the extraction of the reflected radiation.

A second radiometer looks to that region of the sky, which is the
source of the reflected part. The polarization states of both
systems should be the same for the present. In dependence of certain system parameters the correlation process would be carried
out in the IF-regions of the radiometer receivers (IF type, the
signals themselves are correlated) or after the square law envelope detection (Envelope type, only the amplitudes are correlated).
The degree of correlation then immediately would be proportional
to the power reflection coefficient $r(\Theta_i)$ at the ground $(\Theta_i=\Theta_z=\Theta_n)$
for the selected polarization. Imaging the power reflection coeffi-
cient should lead to better results, because the relative contrast
is higher than that for the (apparent) brightness-temperatures. In
the case of a complex terrain medium the look angle (Θ_z) of the
sky-radiometer had to be varied to obtain the power scattering
characteristic $\sigma(\Theta_r, \Theta_i = \Theta_z)$ or the receiving characteristic of
the sky-radiometer had to be changed.

Such a radiometer configuration probable could supply some more
informations, by switching of different polarization states at both
radiometer channels. In terms of Stokes parameters, switching the
polarization state of the sky-radiometer means the extraction
(filtering) of different incident waves with different Stokes vectors and switching the polarization state of the downward looking
radiometer means the respective determination of the Stokes parameters of the reflected wave as already described in the last chapter. By a virtual manipulation of the input field a complete series
of measurements at short intervals should make it possible to determine all independent elements of the Mueller matrix (bistatic case),
which definitely characterizes the object. Thus two advantages make
this method superior to the conventional method.

Principally a correlation method can also be realized for the emitted part of the natural radiation by using two radiometers the lines
of sight of which cross at the object. In doing so, it would be of
great advantage that the emitted radiation is mostly much stronger
than the reflected radiation (see Fig. 3) by which a better signal
to noise ratio is obtained. On the other hand the set of informations would be not so complete as in the upper case, because it is
not possible to seperate and virtually manipulate the radiation
of the source which generates the emitted part. Besides of that
it turns out, that the signal to noise ratio when extracting the
reflected part becomes better for bad weather conditions (see
Fig. 12) or by shifting the frequency f to the absorption bands of
atmosphere (22 GHz, 60 GHz, 120 GHz, 180 GHz ...), conditions which
normally affect radiometric contrast $(T_{PHYS} - T_B)$ very hard, because that profits from the difference $(T_{PHYS} - T_{SKY})$ (compare
equation (3)).

At any case, the realization of this method will require greatest
efforts to be successfull not only for strong but also for weak

reflecting media. A series of system parameters as the preferable
frequency bands, optimum microwave-bandwidths, necessary temperature
resolutions, the efficiency of the IF-type- and Envelope-type correlation methods, optimum antenna characteristics, calibration
methods, possible flight profiles and so on must be investigated.

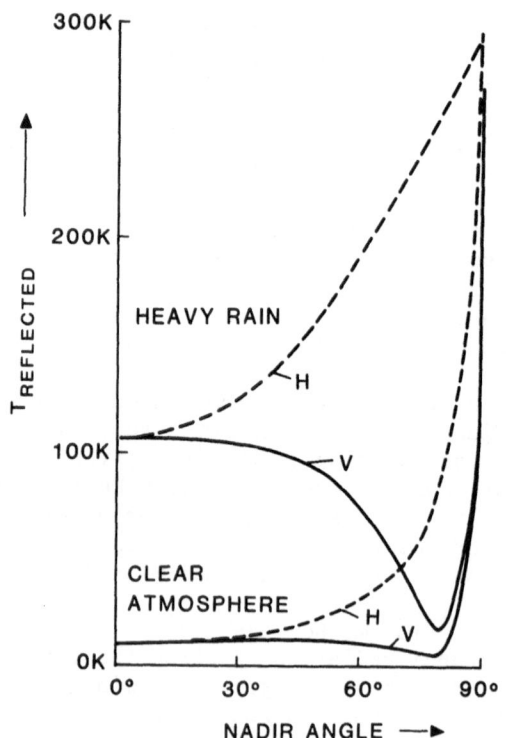

Figure 12. Example for the influence of the weather condition on the reflected part of the natural microwave
radiation
water surface as above, horizontal (---) and vertical (——)
polarization.

4. CONCLUSION

Some examples have shown, that polarization plays a very important
role also in the field of microwave radiometry.

By discussion of the Stokes parameter of a partially polarized wave it was demonstrated that thermal microwave radiation after interaction with differen media should contain more informations, than obtained by only measuring the vertically and horizontally polarized components. This could lead to new signatures or new textures at least and microwave images could have a better quality when all independent informations are summarized.

Finally a correlation method was outlined, which should allow a better solution of the remote sensing problem of microwave radiometry by separation of the reflected sky radiation from the emitted ground radiation. At time, the whole future development with respect to this method cannot be told. A series of system parameters must be optimized.

Thus once more again the prerequisite of all these activities must be the improvement of the sensor technology, to realize an absolute accuracy of measurement, which is almost one order of magnitude better than today.

5. REFERENCES

[1] M. Born and E. Wolf,
     "Principles of Optics", Pergamon Press, London, 1965.

[2] F.T. Ulaby, R.K. Moore and A.K. Fung,
     "Microwave Remote Sensing", Volume 1, Addison-Wesley
     Publishing Company, London, 1981.

[3] J.D. Krans
     "Radio Astronomy",
     Mc Graw-Hill Book Company, New York, p. 108-130, 1966.

[4] W-M. Boerner, M.B. El-Arini et al.
     "Polarization Utilization in Radar Target Reconstruction",
     Electromagnetic Imaging Division, Communications Laboratory,
     University of Illinois at Chicago, Technical (Interim) Report
     UICC, CL-EMID-NANRAR-81-01, 1981.

[5] K. Fujimoto,
     "On the Correlation Radiometer Technique",
     IEEE Trans. Microwave Theory Tech., Vol MTT-2,
     pp 203-211, March 1964.

6. ACKNOWLEDGMENT

The author wishs to thank Professor Boerner, whose enthusiasm for all kinds of polarization phenomena induced me to write this paper. I further thank Mrs. Kappelmaier and Mrs. Malchow for typing this manuscript.

III.5 (SP.3)

MULTISTATIC VECTOR INVERSE SCATTERING

Gregory E. Heath

Massachusetts Institute of Technology
Lincoln Laboratory
Lexington, MA 02173-0073, U.S.A.

ABSTRACT. Two new vector (i.e., polarization dependent) methods
for constructing a three-dimensional radar image of a target from
transient or multi-frequency far field multistatic scattering data
have been formulated. In both methods equivalent current image
functions can be constructed directly from transient or multifrequency far field scattering data by using efficient inverse Radon
or Fourier Transform computations, respectively. Analytic continuation of far field data into the near field is unnecessary.

I. INTRODUCTION

     The first inverse scattering method to utilize complete polar-
ization information in a logical and consistent manner was developed during 1968-1974 [1-5] for opaque scatterers satisfying impedance boundary conditions. However, successful target reconstruction required the computationally difficult and inefficient task of
using analytic continuation to construct near field quantities from
far field measurement data in order to implement inverse boundary
conditions. Later multistatic investigations of perfectly conducting targets (1976 [6], 1979/80 [7-9]) avoided the need for analytic
continuation by using various modifications of physical optics to
express the target image as a computationally efficient inverse
Fourier Transform of far field data. However, the role of polarization was not treated adequately.

     Recently, two new vector (i.e., polarization dependent)
methods for constructing a three-dimensional radar image of a
target from transient or multi-frequency far field multistatic
scattering data have been formulated. Multistatic vector physical
optics far field inverse scattering (POFFIS) extends Bojarski's

599

W.-M. Boerner et al. (eds.), Inverse Methods in Electromagnetic Imaging - Part 1, 599-608.

polarization independent multi-monostatic technique for a perfectly conducting target [10] to the vector case of arbitrarily
polarized multistatic scattering [11]. The inverse method of
solenoidal equivalent currents (IMSEC) replaces the restriction of
perfectly conducting targets and the use of physical optics with
one, less restrictive, assumption: the scattering from an
arbitrary (i.e., transparent or opaque) target can be represented
by reradiation from equivalent magnetic and electric currents
whose solenoidal (divergence-free) components depend on the
incident field through a polarization dependent, linear,
time-invariant, causal relationship that is spatially local and
temporally isotropic (irrotational (curl-free) current components
are ignored because they do not radiate) [12]. As a result of the
IMSEC assumption, the solenoidal current components are required
to vanish outside of the volume enclosed by the external target
surface and, like physical optics equivalent currents, can be used
to construct three-dimensional image functions that determine the
location and shape of the target. In both methods equivalent
current image functions can be constructed directly from transient
or multi-frequency far field scattering data by using efficient
inverse Radon and Fourier Transform computations, respectively.
Analytic continuation of far field data into the near field region
is not necessary.

The new POFFIS formulation obviates some of the objections to
previous work by expressing equivalent magnetic and electric current image functions in terms of measurable, polarization dependent, far field quantities satisfying reciprocity and bistatic-
monostatic equivalence. However, when the polarization reference
plane for each radar is the propagation plane containing the
transmitter and receiver lines of sight, reconstruction requires
only principal polarization measurement data. In contrast, IMSEC
yields exact reconstruction formulae without eliminating depolarization effects. Since actual induced currents may have solenoidal
components that do not vanish in current-free regions (the non-
zero solenoidal and irrotational components cancel) they are replaced by equivalent currents whose solenoidal components satisfy
the IMSEC assumption. However, in order to account for arbitrarily polarized scattered fields, both magnetic and electric equivalent currents must be used. Although solenoidal components and
the corresponding image functions can be constructed directly from
far field scattering data, irrotational components cannot be
determined without additional information regarding the nature of
the fields or currents within the target volume.

II. EQUIVALENT CURRENT SCATTERING THEORY

The incident field is assumed to be a linearly polarized uniform plane wave of the form

$$[\vec{E}^i(\hat{k}_i,\vec{r},t), \eta\vec{H}^i(\hat{k}_i,\vec{r},t)] = [\hat{e}, \hat{h}]f(t-\hat{k}_i\cdot\vec{r}/c) \qquad (1)$$

where $[\vec{E}^i, \vec{H}^i]$ are the incident electric and magnetic field vectors at position $\vec{r}$ and time t, η is the intrinsic impedance of the uniform medium surrounding the target, $\hat{k}_i$ is the unit incident wave normal, c is the speed of light and f(t) is an arbitrary time dependent incident waveform. The unit polarization vectors $[\hat{e}, \hat{h}]$ and $\hat{k}_i$ form the right handed orthogonal triplet $[\hat{e}, \hat{h}, \hat{k}_i]$ so that

$$[\vec{E}^i, \eta\vec{H}^i] = \hat{k}_i \times [-\eta\vec{H}^i, \vec{E}^i], \qquad \hat{k}_i\cdot[\vec{E}^i, \eta\vec{H}^i] = 0 . \qquad (2)$$

The incident field induces electric currents on the target surface and within the target volume. The scattered fields can be interpreted as reradiation, in the presence of the target, from the induced electric current. According to the Equivalence Principle, the scattered fields <u>outside of the target</u> can also be represented by the radiation, in the absence of the target, of nonunique equivalent magnetic and electric currents.

A. POFFIS Currents

POFFIS currents are conveniently described in terms of the target surface describing function $a(\vec{r})$ (< 0 outside of the target volume; $= 0$ on the target surface; > 0 inside of the target volume) and the target characteristic function $\gamma(\vec{r}) = u[a(\vec{r})] = $ (0 outside; 1/2 on the surface; 1 inside) where u is the Heaviside unit step function. Consequently

$$\nabla\gamma(\vec{r}) = \delta(a) \hat{n}_o(\hat{n}_o\cdot\nabla)a \qquad (3)$$

where δ is the Dirac delta distribution and $\hat{n}_o(\vec{r})$ is the unit outward surface normal. The target surface is decomposed into the illuminated surface, S_1 ($\hat{k}_i\cdot\hat{n}_o < 0$), the shadow boundary, C ($\hat{k}_i\cdot\hat{n}_o = 0$), and the shadowed surface, S_2 ($\hat{k}_i\cdot\hat{n}_o > 0$).

Equivalent <u>volume</u> current densities for a perfectly conducting target can be expressed in the form

$$[\eta\vec{J}, \vec{M}] = [\eta\vec{J}_o(\hat{k}_i,\hat{e},\vec{r}), \vec{M}_o(\hat{k}_i,\hat{e},\vec{r})]f(t-\hat{k}_i\cdot\vec{r}/c) \qquad (4)$$

where

$$[\eta\vec{J}_o, \vec{M}_o] = [\mp\nabla\gamma \times \hat{h}, \hat{e} \times \nabla\gamma] \qquad (5)$$

are non-zero only on the target surface because of $\delta(a)$. The upper and lower signs in (5) and (6) apply to S_1 and S_2, respectively. The equivalent <u>surface</u> current densities corresponding to (5) are given by

$$[\eta\vec{J}_s, \vec{M}_s] = [\pm\hat{n}_o\times \eta\vec{H}^i, -\vec{E}^i\times \hat{n}_o] . \qquad (6)$$

<u>All</u> of the equivalent surface currents on S_2 can be replaced by equivalent surface currents on S_1 to obtain the

conventional physical optics result $[n\vec{J}_s, \vec{M}_s] = [2\hat{n}_o \times n\vec{H}^i, 0]$
on S_1, and $[0, 0]$ on S_2. The difference between the two
surface current distributions is $[\hat{n}_o \times n\vec{H}^i, \vec{E}^i \times \hat{n}_o]$ which
produces no field outside of the target surface because the
incident field $[\vec{E}^i, \vec{H}^i]$ satisfies the sourceless Maxwell equations on and within the target surface. It should be obvious, on
physical grounds, that the currents on S_2 in (5) can contain no
information regarding the shape of either S_1 or S_2. Accordingly,
the Maggi-Rubinowicz [13,11] transformation can be applied to replace the currents on S_2 by equivalent line currents on the shadow boundary C.

Only some of the equivalent surface currents on S_1 can be
replaced by equivalent surface currents on S_2 or by equivalent
line currents on C. The remaining currents on S_1 contain information on the illuminated surface shape. Exact (i.e., within the
assumptions of the physical optics approximation) target reconstruction can be obtained from direct illumination of the complete surface by using two oppositely directed transmitters.

The three ($\hat{e}$, $\hat{h}$, $\hat{k}_i$) components of $\nabla\gamma$ are suitable image
functions because they have the characteristics of Dirac
delta distributions located on the target surface and contain the
same shape information as the components of $[n\vec{J}_o, \vec{M}_o]$.

B. IMSEC Currents

All equivalent currents can be uniquely decomposed
into irrotational (curl-free) and solenoidal (divergence-free,
rotational) parts. However, only the solenoidal part of the
equivalent currents radiate. Therefore, neglecting creeping and
traveling wave effects, target reconstruction using scattered
field measurements will only yield images located in the support
of the solenoidal part. Unfortunately, the solenoidal part may
be non-zero in regions where the total equivalent current vanishes (the non-zero solenoidal part is canceled by the non-zero
irrotational part). Therefore assumptions that guarantee vanishing solenoidal parts outside of the target volume are desirable.

The solenoidal part of IMSEC equivalent currents are
assumed to be determined by the local incident field through
the following polarization dependent, linear, time invariant,
causal, temporally isotropic relationship:

$$
\begin{bmatrix} n\vec{J}_R \\ \vec{M}_R \end{bmatrix} = \int_{-\infty}^{t} d\zeta \begin{bmatrix} w_e(t-\zeta)\overset{\leftrightarrow}{\Gamma}_e(\hat{k}_i,\vec{r}) \cdot \vec{E}^i(\hat{k}_i,\vec{r},\zeta) \\ w_m(t-\zeta)\overset{\leftrightarrow}{\Gamma}_m(\hat{k}_i,\vec{r}) \cdot n\vec{H}^i(\hat{k}_i,\vec{r},\zeta) \end{bmatrix}
$$

$$= \begin{bmatrix} \eta \vec{J}_0(\hat{k}_i,\hat{e},\hat{r}) \ g_e(t-\hat{k}_i \cdot \vec{r}/c) \\ \vec{M}_0(\hat{k}_i,\hat{e},\hat{r}) \ g_m(t-\hat{k}_i \cdot \vec{r}/c) \end{bmatrix} \qquad (7)$$

where

$$[\eta \vec{J}_0, \vec{M}_0] = [\ddot{\Gamma}_e \cdot \hat{e}, \ \ddot{\Gamma}_m \cdot \hat{h}] \ , \quad [\ddot{\Gamma}_e, \ \ddot{\Gamma}_m] \cdot \hat{k}_i = 0 \qquad (8)$$

and $g_{e,m}(t)$ is the time convolution of $w_{e,m}(t)$ and $f(t)$. The requirement of solenoidality for <u>arbitrary</u> $f(t)$ results in the following auxiliary constraints:

$$\nabla \cdot [\ddot{\Gamma}_e, \ \ddot{\Gamma}_m] = 0, \quad \hat{k}_i \cdot [\ddot{\Gamma}_e, \ \ddot{\Gamma}_m] = 0 \ . \qquad (9)$$

In addition, the solenoidal currents must be curl-free in the current free region outside of the target. Therefore, outside of the target,

$$\nabla \times [\ddot{\Gamma}_e, \ \ddot{\Gamma}_m] = 0, \quad \hat{k}_i \times [\ddot{\Gamma}_e, \ \ddot{\Gamma}_m] = 0 \ . \qquad (10)$$

Combining (9) and (10) shows that $[\ddot{\Gamma}_e, \ \ddot{\Gamma}_m]$ vanishes outside of the target. Consequently, combinations of the cartesian components of $\ddot{\Gamma}_e$ and $\ddot{\Gamma}_m$ are suitable for image functions.

III. SCATTERED FIELD POLARIZATION STRUCTURE

The temporal Fourier component of the scattered electric field, $\ddot{E}^s$, can be expressed in terms of the range normalized far field scattering dyadic, $\ddot{S}$, through the relations

$$\ddot{E}^s = 2\sqrt{\pi} \ F(jck) \ g(k,r) \ \ddot{S}(k,\hat{k}_i,\hat{k}_s) \cdot \hat{e} \ , \qquad (11)$$

where F is the temporal Fourier transform of the incident pulse at the frequency $ck/2\pi$, and $g = \exp(-jkr)/4\pi r$ $(r=|\vec{r}|)$ is the scalar free space Green's function.

The linear and circular polarization representations of $\ddot{S}$ are given by

$$\ddot{S} = \hat{H}_s(a_{HH}\hat{H}_i + a_{HV}\hat{V}) + \hat{V}(a_{VH}\hat{H}_i + a_{VV}\hat{V})$$

$$= \hat{R}_s(a_{RR}\hat{R}_i^* + a_{RL}\hat{L}_i^*) + \hat{L}_s(a_{LR}\hat{R}_i^* + a_{LL}\hat{L}_i^*) \qquad (12)$$

where the a's are the corresponding scattering matrix coefficients [14,15] and $\hat{V}$ is the vertical polarization unit vector in the $\hat{k}_s \times \hat{k}_i$ direction. The scattered and incident wave horizontal polarization unit vectors are defined by $\hat{H}_s = \hat{V} \times \hat{k}_s$ and $\hat{H}_i = \hat{k}_i \times \hat{V}$, respectively. The corresponding right (upper sign) and left (lower sign) circular polarization unit vectors are given by $\hat{R}_s, \hat{L}_s = (\hat{H}_s \mp j\hat{V})/\sqrt{2}$ and $\hat{R}_i, \hat{L}_i = (\hat{H}_i \pm j\hat{V})/\sqrt{2}$.

A.   POFFIS Polarization

The physical optics scattering dyadic can be <u>uniquely</u> decomposed into reflected and diffracted parts [11], i.e.,

$$\ddot{S} = \ddot{S}_1 + \ddot{S}_2 = \ddot{S}_R + \ddot{S}_D \tag{13}$$

where $\ddot{S}_{1,2}$ are contributions from the surface currents on $S_{1,2}$, respectively. The diffracted component, $\ddot{S}_D$, can <u>always</u> be interpreted as reradiation from shadow boundary line currents. It contains no information concerning the shape of S_1 and S_2 outside of their common boundary, C. Therefore, it is not used in POFFIS target reconstruction. Alternate forms of $\ddot{S}_D$ in terms of $\hat{N}_1$, $\hat{N}_2$ and $\hat{L}$ are given in [11] where $\hat{N}_1(\hat{K}), \hat{N}_2(\hat{K})$ and $\hat{L}(\hat{K})$ $(\hat{K} \equiv k(\hat{k}_s - \hat{k}_i))$ are integrals of $n_o(\vec{r})\exp(j\vec{K}\cdot\vec{r})$ over S_1, S_2 and C, respectively.

The reflected component, $\ddot{S}_R(k, \hat{k}_i, \hat{k}_s)$ <u>must</u> be interpreted as reradiation from illuminated area surface currents. It is given by [11]

$$\ddot{S}_R = a_R(\hat{H}_s\hat{H}_i + \hat{V}\hat{V}) = a_R(\hat{L}_s\hat{R}_i^* + \hat{R}_s\hat{L}_i^*) \tag{14}$$

where

$$a_R = (-j/2\sqrt{\pi}) \, \hat{K} \cdot \hat{N}_1(\hat{K})$$
$$= (a_{HH} + a_{VV})/2 = (a_{LR} + a_{RL})/2 . \tag{15}$$

All eight linear and circular polarization scattering matrix coefficients are decomposed into reflected and diffracted parts in Table 1 of [11].

Although all of the physical optics scattering coefficients satisfy the principle of bistatic-monostatic equivalence [16], the equivalence is satisfied with a lower degree of error by the combinations in (15) that yield the reflected coefficient a_R [11].

B.   IMSEC Polarization

The transverse nature of $\ddot{\Gamma}_e$ and $\ddot{\Gamma}_m$ indicated in (9) results in reradiated electric fields caused by electric and magnetic solenoidal equivalent currents that are polarized perpendicular and parallel, respectively, to the propagation plane containing the transmitter and receiver lines of sight [12]. Consequently

$$\ddot{S} = \ddot{S}_e + \ddot{S}_m \tag{16}$$

where

$$\ddot{S}_e = \hat{V}[a_{VH}\hat{H}_i + a_{VV}\hat{V}]$$

$$= (-jk/2\sqrt{\pi})w_e(jck)\int_V d^3\vec{r}\,'\,\vec{\vec{\Gamma}}_e(\hat{k}_i,\vec{r}\,')\exp(j\vec{K}\cdot\vec{r}\,') \qquad (17)$$

and

$$\vec{\vec{S}}_m = \hat{H}_s[a_{HH}\hat{H}_i + a_{HV}\hat{V}]$$

$$= (-jk/2\sqrt{\pi})w_m(jck)\int_V d^3\vec{r}\,'\,[-\hat{k}_s\times\vec{\vec{\Gamma}}_m\times\hat{k}_i]\exp(j\vec{K}\cdot\vec{r}\,'). \qquad (18)$$

IV. RECONSTRUCTION ALGORITHMS

The results of the last chapter can be used to obtain inversion formulae in terms of spatial inverse Fourier Transforms $[\vec{K}\to\vec{r}]$ of multifrequency far field multistatic scattering data. The conventional relationship between Fourier and Radon Transforms [17] can then be used to obtain alternate inversion formulae in terms of inverse Radon transforms of transient scattering data. Filtered and prefiltered back-projections of both types of data can also be used [12,18,19].

A. POFFIS Inversion Formulae

The divergence theorem can be used to establish the following spatial Fourier Transform $(\vec{r}\to\vec{K})$ pairs [11]:

$$-\nabla\gamma\to(2j\sqrt{\pi}/K^2)\vec{K}[a_{HH}(k,\hat{k}_i,\hat{k}_s)+a_{HH}^*(k,-\hat{k}_i,-\hat{k}_s)] \qquad (19a)$$

$$\to(2j\sqrt{\pi}/K^2)\vec{K}[a_{VV}(k,\hat{k}_i,\hat{k}_s)+a_{VV}^*(k,-\hat{k}_i,-\hat{k}_s)] \qquad (19b)$$

$$\to(2j\sqrt{\pi}/K^2)\vec{K}[a_R(k,\hat{k}_i,\hat{k}_s)+a_R^*(k,-\hat{k}_i,-\hat{k}_s)] \qquad (19c)$$

where a_R can be obtained from linear or circular principal polarization measurements using (15). The three forms on the right hand side of (19) are equivalent only in the physical optics limit. Although each reduces to Bojarski's polarization independent monostatic results [10] when $\hat{k}_s = -\hat{k}_i$, only (19c) is consistent with the polarization dependent monostatic result of Boerner and Ho [17]. The forms in (19a) and (19b) also apply to scalar wave scattering with Neumann and Dirichlet boundary conditions, respectively. For two-dimensional scattering the right hand sides of (19) must be multiplied by $\sqrt{(jk/\pi)}$.

Physical optics scattering coefficients do not, in general, satisfy electromagnetic reciprocity. However, the electromagnetic reciprocal of the physical optics approximation is the image induction approximation [20] which also yields (19) [11]. Therefore (19) is consistent with reciprocity.

The ability to use only HH or VV scattering coefficients in (19a) and (19b) does not imply that multistatic POFFIS inversion

should be attempted with a disregard for polarization effects.
The simple principal polarization forms in (19) are valid only
when the propagation plane is the polarization reference plane.
Consequently, the $\hat{H}$ and $\hat{V}$ polarization directions indicated in
(19) vary with receiver position. More practical polarization
conventions require orthogonal polarization measurements.
However, known target symmetries can be exploited to reduce the
number of measurements [14,15,21].

B.   IMSEC Inversion Formulae

IMSEC inversion formulae follow directly from (17) and (18),
i.e.,

$$\ddot{\Gamma}_e \approx [2j\sqrt{\pi}/kw_e(jck)]\hat{V}[a_{VH}\hat{H}_s + a_{VV}\hat{V}] \tag{20}$$

and

$$\ddot{\Gamma}_m \approx [2j\sqrt{\pi}/kw_m(jck)]\hat{V}[a_{HV}\hat{H}_s - a_{HH}\hat{V}] . \tag{21}$$

V.   EXAMPLES

Four IMSEC images of an infinitely long perfectly conducting circular (radius = a) cylinder in the $\hat{z}$ direction were
obtained for the two-dimensional case of broadside incidence
($\hat{k}_i = -\hat{x}$) and reception ($\hat{k}_s = \cos\beta\ \hat{x} + \sin\beta\ \hat{y}$). The data were
simulated from the exact eigenfunction series expansion and a 64
x 64 two-dimensional FFT (zero padded with 16 x 16 data points)
was used to approximate the inverse Fourier Transform. Sample
points in K space were chosen to yield a 4a ambiguity interval,
i.e., $\Delta K_x = \Delta K_y = 2\pi/4a$. Spatial resolution was $\Delta x = \Delta y =$
$4a/16 = \hat{a}/4$. The resulting range of bistatic angles and
circumference to wavelength ratios was:

$$-133° \leq \beta = 2\tan^{-1}(K_y/K_x) \leq 127° \tag{22}$$

and

$$2.75 \leq ka = (1/2)[K_x + K_y^2/K_x] \ a \leq 17.2. \tag{23}$$

Therefore resonant region data was used for a specular point
range of approximately ±65° about the transmitter line of sight.

The images in Figure 1 show little difference between the
three principal polarization cases with image functions
$|\vec{M}_{oz}|^2$ (HH), $|n\vec{J}_{oz}|^2$ (VV) and $(1/4)|n\vec{J}_{oz} + \vec{M}_{oz}|^2$ (LR).
However, the computer printout indicated that background noise
was three orders of magnitude lower in the circular polarization
case. In all three cases the circular shape is recovered within
≈ ±45° of the transmitter line of sight without severe
distortion. In contrast, the circular orthogonal polarization
image function $(1/4)|n\vec{J}_{oz} - \vec{M}_{oz}|^2$ (RR) yields a distorted shape

IMSEC IMAGES OF A CIRCULAR CYLINDER

and observable creeping wave contamination. Experimentation with spectral weights ($w_e = w_m$) consisting of various powers of jk did not substantially improve the RR image. There was no observable difference between the principal polarization IMSEC images and the POFFIS images with the $(k, -\hat{k}_i, -\hat{k}_s)$ terms omitted (not shown) because of the simplicity of the target.

REFERENCES

1. V. H. Weston, J.J. Bowman and E. Ar, Arch. Rational Mech. Anal., 31, pp. 199-213 (1968).

2. V. H. Weston and W. M. Boerner, Can. J. Phys., 47, pp. 1177-1184 (1969).

3. W. M. Boerner and H.P.S. Ahluwalia, Can. J. Phys., 50, pp. 3023-3061 (1972).

4. H.P.S. Ahluwalia and W. M. Boerner, IEEE Trans. AP-21, pp. 663-672 (1973).

5. _____   IEEE Trans. AP-22, pp. 673-682 (1974).

6. S. Rosenbaum-Raz, IEEE Trans. AP-24, pp. 66-70 (1976).

7. J. Detlefsen, "Abbildung mit Mikrowellen", Fortschrittberichte, der VDI-Zeitschriften, Reihe 10, Nr. 5. VDI-Verlag Dusseldorf (1979).

8. _____   IEEE Trans. AP-28, pp. 377-380 (1980).

9. _____   1980 International U.R.S.I. Symposium, Munich, Germany (August 26-29) pp. 322C/1.

10. N. N. Bojarski, IEEE Trans. AP-30, pp. 980-989 (1982).

11. G. E. Heath, "Direct and Inverse Multistatic Physical Optics: The Significance of Reciprocity, Polarization and Bistatic-Monostatic Equivalence", M.I.T./LL (to be publ.).

12. G. E. Heath, 1981 International IEEE/AP-S Symposium, Los Angeles, CA (June 19) pp. 620-623.

13. A. Rubinowicz, Progress in Optics, 5, E. Wolf, Ed. (John Wiley and Sons, Inc., 1965)pp. 199-240.

14. G. E. Heath, IEEE Trans. AP-29, pp. 523-525 (1981).

15. G. E. Heath, IEEE Trans. AP-29, pp. 429-434 (1981).

16. R. E. Kell, Proc. IEEE, 53, pp. 983-988 (1965).

17. W. Boerner and C. Ho, Wave Motion, 3, pp.311-333 (1981).

18. A. J. Rockmore, R. V. Denton and B. Friedlander, IEEE Trans. AP-27, pp. 239-241 (1979).

19. M. Y. Chiu, H. H. Barrett and R. G. Simpson, J. Opt. Soc. Am., 70, pp. 755-762 (1980).

20. R. F. Harrington, IRE Trans. AP-7, pp. 150-153 (1959).

21. M. Davidovitz and W. M. Boerner, IEEE Trans. AP-31, pp. 237-242 (1983).

The work reported herein was done at Lincoln Laboratory, a center for research operated by Mass. Inst. of Technology. The program is sponsored by Ballistic Missile Defense Prog.Office, Dept. of the Army; supported by the Ballistic Missile Defense Adv. Tech. Ctr. under A.F. Contract F19628-80-C-0002. The views and conclusions contained herein are those of the contractor and should not be interpreted as necessarily representing the official policies, either expressed or implied, of the U.S. Government.

III.6 (RP.2)

EXTENSION OF KENNAUGH'S OPTIMAL POLARIZATION CONCEPT TO THE ASYMMETRIC MATRIX CASE

Marat Davidovitz and Wolfgang-M. Boerner

Electromagnetic Imaging Division, Communications Lab.
Department of Electrical Engineering & Computer Science
University of Ill. at Chicago P.O. Box 4348, M/C 154
Chicago, Illinois 60680

Abstract
The polarization scattering matrix measured by a bistatic radar system generally will be asymmetric. Similarly, when the radar system is monostatic but the intervening propagation medium is non-reciprocal, the scattering matrix is asymmetric. In this paper we generalize Kennaugh's optimal polarization theory to the case of the asymmetric scattering matrix. The radar antenna polarizations to be used for maximum and zero power reception are defined and geometrically interpreted on the Poincaré sphere. These polarizations, termed optimal polarizations, may be used to enhance the level of target echo return, discriminate against undesired interference sources, or to classify radar targets.

I. Introduction
A monochromatic plane wave with a fixed polarization state impinging upon a radar target induces a surface current distribution, which in turn gives rise to a reradiated or scattered field. The polarization state of a wave scattered in the direction of observation will, in general, differ from that of the incident wave. The transformation of the polarization state upon reflection can be represented by a complex scattering matrix, characteristic of the target at a specific aspect and source frequency. The formula used to calculate the target radar cross section or echoing area involves the scattering matrix as well as the transmitting and receiving antenna polarizations. The problem facing the radar operator is the one of selecting the optimal radar polarizations which will, for example, maximize the level of echo signal or minimize the effect of undesired sources, such as clutter return or jamming signals.

W.-M. Boerner et al. (eds.), Inverse Methods in Electromagnetic Imaging - Part 1, 609–628.
© *1985 by D. Reidel Publishing Company.*

Utilizing the principle of electromagnetic reciprocity, one can show that when the radar system is monostatic and the intervening propagation medium is reciprocal, the scattering matrix is symmetric. The optimal polarizations associated with a symmetric scattering matrix were first defined by Kennaugh [1] in 1950. In particular, he found that a co-polarized transmitting and receiving radar antenna can be used to obtain the maximum echo return possible and that there exists only one polarization for which this maximum is achieved. Kennaugh also showed that there are two polarizations for which the back-scattered wave is orthogonal to the co-polarized radar antenna, thus rendering the radar "blind" to the incoming wave.

In a situation where the radar system is bistatic or one in which the radar system is monostatic but the medium of propagation is non-reciprocal, the scattering matrix generally is asymmetric. In this paper we attempt to generalize the optimal polarizations theory to the case of the asymmetric scattering matrix. In particular, we show that the maximum echo return can no longer be obtained with an identically polarized transmitting and receiving antenna. The optimal transmitting and receiving antenna polarizations are defined in terms of the scattering matrix elements. It is subsequently proven that for the case of the asymmetric scattering matrix, the co-polarized echo return will always be less than the absolute maximum obtained with separately polarized transmitting and receiving antennas. Finally, we represent the optimal polarizations on the Poincaré sphere and prove certain geometrical properties of the resulting configuration.

II. Overview and Statement of the Problem

Under the condition that the radar antennas are located sufficiently far from the target, we may consider the wave incident on the target and the scattered wave observed at the receiving antenna to be quasi-planar. In such case the electric field vectors $\underline{E}^i$ and $\underline{E}^s$ of the incident and the scattered wave, respectively, will lie entirely in the planes perpendicular to directions of incidence and observation. The standard two-dimensional orthonormal polarization bases may then be used to represent the electric field vectors [2]. Due to the linearity of Maxwell's equations, the components of the vectors $\underline{E}^i$ and $\underline{E}^s$ can be related by a 2x2 matrix transformation S as follows:

$$\underline{E}^s = S \cdot \underline{E}^i \tag{1}$$

The scattering matrix S serves as a discriptor of the polarizing properties of the target.

The measurable quantity of primary interest to the radar operator is the target radar cross section or echoing area. For a given incident wave $\underline{E}^i$, the echoing area A is a measure of the power contained in the wave scattered toward the receiver. The echoing area A is calculated according to the following standard formula

$$A = 4\pi R^2 \; \frac{\| S \cdot \underline{E}^i \|}{\| \underline{E}^i \|} \tag{2}$$

where $\| \cdot \|$ denotes the norm of the enclosed vector quantity. The norm is defined as $\| \underline{E} \| = \underline{\tilde{E}}{}^* \cdot \underline{E}$, where "~" and "*" denote transposition and conjugation, respectively.

However, when there is a mismatch between the polarization state of the scattered wave and that of the receiving antenna, only a fraction of the reflected wave's power is sampled. The formula for the resulting effective echoing area must be appropriately redefined to account for the dependence on the receiving antenna polarization. The following bilinear matrix form defines the effective echoing area A_e

$$A_e = | \underline{\tilde{h}}{}^r \cdot S \cdot \underline{h}^t |^2 \tag{3}$$

where $\underline{h}^t$ and $\underline{h}^r$ are the normalized ($\| \underline{h}^r \| = 1 = \| \underline{h}^t \|$) vector heights of the transmitting and receiving antenna, respectively [3]. Note, the quantity $A_e = A_e(\underline{h}^t, \underline{h}^r)$ is a function of two independent variables, namely $\underline{h}^t$ and $\underline{h}^r$. The problem which we pose in this paper is the one of finding the maxima and minima of A_e and the corresponding antenna polarizations for which the optimum values of A_e are obtained. Such polarizations are termed optimal.

We first consider the somewhat simplified case of expression (3), namely one in which the transmitting and the receiving antenna are identically polarized, i.e. $\underline{h}^t = \underline{h}^r = \underline{h}$. At this point it is proper to raise the question of just what do we mean when we say identically polarized antennas, since in the general case of a bistatic radar system, the transmitting and the receiving antenna polarizations are referred to two separate local coordinate systems. Evidently there is no unique answer to the posed question, since several definitions of co- and cross-polarizations have been offered in the past [4,5]. However, it is important to note that a unique definition of co-polarization is not crucial to the theory presented in this paper. The main reason is that the optimal polarizations are physically defined by the radar target itself, and are, therefore, invariant to the various definitions of coordinate systems. In other words, even though the mathematical expressions representing the optimal polarizations may change depending on the particular choice of the reference coordinate systems, the physical attributes of the optimal polarizations, such as ellipticity, sense of rotation, and the orientation angle, referred to the plane defined by the directions of incidence and observation, remain invariant. We may, therefore, without any loss of generality, consider the plane defined by the directions of incidence and observation to be the polarization reference plane. We then proceed to construct right-handed coordinate frames at the transmitter and receiver sights, letting the y-axis lie in the reference plane and x-axis be perpendicular to it, while the z-axis will be directed at the target [3].

Returning to expression (3) and setting $\underline{h}^t = \underline{h}^r = \underline{h}$, we obtain the following quadratic matrix form

$$A_e = |\underline{\tilde{h}} \cdot S \cdot \underline{h}|^2 \tag{4}$$

The scattering matrix S appearing in (4) will, in general, be asymmetric. From elementary matrix theory we know that any asymmetric matrix can be decomposed into a symmetric (S_s) and a skew- or anti-symmetric (S_a) component as follows:

$$S_s = (S + \tilde{S})/2 \tag{5a}$$

$$S_a = (S - \tilde{S})/2 \tag{5b}$$

The value of the quadratic matrix form defined by (4) remains unchanged if the scattering matrix S is replaced by its symmetric component S_s.

Taking into consideration the discussion of the previous paragraph, we may subdivide the general problem into two parts. In the first part we will investigate the properties of symmetric scattering matrices under the restriction of identically polarized transmitting and receiving antennas. In the second part this restriction will be removed, thus forcing us to consider a more general problem of the asymmetric scattering matrix and separately polarized radar antennas.

III. Polarization Basis Transformations
Transformation of polarization vectors from one orthonormal basis to another is accomplished through the use of unitary matrices.

Unitary change-of-basis transformations have the important property of preserving the total power in the wave [2], i.e. the norm of the electric field vector $\|\underline{E}\|$ is invariant under unitary transformations. A unitary change-of-basis matrix T must satisfy the following

$$|\det\{t\}| = 1 \tag{6a}$$

$$T^{-1} = \tilde{T}* \tag{6b}$$

The most general form of a unitary matrix, satisfying the above requirements, is given by

$$T = \begin{bmatrix} e^{j\psi_1}\cos(\gamma/2) & -e^{j\psi_2}\sin(\gamma/2) \\ e^{j\psi_3}\sin(\gamma/2) & e^{j\psi_4}\cos(\gamma/2) \end{bmatrix} \tag{7}$$

where, because of condition (6a), the phases of the matrix are interrelated as follows: $\psi_1 + \psi_4 = \psi_2 + \psi_3$. Without any loss of generality we may choose to set $\psi_1 = \psi_4 = 0$ and let $\psi_3 = -\psi_2 = \delta$, in which case the matrix T takes on the following form [6,7]

$$T = \begin{bmatrix} \cos(\gamma/2) & -e^{-j\delta}\sin(\gamma/2) \\ e^{j\delta}\sin(\gamma/2) & \cos(\gamma/2) \end{bmatrix} = \frac{1}{(1+\rho\rho*)^{1/2}} \begin{bmatrix} 1 & -\rho* \\ \rho & 1 \end{bmatrix} \tag{8}$$

where $\rho = \tan(\gamma/2)e^{j\delta}$.

The particular explicit form of T given in (8) was chosen because of its intimate connection to the geometrical representation of unitary transformations on the Poincaré sphere. In fact, when the Poincaré sphere is used to represent two orthonormal bases, namely $[\underline{e}_A, \underline{e}_{A'}]$ and $[\underline{e}_B, \underline{e}_{B'}]$, the geometrical transformation analogous to (8) is accomplished using the angles γ and δ as shown in Fig. 1.

Utilizing the change-of-basis matrix defined above, we proceed to transform the vectors $\underline{E}^i$, $\underline{E}^S$ from the $[\underline{e}_A, \underline{e}_{A'}]$ basis into the $[\underline{e}_B, \underline{e}_{B'}]$ basis in the following manner

$$\underline{E}^i(A,A') = T \cdot \underline{E}^i(B,B') \tag{9a}$$

$$\underline{E}^S(A,A') = T* \cdot \underline{E}^S(B,B') \tag{9b}$$

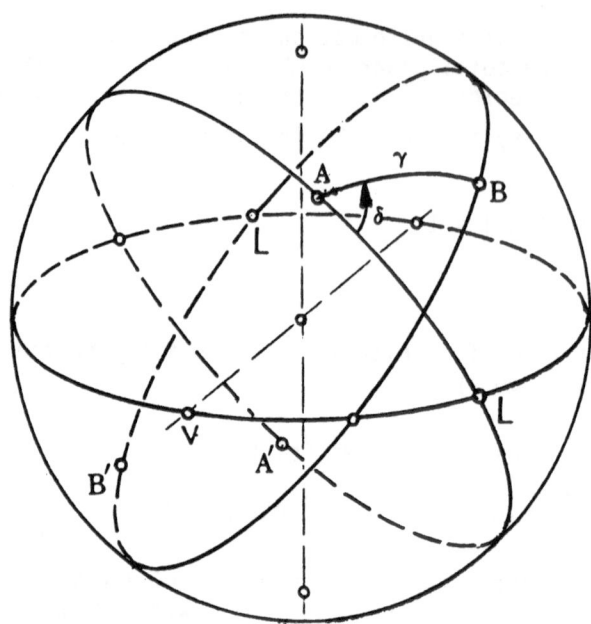

FIGURE 1:   POLARIZATION BASIS TRANSFORMATION ON THE POINCARE
            SPHERE

Note that the incident and the scattered wave are propagating in
opposite directions. Consequently, in order for the transformation to have the same effect on the polarization state of $\underline{E}^i$ and
$\underline{E}^S$, the change-of-basis matrices used to transform $\underline{E}^i$ and $\underline{E}^S$ must
be related through conjugation [2,6]. The latter fact is reflec-
ted in relations (9a,b).

Recalling that the relation between $\underline{E}^i(A,A')$ and $\underline{E}^S(A,A')$ is given
by

$$\underline{E}^S(A,A') = S(A,A') \cdot \underline{E}^i(A,A') \tag{10}$$

we can transform the scattering matrix $S(A,A')$ into the $[\underline{e}_B, \underline{e}_{B'}]$
basis. Substituting the relations in (9a,b) into (10), we obtain
the scattering matrix representation in the $[\underline{e}_B, \underline{e}_{B'}]$ basis. The
resulting expression is given by

$$S(B,B') = \tilde{T} \cdot S(A, A') \cdot T \tag{11}$$

The general situation is conveniently summarized in the following
mneumonic diagram

$$\underline{E}^i(B,B') \qquad S(B,B') \qquad \underline{E}^S(B,B')$$

$$T \downarrow \xrightarrow{\hspace{3cm}} \quad \downarrow T^*$$

$$\underline{E}^i(A,A') \qquad S(A,A') \qquad \underline{E}^S(A,A')$$

Finally, we prove that the type of transformation prescribed by equation (11) preserves the (skew-) symmetry of a matrix. Starting with a symmetric matrix S_s or a skew-symmetric matrix S_a and applying the transformation yields

$$\tilde{T} \cdot S_{s,a} \cdot T = S'_{s,a} \tag{12}$$

Taking note of the fact that $\tilde{S}_{s,a} = \pm S_{s,a}$, leads to following equation

$$\tilde{S}'_{s,a} = (\tilde{T} \cdot S_{s,a} \cdot T) = \tilde{T} \cdot \tilde{S}_{s,a} \cdot T = \pm (\tilde{T} \cdot S_{s,a} \cdot T) = \pm S'_{s,a} \tag{13}$$

where the lower sign corresponds to the lower subscript.

Moreover, it is a well-known fact that the transformation of equation (11) can take a complex symmetric matrix into the complex diagonal form [2]. This important fact is used in the next section, where the optimal polarization properties of symmetric scattering matrices are considered.

IV. Optimal Polarizations Associated with a Symmetric Scattering Matrix

Consider a situation in which the scattering matrix, obtained either through measurement or theoretical computation, is represented in an orthonormal basis $[\underline{e}_A, \underline{e}_{A'}]$ by a complex asymmetric matrix $S(A,A')$. The effective echoing area obtained by transmitting and receiving with the same polarization $\underline{h}(A, A')$ ($\|\underline{h}(A, A')\| = 1$) is given by

$$A_e = |\tilde{\underline{h}}(A,A') \cdot S(A,A') \cdot \underline{h}(A,A')|^2 = |\tilde{\underline{h}}(A,A') \cdot S_s(A,A') \cdot \underline{h}(A,A')|^2 \tag{14}$$

where we have replaced the matrix $S(A,A')$ by its symmetric component $S_s(A,A')$. Note, no loss in generality occurs if the matrix $S(A,A')$ is symmetric from the outset.

The mathematical computations involved in optimization of the quadratic form for A_e can be simplified by finding a basis in which the scattering matrix is diagonal. Recalling that the type of transformation described in the preceeding section can diagonalize a complex symmetric matrix, we pose the problem of finding the change-of-basis matrix T_M such that

$$\tilde{T}_M \cdot S(A,A') \cdot T_M = S_d(M,M') \tag{15}$$

where $S_d(M,M')$ is a complex diagonal matrix. Conjugating and transposing (15) and multiplying both sides of the original equation by the corresponding sides of the new equation, we obtain an equivalent formulation of the problem. The resulting equation is given by

$$\tilde{T}_M^* \cdot P(A,A') \cdot T_M = T_M^{-1} \cdot P(A,A') \cdot T_M = P_d(M,M') \tag{16}$$

where $P(A,A') = S_s^*(A,A') \cdot S_s(A,A')$ and $P_d(M,M') = S_d^*(M,M') \cdot S_d(M,M')$. We recognize that $P(A,A')$, $P_d(M,M')$ are the so-called power scattering matrices [2]. We also note the fact that the diagonal elements of $P_d(M,M')$ are the eigenvalues of $P(A,A')$ and are equal to the squared magnitudes of the respective elements of the matrix $S_d(M,M')$. The basis $[\underline{h}_M, \underline{h}_{M'}]$ consists of the eigenvectors of $P(A,A')$ and can, therefore, be termed the "characteristic" basis. It can easily be proven that such a basis is orthogonal [8]. Moreover, Kennaugh showed [1] that the maximum value of A_e is equal to the larger eigenvalue of $P(A,A')$ and the maximum polarization is the corresponding eigenvector.

The eigenvalues of $P(A,A')$ are found in the usual manner, i.e., by solving the equation $\det \{P(A,A') - |\lambda|^2 I\} = 0$. The solutions satisfying this equation are

$$|\lambda_{1,2}|^2 = \frac{\text{tr}\{P(A,A')\} \pm \sqrt{\text{tr}^2\{P(A,A')\} - 4 \cdot \det\{P(A,A')\}}}{2} \tag{17}$$

where $\lambda_{1,2}$ are the complex diagonal elements of $S_d(M,M')$. It is interesting to note that $\text{tr}\{P(A,A')\} = \text{tr}\{P_d(M,M')\}$ and $\det\{P(A,A')\} = \det\{P_d(M,M')\}$, i.e. the trace and the determinant are invariant under the change-of-basis transformation. Furthermore, we can show that

$$\text{tr}\{P(A,A')\} = \text{span}\{S_s(A,A')\} \tag{18a}$$

and

$$\det\{P(A,A')\} = |\det\{S_s(A,A')\}|^2 \tag{18b}$$

where the span of a complex matrix is defined as the sum of the magnitudes squared of its elements.

To find the change-of-basis matrix T_M and the eigenvectors corresponding to $|\lambda_{1,2}|^2$, we rewrite equation (16) as follows:

$$P(A,A') \cdot T_M = T_M \cdot P_d(M,M') \qquad (19)$$

The matrix equation (19) corresponds to four scalar equations. If we use the explicit form of T_M defined in Section III, any one of the four scalar equations yields the complex parameter ρ_M

$$\rho_M = \frac{|\lambda_1|^2 - P_{AA}}{P_{AA'}} \qquad (20)$$

where P_{AA}, P_{AA}' are the elements of the power scattering matrix $P(A,A')$ and $|\lambda_1|^2$ is given by (17). The eigenvectors $\underline{h}_M$ and $\underline{h}_{M'}$ and the change-of-basis matrix T_M can be defined in terms of ρ_M as follows:

$$\underline{h}_M = C_M \begin{bmatrix} 1 \\ \rho_M \end{bmatrix}, \quad \underline{h}_{M'} = C_M \begin{bmatrix} -\rho_M^* \\ 1 \end{bmatrix}, \quad C_M = \frac{1}{\sqrt{1+\rho_M \rho_M^*}} \qquad (21)$$

and

$$T_M = [\underline{h}_M, \underline{h}_{M'}] = C_M \begin{bmatrix} 1 & -\rho_M^* \\ \rho_M & 1 \end{bmatrix} \qquad (22)$$

From equation (17) it is clear that $|\lambda_1|^2 > |\lambda_2|^2$ and we therefore conclude that $A_{e,max} = |\lambda_1|^2$ and the maximum polarization is specified by $\underline{h}_M$.

We may use the following formula to calculate the effective echoing area A_e

$$A_e = |\underline{\tilde{h}} \cdot (M,M') \cdot S_d(M,M') \cdot \underline{h}(M,M')|^2 \qquad (23)$$

The matrix $S_d(M,M')$ has been determined by straight-forward matrix multiplication in accordance with equation (15) and using the matrix T_M displayed in (22).

We now turn to the problem of finding the so-called "null" polarization in the characteristic basis $[\underline{h}_M, \underline{h}_{M'}]$. The "null", $\underline{h}_0$, is defined by the condition $A_e = 0$ [9], which in turn leads to the following equation

$$\tilde{\underline{h}}_0(M,M') \cdot S_d(M,M') \cdot \underline{h}_0(M,M') = 0 \qquad (24)$$

Two "null" polarizations generally will be found by expanding equation (24). The polarization ratios defining the "null" polarizations are given by [1]

$$\rho_{1,2} = \pm j \sqrt{\frac{S_{MM}}{S_{M'M'}}} = \sqrt{\frac{|S_{MM}|}{|S_{M'M'}|}} \exp\left[j\frac{(\phi_{MM} - \phi_{M'M'})}{2} \pm j(\pi/2)\right] =$$

$$= \tan(\gamma_{1,2}/2) e^{j\delta_{1,2}} \qquad (25)$$

In the characteristic basis, the magnitudes of the polarization ratios $\rho_{1,2}$ are equal, i.e. $\gamma_1 = \gamma_2 = \gamma$, and the phase difference $\delta_1 - \delta_2$ is equal to π. There is an interesting geometrical interpretation of these facts.

We first represent the orthonormal basis $[\underline{h}_M, \underline{h}_{M'}]$ on the Poincaré sphere by antipodal points M,M', as shown in Fig. 2. In accordance with the Poincaré sphere rules [9,10,11], all polarizations with equivalued polarization ratio magnitude lie on a small circle, centered on the diameter joining M and M'. Change in the phase of a polarization ratio, on the other hand, rotates the point representing the polarization about the diameter MM'. Using these rules to map the "null" polarizations $N_{1,2}$ on the Poincaré sphere we come to the following conclusions; since the difference in the rotations of the points N_1, N_2 about MM' is equal to π, they are located on the same great circle. Moreover, the fact that $\gamma_1 = \gamma_2 = \gamma$ implies that the angle N_1ON_2 (O is the center of the sphere) is bisected by MM'. The conclusions are illustrated in Fig. 2.

V.   Optimal Polarizations Associated with the Asymmetric
      Scattering Matrix
In the preceeding section we have constrained our radar system to transmit and receive with the same polarization. In this section we remove this restriction and attempt to find distinct transmitting and receiving antenna polarizations, such that the effective echoing area is maximized.

We will start by transforming the general asymmetric scattering

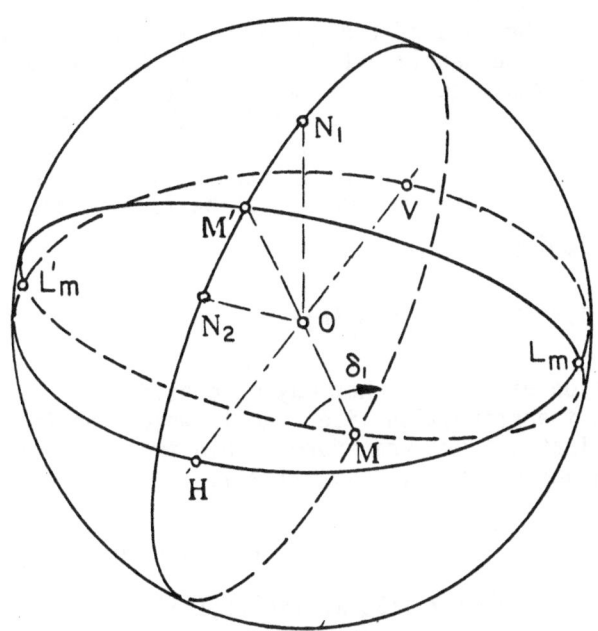

FIGURE 2:   OPTIMAL POLARIZATION FOR CO-POLARIZED
            TRANSITORY AND RECEIVING ANTENNA

matrix $S(A,A')$ into the characteristic basis $[\underline{h}_M, \underline{h}_{M'}]$. As we have
seen, this transformation diagonalizes the symmetric component of
$S(A,A')$. We also recall that this type of transformation preserves
the skew-symmetry of the anti-symmetric component ($S_a(A,A')$) of
$S'(A,A')$. Therefore, the transformed matrix $S(M,M')$ can be ex
pressed as the sum of a diagonal and a skew-symmetric matrix, i.e.

$$S(M,M') = \tilde{T}_M \cdot S(A,A') \cdot T_M = \tilde{T}_M \cdot S_s(A,A') \cdot T_M + \tilde{T}_M \cdot S_a(A,A') \cdot T_M =$$

$$= S_d(M,M') + S_a(M,M') = \begin{bmatrix} S_{MM} & S_{MM'} \\ -S_{MM'} & S_{M'M'} \end{bmatrix} \qquad (26)$$

We next turn to the problem of finding the polarizations $\underline{h}^{t,r}(M,M')$
($\|\underline{h}^{t,r}(M,M')\|=1$) which will maximize the effective echoing area A_e,
computed according to the formula

$$A_e = |\tilde{\underline{h}}^r(M,M') \cdot S(M,M') \cdot \underline{h}^t(M,M')|^2 \qquad (27)$$

Again, just as in the case of the symmetric scattering matrix, diagonalization of the matrix will facilitate the mathematical procedure used to maximize the bilinear form (27). As we know, an asymmetric matrix cannot be diagonalized by the type of transformation defined in Section III. However, from the theory of matrices it is known that an asymmetric scattering matrix $S(M,M')$ can be diagonalized by two distinct change-of-basis matrices T_I and T_R used in the following manner

$$\tilde{T}_R \cdot S(M,M') \cdot T_1 = S_d(R;I) \tag{28}$$

The above equation discribes, what may be considered, a mixed change-of-basis transformation, i.e. one in which the incident and scattered polarizations are transformed into two distinct bases, namely $[\underline{e}_I, \underline{e}_{I'}]$ and $[\underline{e}_R, \underline{e}_{R'}]$, respectively.

Rewriting the formula for A_e in the following fashion

$$A_e = |\tilde{\underline{h}}^r(R,R') \cdot S_d(R;I) \cdot \underline{h}^t(I,I')|^2 \tag{29}$$

We can show that the maximum value of A_e is equal to the squared magnitude of the larger non-zero element of $S_d(R,I)$. The polarizations for which this maximum is achieved are given by either $\underline{h}^r = \underline{e}_R$, $\underline{h}^t = \underline{e}_I$ or $\underline{h}^r = \underline{e}_{R'}$, $\underline{h}^t = \underline{e}_I$, depending on which matrix element has a greater magnitude. Therefore, our next objective is to find the change-of-basis matrices T_I and T_R, the matrix $S_d(R;I)$ and express them in terms of the elements of $S(M,M')$. Conjugating and transposing equation (28) we get

$$\tilde{T}_I^* \cdot \tilde{S}^*(M,M') \cdot T_R^* = \tilde{S}_d^*(R;I) \tag{30}$$

Now, pre- and post- multiplying both sides of the equation (28) by the corresponding sides of equation (30), we reduce the more difficult original problem to two problems whose solutions are derived in Section IV. The resulting equations are given by

$$\tilde{T}_I^* \cdot P(M,M') \cdot T_I = P_d(R;I) \tag{31a}$$

$$\tilde{T}_R \cdot P'(M,M') \cdot T_R^* = P_d(R;I) \tag{31b}$$

where $P(M,M') = \tilde{S}^*(M,M') \cdot S(M,M')$; $P'(M,M') = S(M,M') \cdot \tilde{S}^*(M,M')$;

$$P_d(R;I) = \tilde{S}_d^*(R;I) \cdot S_d(R;I)$$

As we immediately observed, the type of problem posed by equations (31a,b) was solved in the preceeding section. The non-zero elements of $P_d(R;I)$ are the eigenvalues of both $P(M,M')$ and $P'(M,M')$, and are equal to the squared magnitudes of the respective elements of $S_d(R;I)$. The bases $[\underline{e}_I, \underline{e}_{I'}]$ and $[\underline{e}_R, \underline{e}_{R'}]$ consist of the eigenvectors of P and P', respectively. Following the procedure used in the preceeding section, but omitting some of the details, we obtain the following results:

$$|\mu_{1,2}| = \frac{tr\{P\} \pm \sqrt{tr^2\{P\}-4det\{P\}}}{2} \tag{32}$$

where $\mu_{1,2}$ are the diagonal elements of $S_d(R;I)$ and P stands for either $P(M,M')$ or $P'(M,M')$;

$$\rho_I = \frac{|\mu_1|^2 - P_{MM}}{P_{MM'}} \;;\; \rho_R = \left(\frac{|\mu_1|^2 - P'_{MM}}{P'_{MM'}}\right)^* \tag{33a,b}$$

where P_{MM}, $P_{MM'}$ are the elements of $P(M,M')$ and P'_{MM}, $P'_{MM'}$ are the elements of $P'(M,M')$;

$$\underline{e}_I = C_I \begin{bmatrix} 1 \\ \rho_I \end{bmatrix} \;;\; \underline{e}_{I'} = C_I \begin{bmatrix} -\rho_I^* \\ 1 \end{bmatrix} \;;\; T_I = [\underline{e}_I \underline{e}_{I'}] = C_I \begin{bmatrix} 1 & -\rho_I^* \\ \rho_I & 1 \end{bmatrix} \tag{34}$$

$$\underline{e}_R = C_R \begin{bmatrix} 1 \\ \rho_R \end{bmatrix} \;;\; \underline{e}_{R'} = C_{R'} \begin{bmatrix} -\rho_R^* \\ 1 \end{bmatrix} \;;\; T_R = [\underline{e}_R \underline{e}_{R'}] = C_R \begin{bmatrix} 1 & -\rho_R^* \\ \rho_R & 1 \end{bmatrix} \tag{35}$$

and $\quad C_{I,R} = \dfrac{1}{\sqrt{1+\rho_{I,R} \cdot \rho_{I,R}^*}}$

From equation (32) it is clear that $|\mu_1|^2 > |\mu_2|^2$ and we, therefore, conclude that the maximum value of A_e is given by $|\mu_1|^2$ and the polarizations which are used to obtain $A_{e,max} = |\mu_1|^2$ are, in fact, $\underline{h}^t = \underline{e}_I$, $\underline{h}^r = \underline{e}_R$.

In Section IV we have demonstrated that when the radar was under constraint to transmit and receive with the same polarization, the maximum echoing area obtainable was given by $A_{e,max} = |S_{MM}|^2 = |\lambda_1|^2$. In this section we saw that by using distinctly polarized transmitting and receiving antennas we obtain the maximum echoing area $A_{e,max} = |\mu_1|^2$. In what follows, it is proven that $|\mu_1|^2$ is the absolute maximum of the function A_e, exceeding the value of $|\lambda_1|^2$. The proof utilizes the fact that the trace and the determinant of the power cattering matrix are invariant under all unitary change-of-basis transformations, i.e.

$$tr\{P_d(R;I)\} = |\mu_1|^2 + |\mu_2|^2 = tr\{P(M,M')\} = |\lambda_1|^2 + |\lambda_2|^2 + 2|S_{MM'}|^2 \tag{36}$$

and

$$det\{P_d(R;I)\} = |\mu_1|^2 \cdot |\mu_2|^2 = det\{P(M,M')\} =$$

$$= |\lambda_1|^2|\lambda_2|^2 + |S_{MM'}|^4 + 2|\lambda_1||\lambda_2|^2|S_{MM'}|^2 \cdot \cos\alpha \tag{37}$$

where

$$\lambda_1 = S_{MM}, \quad \lambda_2 = S_{M'M'}, \quad \text{and} \quad \alpha = \phi_{MM} + \phi_{M'M'} - 2\phi_{MM'} .$$

From equation (36,37) it follows that

$$(|\mu_1|^2 - |\mu_2|^2)^2 = (|\lambda_1 - |\lambda_2|^2)^2 +$$

$$+4|S_{MM'}|^2 (|\lambda_1|^2 + |\lambda_2|^2 - 2|\lambda_1||\lambda_2|\cos\alpha) \tag{38}$$

and consequently

$$(|\mu_1|^2 - |\mu_2|^2)^2 > (|\lambda_1|^2 - |\lambda_2|^2)^2 \tag{39}$$

From equations (36) and (39) we obtain the following set of inequalities

$$\left. \begin{array}{l} |\mu_1|^2 + |\mu_2|^2 > |\lambda_1|^2 + |\lambda_2|^2 \\ \\ |\mu_1|^2 - |\mu_2|^2 > |\lambda_1|^2 - |\lambda_2|^2 \end{array} \right\} \quad |\mu_1|^2 > |\lambda_1|^2 \tag{40}$$

which yield the sought for proof of the fact that $|\mu_1|^2 > |\lambda_1|^2$.

Finally, we represent the maximum polarizations $\underline{h}^t = \underline{e}_I$ and $\underline{h}^r = \underline{e}_R$

on the Poincaré sphere. The optimal polarizations $h^{t,r}$ may be specified by their complex polarization ratios, as follows:

$$\rho^t = \frac{h_M^t}{h_{M'}^t} = \rho_I = \tan(\frac{\xi_I}{2})e^{j\delta_I} \; ; \text{ and } \rho^r = \frac{h_M^r}{h_{M'}^r} = \rho_R = \tan(\frac{\xi_R}{2})e^{j\delta_R} \quad (41a,b)$$

Taking note of the fact that $P'_{MM} = P_{MM}$ and $P'_{MM'} = -P^*_{MM'}$ and using equations (33a,b), we obtain the following result

$$\frac{\rho^t}{\rho^r} = \frac{|\mu_1|^2 - P_{MM}}{P_{MM'}} \cdot \frac{-P_{MM'}}{|\mu_1|^2 - P_{MM}} = -1 \quad (42)$$

The implications of the above result are that $\xi_R = \xi_I = \xi$ and that $\delta_R - \delta_I = \pi$. In accordance with the Poincaré sphere rules expounded at the end of the preceeding section we see that the optimal polarizations $h^{t,r}$, represented by points I,S respectively, are located on the same great circle and the angle IOS is bisected by the diameter MM'. Furthermore, we can show that the angle β included between the planes of the great circles containing N_1, N_2 and I,S is given by

$$\beta = \text{arccot}\{\frac{|\lambda_1| + |\lambda_2|}{|\lambda_1| - |\lambda_2|} \tan\nu\} \quad (43)$$

where $\nu = \phi_{MM'} - 1/2(\phi_{MM} + \phi_{M'M'})$

The complete optimal polarizations configuration on the Poincaré sphere is illustrated in Fig. 3.

V.   Effect of Faraday Rotation on the Optimal Polarizations

Faraday rotation is the anisotropic effect exhibited by magneto-ionic media, such as the ionosphere. As a result of this effect, the polarization plane of a linearly polarized wave rotates continuously along the direction of propagation.

The purpose of this section is to investigate the effects of Faraday rotation on the optimal polarizations defined in the preceeding sections. In order to keep the exposition intuitively

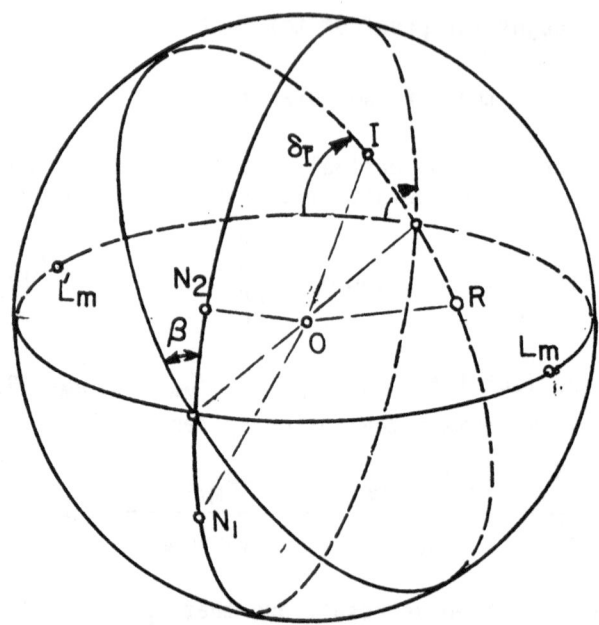

FIGURE 3: OPTIMAL POLARIZATION CONFIGURATION FOR AN
ASYMMETRIC SCATTERING MATRIX

clear, the example considered here will be simple, although not completely devoid of practical value.

Consider a body of revolution or any other plane symmetric scatterer illuminated by a monostatic radar with linearly polarized transmitting and receiving antenna. It is a well-known fact that if the radar's line of sight and the vibration plane of the linearly polarized antenna lie in the target's symmetry plane, no cross-polarization is observed, i.e. the scattering matrix is diagonal [1]. Denoting the linear polarizations by x and y, the scattering matrix for such a target can be written as follows:

$$S_d(x,y) = \begin{bmatrix} S_{xx} & 0 \\ 0 & S_{yy} \end{bmatrix} \qquad (44)$$

As the term Faraday rotation implies, the influence of the intervening magneto-ionic propagation medium on the incident linear polarization can be accounted for by a simple rotation matrix, given by [12]

$$R(\phi) = \begin{bmatrix} \cos\phi & \sin\phi \\ -\sin\phi & \cos\phi \end{bmatrix} \qquad (45)$$

The preceeding example served to illustrate a potential application of the optimal polarization theory. As the results indicate,
under special circumstances the optimal polarizations are inti-
mately related to a physical parameter of the total scattering
problem. In particular we see that the polarization ratios of the
maximum polarizations (equations 47a,b) are simply related to the
physical parameter of the propagation medium, namely the Faraday
rotation angle ϕ.

VI. Conclusion

The optimal polarizations of the scattering matrix are derived for
the general asymmetric case. In particular, it is found that unlike in the symmetric scattering matrix case, the maximum return
is no longer obtained utilizing identically polarized transmit-receive antennas. Therefore, in addition to the optimal polarizations found for the symmetric matrix, there are two more optimal
polarizations associated with the asymmetric scattering matrix.
When these polarizations are used by the transmit and receive antennas maximum target echo is observed.

It is also shown that when the optimal polarizations are mapped
onto the surface of the Poincaré sphere, certain voltage and power
calculations associated with the scattering matrix can be very
conveniently performed by geometrical methods. For example, it is
shown that the product of the chord lengths extending from point
A, representing the transmit-receive antenna polarization to the
null polarizations N_1 and N_2, is proportional to the magnitude of
the voltage recorded at the receiver. Similarly, a certain combination of the three chord lengths associated with the chordal
triangle N_1AN_2 directly determines the phase of the received voltage.

Finally, it is shown that the amount of scattering matrix measurements required to characterize completely bistatic scattering by
targets having inversion symmetry can be significantly reduced.
The necessary relations are derived upon simple geometrical arguments. The relations are subsequently numerically verified for
specific geometries.

It should be noted that the theory presented here, and developed
in greater detail in the M.Sc. thesis of Davidovitz [12], namely
the derivation of optimal polarizations for the asymmetric scattering matrix, constitutes a direct extension of the concepts
first developed by Kennaugh [1], and later expounded by Huynen
[8], for the symmetric scattering matrix. In fact, the generality
of the present theory is verified by showing that Kennaugh's
results can be obtained therefrom in a straight-forward manner.

The Poincaré sphere results obtained here, are derived in part

The angle ϕ is the total amount by which the plane of polarization is rotated while traveling in the anisotropic medium. Since the medium is non-reciprocal, the rotation of the polarization plane resulting from back-propagation to the antenna will be in the same direction and will, therefore, be represented by the same rotation matrix $R(\phi)$. The overall scattering matrix $S'(x,y)$, including the effect of the target as well as that of the magneto-ionic propaga-tion medium can be constructed as follows:

$$S'(x,y) = R(\phi) \cdot S_d(x,y) \cdot R(\phi) =$$

$$= \begin{bmatrix} S_{xx}\cos^2\phi - S_{yy}\sin^2\phi & (S_{xx} + S_{yy})\cos\phi\sin\phi \\ -(S_{xx} + S_{yy})\cos\phi\sin\phi & -S_{xx}\sin^2\phi + S_{yy}\cos^2\phi \end{bmatrix} \quad (46)$$

The resulting matrix is of the form shown in (26). We, therefore, conclude that the basis $[\underline{h}_x, \underline{h}_y]$ is the so-called characteristic basis. This in turn implies that $|S'_{xx}|^2$ is the maximum echoing area obtainable with a co-polarized antenna. The maximum polarization is, of course, specified by $\underline{h}_x$.

However, it is intuitively clear that, in fact, $|S_{xx}|^2$ and not $|S'_{xx}|^2$ is the absolute maximum echo return obtainable from that target. Solution of equation (32) shows that the larger eigenvalue of $P'(x,y) = \tilde{S}'^*(x,y) \cdot S'(x,y)$ is indeed equal to $|S_{xx}|^2$. Further more, the polarization ratios of the maximum polarizations, specified by equations (33a,b), are found to be

$$\rho_I = \tan\phi \; ; \quad \rho_R = -\tan\phi \quad\quad\quad (47a,b)$$

The "null" polarizations associated with the scattering matrix $S'(x,y)$ are given by $\rho'_{1,2}$, where

$$(\rho'_{1,2}) = \frac{\tan^2\phi + \rho^2_{1,2}}{1+\rho^2_{1,2}\tan^2\phi} \quad\quad\quad (48)$$

and $\rho_{1,2} = \pm \sqrt{-\dfrac{S_{xx}}{S_{yy}}}$ are the "null" polarizations of the scatter-

ing matrix $S_d(x,y)$.

from the work of Kennaugh [1], as well as from the papers on the theory of interference of polarized light of Pancharatnam [11].

In light of the present results, several recommendations can be made:

1) Phenomenological approach to target analysis utilized by Huynen in his thesis [8] should be applied to the bistatic scattering matrix case, in order to establish what useful information, such as target shape, compositon, etc. can be extracted from the scattering matrix parameters.

2) The Poincaré sphere theory should be extended further by devising a method of representing complex bilinear transformations on the surface of the sphere. This will provide a geometrical means of determining polarization of the scattered wave, given the incident polarization and the Poincaré sphere representation of the scattering matrix.

   A great wealth of the Poincaré sphere theory in both the fully, as well as the partially-polarized wave cases can be found in the papers of Pancharatnam referenced in [11].

3) Finallly, it is proposed, that the theory of matrices be applied to the problem of scattering matrix analysis in order to justify the already existing results by rigorous mathematical methods.

VII. Acknowledgements
The research reported in this paper was supported under Contract No. ONR N00001480-C-0773, and we thank Dr. Richard G. Brandt for the continued interest he has shown in our research. We also thank Mrs. Deborah Foster and Mr. Richard Foster for skillfully typing this manuscript and for preparing the illustrations.

VIII. References
[1] E.M. Kennaugh, "Polarization Properties of Radar Reflections", M. Sc. thesis, Dept. Elec. Engr., Ohio State Univ., Columbus, OH., 1952 (Antenna Laboratory, Rept. 389-12, March 1, 1952).

[2] C.D. Graves, "Radar Polarization Power Scattering Matrix", Proc. IRE, Vol. 44, pp. 248-252, Feb. 1956.

[3] E.M. Kennaugh, "Contributions to the Polarization Properties of Radar Targets", Commemorative Collection of Unpublished Notes and Reports in Four Volumes", ElectroScience Lab., Dept. of Electrical Engineering, Ohio State University, Columbus, OH 43212, June, 1984.

[4] A.C. Ludwig, "The Definition of Cross-Polarization", IEEE, Trans A&P-1, (Comm.) Jan. 1973,

[5] G.A. Deschamps, "Polarization Descriptors - Utilization of
     Symmetries", These Proceedings, Part 1, Topic III.1, pp.
     484-511, 1984.

[6] A.L. Maffet, "Scattering Matrices in Methods of Radar Cross
     Section Analysis", Ed. by J.W. Crispin, K.M. Siegal, Academic
     Press, NY, 1968.

[7] W-M. Boerner, "Use of Polarization in Electromagnetic Inverse
     Scattering", Radio Sci., Vol. 16(6), (Special Issue including
     Papers: 1980 Munich Symposium on Electromagnetic Waves), pp.
     1037-1045, Nov./Dec. 1981.

[8] J.R. Huynen, "Phenomenological Theory of Radar Targets", Ph.D.
     Dissertation, Technical University of Delft, The Netherlands,
     1970.

[9] D.B. Kanareykin, N.F. Pavlov, and V.A. Potekhin, The Polariza-
     tion of Radar Signals. Moscow: Sovyetskoye Radio, (Russian
     Version) 1966.

[10] G.A. Deschamps, "Geometrical Representation of the Polariza-
     tion of a Plane Electromagnetic Wave", Proc. IRE, Vol. 39(5),
     pp. 543-548, 1951.

[11] S. Pancharatnam, "Generalized Theory of Interference and its
     Applications", Proc. Ind. Acad. Sci., A, Vol.44, 1956,
     pp.247-262.

[12] M. Davidovitz, "Analysis of Certain Characteristic Properties
     of the Bistatic, Asymmetric Scattering Matrix, M.Sc. thesis,
     Univ. of Ill. at Chgo., Electr. Engr. & Comp. Sci. Dept.,
     Communications Lab. Rept. No. 83-04-15, April 15, 1983.

III.7 (RP.3)

ON THE NULL REPRESENTATION OF POLARIZATION PARAMETERS

Bob Raven

Westinghouse DEC, Baltimore, MD

Abstract-A sampling of simulated, complex RCS scattering matrix
data are represented as null pairs and plotted on an equal-area
projection of the Poincaré sphere. Examples include randomly
oriented collections of conducting thin wires, dihedrals, and
trihedrals and rotating composite dihedrals.

I. INTRODUCTION

This paper is concerned with the graphical display of polarization parameters of monostatic RCS scattering matrices, specifically, with the display of the polarization nulls of simulated,
complex reflectors. Graphical displays of empirical data are
desireable to exhibit relationships and for preliminary data analysis. For scattering matrix parameters, the problem is the common
one of showing an object of more than three dimensions in a two-
dimensional format. A monostatic scattering matrix is six-dimen-
sional (three complex numbers). Typically, one or two of these,
representing the distance to the target in wavelengths and a measure of the reflected signal magnitude are suppressed to reduce the
dimensionality. In the following discussion, both of these parameters are suppressed leaving the "pure" polarization matrix as a
four-dimensional object to be displayed. Several canonical forms
have been employed: (1) standard transmit and receive polarizations
usually horizontal and vertical or right and left circular, (2) the
eigenpolarization and values, and (3) the null polarizations of E.
M. Kennaugh [1]. For present purposes, the last form was selected
because it lends itself to a representation as point pairs on the
surface of the Poincaré sphere which is suitable for a finite,
equal-area projection. Null pair diagrams have beem previously used

629

W.-M. Boerner et al. (eds.), Inverse Methods in Electromagnetic Imaging - Part 1, 629–641.
© *1985 by D. Reidel Publishing Company.*

for assessment of polarimetric data [2],[3],[4] but with orthogra-
phic or rectangular coordinates which produce severe distortion on
the equator or at the poles. An equal-area projection preserves
areas while introducing modest shape distortion at the edges. It
is useful for preliminary analysis of null distribution and clustering. A number of equal-area projections have been used by car-
tographers. For this study, the projection selected is a modification of the Lambert meridional projection applied to the whole
sphere by Aitoff [5].

II. COMPUTATIONS

A. General Description

Two groups of reflector models were examined: (1) combinations
of multiple conducting thin wires, dihedrals, and trihedrals with
random amplitudes, phases, and orientations, and (2) specular reflections from single rotating composite dihedrals. In the first cat-
egory, scattering matrices, S, were constructed as sums of reflector matrices, S_{rk}, multiplied by a complex Gaussian variate, Z_{rk},
plus, in some cases, a complex Gaussian noise matrix, S_n.

$$S = \sum_{k=1}^{N} Z_{rk} S_{rk} + S_n \qquad (1)$$

Nulls were then calculated for each S and their coordinates plotted on a projection of the surface of the Poincaré sphere. Six-
teen such cases are displayed in following figures.

Composite material dihedrals illustrate the marked deviation
from the case of conducting surfaces when a finite Brewster angle
is present. The five material combinations listed in table 1 were
considered:

Table 1.

Dihedral Material Combinations

Steel/Sea water	Wood/Sea water
Steel/Dry ground	Wood/Moist ground
Steel/Moist ground	

Large smooth surfaces giving specular reflections with no diffraction or surface roughness effects were assumed with Fresnel's expressions, given below, used to determine reflected polarization
as a function of incidence angle.

$$R_H = \frac{\sin\psi - \sqrt{e_c - \cos^2\psi}}{\sin\psi + \sqrt{e_c - \cos^2\psi}} \qquad (2)$$

$$R_V = \frac{e_c\sin\psi - \sqrt{e_c - \cos^2\psi}}{e_c\sin\psi + \sqrt{e_c - \cos^2\psi}} \qquad (3)$$

$$e_c = e_1 - 160\lambda\sigma \quad (\lambda m \ \& \ \sigma mhos) \tag{4}$$

A K_a-band frequency was assumed. Published E-M material properties at this frequency are fragmentary. The following data were
adapted from several sources for this study [6],[7].

Table 2
E-M Properties Assumed at K_a-band

	e_1	σ-mhos/m
Steel	1.0	10^7
Sea water	65.0	4.3
Moist grnd	20.0	.5
Dry ground	3.0	.01
Wood	2.0	.2

Variation of $|R_V|$ and $|R_H|$ with incidence angle is plotted
in fig. 1.

Figure 1. Reflection Coefficient Magnitudes at K_a-band

For the composite dihedrals, the dihedral was horizontal and
normal to the incident direction and rotated about the line of
intersection such that the dielectric surface traversed incidence
angles of from 2.5° to 85°. The null trajectories were then
plotted to show the marked change in reflection characteristics as
the incidence angle traversed the Brewster angle of the dielectric
surface. A polarization transformation subroutine which computed
the reflected polarization from arbitrarily oriented surfaces with
arbitrary incidence, polarization, and complex dielectric constant
was used to track the direction of propagation and polarization
from surface to surface.

B. The Aitoff Projection

The Aitoff projection used to display nulls on the Poincaré
sphere is a Lambert meridional projection stretched by a factor of
two in the horizontal direction. It is implemented as follows:

A_S = azimuth angle
E_S = elevation angle
B = chord angle
C = chord length
D = polar angle from the equator to the chord plane
K = scale factor
X,Y = rectangular coordinates

1. Transform spherical angles to hemispherical angles by halving the azimuth angle.

$$A_H = .5A_S$$
$$E_H = E_S \tag{5}$$

2. Compute the chord angle, chord, and polar angle from the equatorial point $A_H, E_H = 0, 0$ to A_H, E_H.

$$B = \cos^{-1}(\cos .5A_S \cos E_S)$$
$$C = 2^{1/2} \sin .5B$$
$$D = \sin^{-1}(\sin E_S / \sin B) \tag{6}$$

3. Transform to rectangular coordinates doubling the X component.

$$X = 2KC\cos D$$
$$Y = KC\sin D \tag{7}$$

C. Polarization Transformation by a Surface

Because the transformation of a polarization vector by a surface may not be readily available, the approach used is summarized below. Figure 2 shows the seven directional vectors and two polarization vectors involved.

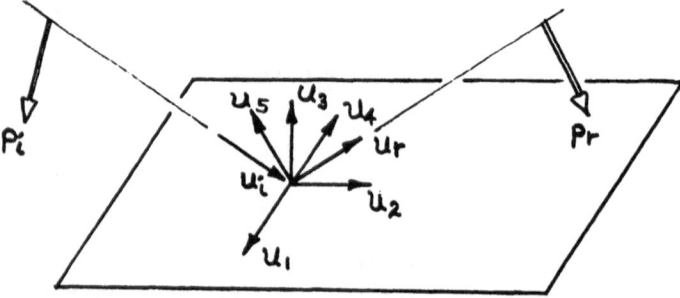

Figure 2. Surface Reflection Vectors

u_i, u_r = incident & reflected ray unit direction vectors
u_1, u_2, u_3 = perpendicular, parallel, & normal surface unit vectors
u_4, u_5 = incident & reflected vertical polarization unit vectors
p_i, p_r = incident & reflected polarization vectors (complex)
T = transformation to surface coordinates $(u_1, u_2, u_3)^T$
S_i = transformation to incident polarization coordinates $(u_1, u_4, u_i)^T$
S_r = transformation to reflected polarization coordinates $(u_1, u_5, u_r)^T$

R = polarization reflectance matrix $\begin{bmatrix} R_V & O \\ O & R_H \end{bmatrix}_0$

J =ray reflection matrix $\begin{bmatrix} 1 & O \\ O & -1 \end{bmatrix}$

Given: u_i, u_3, p_i

$$u_1 = \frac{u_i \times u_3}{|u_i \times u_3|} \qquad u_2 = u_3 \times u_1 \qquad u_4 = u_1 \times u_i \qquad u_5 = u_1 \times u_r \qquad (8)$$

With these definitions, transformation of the ray and the polarization can be expressed compactly. The ray is transformed to surface coordinates by T, reflected by J, and transformed back to the
initial coordinate system by T^T.

$$u_r = T^T J T u_i \qquad (9)$$

Similarly, the polarization is transformed to incident polarization coordinates by S_i, multiplied by the complex reflectance
matrix R, and transformed back to the initial coordinate system
by S_r^T.

$$p_r = S_r^T R S_i p_i \qquad (10)$$

III. RANDOM ELEMENT COLLECTIONS

    Sixteen plots of random collections of elements are shown in
figs. 3-18. These examples are not intended to model real clutter
but to illustrate phenomena which can occur in idealized scattering
models. Figure 3 shows the distribution of nulls for random noise
matrices. The nulls appear to be uniformly distributed. This is an
important property if the representation is to be useful and suggest-
ive. The author is unaware of a mathematical demonstration of this
property. In fig. 4, the null pairs in fig. 4 are connected. Fig. 5
shows the nulls of single conducting thin wires or dipoles with random orientations, amplitudes, and phases. The nulls of such reflectors are coincident and on the equator with their azimuth equal to
twice the rotation angle. In figs. 6 and 7, collections of two and
five wires with random orientations, amplitudes, and phases are displayed. Only two elements are sufficient mask the linear element
characteristic of fig. 5, while five elements already seem an adequate model for random noise.
    In figs. 8-18, combinations of conducting dihedrals with ran-
dom orientations, amplitudes, and phases in combination with noise
and trihedrals are shown. Figure 8 shows only single dihedrals.
The nulls plot as equatorial pairs spaced by 180° with their azimuth
given by twice the rotation angle. In figs. 9 and 10, collections
of two and five such elements are plotted. A dihedral characteristic
is preserved in that all null pairs occur at the same elevation.
Only two elements, however, are sufficient to disturb the nulls substantially from the equator. Further, there is little to distinguish
the two and five element cases. In fig. 11, the effect of a signal-
-to-noise ratio of 10dB on the case in fig. 10 is shown. Signal-to-
noise is defined here as the ratio of matrix components. In fig. 12,

the rotation angle of the dihedrals were limited to 30° to show the
sensitivity to this variable. Figures 13, 14, and 15 show the effect of various mixtures of dihedrals and trihedrals. Not surprisingly, an equal power ratio has considerable resemblance to random
noise. Power ratios in these cases were defined in terms of the
total power of all matrix components for the two constituents. The
last three plots, figs. 16, 17, and 18, show single trihedrals and
dihedrals for three signal-to-noise ratios. Significant deteriation
of polarization features is apparent at a 10dB signal-to-noise ratio.

VI. SPECULAR REFLECTIONS FROM COMPOSITE DIHEDRALS

     Figures 19-23 show five examples of specular reflections from
composite dihedrals. In these examples, horizontal dihedrals were
rotated about their line of intersection such that the angle of
incidence on the first plane varied from 2.5° to 85° and the null
migration plotted. In the first three cases, one surface is conducting and the other is dielectric modeling, for instance, the side
of a ship and the sea or a van and the ground. In these examples,
the null migration is such that the object first appears to be a
conducting dihedral, then a wire, and lastly a trihedral or plate
as the Brewster angle of the dielectric surface is traversed. Figure 21 is noteworthy in that, by chance, one of the sample incidence
angles was very close to the Brewster angle for the dielectric surface, and the reflection is almost identical to that from a thin
horizontal wire. Figures 22 and 23 show null migrations for a similar geometry where both surfaces are dielectrics, and there are two
Brewster angles. In these cases, the nulls perform loops first
aping trihedrals, then dihedrals, and lastly back to trihedrals as
the two Brewster angles are traversed.

IV REFERENCES

[1] E.M. Kennaugh, "Effects of type of polarization on echo characteristics," Antenna Lab Rpts. 389-1/24, Ohio State Univ., 1949-1954.
[2] J. R. Huynen, "Phenomenological theory of radar targets," Dissertation Univ. of Delft, Drukkerij BronderOffset, Rotterdam, 1970.
[3] L. A. Morgan & S. Weisbrod, "RCS matrix studies of sea clutter,"
Workshop on Polarization Technology, Vol. 1, GACIAC, IIT Res. Inst.,
Chicago, IL, 1980, pp. 103-120.
[4] J. C. Daley, "Radar target discrimination based on polarization
effects," Workshop on Polarization Technology, Vol. 1, GACIAC, IIT
Res. Inst., Chicago, IL, 1980, pp. 185-213.
[5] C. H. Deetz & O. S. Adams, "Elements of map projection," Spec.
Pub. No. 68, U.S. Gov't Printing Off., 1945, pp. 160-163.
[6] D. J. Kerr, Ed., "Propagation of short radio waves," Vol. 13,
MIT Rad Lab Series, McGraw Hill, 1951, p. 398.
[7] H. Jasik, Ed., "Antenna engineering handbook," McGraw Hill, 1961.

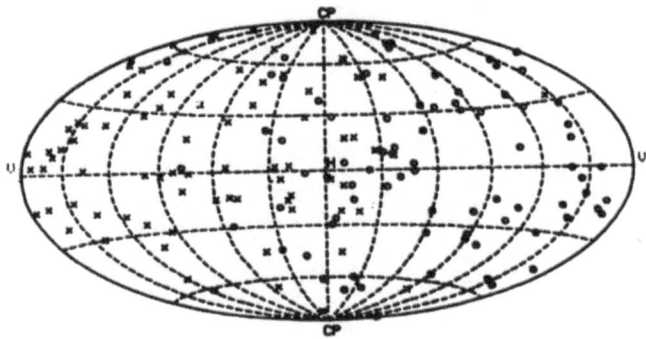

Figure 3. Aitoff Equal-area Projection of the Poincaré Sphere.
Co-pol Nulls of 80 Random Noise Samples.

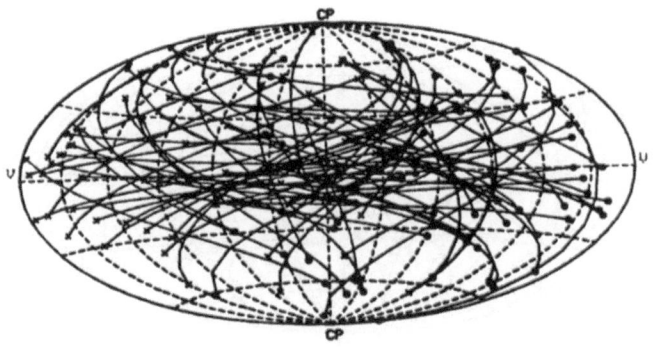

Figure 4. Aitoff Equal-area Projection of the Poincaré Sphere.
Co-pol Nulls of 80 Random Noise Samples (connected).

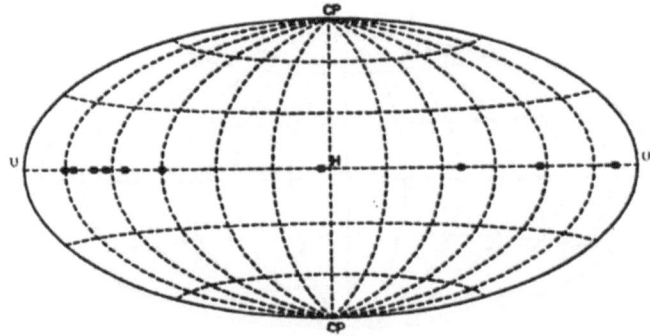

Figure 5. Aitoff Equal-area Projection of the Poincaré Sphere.
Co-pol Nulls of 10 Wire Reflectors. Single wires with Random Amplitudes, Phases, & Attitudes.

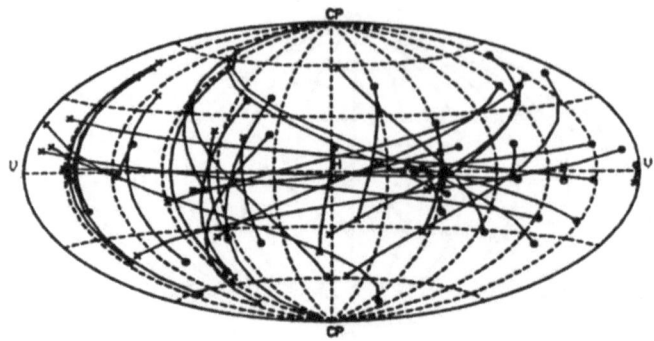

Figure 6.   Aitoff Equal-area Projection of the Poincaré Sphere.
Co-pol Nulls of 40 Wire Reflectors.   Two Wire Elements with Random
Amplitudes, Phases, & Attitudes.

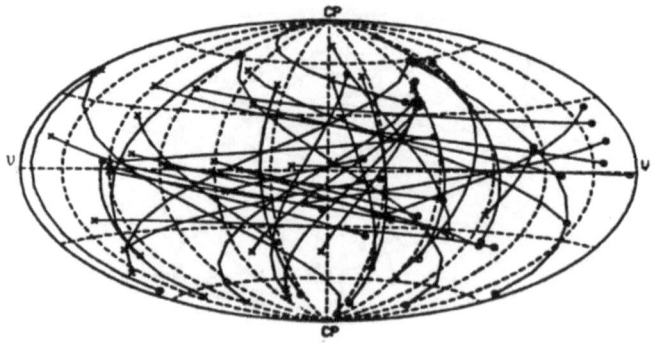

Figure 7.   Aitoff Equal-area Projection of the Poincaré Sphere.
Co-pol Nulls of 40 Wire Reflectors.   Five Wire Elements with Random
Amplitudes, Phases, & Attitudes.

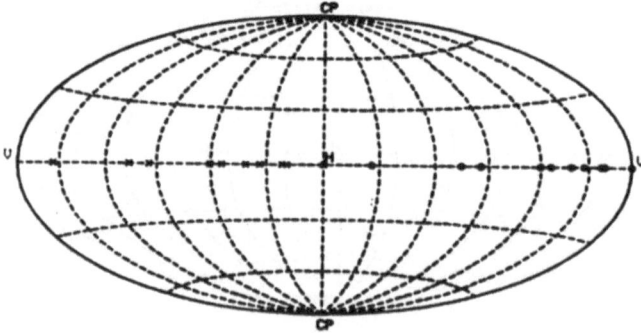

Figure 8.   Aitoff Equal-area Projection of the Poincaré Sphere.
Co-pol Nulls of 40 Dihedral Reflectors. Single Dihedral Elements
with Random Amplitudes, Phases, & Attitudes.

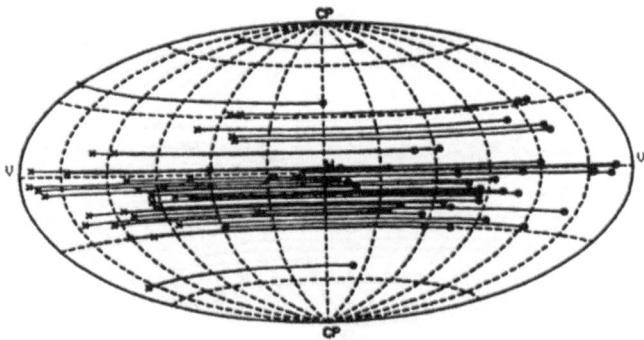

Figure 9. Aitoff Equal-area Projection of the Poincaré Sphere.
Co-pol Nulls of 40 Dihedral Reflectors. Two Dihedral Elements
with Random Amplitudes, Phases, & Attitudes.

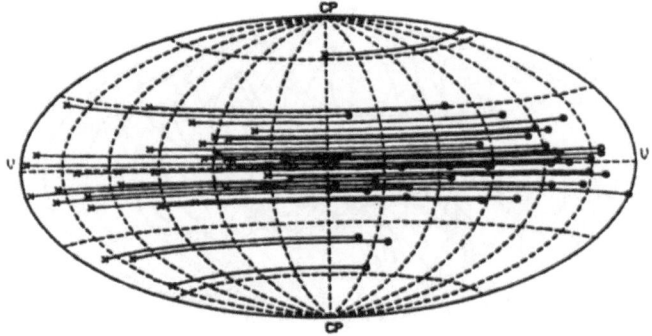

Figure 10. Aitoff Equal-area Projection of the Poincaré Sphere.
Co-pol Nulls of 40 Dihedral Reflectors. Five Dihedral Elements
with Random Amplitudes, Phases, & Attitudes.

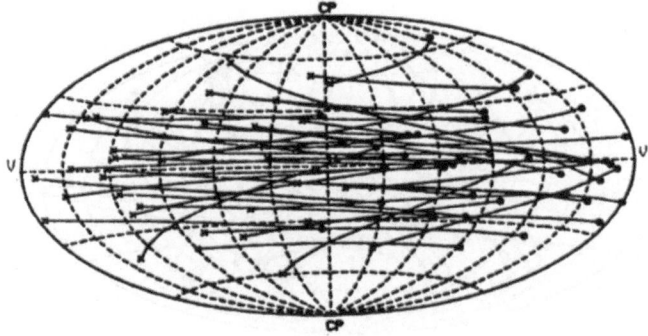

Figure 11. Aitoff Equal-area Projection of the Poincaré Sphere.
Co-pol Nulls of 40 Dihedral Reflectors. Five Dihedral Elements
with Random Amplitudes, Phases, & Attitudes plus Noise. SNR=10dB.

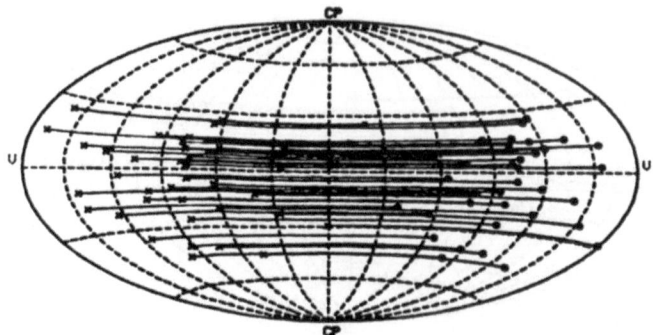

Figure 12.   Aitoff Equal-area Projection of the Poincaré Sphere.
Co-pol Nulls of 40 Dihedral Reflectors.   Five Dihedral Elements
with Random Amplitudes, Phases, & Attitudes. $|\Theta|<30°$.

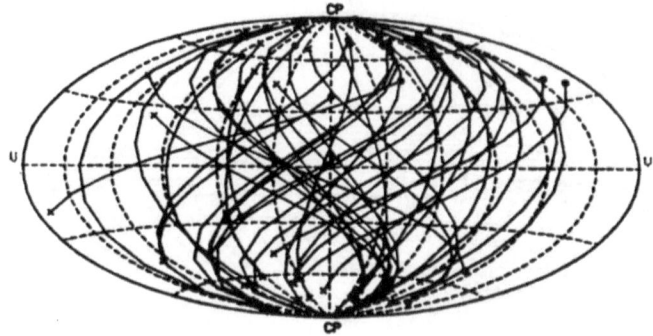

Figure 13.   Aitoff Equal-area Projection of the Poincaré Sphere.
Co-pol Nulls of 40 Dihedral/Trihedral Reflectors.   Five Dihedrals &
One Trihedral with Random Amplitudes, Phases, & Attitudes. TDR=10dB.

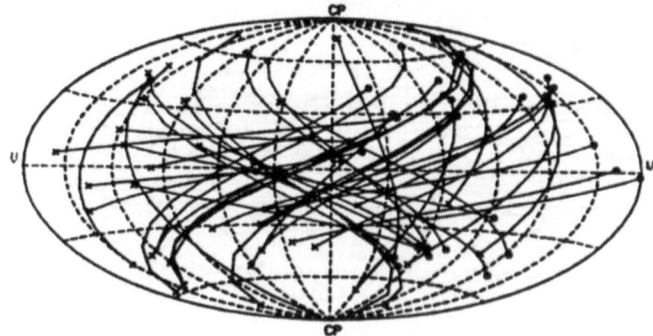

Figure 14.   Aitoff Equal-area Projection of the Poincaré Sphere.
Co-pol Nulls of 40 Dihedral/Trihedral Reflectors. Five Dihedrals &
One Trihedral with Random Amplitudes, Phases, & Attitudes. TDR=0dB.

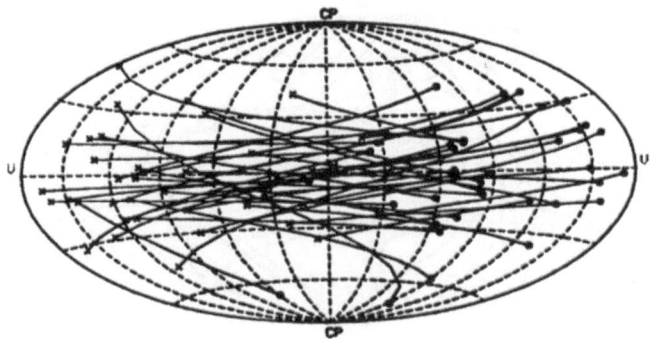

Figure 15. Aitoff Equal-area Projection of the Poincaré Sphere. Co-pol Nulls of 40 Dihedral/Trihedral Reflectors. Five Dihedrals & One Trihedral with Random Amplitudes, Phases, & Attitudes. TDR=-10dB.

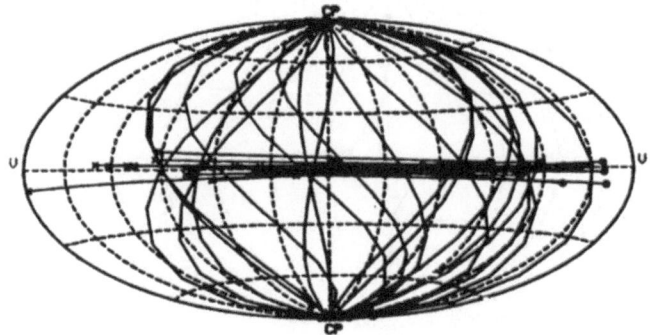

Figure 16. Aitoff Equal-area Projection of the Poincaré Sphere. Co-pol Nulls of 40 Dihedral & Trihedral Reflectors. Single Elements with Random Amplitudes, Phases, & Attitudes plus Noise. SNR=30dB.

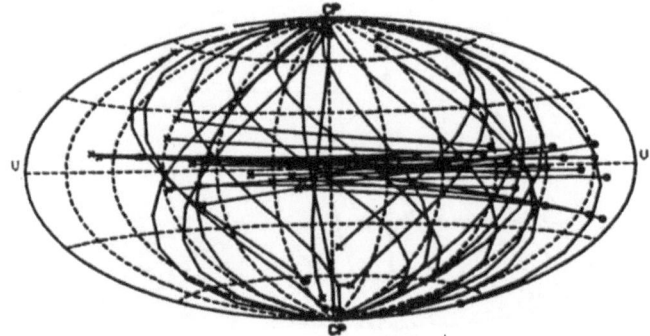

Figure 17. Aitoff Equal-area Projection of the Poincaré Sphere. Co-pol Nulls of 40 Dihedral & Trihedral Reflectors. Single Elements with Random Amplitudes, Phases, & Attitudes plus Noise. SNR=20dB.

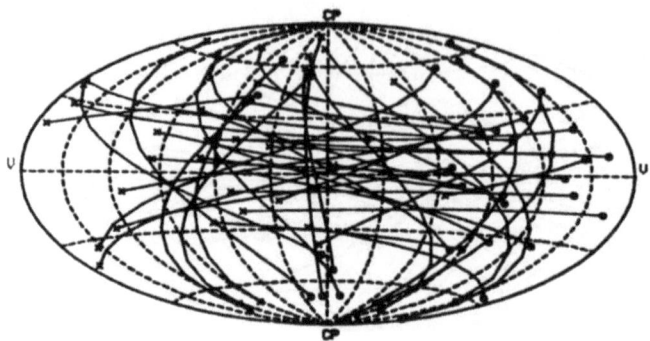

Figure 18.   Aitoff Equal-area Projection of the Poincaré Sphere.
Co-pol Nulls of 40 Dihedral & Trihedral Reflectors. Single Elements
with Random Amplitudes, Phases, & Attitudes plus Noise. SNR=10dB.

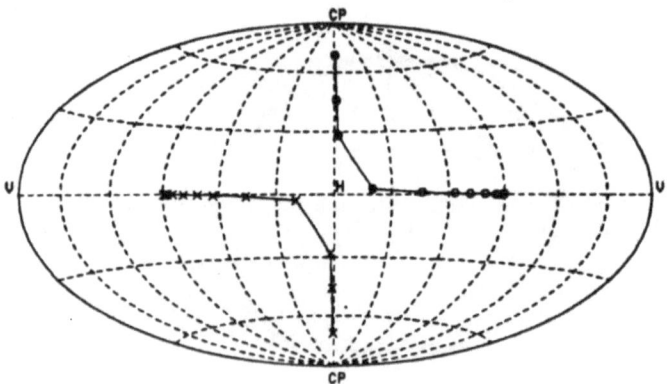

Figure 19.   Aitoff Equal-area Projection of the Poincaré Sphere.
Co-pol Null Migrations of Composite Dihedral with Incidence from
2.5^o to 85°.   Steel Cond./Sea Water. K_a-band.

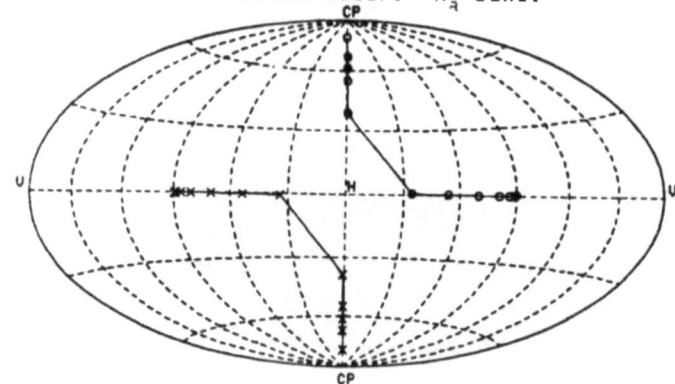

Figure 20.   Aitoff Equal-area Projection of the Poincaré Sphere.
Co-pol Null Migrations of Composite Dihedral with Incidence from
2.5^o to 85°.   Steel Cond./Moist Ground. K_a-band.

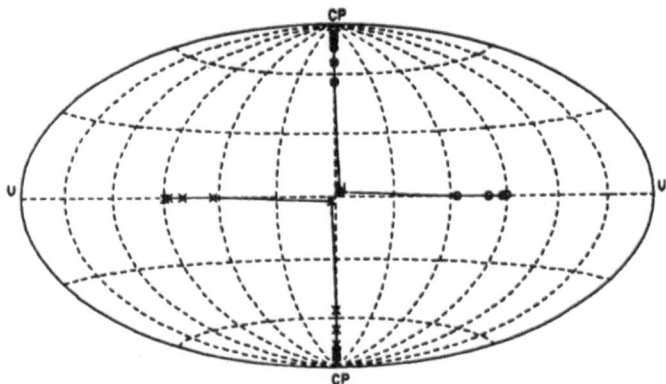

Figure 21. Aitoff Equal-area Projection of the Poincaré Sphere.
Co-pol Null Migrations of Composite Dihedral with Incidence from
2.5° to 85°. Steel Cond./Dry Ground. K_a-band.

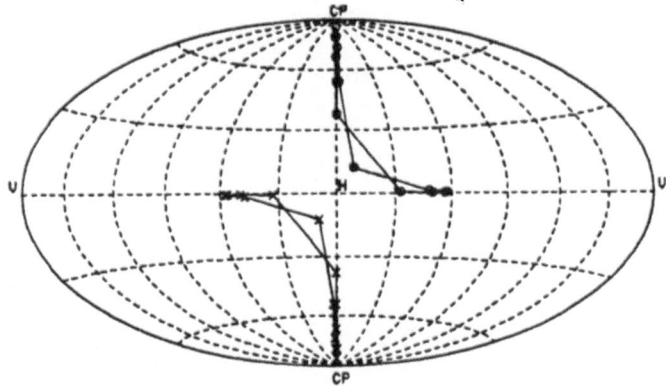

Figure 22. Aitoff Equal-area Projection of the Poincaré Sphere.
Co-pol Null Migrations of Composite Dihedral with Incidence from
2.5° to 85°. Wood/Sea Water. K_a-band.

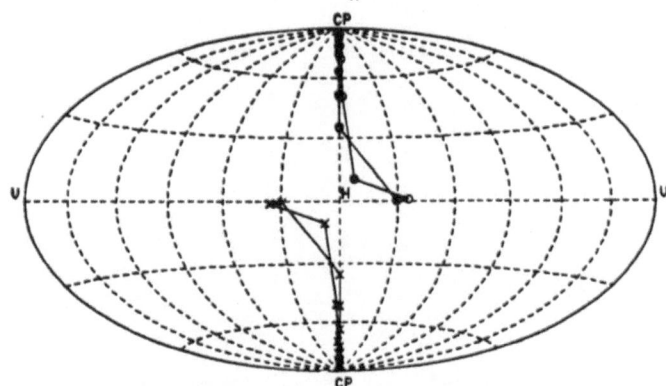

Figure 23. Aitoff Equal-area Projection of the Poincaré Sphere.
Co-pol Null Migrations of Composite Dihedral with Incidence from
2.5° to 85°. Wood/Moist Ground. K_a-band.

THEORY AND DESIGN OF A DUAL POLARIZATION RADAR FOR CLUTTER
ANALYSIS

Jerald D. Nespor, Amit P. Agrawal, Wolfgang-M. Boerner

Communications Laboratory, Polarization Radar Data Processing
Division, Department of Electrical Engineering & Computer Science,
Mail Stop M/C 154, 851 S. Morgan St., P.O. Box 4348, 1141-SEO,
4210-11-SEL, Chicago, IL 60680, USA

ABSTRACT
Solutions of the inverse problem, that of extracting orientation
and shape information from polarization measurements, provide useful meteorological data.

Currently, such information is obtained from circular and linear
dual-polarization measurables derived from auto and cross correlations of the main and orthogonal channels.

In this paper, an alternative integrated approach is used to deduce meteorological information from measurements of the relative
backscattering matrix derived from polarimetric techniques of
Kennaugh. For meteorological scatterers, such measurements would
be gathered in a time series.

The time elements gathered from a single measurement of the relative backscattering matrix is used to determine the co and cross
pol nulls on the Poincaré sphere or polarization chart. The optimal polarizations are then given in terms of elliptical and/or
circular dual polarization measurables where the equivalency of
the coherency and Stokes parameter measuring radar is given. Finally, the statistics of the co-pol null distributions on the
Poincaré sphere are given.

1. Introduction
Polarization, along with amplitude, frequency, phase and doppler,
are the five parameters that completely describe an electromagnetic wave. In propagation through a precipitation medium, the polarization state can be the parameter most significantly changed.
Due to engineering limitations prior to the last decade, such as

643

W.-M. Boerner et al. (eds.), Inverse Methods in Electromagnetic Imaging – Part 1, 643–659.

achieving high polarization purity within an antenna beam, coupled with the lack of theoretical development, polarization diversity techniques applied to radar meteorology have had a low level of research effort.

There has, however, been substantial progress within the last decade with the development of theory for predicting and interpreting backscatter and propagation measurements with the construction of antennas with good polarization characteristics, as well as, fast polarization switching techniques.

As the scope of research relating to polarization has increased, new ideas for meteorological applications have been proposed. It is the objective of this paper to suggest an integrated approach to polarization diversity techniques which have the possibilities of adding greater insight and reliability to meteorological identification and classification of different precipitation states.

2. The Scattering Matrix

It was first shown by Sinclair [1950] that a radar target acts like a polarization transformer which is described by its associated scattering matrix [S]. At the target scatterer, then, the target scattering matrix characterizes the scattering properties of the target such that the scattered electric field can be related to the incident electric field by:

$$
\begin{bmatrix} E_{s1} \\ E_{s2} \end{bmatrix} = \frac{\exp(-j\,k_0 r)}{r} \begin{bmatrix} S_{11} & S_{12} \\ S_{21} & S_{22} \end{bmatrix} \begin{bmatrix} E_{i1} \\ E_{i2} \end{bmatrix} . \tag{1a}
$$

where the subscripts s and i refer to the scattered and incident electric fields, respectively. The subscripts 1 and 2 refer to any two general orthogonal polarizations.

In vector notation, equation (1a) can be written as:

$$
\underline{E}_s = \frac{\exp(-jk_0 r)}{r} [S]\,\underline{E}_i \tag{1b}
$$

Assuming that the propagation space is reciprocal, one can now represent the scattering matrix in terms of amplitudes and phases where for a monostatic radar the relative scattering matrix becomes symmetric and is given by:

$$
[S]_{SMR} = \begin{bmatrix} |S_{11}|\,e^{j(\phi_{11} - \phi_{12})} & |S_{12}| \\ |S_{21}| & |S_{22}|e^{j(\phi_{22} - \phi_{12})} \end{bmatrix} \tag{2}
$$

where the $||$ brackets denote absolute value.

For the monostatic reciprocal radar case, it can be shown from the Reciprocity Theorem that $|S_{21}| = |S_{12}|$ and $\phi_{21} = \phi_{12}$. Throughout the remaining discussion, it will be assumed that the [S] implies the monostatic $[S]_{SMR}$ case unless specifically stated otherwise.

The radar cross section of a target is related to the scattering matrix by:

$$\sigma_{ij} = |S_{ij}|^2 \tag{3}$$

where i, j = 1 or 2

The radar cross section of a random array of objects (such as precipitation particles) fluctuates with pulse and range time. The fluctuations are not caused by changes in the radar cross sections of the individual objects; but rather they are due almost entirely to phase effects. Each object in the contributing region gives rise to a scattered wavelet with a certain phase, and the total scattered electric field is the sum of the interference effects of the individual wavelets. Since the relative phases of the wavelets change as the objects move about during the interval between radar pulses, the total echo power varies with time. The relative phases also vary with time in a random fashion, so the fluctuations are called "random phase" fluctuations. The amplitudes of the wavelets (or equivalently, the random cross sections of individual objects) sometimes vary as well. However, any echo fluctuations due to such amplitude changes will be superimposed or masked out by the "random phase" fluctuations.

Random-phase fluctuations are a characteristic feature of precipitation echoes. Because of the random-phase fluctuations, the radar cross-section for a distributed target can be expressed as:

$$\overline{\sigma}_{ij} = \overline{|S_{ij}|^2} \tag{4}$$

where i, j = 1 or 2 and the bar above the elements denote averaging for an ensemble of particles within the radar pulse.

Random-phase fluctuations also occur in range-time as the contributing region moves out from the radar; thus, the contributing region contains a different set of hydrometeors each time it moves out by its own length. Fluctuations of a similar nature can also occur as the antenna is rotated because the contributing region moves with the antenna beam.

The echo power $[P_R]$ received from a contributing region at range r for any one radar pulse is not, in general, equal to the average power <[P_R]> because of the random phase fluctuations. Thus, the power, [P_R], may be higher or lower than <[P_R]> by an amount that can only be predicted statistically. Measuring the average echo power <[P_R]> from any contributing region exactly would require averaging over an infinite time interval. In practice, however, it is therefore necessary to estimate the average echo power <[P_R]> by observing only a limited number of echoes from the contributing region.

From Figure 1, the following equation for the ensemble average radar cross-section is:

$$\overline{\sigma} = \sum_{i=1}^{N} \overline{|S_i|^2} \qquad (5)$$

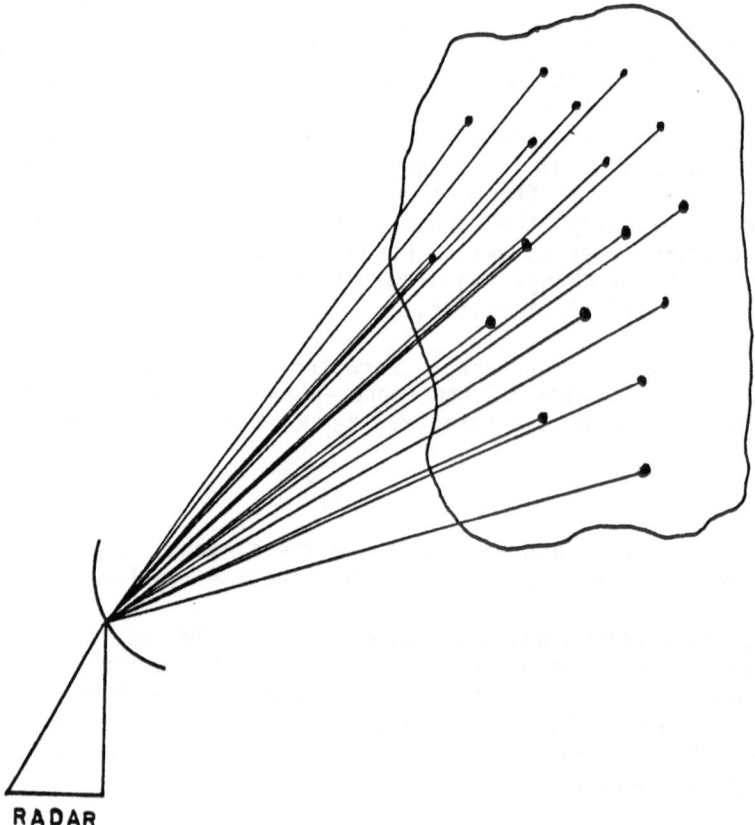

RADAR

FIGURE 1: AMPLITUDE STATISTICS OF THE RADAR CROSS SECTION

Thus, by the Central Limit Theorem which states that the mean of N samples follows a Gaussian distribution whatever the distribution of the individual measurements. It can then be assumed that the scattering matrix elements are zero-mean Gaussian and statistically independent when N is large. Consequently, the radar cross-section is Rayleigh distributed, because it is the product of two Gaussian distributions. It can be seen that the above assumptions are in total agreement with the earlier work of Marshall and Hitschfeld [1953].

3. Transformations From Linear To Circular And Elliptical Polarizations

Thus far, no stipulation has been made about a specific polarization basis in the above sections. Up to this point, the only requirement has been that the transmitted polarizations be orthogonal. Through the appropriate congruent transformations, it has been shown in Crispin and Siegal [1969] that the linear scattering matrix (LSM), $[S]_\ell$, can be transformed to a circular scattering matrix (CSM), $[S]_c$, or, in general, to an elliptical scattering matrix (ESM), $[S]_e$.

To transform $[S]$ to different polarization states, the following matrix transformation can be used:

$$[S']_i = [T][S][T]^T \tag{6}$$

where the subscript $i \to \ell$ represents the scattering matrix in a linear polarization basis where ℓ is usually implied when no subscript is present. For the scattering matrix in circular or elliptical polarization basis, $I \to c$ or e, respectively. The superscript T denotes the transpose of the transformation matrix $[T]$.

The most general form of $[T]$ can be found to be:

$$[T] = \begin{bmatrix} e^{j\phi_1} \cos\beta & -e^{j\phi_2} \sin\beta \\ e^{j\phi_3} \sin\beta & e^{j\phi_4} \cos\beta \end{bmatrix} \tag{7}$$

where $\phi_2 - \phi_1 = \phi_4 - \phi_3$. It should be noted that the most general basis is elliptical; but when all the phases are set to zero, $[T]$ reduces to an ordinary rotational matrix which rotates a particular linear polarization base by an angle β. There exist numerous transformation matrices for transforming between unit vectors of any polarization basis depending on one's choice of ϕ_1, ϕ_2, ϕ_3, ϕ_4 and β.

In this work, the transformation matrix will be taken as:

$$[T] = \frac{1}{\sqrt{2}}\begin{bmatrix} 1 & -j \\ 1 & +j \end{bmatrix} \tag{8}$$

to transform between the linear and the circular scattering matrices. Hence, from equation (6),

$$[S]_c = \frac{1}{2}\begin{bmatrix} 1 & -j \\ 1 & +j \end{bmatrix}\begin{bmatrix} S_{11} & S_{12} \\ S_{12} & S_{22} \end{bmatrix}\begin{bmatrix} 1 & 1 \\ -j & +j \end{bmatrix} \tag{9}$$

Finally, the scattering matrix in circular polarization basis becomes:

$$[S]_c = \frac{1}{2}\begin{bmatrix} S_{11} - S_{22} - j2S_{12} & S_{11} + S_{22} \\ S_{11} + S_{22} & S_{11} - S_{22} + j2S_{12} \end{bmatrix} \tag{10}$$

The transformation given in equation (7) is necessary if the non-ideal nature of the radar antenna is to be taken into account. Whether one desires to use linear or circular polarizations, in practice, the received polarizations will be elliptical. For any practical analysis for a dual-polarization system, antenna effects should be included by using the rotational and elliptical transformations presented above.

4.   The Backscattered Coherency Matrix

For a radar volume filled with scatterers, the received power is due to the backscattered components of all N scatterers within the radar volume. Thus, the partially polarized signals can be represented using the coherency matrix $[J]$ as:

$$[J]_s = \langle \underline{E}_s \underline{E}_s^{*T} \rangle = \begin{bmatrix} \langle E_{s1}E_{s1}^* \rangle & \langle E_{s1}E_{s2}^* \rangle \\ \langle E_{s2}E_{s1}^* \rangle & \langle E_{s2}E_{s2}^* \rangle \end{bmatrix} \tag{11}$$

$$= \langle \frac{1}{r^2} [S]\ \underline{E}_i\underline{E}_i^{*T}\ [S]^{*T} \rangle \tag{12}$$

where $*$ and T denote the complex conjugate and the transpose, respectively and the subscripts s and i denote the scattered and incident coherency matrix, respectively.

Expanding the above equation (12) for the monostatic reciprocal case $(S_{12} = S_{21})$ yields:

$$J_{s11} = \langle |S_{11}|^2 |E_{i1}|^2 + 2Re[S_{11}S_{12}^* E_{i1}E_{i2}^*] + |S_{12}|^2 |E_{i2}|^2 \rangle$$

$$J_{s12} = <S_{11}S_{12}{}^\star \ |E_{i1}|^2 + S_{11}S_{22}{}^\star \ E_{i1}E_{12}{}^\star + |S_{12}|^2 \ E_{i2}E_{i1}{}^\star +$$
$$+ \ S_{22}^\star S_{12} \ |E_{i2}|^2 > \qquad\qquad (13)$$

$$J_{s21} = J_{s12}{}^\star$$

$$J_{s22} = < \ |S_{22}|^2 \ |E_{i2}|^2 + 2Re[S_{22}S_{12}{}^\star \ E_{i2}E_{i1}{}^\star] + |S_{12}|^2|E_{i1}|^2 >$$

5.   The Stokes Vector And The Average Mueller Matrix

up to this point, the polarization properties of distributed targets such as precipitation have been represented in terms of the
incident and scattered coherency matrices and the 2x2 scattering
matrix. The partially polarized wave backscattered from the distributed target can also be represented in terms of the 4x1 Stokes
vector and the average Mueller matrix [M̄] can be defined (Kennaugh
1949-1954) as:

$$\underline{g}_s = [\bar{M}]\underline{g}_i \qquad\qquad (14)$$

where $\underline{g}_s$ and $\underline{g}_i$ are, respectively, the scattered (partially polarized) and incident (completely polarized) waves, expressed as
Stokes vectors and [M̄] is the 4x4 real and symmetric average
Mueller matrix used to represent partially polarized signals. It
has been shown by Chan [1981] that a transformation exists between
the scattering matrix [S] and the coherent Mueller matrix [M].
Equation (14) can then be expressed as:

$$\begin{bmatrix} g_0 \\ g_1 \\ g_2 \\ g_3 \end{bmatrix}_s = \begin{bmatrix} \bar{m}_{11} & \bar{m}_{12} & \bar{m}_{13} & \bar{m}_{14} \\ \bar{m}_{21} & \bar{m}_{22} & \bar{m}_{23} & \bar{m}_{24} \\ \bar{m}_{31} & \bar{m}_{32} & \bar{m}_{33} & \bar{m}_{34} \\ \bar{m}_{41} & \bar{m}_{42} & \bar{m}_{43} & \bar{m}_{44} \end{bmatrix} \begin{bmatrix} g_0 \\ g_1 \\ g_2 \\ g_3 \end{bmatrix}_i \qquad\qquad (15)$$

The Stokes parameters and the elements of the coherency matrix are
related by:

$$g_0 = J_{11} + J_{22}$$

$$g_1 = J_{11} - J_{22}$$

$$g_2 = J_{12} + J_{21}$$

$$g_3 = j(J_{12} - J_{21}) \qquad\qquad (16)$$

6.   Optimal Polarizations Concept For Dual Polarization Radar

Optimal polarizations are those which produce nulls in the co-polarized and the cross-polarized components of backscattered radiation.   The nulls in the co-polarized and the cross-polarized channels have been termed as 'co-pol' and 'x-pol' nulls, respectively,
in this paper. These are the polarizations where minimum and maximum voltages are recieved across the antenna terminals of a radar.

In order to obtain all the backscatter properties of an ensemble
of scatterers, it is necessary to make a complete measurement of
the polarization, amplitude and relative phase characteristics of
the backscattered signals.   This can be done, for example, by
means of measuring either the Stokes parameters or the coherency
matrix.   The two methods have been seen to be equivalent in the
previous section. Such measurements for a dual polarization radar
require two separate sequential orthogonal transmissions. Part of
the information contained in this system of measurements can also
be obtained from a knowledge of the two optimal polarizations, say
one co-pol and one cross-pol null. Optimal polarizations do not
in themselves constitute a complete backscatter description, but
they can be made complete by a measurement of the cross-section
and the degree of polarization. Alternatively, the measurement
can be completed by obtaining the power in the co-pol channel and
the orthogonal channel [G.C. McCormick, 1982].

7.   Optimal Polarizations For Completely Polarized Signals

The backscattered signal from a single particle is completely polarized.   First of all, the optimal polarizations for a single
particle will be formulated and then these formulations will be
extended for an ensemble of particles.   Making the appropriate
substitutions, equation (7) yields:

$$[T] = \frac{1}{(1+\rho\rho\star)^{1/2}} \begin{bmatrix} 1 & -\rho \\ \rho\star & 1 \end{bmatrix} \tag{17}$$

where $\rho = e^{j\alpha}\tan{\gamma}/2$. It should be noted that the transformation
matrix [T] is also uniitary.

Recalling from equation (1b) that the incident electric field was
represented in matrix notation by $\underline{E}_i$, and the scattered electric

field by $\underline{E}_s$; then the transformations of the electric field to the
new optimal polarization basis are given by:

$$\underline{E}_i' = [T] \underline{E}_i \tag{18}$$

$$\underline{E}_s' = [T\star] \underline{E}_s \tag{19}$$

where the prime denotes the new change of basis and the * denotes the complex conjugate.

When equations (18) and (19) are substituted into equation (1a), the electric fields can be expressed in the new optimal basis as:

$$\begin{bmatrix} E'_{s1} \\ E'_{s2} \end{bmatrix} = \frac{\exp(-jk_0 r)}{r} [T*] \begin{bmatrix} S_{11} & S_{12} \\ S_{12} & S_{22} \end{bmatrix} [T*]^T \begin{bmatrix} E'_{i1} \\ E'_{i1} \end{bmatrix} \tag{20}$$

Alternatively, in matrix notation equation (1b) becomes:

$$\underline{E}'_s = \frac{\exp(-jk_0 r)}{r} [S'] \underline{E}'_i \tag{21}$$

where

$$[S'] = [T*][S][T*]^T \tag{22}$$

and is the scattering matrix in the optimal polarization base.

Expanding equation (22), the elements of [S'] become:

$$S'_{11} = (1 + \rho\rho*)^{-1} (\rho^2 S_{22} + 2\rho S_{12} + S_{11}) \tag{23a}$$

$$S'_{12} = (1 + \rho\rho*)^{-1} (\rho^2 S_{22} + (1 - \rho\rho*)S_{12} - \rho* S_{11}) \tag{23b}$$

$$S'_{21} = S'_{12} \tag{23c}$$

$$S'_{22} = (1 + \rho\rho*)^{-1} (\rho*^2 S_{11} - 2\rho* S_{12} + S_{22}) \tag{23d}$$

Expressions for the co and cross pol nulls can be calculated by assuming the first normalized sequential pulse (polarization 1) is transmitted. Equation (20) gives:

$$\begin{bmatrix} E'_{s1} \\ E'_{s2} \end{bmatrix} = \frac{\exp(-jk_0 r)}{r} [S'] \begin{bmatrix} 1 \\ 0 \end{bmatrix} \tag{24}$$

Setting $E'_{s1} = 0$ in equation (20) yields the following expression for the co-pol nulls:

$$\rho_{1co} = \frac{-S_{12} \pm (S_{12}^2 - S_{11}S_{22})^{1/2}}{S_{22}} \tag{25}$$

Setting $E'_{s2} = 0$ in equation (20) yields the following expression for the cross-pol nulls:

$$\rho_{1x} = \frac{-B \pm (B^2 - 4AC)^{1/2}}{2A} \qquad (26)$$

where

$$A = S_{22}S^*_{12} + S^*_{11}S_{12}$$

$$B = |S_{11}|^2 - |S_{22}|^2$$

$$C = -A^*$$

Alternatively, one can obtain expressions for the co and cross-pol nulls in polarization 2 by assuming that the normalized second sequential pulse is transmitted:

$$\begin{bmatrix} E'_{s1} \\ E'_{s2} \end{bmatrix} = \frac{\exp(-jk_o r)}{r} [S'] \begin{bmatrix} 0 \\ 1 \end{bmatrix} \qquad (27)$$

8. Optimal Polarizations for Completely Polarized Signals and Their Relationship to the Elliptical Depolarization Ratio

In general, the backscattered signal from a target will be elliptically polarized, and it may be desirable for an adaptive dual-polarization radar to receive an elliptical mode. An elliptical depolarization ratio (EDR) may be defined where CDR is just a special case.

Recalling from Section 3 that the scattering m matrix in an transformed elliptical polarization base can be expressed as:

$$[S']_e = \begin{bmatrix} S'_{e11} & S'_{e12} \\ S'_{e12} & S'_{e22} \end{bmatrix} \qquad (28)$$

or

$$[S']_e = S'_{e12} = \begin{bmatrix} \eta'_{e11} & 1 \\ 1 & \eta'_{e22} \end{bmatrix} \qquad (29)$$

Following the same procedure as in the previous section, optimal polarizations can be expressed in terms of the non-logarithmic parameters used to calculate EDR as:

$$\rho_{1co} = \frac{-1 \pm \sqrt{1 - \eta'_{e11}\eta'_{e22}}}{\eta'_{e11}} \qquad (30)$$

$$\rho_{1x} = \frac{1}{2} \ \frac{|\eta'_{e11}|^2 - |\eta'_{e22}|^2}{|\eta'^\star_{e11} + \eta'_{e22}|} \ \pm \ \sqrt{1 + \frac{(|\eta'_{e11}|^2 - |\eta'_{e22}|^2)^2}{4|\eta'^\star_{e11} + \eta'_{e22}|^2}} \tag{31}$$

9. Optimal Polarizations For Partially Polarized Signals

The backscattered signal from an ensemble of particles (e.g. precipitation) is always partially polarized. Thus, expressions for the optimal polarizations for the partially polarized signals will be developed in this section.

The backscattered coherency matrix can now be expressed in terms of the optimal scattering matrix elements, as given in equations (23 a-d). When polarization 1 is transmitted, the backscattered coherency matrix elements in the optimal polarization base simplify to:

$$J'_{s11} = \langle |S'_{11}|^2 \ |E'_{i1}|^2 \rangle$$

$$J'_{s12} = \langle S'_{11} S'^\star_{12} \ |E'_{i1}|^2 \rangle \tag{32}$$

$$J'_{s21} = J'^\star_{s12}$$

$$J'_{s22} = \langle |S'_{12}|^2 \ |E'_{i1}|^2 \rangle$$

For either J'_{s11} or J'_{s22} a minimum, the optimal polarizations can be seen to be formally the same as for the completely polarized case presented in Section 7 with S_{11}, S_{12} and S_{22} being replaced by their ensemble averages:

$$\rho_{1co} = \frac{-\overline{S_{12}} \pm \sqrt{\overline{S_{12}}^2 - \overline{S_{11}} \ \overline{S_{22}}}}{\overline{S_{22}}} \tag{33}$$

$$\rho_{1x} = \frac{-\overline{B} \pm \sqrt{\overline{B}^2 - 4\overline{A} \ \overline{C}}}{2\overline{A}} \tag{34}$$

where

$$\overline{A} = \overline{S_{22} S_{12}^\star} + \overline{S_{11}^\star S_{12}}$$

$$\overline{B} = |\overline{S}_{11}|^2 - |\overline{S}_{22}|^2$$

$$\overline{C} = \overline{S_{11} S_{12}^\star} - \overline{S_{22}^\star S_{12}}$$

where the bar denotes ensemble averaging.

Similarly, expressions for the optimal polarizations can be obtained for polarization 2.

10. Optimal Polarizations For Partially Polarized Signals And Their Relationship To The Elliptical Depolarization Ratio

It can be seen that the optimal polarizations for partially polarized signals in terms of the non-logarithmic parameter used to calculate EDR are:

$$
\rho_{1co} = \frac{-1 \pm \sqrt{1 - \overline{\eta}'_{e11}\ \overline{\eta}'_{e22}}}{\overline{\eta}'_{e11}} \tag{35}
$$

$$
\rho_{1x} = \frac{1}{2} \frac{|\overline{\eta}'_{e11}|^2 - |\overline{\eta}'_{e22}|^2}{|\overline{\eta}'^*_{e11} + \overline{\eta}'_{e22}|^2} \pm \sqrt{1 + \frac{(|\overline{\eta}'_{e11}|^2 - |\overline{\eta}'_{e22}|^2)^2}{4|\overline{\eta}'^*_{e11} + \overline{\eta}'_{e22}|^2}} \tag{36}
$$

11. The Degree of Polarization

The degree of polarization, P, is defined as the ratio of the intensity of the polarized portion of the wave to the total intensity of the wave and is given by:

$$
P = \sqrt{1 - \frac{4\det[J]}{(Tr[J])^2}} \tag{37}
$$

When polarization 1 is transmitted, the $\det[J'_s]$ and $Tr[J'_s]$ can be calculated from equation (32) as:

$$
\det[J'_s] = <|S'_{11}|^2\ |E'_{i1}|^2> <|S'_{12}|^2\ |E'_{i1}|^2> -
$$
$$
- <S'_{11}S'^*_{12}\ |E_{i1}|^2> <S'_{11}{}^*S'_{12}\ |E_{i1}|^2> \tag{38}
$$

$$
Tr[J'_s] = <|S'_{11}|^2\ |E'_{i1}|^2> + <|S'_{12}|^2 <|E'_{i1}|^2> \tag{39}
$$

The degree of polarization, therefore, is obtained by substituting equations (38) and (39) into the above expression for P, yielding:

$$
P_1 = \sqrt{1 - \frac{4[\overline{|S_{11}|^2}\ \overline{|S_{12}|^2} - \overline{|S_{11}S_{12}|^2}]}{(\overline{|S_{11}|^2} + \overline{|S_{12}|^2})^2}} \tag{40}
$$

Similarly, P_2 can be found to be:

$$P_2 = \sqrt{1 - \frac{4[\overline{|S_{22}|^2 \, |S_{12}|^2} - \overline{|S_{22} \, S_{12}|^2}]}{(\overline{|S_{22}|^2} + \overline{|S_{12}|^2})^2}} \tag{41}$$

12. The Degree Of Coherence (Orientation)

The degree of coherence, μ, is the complex correlation factor between intensities in two recieving channels and it can be expressed as:

$$\mu = \frac{J_{12}}{\sqrt{J_{11}J_{22}}} \tag{42}$$

Thus, when polarization 1 is transmitted the degree of coherence is given by:

$$\mu_1 = \frac{\overline{S_{11}S_{12}^*}}{[\overline{|S_{11}|^2} \cdot \overline{|S_{12}|^2}]^{1/2}} \tag{43}$$

Similarly, when polarization 2 is transmitted, the degree of coherence becomes:

$$\mu_2 = \frac{\overline{S_{22} \, S_{12}^*}}{[\overline{|S_{12}|^2} \cdot \overline{|S_{22}|^2}]^{1/2}} \tag{44}$$

13. Mean And Spread Of The Optimal Polarizations For Partially Polarized Signals

When optimal polarizations are calculated as in Section 9, a distribution of co-pol and cross-pol nulls will then result from fluctuations due to the motion of the scatterers in the distributed target.

The mean values of the optimal polarizations are σ_{co} and σ_x, as calculated in Section 9. The variances of the clustering about the mean values are derived below.

In Section 2, it was assumed that S_{11}, S_{12} and S_{22} were Gaussian. Thus, η_{11}' is also Gaussian.

Now, from equation (32), it can be seen for the polarization 1 being transmitted, that the mean of the square of EDR becomes:

$$\langle |\eta_{11}'|^2 \rangle = \frac{\langle |S_{e11}'|^2 \rangle}{\langle |S_{e12}'|^2 \rangle} = \frac{J_{11min}'}{J_{22max}'} \approx \frac{\det[J']}{(Tr[J'])^2} \tag{45}$$

Since $<|\eta_{11}'|^2>$ is assumed to be a zero-mean Rayleigh distribution (i.e., $\eta_{11}' = 0$), $<|\eta_{11}'|^2>$ becomes the mean square deviation, and the root mean square deviation (spread) of the co-pol nulls distributed on the Poincaré Sphere is given by:

$$\sigma_{co} \approx \sqrt{\frac{det[J']}{(Tr[J'])^2}} \approx \frac{1}{2}\sqrt{1 - p^2} \tag{46}$$

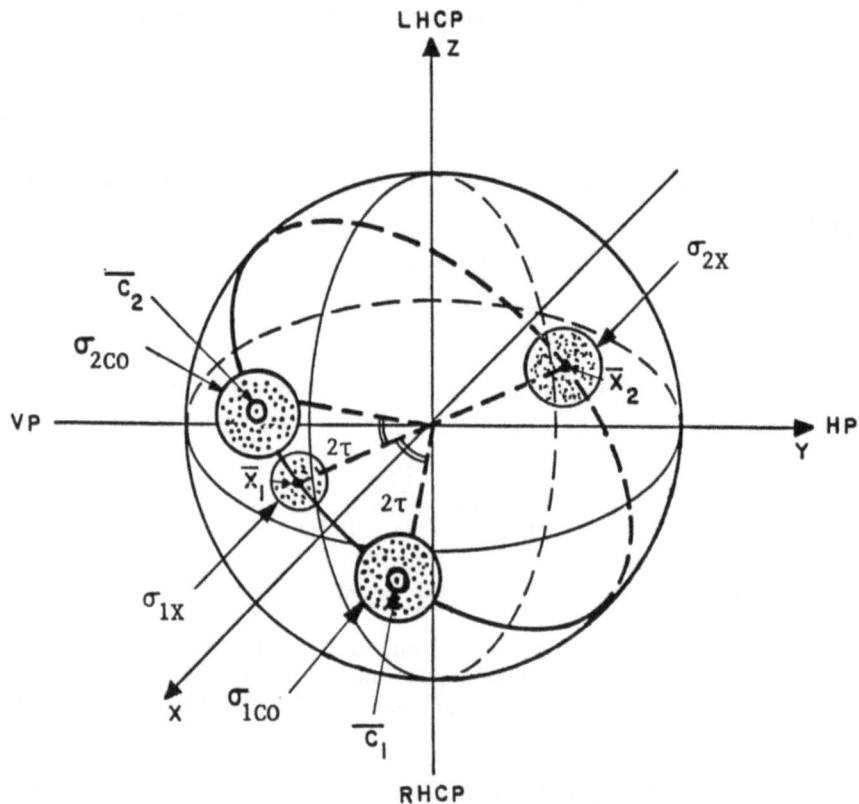

FIGURE 2: THEORETICAL REPRESENTATION OF OPTIMAL POLARIZATION ON THE POINCARE SPHERE

Similarly, the spread of the cross-pol nulls can be calculated and is given by:

$$\sigma_x \approx \frac{1}{2}\sqrt{1 - p^2} \tag{47}$$

The spread has general validity because all optimal polarizations represented on the Poincaré Sphere are unique.

14. Graphical Representation Of The Optimal Polarizations

The optimal polarizations for completely and partially polarized backscattered signals can be represented on the Poincaré Sphere.

For ensemble of particles (precipitation), the optimal polarizations will distribute themselves on the sphere because of the fluctuations due to the motion of the scatterers as described in Section 2. This distribution of optimal polarizations represented on the Poincaré Sphere along with mean and spread is demonstrated in Figure 2.

In Figure 3, plots of the optimal polarizations are given for rain. These plots for rain on the polarization chart were obtained from Poelman [1983] of the SHAPE-TC. The prescribed clustering distribution is again evident. The theory developed above can be applied for classifying and identifying the types of the clutter.

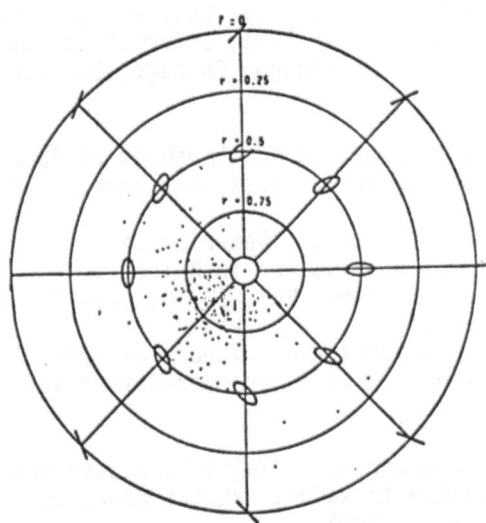

FIGURE 3:   OPTIMAL POLARIZATIONS REPRESENTED ON THE POLARIZATION CHART FOR RAIN TRANSMITTING RHCP (Poelman, 1983)

16. Acknowledgements

This research was supported, in part, by Army Research Office contract No. DAAG-20-80-K0027, and Office of Naval Research Grant No. N00001480-C-07730, and we wish to express our sincere gratitude to Drs. James W. Mink, Walter Flood, and Richard J. Brandt for their continued interest in our research. The authors wish to express special thanks to Mr. Richard and Mrs. Deborah Foster for the efficient typing and the supplying of the figures used in this manuscript.

19. References

[1] Boerner, W-M., El-Arini, M.B., Chan, C-Y., and Mastoris, P.M., (1981), "Polarization Dependence in Electromagnetic Inverse Problems", IEEE Trans. A & P, Vol. AP-29, No. 2, pp. 262-271, March, 1981.

[2] Boerner, W-M., and El-Arini, M.B., "Utilization of the Optimal Polarization Concept in Radar Meteorology", 20th Conf. on Radar Meteorology, Boston, pp. 656-665, 1981.

[3] Bolinder, E.F., "Geometric Analysis of Partially Polarized Electromagnetic Waves", IEEE, Trans A&P, Vol. 15(1), pp. 37-40, 1967.

[4] Born, E., and Wolf, E., "Principles of Optics", 2nd Edition, pp. 552. Pergamon, NY, 1964.

[5] Chan, C-W., "Studies on the Power Scattering Matrix of Radar Targets", M.Sc. Thesis, May 1981, Information Engineering Dept., Univ. of Ill. at Chicago, Chicago, IL 60680, #CL-EMID-81-02.

[6] Crispin, J.W., and Siegal, K.M., "Methods of Radar Cross Section Analysis", Acadenic Press, New York, 1969.

[7] Hendry, A., McCormick, G.C., "Deterioration of Circular Polarization Clutter Cancellation in Anisotropic Precipitation Media", Electronic Letters, Vol. 10, pp.165-166, 1974.

[8] Huynen, J-R., "Phenomenological Theory of Radar Targets", Ph.D. Dissertation, Drukkerij Bronder-Offset N.V., Rotterdam, 1970.

[9] Ishimaru, A., and Chung, R., "Multiple Scattering Effects on Wave Propagation Due to Rain", Extract Annuales DesTelecommunications, pp. 11-12, 1980.

[10] Jameson, A.R., "Microphysical Interpretation of Multi-Parameter Radar Measurements in Rain", Part 1, Interpretation of Polarization Measurements and Estimation of Raindrop Shapes, Submitted to the Jornal of Atmos. Sci., 1983.

[11] Jameson, A.R., and Beard, K.V., "Raindrop Axial Ratios", J. Applied Meteorology, Vol. 21, pp. 257-259, 1982.

[12] Kennaugh, E.M., "Effects of Type of Polarization on Echo Characteristics", Reports 389-1 to 389-24, Antenna Lab., Ohio State Univeristy, Columbus OH, 1949 to 1954.

[13] Marshall, J.S., and Hitschfeld, W., "Interpretation of the Fluctuating Echo from Randomly Distributed Scatterers", Canadian J. Physics, Part 1, pp. 962-995, 1953.

[14] McCormick, G.C. "Private Communications", May, 1982.

[15] McCormick, G.C. and Hendry, A., "Polarization-Related Parameters for Rain: Measurements Obtained by Radar", Radio Sci., Vol. 11, pp. 731-740, 1976.

[16] McCormick, G.C. and Hendry, A., "Principles for the Radar Determination of the Polarization Properties of Precipitation", Radio Science, Vol. 10(4), pp. 421-434, 1975.

[17] McCormick, G.C. and Hendry, A., and Barge, B.L., "The Anisotropy of Precipitation Media", Nature, 238, pp. 214-216, 1972.

[18] Nespor, J.D., "Theory and Design of a Dual Polarized Radar for Meteorological Studies", M.Sc. Thesis, University of Illinois at Chicago, Chicago, IL   60680, July, 1983.

[19] Nespor, J.D., Jameson, A.R., Boerner, W-M., "Sensitivity of Optimal Dual Polarization Measurements to the Estimation of Raindrop Shapes", Proc. 21st Conf. on Radar Meteor., American Meteor. Soc., Edmonton, Alberta, Canada, Sept. 1983.

[20] Poelman, A.J., "Multi-Notch Logic-Product Polarization Suppression Filters", Proc. of the 2nd Workshop on Polarimetric Radar Technology, May, 1983.

[21] Poelman, A.J.. "Cross Correlation of Orthogonolly Polarized Backscatter Components", IEEE Trans AES, Vol. 12(6), pp. 674-682, 1976.

[22] Rosien, R., Hammers, D., Ioannidis, G., Bell, J., and Nemit, J., "Implementation Techniques for Polarization Control for ECCM", RADC-TR-79-4, Final Tech. Rept., Feb., 1979.

[23] Seliga, T.A., and Bringi, V.N., "Potential Use of Radar Differential Reflectivity Measurements at Orthogonal Polarizations for Measuring Precipitation", J. Applied Meteor., Vol. 15, pp. 69-76, 1976.

[24] Sinclair, G., "The Transformation and Reception of Ellipitically Polarized Waves", Proc. IRE, Vol. 38, pp. 148-151, Feb., 1950.

III.9 (SR.3*)

RADAR TARGET HANDLING IN CLUTTER, CONSIDERATION OF
THE MEASUREMENT SYSTEM.

D. J. R. Stock

AEG-Telefunken

Introduction

Some ideas of Dunn and Howard (1) about glint can be
extended to polarisation-glint as concept.
When an antenna illuminates a fixed target in vacuum,
the waves reflected from the target combine on the
aperture as a vectorsum. Since the target-antenna
distance is fixed, no statistical amplitude-changes
are detected as a function of time (excluding noise).

   If the target is moving, the waves reflected from
various parts of the target show distance variations
with time which are very small compared to the distance between target and antenna, but may be multiples
of $\lambda/4$. The vectorsum of the waves on the antenna-
aperture shows a "glint" component.
Poelman (2) based on Born + Wolf (3) gives a similar
argument for what may be called polarisation-glint.
It is the purpose of this paper to consider this
problem with special emphasis on the antenna system
rather than on special properties of the target.

We assume following Poelman that a transmitting antenna radiates a wave given by the vector.[1]

$$e_T = \begin{bmatrix} \cos \varepsilon_T \\ \sin \varepsilon_T \ e^{j\delta_T} \end{bmatrix} \qquad /1/$$

For convenience we suppose that there are two receiving antennas, orthogonally polarized to each other.

W.-M. Boerner et al. (eds.), Inverse Methods in Electromagnetic Imaging - Part 1, 661–671.

$$e_R = \begin{bmatrix} \cos \varepsilon_R \\ \sin \varepsilon_R \ e^{j\delta_R} \end{bmatrix} \qquad e_{R_\perp} = \begin{bmatrix} \sin \varepsilon_R \ e^{j\delta_R} \\ -\cos \varepsilon_R \end{bmatrix} \qquad /2/$$

If the target be represented by a 2x2 scattering Matrix T, then the backscatter from the target is:

$$T \ e_T \qquad\qquad\qquad\qquad /3/$$

The average backscatter power is then:

$$P_T = \langle (T \ e_T)^* \ T \ e_T \rangle = \langle \widetilde{e_T}^* \ \widetilde{T}^* \ T \ e_T \rangle \quad /4/$$

a hermitian form.[2)]
In addition to the matrix

$$\widetilde{T}^* \ T$$

called "the power polarization matrix of Graves " in the literature, one can introduce the correlation matrix of (2), (3):

$$J = \begin{bmatrix} J_{yy} & J_{xy} \\ J_{yx} & J_{xx} \end{bmatrix} \qquad /5/$$

Both matrices appear in quadratic forms which lead to the following polynomials:

$$\det (\widetilde{T}^* \ T - k_0^2) = 0$$
$$\qquad\qquad\qquad\qquad\qquad /6/$$
$$\det (J - k^2) \quad\ = 0$$

The solutions are easily found by expansion:

$$2 \ k^2 = \mathrm{Tr} \ J \ (1 \overset{+}{}_{-} \sqrt{1 - \frac{4 \ \det J}{(\mathrm{Tr} \ J)^2}} \ ) \quad /7/$$

The polarizationfactor /2/, /3/ is defined as:[3)]

$$p = \sqrt{1 - \frac{4 \ \det J}{(\mathrm{Tr} \ J)^2}} \qquad /8/$$

It may be easily shown that:

$$\langle \widetilde{e_T}^* \ \widetilde{T}^* \ T \ e_T \rangle = \mathrm{Tr} \ J \qquad /10/$$

and when p = 1

$$\text{Tr } J = k^2_{\text{max}}$$

A cross-correlation function (2), (3) in form

$$\mu = \frac{J_{xy}}{\sqrt{J_{xx} J_{yy}}}$$

plays an important part in the work that follows, since the various receiver systems discussed all make use of two (othogonal) channels. The backscatter vector sums of waves falling on the two receive-apertures give rise to the bilinear forms:

$$V_R = (\widetilde{T e_T}) e_R$$

$$V_R = (\widetilde{T e_T}) e_{R\perp}$$

with associated powers:

$$P_R = V_R V_R^*$$

$$P_{R\perp} = V_{R\perp} V_{R\perp}^*$$

Discussion of the Radarsystem

As discussed above and shown in Fig. 1, the antenna system consists of three antennas, a configuration which may be easily realized for multistatic radar. In pratice today, however, the weather radars for example (4) have only one antenna, so that such sy-stems will be preferred here[4]
Such a system is switchable:
>   1. The polarization e_T is radiated, and the copolar polarization e_R received.
>   2. The crosspolarization $e_{T\perp}$ is radiated and $e_{R\perp}$ received.
superposition is assumed.

The measurement Quantities

As noted above, the backscatter power is:

$$P_T = \langle \widetilde{e_T}^* \widetilde{T}^* T e_T \rangle$$

since $\widetilde{T}^* T$ is hermitian, P_T is a hermitian[5] form. Since $P_T = 0$ not only for the trivial case $e_T = 0$,

it is positive semi-definite (PSD).
The received voltages give rise to the power forms:

$$P_R = \widetilde{e}_T \, \widetilde{T} \, e_R \cdot \widetilde{e}_T{}^* \, \widetilde{T}^* \, e_R{}^*$$

$$P_{R\perp} = \widetilde{e}_T \, \widetilde{T} \, e_{R\perp} \cdot \widetilde{e}_T{}^* \, \widetilde{T}^* \, e_{R\perp}{}^* \qquad /11/$$

$$V_R \, V_{R\perp}{}^* = \widetilde{e}_T \, \widetilde{T} \, e_R \cdot \widetilde{e}_T{}^* \, \widetilde{T}^* \, e_{R\perp}{}^*$$

We rewrite $e_{R\perp}$ as:

$$e_{R\perp} = e^{-j\delta} \begin{bmatrix} e^{-j\delta_R} & 0 \\ 0 & e^{j\delta_R} \end{bmatrix} \begin{bmatrix} 0 & 1 \\ -1 & 0 \end{bmatrix}$$

$$= e^{-j\delta_R} \Lambda \sigma_1 \, e_R \qquad /12/$$

where

$$\widetilde{\Lambda}^* \Lambda = 1$$

$$\widetilde{e}_R \, \widetilde{\sigma}_1 \, e_{R\perp} = 1$$

so that

$$P_{R\perp} = \widetilde{e}_T \, \widetilde{T} \, \sigma_1 \, e_R \cdot \widetilde{e}_T{}^* \, \widetilde{T}^* \, \sigma_1 \, e_R{}^*$$

$$V_R \, V_{R\perp}{}^* = (\widetilde{e}_T \, T \, e_R \cdot \widetilde{e}_T{}^* \, \widetilde{T}^* \, \Lambda^* \, \sigma_1 \, e_R{}^*) \, e^{j\delta_R} \qquad /13/$$

from /6/ one obtains

$$2 \, k_0{}^2 = \mathrm{Tr} \, (\widetilde{T}^* \, T)(1 \pm P_0)$$

where

$$P_0 = \sqrt{1 - \frac{4 \, \det \, \widetilde{T}^* \, T}{(\mathrm{Tr} \, \widetilde{T}^* \, T)^2}}$$

Bickel (5) shows that $\mathrm{Tr} \, \widetilde{T}^* \, T$ is invariant for
polarizations-rotation of the receiving antenna,
and Graves (6) shows the invariance of $\det T^* \, T$
under this condition.
According to Mirsky (7)

$$k_0{}^2{}_{min} \, \widetilde{e}_T{}^* \, e_T \leq \widetilde{e}_T{}^* \, \widetilde{T}^* \, T \, e_T \leq k_0{}^2{}_{max} \, \widetilde{e}_T{}^* \, e_T$$

since

$$\widetilde{e_T}^* \, e_T = 1$$

$$k_{0\ min}^{2} \; \leq \; \widetilde{e_T}^* \; \widetilde{T}^* \; T \; e_T \; \leq \; k_{0\ max}^{2} \qquad /14/$$

Then $k_{0\ max}^{2}$ must be the entire radiated power. Equality right in the inqualitiy results when the target reflects the entire power, i.e., it is loss-less.

The "orthogonal" target
_____

To show the simplicity of the measurement systems, we define an orthogonal target:

$$T_\perp$$

using:

$$P_{R\perp} = \widetilde{e}_T \, T_\perp \, e_R \cdot \widetilde{e}_T^* \, T_\perp^* \, e_R$$

where

$$T_\perp = \sigma_1 \Lambda_T \, T \, \Lambda \sigma_1 \qquad \text{and } \Lambda_T = \begin{bmatrix} e^{-j\delta_T} & 0 \\ 0 & e^{j\delta_T} \end{bmatrix}$$

is the backscattermatrix of the orthogonal target. Noting that

$$P_{T\perp} = \widetilde{e}_T^* \, \widetilde{\sigma}_1 \, \Lambda_T^* \, \widetilde{T}^* \, \Lambda^* \, \sigma_1 \, \widetilde{\sigma}_1 \, \Lambda_T \, \widetilde{T} \, \Lambda \, \sigma_1 \, e_T$$

and that

$$\Lambda_T^* \, \sigma_1 \, \widetilde{\sigma}_1 \, \Lambda = 1 \qquad \sigma_T = \sigma_R$$

we have

$$P_{T\perp} = \widetilde{e}_T^* \, T_\perp^* \, T_\perp \, e_T$$

analogous to /4/.
Following Bickel and Graves one can also show that:

$$\det \widetilde{T_\perp^*} \; T_\perp \quad \text{and} \quad \mathrm{Tr} \; \widetilde{T_\perp^*} \, T_\perp$$

are invariant under rotation.
McCormick and Hendry (8) show a system in which a network of couplers and phase shifters behind the antenna separate the signal into two orthogonal components. Each component feeds a separate receiver, and the receiver outputs are fed to a correlator. They compare this method (9) with a method of Seli-ga and Bringi (10) where the polarizations are

switched.
The two methods are really the same when one uses
the concept of orthogonal targets. As shown above,
the orthogonal splitting can take place in the target
plane, or in the network as shown by McCormick and
Hendry.

Antenna adjustment for measurement

McCormick and Hendry (11) show that when the antenna
polarization is adjusted so that a minimum exists in
one of the orthogonal channels, one obtains:

$$\frac{k^2_-}{k^2_+} = \frac{Tr\ J - \sqrt{Tr\ J - 4\ det\ J}}{Tr\ J + \sqrt{Tr\ J - 4\ det\ J}} \approx \frac{det\ J}{(\ Tr\ J)^2}$$

(our notation)
They further note that a non-zero value in one
channel $(k^2_- \neq 0)$ results from a non-coherent generator[6].
The concept of orthogonal targets allows this antenna
adjustment to be made, although from the following
considerations this may not be necessary.
One defines the orthogonal matrix J by a derivation
using $e_{T\perp}$ instead of e_T.
It is then easy to show in general without specifi-
cation of the target that although Tr J is not an
invariant under polarisation-rotation,

$$Tr\ (\ J + J_\perp\ )$$

is independent of ε_T, δ_T. That is, Tr (J + J_\perp)
may be measured with two arbitrary orthogonal pola-
rizations.
Further this target invariant is related to Graves'
invariant:

$$Tr\ (\ \widetilde{T^*}\ T\ ) = Tr\ (\ J + J_\perp\ )$$

An alternative statement of the McCormick and Hendry-
method is: using /14/ the extreme value of $\widetilde{T^*}\ T$ can
be obtained by antenna adjustment.

$$k_0{}^2_- \leq Tr\ J \leq k_0{}^2_+$$

$$k_0{}^2_{-\perp} \leq Tr\ J_\perp \leq k_0{}^2_{+\perp}$$

and the constant values are:

$$\text{Tr} \ (\ J \ + \ J_\perp) \ = \ \text{Tr} \ \widetilde{T}^* \ T \ = \ k_0^2 \ _+ + \ k_0^2 \ _-$$

of special interest here is the result that the properties of the polarization-switchable elements of a
phased-array antenna can be used (12). The measurement steps are:
1. The arbitrary polarization e_T is radiated.
2. The copolarization e_R is received.
3. The antenna is switched   to the orthogonal
   polarization.
4. The polarization $e_{T\perp}$ is radiated.
5. The "cross"-polarization $e_{R\perp}$ is received.
see Fig. 2 and 3.

Summary

The problem of target glint as expressed by the J-
matrix was considered here more from the standpoint
of the measuring system than from target properties.
Only pairs of orthogonal targets were considered in
order to throw emphasis on the antenna-receiver system. It is then apparent that the various systems
in the literature are conceptually identical.

Acknowledgement

Discussion with H. Trogus, AEG-TELEFUNKEN concerning
the KR-75 coastel Radar with phased-array antenna
is gratefully acknowledged.

Index:

1.) This form is chosen to emphasize the polarisation
properties. The wave is described fully by:

$$\vec{E} = \vec{E}_T \ e^{j(\omega t - \beta z)}$$

where $\quad \vec{E}_T = E_0 \ \vec{e}_T$

and $\quad \vec{e}_T = \vec{1}_x \cos \varepsilon_T + \vec{1}_y \sin \varepsilon_T \ e^{j\delta_T}$

2.) If T is a n x n-matrix, $\widetilde{T}$ is its transpose, T^* its
complex conjugate, T^{-1} the inverse, det T the determinant, and Tr (T) the trace of T. T is hermitian, when

FIG. 1

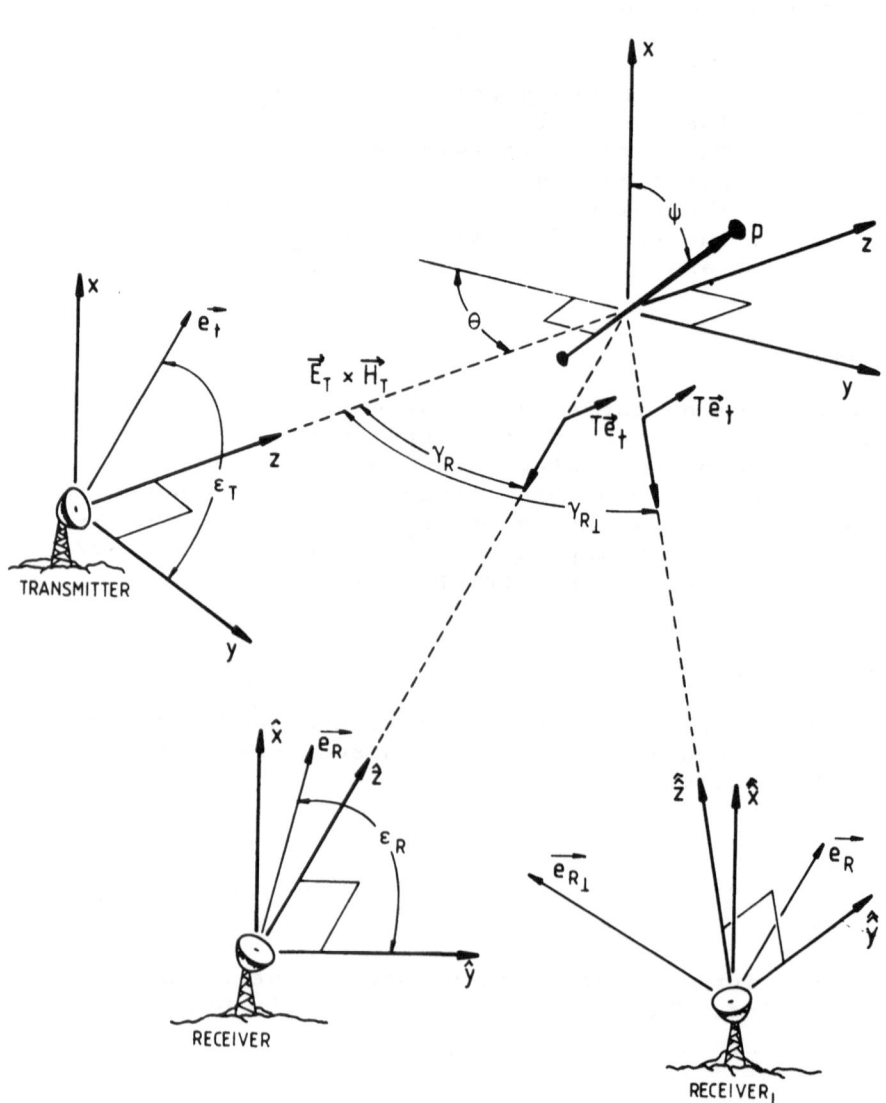

FIG. 2
Receiving antenna   orthogonal polarized
                                           receiving antenna

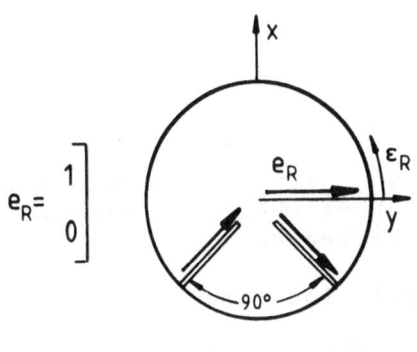

$$e_R = \begin{bmatrix} 1 \\ 0 \end{bmatrix}$$

$$e_{R\perp} = \sigma_1 \begin{bmatrix} 1 \\ 0 \end{bmatrix} = \begin{bmatrix} 0 \\ -1 \end{bmatrix}$$

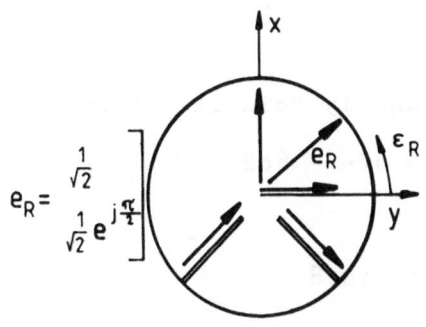

$$e_R = \begin{bmatrix} 1 \\ 0 \end{bmatrix}$$

$$e_{R\perp} = \sigma_1 \begin{bmatrix} 0 \\ 1 \end{bmatrix} = \begin{bmatrix} 1 \\ 0 \end{bmatrix}$$

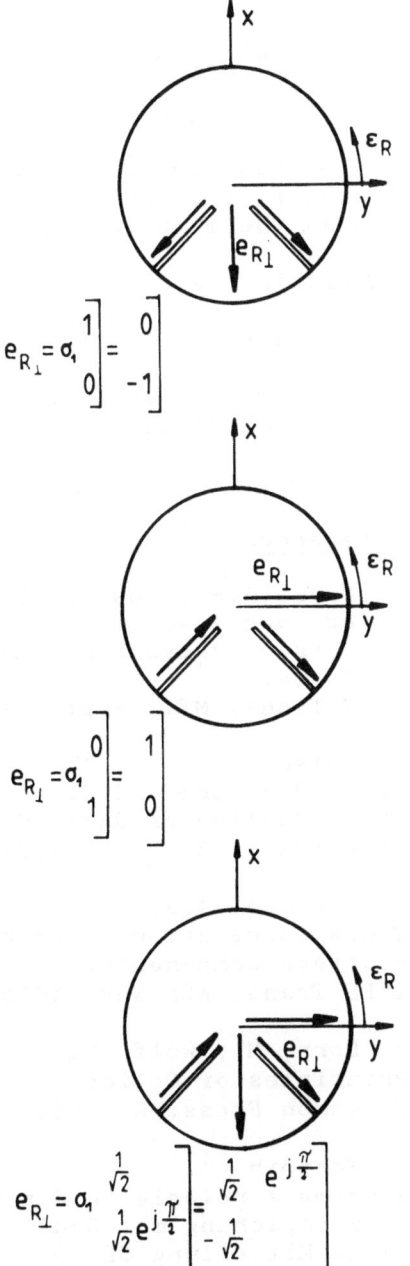

$$e_R = \begin{bmatrix} \frac{1}{\sqrt{2}} \\ \frac{1}{\sqrt{2}} e^{j\frac{\pi}{2}} \end{bmatrix}$$

$$e_{R\perp} = \sigma_1 \begin{bmatrix} \frac{1}{\sqrt{2}} \\ \frac{1}{\sqrt{2}} e^{j\frac{\pi}{2}} \end{bmatrix} = \begin{bmatrix} \frac{1}{\sqrt{2}} e^{j\frac{\pi}{2}} \\ -\frac{1}{\sqrt{2}} \end{bmatrix}$$

$T^* = \widetilde{T}$, T is unitary when $\widetilde{T}^* \cdot T = 1$. For a complex matrix T, $\widetilde{T}^* \cdot T$ is hermitian.

3.) Since

$$k^2_{max} = 1/2 \ Tr \ J \ (\ 1 + p \ )$$
$$k^2_{min} = 1/2 \ Tr \ J \ (\ 1 - p \ ) \qquad /9/$$

it is possible to represent k^2_{max}/k^2_{min} as a bilinear transformation on the smith chart when p is considered as the magnitude of the reflectance.

4.) within the clutter decorrelation time.

5.) for a single dipole, T and $\widetilde{T}^* \cdot T$ have rank 1.

6.) this may be seen for example for the dipole cloud of Poelman (2) when $\theta_0 = 0$.

References:

1. Dunn, J.H., Howard, D.D.
Radar Target Amplitude, Angle and Doppler Scintillation from Analysis of the Echo Signal Propagation in Space.
IEEE Trans. MTT, Sept. 1968, S 715 - 728

see also
Sims, R.J. Graf, E.R.
The Reduction of Radar Glint by Diversity Techniques.
IEEE Trans. AP, July 1971, S 462 - 468

2. Poelman, A.J.
Cross Correlation of Orthogonally Polarized Back-scattern Components.
IEEE Trans. AES Nov. 1976, S 674 - 682

3. Born, M., Wolf, E.
Principles of Optics
Pergamon Press, NY 1959, Sec. 10.8

4. Schroth, A.
Radarmeteorologie in der Wolkenphysik und Ausbreitungsforschung mit Radiowellen.
DFVLR-Mitteilung 82-09

5. Bickel, S.H.
Some Invariant Properties of the Polarization Scat-
tering Matrix.
Proc. IEEE Aug. 1965, S 1070 - 1072

6. Graves, C.D.
Radar Polarization Power Scattering Matrix
Proc. IRE Feb. 1956, S 246 - 252

7. Mirsky, L.
An Introduction to Linear Algebra
Clarendon Press, Oxford 1961, S 388

8. McCormick, G.C., Hendry, A.
Principles for the Radar Determination of the Polari-
zation Properties of Precipitation
Radio Science, April 1975, S 421 - 434

9. McCormick, G.C.
Relationship of differential Reflectivity to Correla-
tion in Dual-Polarization Radar.
Electronic Letters, 10 May 1979, S 265 - 266
10. Seliga, T.A., Bringi, V.N.
Potential Use of Radar Differential Reflectivity
Measurements of Orthogonal Polarization for Measu-
ring Precipitation
J. Appl. Meteorol. 1976, S 69 - 76

11. McCormick, G.C. Hendry, A.
Techniques for the Determination of the Polarization
Properties of Precipitation
Radio Science, Nov. - Dec. 1979, S 1027 - 1040

12.Trogus, H.
Costal Radar KR-75 a Contribution Towards the Preven-
tion of Collisions.
AEG-TELEFUNKEN Progress 1980, S 23 - 30